Test statistic for testing H_0: $p = p_0$

$$z = \frac{\hat{p} - p_0}{\sqrt{\frac{p_0(1 - p_0)}{n}}}$$

Pooled estimate of a population proportion using data from two independent samples

$$\bar{p} = \frac{x_1 + x_2}{n_1 + n_2}$$

Test statistic for testing H_0: $p_1 = p_2$

$$z = \frac{\hat{p}_1 - \hat{p}_2}{\sqrt{\frac{\bar{p}(1 - \bar{p})}{n_1} + \frac{\bar{p}(1 - \bar{p})}{n_2}}}$$

Equation for sample regression line

$$y_c = b_0 + b_1 x$$

Slope of sample regression line

$$b_1 = \frac{\Sigma\, xy - \frac{\Sigma\, x\, \Sigma\, y}{n}}{\Sigma\, x^2 - \frac{(\Sigma\, x)^2}{n}}$$

y-intercept of sample regression line

$$b_0 = \bar{y} - b_1 \bar{x}$$

Sample coefficient of determination

$$r^2 = \frac{b_1 \left[\Sigma\, xy - \frac{\Sigma\, x\, \Sigma\, y}{n}\right]}{\Sigma\, y^2 - \frac{(\Sigma\, y)^2}{n}}$$

Sample correlation coefficient

$$r = \frac{n\, \Sigma\, xy - (\Sigma\, x)(\Sigma\, y)}{\sqrt{n\, \Sigma\, x^2 - (\Sigma\, x)^2}\, \sqrt{n\, \Sigma\, y^2 - (\Sigma\, y)^2}}$$

Test statistic for χ^2 tests of independence and homogeneity, $r \times c$ contingency table

$$X^2 = \sum_{i=1}^{r} \sum_{j=1}^{c} \left[\frac{(O_{ij} - E_{ij})^2}{E_{ij}}\right]$$

Test statistic for χ^2 tests of independence and homogeneity, 2×2 contingency table

$$X^2 = \frac{n(ad - bc)^2}{(a + c)(b + d)(a + b)(c + d)}$$

Spearman rank correlation coefficient

$$r_s = 1 - \frac{6\, \Sigma\, d^2}{n(n^2 - 1)}$$

Introductory Statistics with Applications

Introductory Statistics with Applications

Wayne W. Daniel

Georgia State University

HOUGHTON MIFFLIN COMPANY Boston
Atlanta Dallas Geneva, Illinois Hopewell, New Jersey Palo Alto London

Chapter-opening art rendered by Herb Rogalski

Printed in the U.S.A.

Library of Congress Catalog Card Number: 76-10897

ISBN: 0-395-24430-7

To Shirley and Kay and their families

Contents

Preface

Introductory Statistics with Applications was written for use in applied statistics classes that draw students from a wide variety of disciplines. The exercises and worked examples that illustrate the concepts and methodology are drawn from the fields of education, psychology, sociology, social work, biology, business, anthropology, agriculture, and other disciplines. Although these exercises and examples are fictitious, I tried to make them as interesting and realistic as possible. I believe that a seemingly tedious statistical technique appears more interesting and relevant when the student is shown how it can be used to solve a real problem. Consequently, the exercises in this book are not just sterile arrays of numbers, with instructions to perform such-and-such an analysis of them.

I have been generous with the exercises. Under most topics I have included enough of them to enable the instructor to have a different one for classroom presentation, homework assignment, and tests. There are exercises at the end of each major section in order to provide immediate reinforcement of the concepts and techniques covered in the section. There are additional exercises at the ends of the chapters to enable students to test their recall of important techniques and concepts after they have completed a chapter.

By design, the mathematical level of this book has been kept at a precalculus level. For most students not majoring in statistics or mathematics, the introductory statistics course should emphasize statistical applications rather than statistical theory and mathematics. In presenting concepts and techniques, therefore, I have tried to capitalize on the student's intuition rather than on an ability to execute a proof or to manipulate a complex algebraic expression.

I have introduced the various topics covered by this text in a logical sequence. Each new topic is illustrated by means of at least one example worked out in sufficient detail for the student to follow without difficulty. In the first six chapters, a given topical section to a large extent builds on the onc that precedes it. Chapters 7 through 11 are independent of each other. The first six chapters, however, provide the conceptual foundation for understanding them.

I designed this text to facilitate innovation and flexibility in the use of it. Although I recommend that an instructor cover most of the topics in the first six chapters as a minimum, all of them do not have to be covered; nor do they all have to be covered in the same detail. In Chapter 1, for example, some instructors may wish to omit the section on computing descriptive measures from grouped data. Others may wish to cover only the case of the single mean and the single proportion in the chapters on estimation and hypothesis testing. In some cases students will already have had an adequate exposure to the concepts and algebra of probability, so that this chapter may be omitted.

Chapter 4 discusses the sampling distributions needed for Chapters 5 and 6. Some instructors, however, may not wish to cover all of Chapter 4 before introducing students to estimation (Chapter 5) and hypothesis testing (Chapter 6). These instructors may wish to go directly from a discussion of the normal distribution in Chapter 3 to a discussion of confidence intervals for a single mean, referring to the material in Chapter 4 on the sampling distribution of the sample mean as these concepts are needed to facilitate the understanding of confidence intervals. Similarly, as a class takes up each new topic in Chaper 5, the instructor can bring up the relevant sampling distribution discussion in Chapter 4 at the same time. This same format can be followed in teaching experimental design (Chapter 7) and analysis of variance (Chapter 8). That is, Chapter 7 does not have to be taught before Chapter 8. The material in the two chapters may be taught simultaneously. For example, when the instructor is teaching one-way analysis of variance (Chapter 8), the material on the completely randomized design can be integrated into the discussion.

The material on estimating and testing hypotheses about variances and the ratio of two variances in Chapters 5 and 6 and the relevant sampling distributions in Chapter 4 may be omitted without losing continuity. Likewise, the instructor may omit the section on Bayes's theorem in Chapter 2, even though covering the other sections of that chapter.

Because of the step-by-step details of the worked examples, I feel that students will be able, on their own, to read and comprehend many topics. Thus the innovative instructor should be able, in a one-term course, to cover one or more chapters beyond the basic first six. I have included four of the more important and popular topics—experimental design, analysis of variance, regression and correlation, chi-square analysis, and other nonparametric tests—from which a selection may be made.

When the book is used for a two-term course, the material in Chapters 1 through 6 can be covered during the first term, followed by Chapters 7 through 11 the second term.

I would like to take this opportunity to express appreciation to my wife Mary, who typed the manuscript and helped me in other aspects of producing this book. I am grateful to Professor Carroll Mohn, of Georgia State University, who checked the accuracy of the solutions to the examples and exercises. I am also grateful to Professors Pickett Riggs and Charles Hopkins, who wrote the *Study Guide* to accompany this text.

Various drafts of the manuscript were read and critiqued by Professors Ralph D'Agostino, of Boston University, David S. Moore, of Purdue University, John Wiorkowski, of the University of Texas at Dallas, and Richard A. Fritz, of Moraine Valley Community College. Their comments and suggestions contributed greatly to the improvement of the final product, and I owe them special thanks. They must, however, be absolved of all responsibility for any shortcomings that the book may still possess.

Wayne W. Daniel

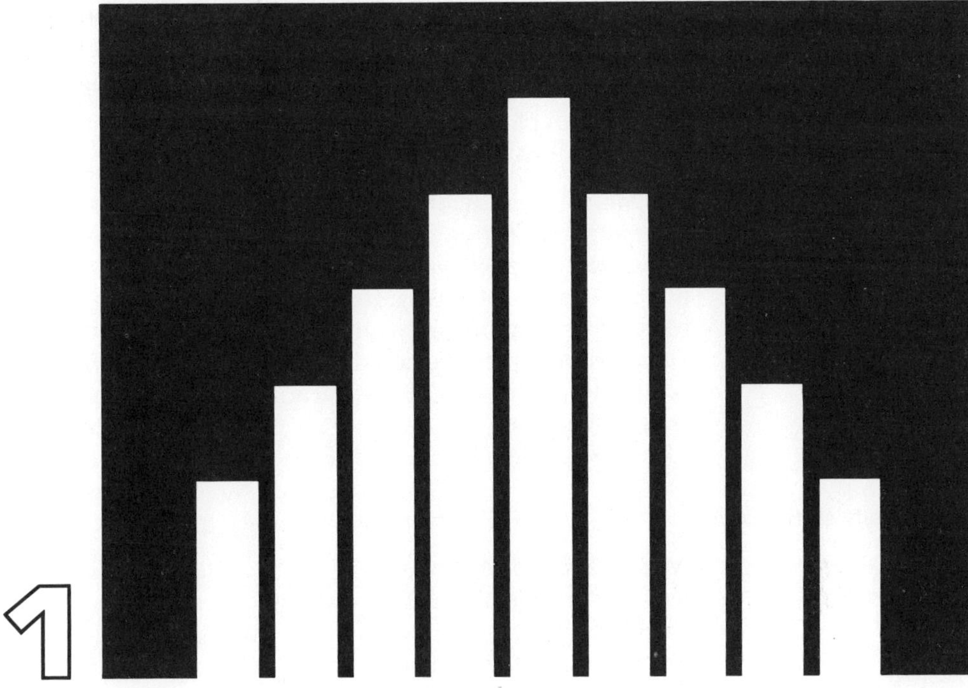

Descriptive Statistics

To the typical person on the street, statistics means numbers. From the morning paper he or she can find out the latest crime statistics for the city, such as the number of murders, car thefts, break-ins, and other crimes that have been reported over a period of time. Or the latest labor statistics for the country, such as the number unemployed. Or the city's latest vital statistics, such as the number of births and deaths that have occurred during some reporting period. Or the latest sports statistics, such as the number of games won and lost by locally favorite teams.

Although these examples are indeed a part of the total concept of statistics, the word has a broader meaning for those whose jobs require a knowledge (though sometimes a minimal one) of the more technical aspects of statistics. To this person, the word statistics refers to those concepts and techniques that are employed in the collection, organization, summarization, analysis, interpretation, and communication of numerical information. These concepts and techniques consequently play an important role in the activities engaged in by practitioners of all the sciences.

A particular series of statistical endeavors is usually designed to achieve one (sometimes both) of the following objectives.

1 To quantitatively describe some collection of persons, places, or things

2 To provide information from which one can draw inferences about some large collection of persons, places, or things through observation of only a small part of the larger collection

Statistical activities aimed at achieving the first goal are referred to as *descriptive statistics*, while those activities designed to meet the second objective are called *inferential statistics*. This chapter covers the most important aspects of descriptive statistics. Chapters 2, 3, and 4 present the basic ideas of probability, probability distributions, and sampling distributions. There topics provide the foundation for an understanding of the concepts and procedures of statistical inference, the subject matter of the remaining chapters.

Why study statistics?

Most readers of this book will not become statisticians. You may therefore ask why you should have to study statistics. The reason, in a nutshell, is that the concepts and techniques of statistics are now used in a wide variety of occupations. Statistical ideas are an integral part of research activities, data-collecting surveys, and the analysis of data generated by the ongoing activities of institutions and organizations.

A worker may need to know only enough about statistics to enable him or her to know when the services of a statistician are needed, and to communicate effectively with the statistician as they work together in planning, conducting, and interpreting the results of the activity requiring the methodology of statistics.

The person who has an understanding of statistical concepts and methodology will be a better consumer of statistics. This person will be better prepared to evaluate the results of research and other reported information. The professional who understands statistics will be able to read with greater understanding the journal literature in his or her chosen field.

Finally, you will find that a knowledge of statistics is of considerable help to you in your other courses. Many textbooks in other fields are written on the assumption that the student has at least a basic understanding of statistical ideas and techniques, and many upper-division courses have as a prerequisite a course in statistics.

Areas of application of statistical methodology

We have noted that the concepts and methodology of statistics are used in many fields. The following are just a few of the areas in which statistics is used.

Agriculture Statistical techniques are used in such activities as plant- and and animal-breeding experiments; the study of the relative merits of different fertilizers, insecticides, and so on; and studies of methods for increasing crop yields.

Biology Statistical methods are used in biology to study the reactions of plants and animals to various environmental stresses, and in the study of heredity.

Business By employing the methods of statistics, business people predict sales volumes, measure consumer reactions to new products, decide how to spend their advertising budgets, and determine how best to use the skills and talents of their employees.

Health and medicine The results of drug research are analyzed by statistical techniques. Health officials use statistics in planning the location and size of hospitals and other health facilities. Research physicians use statistical analysis to help them evaluate the effectiveness of different treatments.

Industry Most manufacturers use some form of quality control. Statistical concepts and techniques form the basis of most of these quality-control programs.

Psychology Psychologists use the concepts and techniques of statistics to measure and compare human behavior, attitudes, intelligence, and aptitudes.

Sociology In sociology, statistical techniques are used in comparative studies of different socioeconomic and cultural groups, and in the study of group behavior and attitudes.

1.1 SOME STATISTICAL TERMINOLOGY

At this point we must define some basic terms that will be used in this text. We shall present only a basic statistical vocabulary in this section, and define additional terms as they occur later on.

Entity The object of attention in statistical analysis is some collection of persons, places, or things. A biologist may be interested in the squirrels inhabiting a certain area. A physician may be interested in patients with a certain set of symptoms. An educator may be interested in students who have been taught to read by a certain method. An agricultural researcher may be interested in a certain variety of wheat. A meteorologist may be interested in the precipitation in a certain area. We shall use the word *entity* as a general term to refer to an individual member of such collections of persons, places, and things.

Variable It is the characteristics of entities that are of interest in a statistical investigation. The biologist may have an interest in the size of the foramen magnum of squirrels. The physician may wish to investigate the cholesterol level of certain patients. The educator may be concerned with the reading achievement of students taught to read by a certain method. The agricultural researcher may be interested in the resistance of a variety of wheat to a certain disease. The meteorologist may be interested in snow as a proportion of total precipitation. Since any such characteristic, as a rule, exhibits a different value when observed in different entities, it is referred to as a *variable*. In addition to the variables already mentioned, others that readily come to mind are heights of human beings, the lifetimes of automobile tires, the coat colors of dogs, and the presence of left-handedness among school children.

Random variable If the numerical values assumed by a variable are the result of chance factors, so that a particular value cannot be exactly predicted in advance, the variable is called a *random variable*.

We shall use capital letters such as X, Y, and Z to represent random variables. Thus we may wish to refer to the random variable "age" as X or the random variable "height" as Y. Individual values of a random variable will be represented by lower-case letters such as x, y, and z. If the random variable X, for example, has six values, we shall refer to these values as x_1, x_2, x_3, x_4, x_5, and x_6. The subscripts serve to distinguish one value of the random variable from another.

Variables may be further classified according to whether they are *continuous* or *discrete*, and whether they are *quantitative* or *qualitative*.

Continuous variable A *continuous variable* is one which theoretically can assume any value within an interval of values. That is, a continuous variable is measured on a continuum. Another way to describe a continuous variable is to point out that no matter how close together two values of a variable may be, it is theoretically possible to find another value that falls between them. An example of a continuous variable is human height. No matter how close in height two people may be, it is theoretically possible to find another person who is taller than the shorter, yet shorter than the taller of the two. Our ability to identify such a person in practice may be hampered because of the limitations of available measuring devices. Other continuous variables include those that are measured on some weight, time, or temperature scale.

Discrete variable When the values which a variable may assume are separated from each other by some amount, the variable is called a *discrete variable*. A feature of a discrete variable is the presence of "gaps" or "interruptions" in the values which it can assume. Examples of discrete variables include the number of admissions to a certain hospital on a given day, the number of automobile accidents occurring within some city limits in a month, the number of colonies of bacteria on an agar plate, and the number of students in first-grade classrooms in a certain school system.

Quantitative variable A variable is said to be a *quantitative variable* whenever the values which it can assume are the results of numerical measurement. Examples of quantitative variables include height, weight, temperature, IQ, test scores, blood pressure, the number of students in first-grade classrooms, and the number of accidents occurring in some geographic area within a given time interval.

Qualitative variable In the case of many variables, it is not possible to make numerical measurements. Many variables are capable only of being classified. The variable "marital status," for example, can be assigned the qualitative values Single, Married, Divorced, Widowed, and perhaps All other. If the entities of interest are undergraduate college students, and if the variable of interest is "class standing," the variable may assume the qualitative values Freshman, Sophomore, Junior, and Senior. A variable whose values consist of categories of classification is called a *qualitative variable.*

Since the determination of the value of a quantitative variable is accomplished by means of some measurement procedure, the result is called a *measurement.* A collection of such measurements is called *measurement data.* When entities are classified on the basis of a qualitative variable, it is usually of interest to *count* the number of entities falling into each category. Data of this type are called *count data.* The word *observation* is frequently used to refer to either a measurment or a count datum.

Population The average person's concept of a population is that it is some aggregate of people, such as the population of a city, a state, or a nation. Although we shall sometimes use the term to refer to a collection of entities, we shall more often use it to refer to a collection of values of some random variable associated with a collection of entities. For example, we may speak of a population of weights, a population of test scores, or a population of cholesterol levels. We shall define a *population,* then, as the largest collection of values (of a variable) for which there is an interest.

This definition indicates that populations are defined by the investigator. They are not predetermined by some process beyond the investigator's control. Suppose, for example, that we are interested in the reading achievement scores of all elementary school students in some school system. Our population, then, consists of all these reading achievement scores. If, however, we are interested only in the scores of fifth-grade students in the system, we have quite a different population. By defining our sphere of interest, we define our population.

Populations may be either *finite* or *infinite*. If indeed our interest is confined to the scores of students in some school system at some point in time, we are defining a finite population. If, however, we are interested in the scores of all elementary school students of the past, present, and future, the population of interest is, for all practical purposes, infinite.

Sample A *sample* is part of a population. The sheer size of a population of interest, even though it is finite, may discourage one from attempting to investigate it in its entirety. It may be necessary or desirable to examine only a fraction (sample) of the population. Although we have already defined statistical inference in general terms, we may be more specific now and say that statistical inference is the procedure whereby inferences about a population are made on the basis of information about a sample from that population.

A note on rounding When recording final or intermediate results in this text, we shall abide by the following rule. If the digit to the right of the position of the last digit to be retained (and recorded) is *less than* 5, record the digit occupying the position of the last digit to be retained. If the digit to the right of the position of the last digit to be retained is *greater than* 5, increase by 1 the digit occupying the position of the last digit to be retained, and record that digit. If the digit to the right of the position of the last digit to be retained is 5, record the digit occupying the position of the last digit to be retained *as is* if it were even, and increase it by 1 before recording it if it is odd. The following examples illustrate this rule.

Final or intermediate result	Recorded result
175.787	175.79
175.783	175.78
175.785	175.78
175.775	175.78

1.2 SUMMARIZING DATA: THE FREQUENCY DISTRIBUTION

Data that are collected in the course of a statistical investigation usually are not amenable to analysis and interpretation in the form in which they are available. To facilitate numerical calculations, one has to extract information from records or questionnaires and put it in some convenient form. The amount of data to be processed may be so great that organization and analysis are impractical without the aid of high-speed computers and other machinery. For moderate amounts of data, the work may be done by hand, and we shall discuss hand procedures here. Bear in mind, however, that most of what can be done by hand can be done by machine. When the data are voluminous, machine methods are preferable.

A very useful device for summarizing large sets of data is the *frequency distribution*, which is a presentation of numerical categories of the variable along with the number of entities falling into each category. The categories, which are non-overlapping and contiguous, are called *class intervals*. Each class interval is identified by its *upper class limit* and *lower class limit*. The class limits specify the magnitude of values that can go into a class interval. That is, a class interval will contain no value smaller than the lower class limit and no value larger than the upper class limit. Frequency distributions may be represented in either tabular or graphic form. You can find examples of frequency distributions in the form of tables and graphs in many published

Figure 1.1 Age at onset of disability of male disabled workers

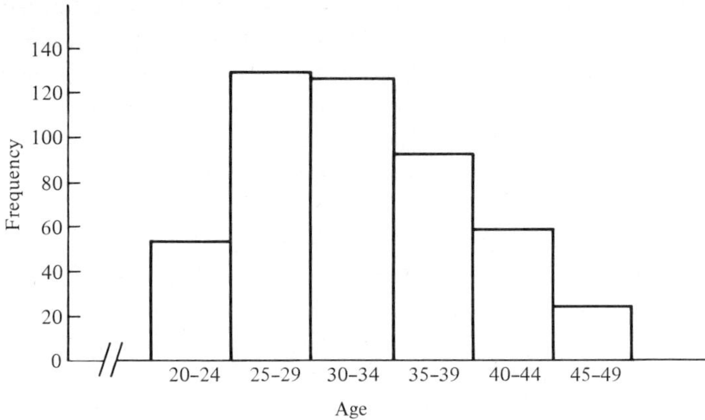

sources, such as newspapers, magazines, and the annual reports of business firms and government agencies. Table 1.1 and Figure 1.1 are examples of frequency distributions displayed in tabular and graphic form, respectively.

Before constructing a frequency distribution, we must decide how many class intervals to use and how wide to make each interval. No hard-and-fast rules can be set down. The person in the best position to decide on such matters is one who is thoroughly familiar with the data and the uses that are to be made of them. One should strive for clarity of presentation without obscuring any of the salient features of the data. Too few intervals may result in a loss of too much detail, while too many may not satisfactorily condense the original data. Generally, large amounts of data require more class intervals than small amounts. A rule of thumb frequently followed is that no fewer than 6 and no more than 15 class intervals should be used with a given set of data.

Table 1.2, based on the data of Table 1.1, illustrated the effect of too few class intervals. The severe condensation shows less detail than Table 1.1

Table 1.1 Age at onset of disability of male disabled workers

Age	Frequency
20–24	53
25–29	129
30–34	125
35–39	91
40–44	57
45–49	24
	479

Source: Derived from Table 12 of *The Disabled Worker Under OASDI*, Research Report No. 6, U.S. Department of Health, Education, and Welfare, Social Security Administration, Washington, D.C., 1964

Table 1.2 Age at onset of disability of male disabled workers

Age	Number
20–29	182
30–39	216
40–49	81
	479

communicates. If class intervals of 20–22, 23–25, . . . , 47–49 had been used, however, the resulting table would have suffered from too much detail. It is usually good to have intervals of equal width, although this may be inconvenient or undesirable in some instances, for example, cases in which there are one or two extremely small and/or extremely large values. Once one has decided on the limits of the class intervals, one records the number of observations falling into each interval. A given observation may be placed in one, and only one, class interval.

With a frequency distribution at hand, you can determine at a glance the frequency of occurrence of values falling into the various class intervals. Or, by appropriately adding frequencies, you can determine the number of values falling within the limits of two or more intervals. The following example will serve to illustrate these points.

Example 1.1 Table 1.3 gives the IQ scores of 150 third-grade students in a certain school system. The teachers would like to prepare a frequency distribution from the scores. Examination of the data reveals that the lowest and highest scores are 85 and 129, respectively. The lower class interval, then, must begin at least as low as 85, and the upper class interval must have an upper limit of at least 129. Suppose a class interval width of ten units is considered. Beginning with 85, this choice of interval width would yield the following equal-width class intervals.

85– 94
95–104
105–114
115–124
125–134

This choice of class interval width results in five intervals, thereby violating the previously stated rule of thumb which says that there should be at least six intervals. Table 1.4 shows the resulting frequency distribution.

Table 1.3 IQ scores of 150 third-grade students

88	91	104	113	125	101	114	105	101	88	126	118	100	111	125	109
119	91	106	120	129	120	109	104	112	101	113	100	106	105	121	128
93	89	124	96	105	95	91	106	93	88	89	100	115	98	108	88
99	120	101	108	118	118	113	114	109	91	104	109	110	113	119	119
106	106	97	104	105	122	112	124	108	121	96	97	99	101	116	118
102	127	121	116	100	95	89	103	115	113	129	91	85	108	103	116
108	98	108	114	102	96	99	108	114	121	107	122	100	116	111	113
109	104	113	118	110	129	124	105	93	115	120	97	112	94	113	122
114	106	105	115	98	112	103	92	125	107	115	118	128	92	85	126
108	114	125	121	122	117										

Let us suppose that the third-grade teachers feel that such a choice of interval widths would condense the data more than they wish. They decide instead to use class intervals that are five units wide. Such a choice leads to the frequency distribution shown in Table 1.5. As one can see, Table 1.5 contains a greater amount of detail than Table 1.4.

The choice of class limits reflects the extent to which the numbers to be grouped have been rounded. If we are grouping numbers that have been rounded to the nearest whole number, a class interval with limits of, say, 20–24 actually would contain all numbers between 19.5 and 24.5. These numbers are sometimes called the *class boundaries* or *true class limits*. The true class limits for the class intervals shown in Table 1.5 are as follows.

84.5– 89.5	109.5–114.5
89.5– 94.5	114.5–119.5
94.5– 99.5	119.5–124.5
99.5–104.5	124.5–129.5
104.5–109.5	

With a frequency distribution as a point of reference, we may quickly determine the number of observations falling into the various categories. By referring to Table 1.5, for example, we see that 20 observations are between 100 and 104 inclusive, 22 observations are in the interval 110–114, and so on. By adding adjacent intervals, we may determine the number of observations falling between two widely spaced values. Thus we see that there are 47 observations between 100 and 109 inclusive.

The cumulative frequency distribution

It is often convenient to have the data of a frequency distribution in cumulative form. Considering the IQ data of third-grade students, for example, we might wish to be able to quickly determine from a table the number of students with IQs below 110. We can do this by means of a table that displays the data in the form of a cumulative distribution. We can obtain a *cumulative*

Table 1.4 Frequency distribution of IQ scores of 150 third-grade students (class intervals 10 units wide)

Class interval	Frequency
85– 94	20
95–104	34
105–114	49
115–124	35
125–134	12
	150

Table 1.5 Frequency distribution of IQ scores of 150 third-grade students

Class interval	Frequency
85– 89	9
90– 94	11
95– 99	14
100–104	20
105–109	27
110–114	22
115–119	19
120–124	16
125–129	12
	150

frequency distribution from a frequency distribution such as that in Table 1.5 by showing, for each class interval, the frequency of that interval plus the frequencies of all preceding intervals. Cumulation may begin with either the smallest or the largest class interval. A cumulative frequency distribution that starts with the smallest class interval shows, for each class interval, the frequency of values that are *less than* or equal to the upper limit of that interval. A cumulative frequency distribution that begins with the largest class interval shows, for each class interval, the frequency of values *greater than* or equal to the lower limit of that interval. An example of a cumulative frequency distribution of the first type ("less than") is shown as Table 1.6.

A table such as Table 1.6 enables us to determine at a glance the number of observations that are less than or equal to any upper class limit. Thus we see that 81 of the 150 observations are less than or equal to 109. There are 122 observations less than or equal to 119. And so on.

The relative frequency distribution

For some purposes one may want to show the proportion, or percentage, of values falling into the various class intervals. The display of data in this fashion is called a *relative frequency distribution.* One can obtain the proportion for a given interval by dividing the frequency of that interval, as shown in Table 1.5, for example, by the total number of observations. Multiplying a proportion by 100 converts it to a percentage of total cases.

One may also construct a *cumulative relative frequency* distribution. One may add successive proportions in the same way as observed frequencies or, alternatively, one may divide cumulative frequencies by the total number of observations. Table 1.7 shows relative frequency and cumulative relative frequency distributions for the data of Table 1.5.

We can use relative frequency distributions in much the same way as frequency distributions. However, they convey information about the rela-

Table 1.6 Cumulative frequency distribution of IQ scores of 150 third-grade students

Class interval	Cumulative frequency
85–89	9
90–94	20
95–99	34
100–104	54
105–109	81
110–114	103
115–119	122
120–124	138
125–129	150

Table 1.7 Relative frequency and cumulative relative frequency distributions of IQ scores of 150 third-grade students

Class interval	Relative frequency	Cumulative relative frequency
85– 89	0.06	0.06
90– 94	0.07	0.13
95– 99	0.09	0.22
100–104	0.13	0.35
105–109	0.18	0.53
110–114	0.15	0.68
115–119	0.13	0.81
120–124	0.11	0.92
125–129	0.08	1.00
	1.00	

tive frequency or proportion of observations of varying magnitudes. From the relative frequency column of Table 1.7, for example, we see that 0.13, or 13%, of the observations are between 100 and 104 inclusive. From the cumulative relative frequency distribution column, we see that 0.35, or 35%, of the observations are less than or equal to 104.

EXERCISES

For each of the exercises below, construct (a) a frequency distribution, (b) a cumulative frequency distribution, (c) a relative frequency distribution, (d) a cumulative relative frequency distribution.

1 Table 1.8 gives the weights, in pounds, of 150 adult males.

2 Table 1.9 shows the heights, in feet, of 100 trees growing on a farm.

3 In a psychology experiment, subjects were required to memorize a certain sequence of words. Table 1.10 lists the times, in seconds, that the participants in the experiment required for memorization.

Table 1.8 Data for Exercise 1

158	176	165	179	168	159	179	162	176	168	184	173	175	169	173	170
179	177	178	175	176	174	173	184	191	179	174	177	163	179	187	168
158	180	181	160	176	171	179	160	163	167	178	178	170	175	173	181
175	168	177	176	176	178	169	176	186	175	184	180	162	178	188	165
171	176	189	178	161	183	175	171	171	187	177	172	168	186	174	180
178	174	181	163	182	177	165	176	186	177	189	168	188	181	177	175
189	171	177	166	184	189	175	183	180	181	166	179	188	185	178	176
164	185	179	178	176	176	186	171	176	175	177	179	176	180	183	184
180	172	188	165	179	184	186	187	170	167	176	182	188	186	170	178
171	181	190	172	165	193										

Table 1.9 Data for Exercise 2

2	2	9	5	9	13	10	16	19	20	3	2	8	5	8
13	11	17	17	24	1	2	9	9	5	14	11	17	15	23
4	2	8	5	5	14	13	16	16	24	4	4	8	7	6
10	10	15	16	21	1	4	9	6	9	11	13	15	22	28
3	1	9	6	5	13	13	16	20	26	1	1	8	8	8
12	13	19	21	28	3	6	6	5	13	14	18	15	21	29
3	5	7	8	13	12	16	16	24	26					

Table 1.10 Data for Exercise 3

100	107	34	57	66	30	79	84	118	77	135	95	130	138	52	126
89	128	100	88	61	108	79	37	93	116	45	57	112	73	129	46
107	109	32	106	122	41	70	96	98	117	97	99	62	88	85	149
75	105	50	99	50	79	43	90	114	53	123	100	69	87	64	85
126	100	102	112	78	118	135	110	64	62	107	127	102	129	88	123
98	110	93	135	58	73	80	125	88	142	103	149	90	145	96	146
119	76	93	99												

Table 1.11 Data for Exercise 4

104	112	128	139	118	132	132	112	106	107	129	125	103	125	104	129
126	126	115	118	117	116	113	122	123	107	122	105	110	133	142	143
116	114	129	117	106	124	115	118	123	101	123	121	124	120	116	117
105	120	146	121	120	102	138	106	113	130	111	123	124	120	113	115
114	122	116	108	122	112	112	123	116	115	116	111	120	119	122	123
124	111	121	111	114	123	107	120	120	106	118	116	135	121	123	117
124	122	134	131												

4 Table 1.11 shows the systolic blood-pressure readings recorded for 100 adults presenting themselves for pre-employment physical examinations.

1.3 SUMMARIZING DATA: THE HISTOGRAM AND THE FREQUENCY POLYGON

One can often communicate the meaning of data more effectively by means of graphs than by the use of tables. There are a variety of ways in which one can present data graphically. You are no doubt familiar with such frequently used devices as the bar graph and the pie or circle chart. We shall limit discussion in this book to two particular types of graph, the histogram and the frequency polygon.

The histogram

The *histogram*, a special type of bar graph, is used to represent a frequency distribution or a relative frequency distribution. Figure 1.2 is a histogram of the frequency distribution of Table 1.5.

Note the following characteristics of the histogram.

1 The possible values of the variable under consideration are assigned to the horizontal axis. The frequencies of occurrence of values of the variable are represented on the vertical axis.

2 Each class interval of the frequency distribution is represented by a bar of the histogram.

3 The bars are of the same width as the corresponding class intervals.

4 The height of a given bar corresponds to the frequency of occurrence of values in the corresponding class interval. That is, for a given set of data, class intervals with large frequencies will be represented in the histogram by tall bars, and class intervals with small frequencies will be represented by short bars.

5 The bars are drawn adjacent to one another. The purpose of this is to portray the continuous nature of the data under consideration. Intelligence, presumably, is a continuous variable. Values of the variable are presented as

Figure 1.2 Histogram of IQ scores of 150 third-grade children

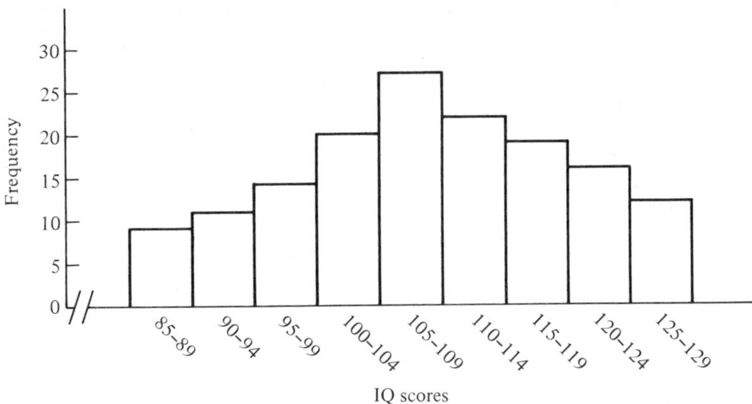

whole numbers because of the limitation of the measuring instrument. Even discrete data, when represented by a histogram, are treated as though they were continuous.

6 Of the total area under the histogram, that proportion enclosed by a given bar is equal to the proportion that the frequency of the corresponding class interval is of the total number of observations. In Figure 1.2, for example, the area enclosed by the first cell is equal to $\frac{9}{150} = 0.06$ of the total area under the histogram.

The frequency polygon

The data of a frequency distribution may be depicted by means of another graphic device known as a *frequency polygon.* One can construct a frequency polygon by first drawing a histogram and then connecting the midpoints of the tops of the cells by a series of straight lines. Figure 1.3 shows a

Figure 1.3 Frequency polygon of IQ scores of 150 third-grade children

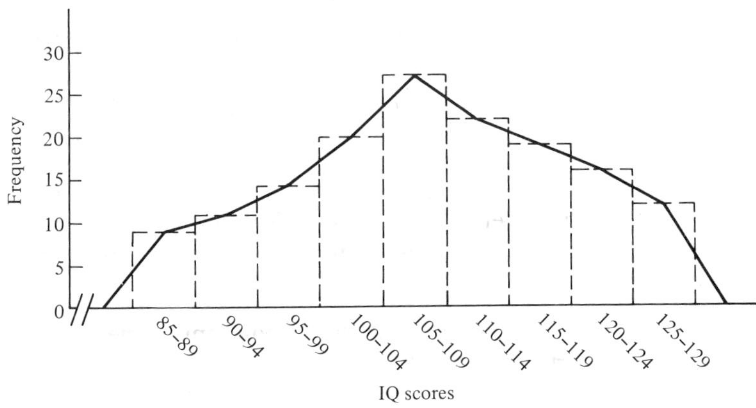

frequency polygon for the IQ data of Table 1.5. The outline of the histogram is usually deleted from the figure of a frequency polygon. The dashed lines in Figure 1.3 show the histogram so that you may see the relationship between the two types of graph.

Some important characteristics of a frequency polygon are the following.

1 The end points of the histogram, unless there is some logical reason to the contrary, are brought down to the horizontal axis at a point corresponding to the midpoint of an imaginary class interval of the same width as those used, and adjacent to the histogram to the right and left.

2 The total area under the curve is equal to the total area under the corresponding histogram.

EXERCISES

5 Refer to Exercise 1. Construct a histogram and a frequency polygon from the data.

6 Refer to Exercise 2. Construct a histogram and a frequency polygon from the data.

7 Refer to Exercise 3. Construct a histogram and a frequency polygon from the data.

8 Refer to Exercise 4. Construct a histogram and a frequency polygon from the data.

1.4 SUMMARIZING DATA: DESCRIPTIVE MEASURES

For many purposes in statistical analysis, the summarization of data by the construction of frequency distributions does not sufficiently condense the available information. More often than not, the analysis calls for a single numerical value which summarizes a facet of the data and, at the same time, conveys information regarding some characteristic of the data. In other words, the numerical value we seek is one which in some sense describes the data. Hence it is called a *descriptive measure*. We can describe a set of data in several ways. In this chapter we shall be concerned with descriptive measures which indicate the *central tendency* and extent of *variability* present in a set of data. Before we take up these topics, however, we shall devote our attention to a subject that is of great importance in statistical analysis.

Simple random sampling

In most activities requiring statistical analysis, the ultimate objective is to learn something about a population. However, it may be impossible or impractical to examine a population in its entirety. For example, a public health agency may be interested in the serum cholesterol levels of adult males living in some large metropolitan area. The size of such an agency's budget, however, usually prohibits the collection of serum cholesterol values for the entire population. The solution is to obtain data on a sample, or part, of the population. The method whereby information about a population is derived from a sample of the population is known as *statistical inference*. We shall discuss it in detail in later chapters. For the time being, let us concern ourselves with the problem of obtaining the sample.

There are a variety of ways to obtain a part of a population. For the statistical inference procedures which constitute the major portion of this book to be valid, however, we assume special types of samples, which we shall call *scientific samples*. One type of sample that is frequently required is a *simple random sample*. In most of the examples and exercises in the rest of this book, it will be either implied or explicitly stated that the data under consideration constitute a simple random sample.

DEFINITION *A* ***simple random sample*** *of size n is a sample selected from a population in such a way that every sample of size n that can be drawn from the population has the same chance of being selected as the sample actually selected.*

In order to satisfy the requirements of a simple random sample, the investigator selects the sample in a special way known as *simple random sampling*.

One way to draw a simple random sample from a finite population is first to make a list of every entity (subject, value, or whatever) in the population and to identify each one by a unique number. Then one determines the particular entities that are to be selected for the sample by means of a *table of random numbers*. A table of random numbers is a table of numbers that have been produced and displayed in such a way that the numbers 0 through 9 occur with approximately equal frequency. The arrangement does not reflect any conscious pattern or system. In other words, the numbers appear in the table in random order. Table A of the Appendix is one page of random numbers selected from a table containing a million random digits (1). Let us illustrate the procedure of drawing a simple random sample by means of an example.

Example 1.2 There are 180 freshmen at a rural high school. To get information about the television-viewing habits of all freshmen, a guidance counselor wishes to select a simple random sample of ten students to fill out a questionnaire. An alphabetical list of the students, numbered consecutively from 1 through 180, is available from the principal's office. The guidance counselor uses Appendix Table A to decide which students will be in the sample.

Since the number of students in the population is 180 (a three-digit number), it is convenient to think of the numbers from 1 through 180 as the three-digit numbers 001, 002, 003, . . . , 180. Only three-digit numbers between 001 and 180 are usable.

The counselor selects a random starting point on the page of random numbers by closing her (or his) eyes and letting the point of a pencil touch the page. The number closest to the pencil point is the starting point. The pencil point comes closest to the number 1 at the intersection of row 36 and column 7. Since the first three-digit number from this position is 131, student number 131 on the list is in the sample. The counselor moves down (the direction of movement is arbitrary, it could have been up, diagonally, etc.) to the next three-digit number, which, since it is 255, is unusable. Continuing down, the next usable number is 063, so student number 63 is in the sample. When the counselor gets to the bottom of the page, she/he merely moves to the right one digit and considers the three-digit numbers going up. (Again the choice of direction is arbitrary. For example, the counselor could have moved to the left, or started at the top of the page.) The procedure followed by the counselor yields the following random numbers:

131, 063, 120, 065, 154, 117, 002, 166, 031, 101

The simple random sample, then, consists of the 10 students who are identified by these numbers on the list.

Simple random sampling may be carried out in one of two ways: *without replacement* or *with replacement.* When sampling *without replacement*, one allows a given entity to appear only once in the sample. When one uses random numbers to select the sample, one discards duplicate numbers when they appear. When one is sampling *with replacement,* there is no limit to the number of times an entity can appear in the sample. In most practical applications, one uses sampling *without replacement.*

[*Note*: As a rule, small populations are not sampled; instead, they are studied in their entirety. In light of this, the above example may be unrealistic.]

Measures of central tendency

A *measure of central tendency* is a single number that designates the center of the collection of numbers from which it is computed. Measures of central tendency are also called *measures of location.* If we think of the values in a set of data as points on a horizontal axis, the measure of location tells where on the axis the "center" of the distribution is located. A measure of central tendency conveys the concept of "average." In this book we shall use the word average in this general sense. When we want to be more specific, we shall use one of the specific terms defined below.

There are several measures of central tendency in common use. We shall limit the discussion in this book to three: the arithmetic mean, the median, and the mode.

The arithmetic mean The most frequently encountered measure of central tendency is the *arithmetic mean*, or *mean*, as it is usually called. It is this measure of central tendency that the typical person has in mind when speaking of the "average."

One computes the mean by adding up the values for which a mean is desired and dividing the resulting sum by the number of values entering into the sum.

Consider some random variable X on which n measurements $x_1, x_2, \ldots, x_n$ have been taken. The mean of the n measurements is found as follows.

$$\text{Mean} = \frac{x_1 + x_2 + \cdots + x_n}{n}$$

If we represent the mean by the symbol $\bar{x}$, we may indicate its calculation in more compact form as

$$\bar{x} = \frac{\sum_{i=1}^{n} x_i}{n}$$

where $\sum_{i=1}^{n} x_i$ indicates that all available x's from x_1 through x_n are to be added. The $i = 1$ and n below and above the summation sign Σ indicate which sequence of values is to be summed. These are called the *limits of summation*. Whenever it is clear from the context what the *limits of summation* are, one may omit these symbols and use Σ alone to indicate summation.

For example, suppose a sample consists of $x_1 = 7$, $x_2 = 4$, $x_3 = 3$, and $x_4 = 2$. The sample mean is computed as follows.

$$\bar{x} = \frac{\sum_{i=1}^{4} x_i}{4} = \frac{7 + 4 + 3 + 2}{4} = 4$$

The Appendix gives a further discussion of summation notation which you may find helpful.

Note the difference between the use of capital X and lower-case or small x. We shall use capital X to refer to random variables such as height, weight, and test score. Later we shall use X_1, X_2, and so on, to distinguish one random variable from another. We shall also use other capital letters such as Y to designate a random variable. We shall designate specific observed values of the random variable X by small x's. Subscripts on small x's serve to distinguish one value from another. Suppose, for example, that IQ tests are administered to a sample of five children. The observed IQ scores might be $x_1 = 105$, $x_2 = 103$, $x_3 = 120$, $x_4 = 115$, and $x_5 = 100$. The subscript i is used to designate a typical value of X. We shall follow the common practice of using n to indicate the number of values in a sample and $\bar{x}$ to designate a sample mean.

We shall designate the mean of a population by the Greek letter μ (mu). When the population is finite, we can obtain the population mean from

$$\mu = \frac{\sum_{i=1}^{N} x_i}{N}$$

where N designates the size of the population.

Example 1.3 Let us refer to Example 1.2 and suppose that X, the random variable of interest, is the number of hours the students spent watching television during the week preceding the interview. Suppose further that the 10 students in the sample reported that they watched television the following numbers of hours:

24, 25, 22, 20, 15, 25, 17, 16, 15, 17

The mean number of hours of viewing during the week by the 10 students is

$$\bar{x} = \frac{24 + 25 + \cdots + 17}{10} = \frac{196}{10} = 19.6$$

A disadvantage of the mean as a measure of central tendency is that it may be so heavily influenced by a single extreme value that it gives a distorted impression of the data. Suppose, for example, that 10 students in a class make scores of 91, 95, 95, 94, 92, 93, 98, 97, 96, 0 on a test. The mean of 85.1 does not seem to be a good representation of the performance of the class as a whole.

The median The second measure of central tendency is the *median*. The *median* is that value which lies in the middle of a sample or population of values when they are arranged in order of magnitude. If the number of values is odd, the median is equal to the middle value. If the number of values is even, the median is equal to the mean of the two middle values. Thus the median divides the observations into two halves such that half the values are less than or equal to the median value and the other half are equal to or greater than the median. Before computing the median, one must arrange the observations in a sample or population in order of magnitude.

Example 1.4 Let us refer to Example 1.3 and compute the median number of hours of television viewing by the 10 students. Arranging the values in order of magnitude from lowest to highest, we have

15, 15, 16, 17, 17, 20, 22, 24, 25, 25

and the median is equal to $(17 + 20)/2 = 18.5$.

The median is not as greatly affected by extreme values as is the mean. To illustrate, consider the data on test scores presented earlier. When the observations are arranged in ascending order, we have

0, 91, 92, 93, 94, 95, 95, 96, 97, 98

and the median is seen to be $(94 + 95)/2 = 94.5$. As the teacher of the class, which "average" grade would you prefer to report, the median of 94.5 or the mean of 85.1?

The mode The *mode* is the value that appears most frequently in a set of data.

Example 1.5 Eleven high school freshmen made the following scores on a manual dexterity test.

70, 83, 74, 75, 81, 75, 92, 75, 90, 94, 75

The mode for these data is 75, since this score occurs more frequently than any other.

A set of data may have either no mode or more than one mode. This is not true for the mean and median, which, for a given set of data, always exist and are unique. The mode is not a widely used measure.

Measures of variability

A measure of central tendency alone usually does not provide a satisfactory description of a set of data. Those interested in the data usually desire, in addition, a measure of the extent to which individual values deviate from the "average." Such measures are called *measures of variability*. Synonyms for variability include *dispersion*, *spread*, and *scatter*.

For example, showing only the mean of a sample of values may not convey sufficient information to clearly portray the nature of the data. Suppose we know that the mean age of five persons at a birthday party is 18 years. If no information regarding the variability of the data is given, the unsuspecting reader might conclude that the party was made up of five teen-agers. The same mean age would be computed from the ages of a 73-year-old grandmother giving a birthday party for her five-year-old grandson and his three cousins, aged 3, 5, and 4.

The range The simplest measure of variability is the *range*, which is the difference between the largest and smallest values in a set of data.

Example 1.6 Let us refer to Example 1.3 and compute the range for the number of hours of television viewing by the sample of ten high school freshmen. The largest and smallest values were 15 and 25, so we have

$$\text{Range} = 25 - 15 = 10$$

The range is of limited value as a measure of variability. In the first place, it takes into account only the extreme values in a set of data, and gives no hint as to how the values in between are dispersed over the interval. Second, the sample range depends on the number of observations in the sample. Extreme values of a population, being fewer in number, are not as likely to appear in small samples as in large ones, and consequently small samples tend to have small ranges and large samples large ranges.

The variance The shortcomings of the range are overcome by another measure of variability known as the *variance*. One obtains the variance of a set of data by subtracting each value from the mean of all the values, squaring each of the resulting differences, adding the squared differences, and dividing this total by the number of values less 1. If we condense these words into compact notation, we have the following formula for the sample variance, denoted by the symbol s^2.

$$s^2 = \frac{\sum_{i=1}^{n} (x_i - \bar{x})^2}{n - 1}$$

Example 1.7 Let us again use the television-viewing data of Example 1.3 and compute the sample variance. Using the formula, we have

$$s^2 = \frac{(24 - 19.6)^2 + (25 - 19.6)^2 + \cdots + (17 - 19.6)^2}{9}$$

$$= \frac{152.4}{9} = 16.93$$

We can see from the formula for the variance that, except for the fact that division is by $n - 1$ rather than n, the variance would be the mean of the squared deviations of the observations from the sample mean. You may well ask why the denominator is $n - 1$ rather than n. The simple answer, which will be expanded on in a later chapter, is that division by $n - 1$ yields a measure that is more useful for inferential purposes. If the analytic objective is merely to describe the variability exhibited by a sample, it is perfectly satisfactory to compute the sample variance by dividing by n. Since, however, most samples are analyzed for the purpose of inferring to a population, division by $n - 1$ is usually used to compute the sample variance.

The denominator, $n - 1$, is called the *degrees of freedom*. A rigorous explanation of degrees of freedom would go beyond the level of this book. We can explain the idea intuitively, however, as follows. Suppose that we use the data of Example 1.3, subtract the mean from each value, and add the difference. We get

$$(24 - 19.6) + (25 - 19.6) + (22 - 19.6) + (20 - 19.6) + (15 - 19.6)$$
$$+ (25 - 19.6) + (17 - 19.6) + (16 - 19.6) + (15 - 19.6)$$
$$+ (17 - 19.6)$$
$$= +4.4 + 5.4 + 2.4 + 0.4 - 4.6 + 5.4 - 2.6 - 3.6 - 4.6 - 2.6$$
$$= 0$$

When we perform similar calculations on any set of data, the sum obtained will be zero. We may express the result symbolically by $\sum_{i=1}^{n} (x_i - \bar{x}) = 0$, which is called an *identity*, since it is true for all allowable values of the variable. Because of the identity, then, once $n - 1$ of the differences, or deviations, are known, the value of the last is automatically determined. Thus we say that there are $n - 1$ degrees of freedom available for computing the sample variance.

The above formula is called the *definitional* or *conceptual formula* for the sample variance, since it is a symbolic representation of the definition of the variance. When a desk or pocket calculator is used for the computations, it is usually more convenient to use an alternative formula, known as a *computational formula*. A computational formula for the sample variance is

$$s^2 = \frac{n \sum x_i^2 - (\sum x_i)^2}{n(n-1)}$$

To illustrate the use of this formula, let us use the data of Example 1.3 to compute

$$s^2 = \frac{10(24^2 + 25^2 + \cdots + 17^2) - (24 + 25 + \cdots + 17)^2}{10(9)}$$

$$= 16.93$$

This is the result we obtained previously. Note that the computational formula is not an approximation of the conceptual formula. The two formulas always give identical results.

The standard deviation The positive square root of the variance is called the *standard deviation.* For many purposes it is a more useful measure of variability than the variance. For one thing, the standard deviation is expressed in the same units as the original observations and the mean, whereas the variance is expressed as units squared. Referring to the previous example, where the unit of measurement is hours, we can say that the mean amount of time the ten students spent watching television is 19.6 hours, the standard deviation is $\sqrt{16.93} = 4.1$ hours, and the variance is 16.93 hours squared.

The following formulas yield the sample standard deviation s.

$$s = \sqrt{\frac{\sum (x_i - \bar{x})^2}{n-1}}, \qquad s = \sqrt{\frac{n \sum x_i^2 - (\sum x_i)^2}{n(n-1)}}$$

The variance and standard deviation of a population are designated by the symbols σ^2 and σ, respectively. The Greek letter σ is pronounced sigma. The variance of a finite population of size N is computed from

$$\sigma^2 = \frac{\sum_{i=1}^{N} (x_i - \mu)^2}{N}$$

The computational formula for σ^2, analogous to that for the sample variance, is

$$\sigma^2 = \frac{N \sum x_i^2 - (\sum x_i)^2}{N^2}$$

To obtain a formula for σ, take the square of the formulas for σ^2.

EXERCISES

9 The population of 180 high school freshmen in Example 1.2 is represented in Table 1.12, which shows the number of hours each subject watched television during the week preceding the interview. Draw a simple random sample of size ten and compute the mean, median, mode, range, variance, and standard deviation.

For each of the following exercises, compute the sample mean, median, mode, range, variance, and standard deviation.

10 The amounts of ascorbic acid (100 μg/ml) in ten solutions are:

1.0, 1.5, 3.0, 2.5, 3.5, 3.6, 4.0, 2.5, 6.0, 5.0

11 The following data show the daily caloric intake of 11 adolescent males. Original data have been divided by 1000 and rounded to the nearest tenth.

2.5, 2.3, 2.4, 2.3, 2.3, 2.5, 2.7, 2.5, 2.6, 2.6, 2.7

12 The gains in weight (in pounds) of ten dairy calves placed on a diet supplement are:

121, 101, 110, 108, 107, 95, 89, 120, 109, 117

13 The number of days of school missed by 15 first-grade students during the school year are:

1, 8, 2, 2, 3, 9, 2, 3, 5, 2, 4, 7, 9, 8, 5

14 The distances (in blocks) that 10 employees of a certain firm live from work are:

11, 20, 12, 11, 14, 8, 16, 5, 7, 13

15 Eleven subjects participating in a psychology experiment are asked to memorize a list of ten words. Later the subjects are asked to recall as many of the words as possible. The results are as follows.

6, 4, 8, 9, 10, 6, 5, 8, 9, 4, 8

Table 1.12 Number of hours of television viewing in a week by 180 students

Subject	Number of hours	Subject	Number of hours	Subject	Number of hours	Subject	Number of hours	Subject	Number of hours
1	23	37	23	73	17	109	21	145	18
2	17	38	20	74	23	110	17	146	19
3	23	39	23	75	22	111	19	147	17
4	18	40	20	76	18	112	18	148	21
5	21	41	19	77	20	113	20	149	16
6	22	42	21	78	18	114	16	150	18
7	18	43	17	79	22	115	19	151	17
8	20	44	22	80	20	116	17	152	21
9	24	45	22	81	21	117	25	153	20
10	22	46	20	82	18	118	19	154	15
11	19	47	19	83	24	119	18	155	19
12	18	48	23	84	19	120	22	156	22
13	21	49	20	85	21	121	19	157	25
14	20	50	19	86	22	122	24	158	20
15	20	51	25	87	19	123	20	159	19
16	17	52	18	88	23	124	17	160	21
17	18	53	22	89	19	125	21	161	18
18	20	54	19	90	21	126	15	162	20
19	18	55	21	91	17	127	23	163	22
20	21	56	20	92	24	128	17	164	19
21	17	57	24	93	21	129	17	165	20
22	20	58	20	94	17	130	20	166	16
23	17	59	19	95	23	131	24	167	19
24	20	60	20	96	19	132	16	168	21
25	23	61	23	97	22	133	17	169	24
26	18	62	18	98	17	134	22	170	20
27	18	63	25	99	20	135	16	171	19
28	21	64	21	100	16	136	21	172	21
29	19	65	20	101	17	137	19	173	23
30	19	66	17	102	17	138	17	174	19
31	15	67	25	103	15	139	16	175	21
32	21	68	20	104	21	140	20	176	20
33	25	69	25	105	23	141	16	177	23
34	22	70	21	106	16	142	18	178	25
35	24	71	18	107	18	143	16	179	18
36	21	72	22	108	18	144	22	180	24

1.5 DESCRIPTIVE MEASURES COMPUTED FROM GROUPED DATA

A good reason for grouping data and constructing frequency distributions as described earlier in this chapter is that the calculation of the various descriptive measures may be greatly facilitated. This is an important consideration when only desk or pocket calculators are available and when the size of the sample or population is quite large. Generally, the importance of knowing how to compute descriptive measures from grouped data is diminishing as a result of the widespread availability of electronic computers. However, it is sometimes necessary to compute descriptive measures from published data that are available only in the form of a frequency distribution.

The arithmetic mean When a set of values are grouped into class intervals, as in Table 1.5, individual observations lose their identity. It is possible only to specify the number of observations that fall within the indicated class limits. In calculating the mean, we assume that the values falling into a particular class interval are all equal to the midpoint of that interval. We may obtain the midpoint of an interval by taking the average of the two class limits. The midpoint of the first interval in Table 1.5, for example, is equal to $(85+89)/2=87$. We assume that the 9 values in the first interval of Table 1.5 are all equal to 87, the 11 values in the second interval are equal to 92, and so on. To calculate the mean, we simply compute the sum of the products obtained by multiplying each frequency by the corresponding midpoint and dividing by the sum of the frequencies. Let m_i represent the midpoint of the ith class interval. The formula for the mean of a sample computed from grouped data is

$$\bar{x} = \frac{\sum_{i=1}^{k} m_i f_i}{n} \tag{1.1}$$

where k is the number of class intervals and $n = \Sigma f_i$.

Example 1.8 Let us refer to Example 1.1, assume that the data are a sample, and compute the mean IQ score for the 150 third-grade students by using the data as grouped in Table 1.5. Intermediate steps in the calculation are shown in columns 3 and 4 of Table 1.13.

From Equation 1.1, we have

$$\bar{x} = \frac{16{,}245}{150} = 108.3$$

Table 1.13 Work table for computing descriptive measures from grouped data of Example 1.1

(1) Class interval	(2) Frequency, (f_i)	(3) Midpoint, (m_i)	(4) $m_i f_i$	(5) $m_i^2 f_i$
85– 89	9	87	783	68,121
90– 94	11	92	1,012	93,104
95– 99	14	97	1,358	131,726
100–104	20	102	2,040	208,080
105–109	27	107	2,889	309,123
110–114	22	112	2,464	275,968
115–119	19	117	2,223	260,091
120–124	16	122	1,952	238,144
125–129	12	127	1,524	193,548
	150		16,245	1,777,905

The formula for the mean of a finite population from grouped data is the same as Equation 1.1, except that μ replaces $\bar{x}$ and N replaces n.

The mean computed from grouped data is only an approximation of the true mean. The mean IQ score for the data of Example 1.8 from the original ungrouped observations is 108.4. (Do not expect the two methods always to give results that agree so closely.)

The median In computing the median from grouped data, we recall that this descriptive measure may be defined as the value that lies in the middle of an ordered series of values. We may also define the median of a distribution as that point on the horizontal axis of the corresponding histogram at which, if a vertical line is drawn, the area under the histogram will be divided into two equal parts. According to this definition, the median is the $(n/2)$th value when the values have been arranged in order of magnitude. In calculating the median, we find that value by using the cumulative frequency distribution and making the assumption that the values within each interval are spread evenly throughout the interval. Before giving a general formula for the calculation of the median, let us look at the calculations by means of an example.

Example 1.9 Let us compute the median IQ score for the 150 scores of Example 1.1. We seek the $n/2 = 150/2 = 75$th value. From Table 1.6, we see that 54 of the observations lie within the first four class intervals and 81 lie within the first five intervals. The median, then, is in the fifth class interval. That is, the 75th observation is in the fifth interval, and we need to reach the 21st observation in the interval in order to reach the 75th observation for the complete set of data. Under the assumption that the 27 observations in the fifth interval are evenly distributed through the interval, we must go $\frac{21}{27}$ths of the way into the interval in order to reach the 75th observation. Since the interval is five units wide, we must add $\frac{21}{27}$ths of 5 to 104.5, the lower true limit of the fifth class interval, to obtain the value of the 75th observation. In other words, for the IQ data, we have

$$\text{Median} = 104.5 + \tfrac{21}{27}(5) = 108.4$$

The procedure described above may be summarized by the following general formula.

$$\text{Median} = L + \frac{j}{f}w$$

where

L = true lower limit of the class interval in which the median is located
j = number of observations still needed to reach the median after the class interval containing the median has been reached
f = number of observations in the interval containing the median
w = width of the class interval containing the median

The mode When a measure for grouped data comparable to the mode of ungrouped data is required, it is usually satisfactory to speak of the *modal class*, which is that class interval containing the largest number of observations. The modal class for the IQ data of Table 1.5, for example, is the fifth interval, whose class limits are 105–109.

The variance and standard deviation In order to compute the variance and standard deviation from grouped data, we make the assumption, as we did in computing the mean, that the observations in a given class interval are located at the midpoint of the interval. The conceptual or definitional formula for the sample variance computed from grouped data is

$$s^2 = \frac{\sum_{i=1}^{k} (m_i - \bar{x})^2 f_i}{n - 1}$$

and the computational formula is

$$s^2 = \frac{n \sum_{i=1}^{k} m_i^2 f_i - (\sum m_i f_i)^2}{n(n - 1)}$$

We can obtain comparable formulas for the variance of a finite population, σ^2, by replacing $\bar{x}$ with μ, n with N, and $n(n - 1)$ with $N \cdot N$.

Example 1.10 Let us compute the variance and standard deviation for the IQ scores of Example 1.1, using grouped data. Intermediate calculations are shown in column 5 of Table 1.13. Using the computational formula, we have

$$s^2 = \frac{150(1{,}777{,}905) - (16{,}245)^2}{150(149)} = 124.64$$

and the standard deviation is

$$s = \sqrt{124.64} = 11.16$$

EXERCISES

16 Refer to Exercise 1 and compute the mean, median, mode, variance, and standard deviation, using grouped data.

17 Refer to Exercise 2 and compute the mean, median, mode, variance, and standard deviation, using grouped data.

18 Refer to Exercise 3 and compute the mean, median, mode, variance, and standard deviation, using grouped data.

19 Refer to Exercise 4 and compute the mean, median, mode, variance, and standard deviation, using grouped data.

CHAPTER SUMMARY

In this chapter we have introduced the subject of statistics and given some basic statistical terminology. We have seen how large masses of data may be summarized for ease of comprehension by means of frequency distributions, histograms, and frequency polygons. We have shown the usefulness of measures of central tendency and measures of variability to describe data. We have also talked about how to calculate the mean, median, mode, variance, and standard deviation for both grouped and ungrouped data. The topics covered in this chapter come under the heading of *descriptive statistics*.

SOME IMPORTANT CONCEPTS AND TERMINOLOGY

Statistics
Descriptive statistics
Inferential statistics
Variable
Random variable
Continuous variable
Discrete variable
Quantitative variable
Qualitative variable
Population
Sample
Frequency distribution
Cumulative frequency distribution
Relative frequency distribution
Frequency polygon
Simple random sample
Mean
Median
Mode
Variance
Standard deviation
Range

REVIEW EXERCISES

Compute the sample mean, median, variance, and standard deviation for Exercises 20 through 24.

20 The number of defective items produced per hour at a manufacturing plant was recorded for ten successive hours. The results were as follows.

5, 5, 6, 5, 6, 10, 5, 4, 4, 3

21 The following are the number of children per family for 15 families living in a low-rent housing development.

2, 5, 7, 6, 5, 3, 3, 4, 4, 8, 4, 2, 6, 4, 7

22 Twelve clerical employees of an insurance firm were given job-satisfaction tests. Here are their scores.

3, 8, 2, 6, 8, 3, 3, 9, 4, 4, 3, 7

Table 1.14 Data for Exercise 25

40	36	41	40	39	34	61	42	47	53	43	39	93	46	32	44	71	45	62	36
49	31	35	36	84	81	51	51	52	66	55	44	38	33	38	38	42	61	48	55
65	54	97	67	88	44	39	42	35	50	73	60	41	40	39	44	58	47	53	45
30	31	32	34	48	76	38	52	63	41	36	50	31	56	35	45	36	48	41	45
32	37	75	30	68	54	37	30	50	50	40	65	52	50	36	38	38	43	51	55

23 The fasting blood-glucose levels of ten children are:

71, 62, 75, 71, 55, 71, 63, 50, 62, 60

24 A sample of ten trucking firms indicated the volume of their business for the past year, in terms of thousands of tons hauled. The results were:

9, 15, 11, 17, 16, 16, 10, 18, 16, 10

25 Table 1.14 shows the typing speeds, in words per minute, of 100 experienced secretaries. From these data, prepare a frequency distribution, cumulative frequency distribution, relative frequency distribution, cumulative relative frequency distribution, histogram, and frequency polygon. Use grouped-data formulas to compute the mean, median, variance, and standard deviation.

REFERENCE

1 The Rand Corporation, *A Million Random Digits with 100,000 Normal Deviates*, Glencoe, Ill.: The Free Press, 1955

2 Probability

We are accustomed to the use of statements involving probabilities in everyday communication. The weatherman informs us of the probability of rain. The doctor tells us what the chances are that our diseases will be cured by certain therapeutic procedures. High school guidance counselors speculate on our chances of succeeding in college. Pollsters tell us the chance our favorite candidate has of winning an election.

Statistical inference, the subject of most of the remainder of this book, rests on the foundation of the theory of *probability*, a branch of mathematics dealing with random or chance phenomena. Probability theory is a subject on which many books have been written. Therefore a complete treatment of the subject in a single chapter is both impossible and undesirable. The objective here, rather, is to introduce you to the basic concepts and techniques necessary for an understanding of statistical inference. Those who wish to expand on what they learn in this chapter will find the programmed texts by Dixon (1) and Earl *et al.* (2) helpful.

2.1 THE ESSENCE OF PROBABILITY

A concern for probability concepts and calculations dates back to the seventeenth century, when, in 1654, the French mathematicians Blaise Pascal (1623–1662) and Pierre de Fermat (1601–1665) formulated the theory of probability through an exchange of numerous letters. This important correspondence was prompted by the questions put to Pascal by the French philosopher and reputed gambler, Antoine Gombaud Chevalier de Méré.

We may define probability very simply as a number, between 0 and 1, that we assign to a phenomenon in order to indicate the likelihood of its occurrence. Probabilities are often expressed as percentages, as when the weatherman informs us that the probability of rain is 30%. Alternatively, he could say that the probability is 0.30. We assign a probability of 0 to any phenomenon that cannot occur, and we assign a probability of 1 to any phenomenon that is certain to occur. Most people would agree that if a ball is tossed in the air, the probability that it will come down is 1. On the other hand, the probability that an unprotected person could survive on the planet Mercury is 0. We assign a probability of 0.5 to any phenomenon that is just as likely not to occur as it is to occur. We assign a probability of 0 or greater, but less than 0.5, to any phenomenon that has a less-than-even chance of occurring. And we assign a probability greater than 0.5, but less than 1, to a phenomenon that has a better-than-even chance of occurring.

In order to facilitate communication in the discussion that follows, we need to define some terms. The first of these is the word *experiment*, which, for our purposes here, we define as any process or activity that yields an *outcome* or *observation*. By this definition, an experiment may be as simple as flipping a coin to see whether it comes up heads or tails, or as complicated as a formal experiment designed to ascertain which of several teaching methods is most effective.

We call each of the possible outcomes of an experiment an *event*. Applying a stimulus to some organism, for example, may constitute an experiment in which the event produced is the time it takes the organism to react to the stimulus. Another example of an experiment is the act of giving a student a final examination. In this case the event of interest might be a numerical score between 0 and 100, say. Or the possible outcomes might consist of "pass" and "fail," in which case one or the other of these events would be observed.

We say that two or more events are *mutually exclusive* if, when one occurs, the others cannot occur. The events "pass" and "fail" in reference to a particular examination are said to be mutually exclusive, since a given person cannot simultaneously both pass and fail the examination.

With these basic terms defined, we are now ready to examine more carefully the concepts of probability. Specifically, in the paragraphs that follow, we shall present the concept of probability from three different points of view; (1) *a priori* probability, (2) the relative frequency of occurrence concept, and (3) subjective probability.

A priori or classical probability

We may define the *a priori* or *classical* concept of probability, which was the concept held by Pascal, Fermat, and their successors up to the present century, as follows.

DEFINITION *If an experiment can result in N equally likely and mutually exclusive outcomes, and if, out of these N outcomes, event E occurs N_E times, the probability of event E, written P(E), is given by*

$$P(E) = \frac{N_E}{N}$$

Such a definition is useful for solving problems related to games of chance of the type for which probability was first discussed. If, for example, we roll a fair six-sided die, we can calculate the probability of observing a given event, say a 1, as 1/6. Or suppose we toss a fair coin; the probability of its landing heads up is 1/2. Finally, consider a well-shuffled deck of ordinary playing cards in which, for the experiment "draw a card," there are 52 possible outcomes. The probability of the event "draw an ace" is 4/52.

Note in these examples that the described experiments did not have to be performed in order for us to calculate the desired probabilities. It was necessary only that we employ logical prior reasoning; hence the term *a priori.*

A disadvantage of the *a priori* concept of probability is the assumption of *equally likely* outcomes, as specified in the definition. This assumption often is not met in practical problems. As a result of this and other limitations, the classical concept of probability has been extended in such a way that solutions to more practical problems are possible.

The relative-frequency concept of probability

The most widely held view of probability is the *relative frequency of occurrence* concept. We may describe this concept in a nonrigorous manner as follows. Consider some event E resulting from n repetitions, or trials, of some experiment. According to the relative-frequency concept of probability, $P(E)$, the probability of event E is equal to the relative frequency of occurrence of type E events as n approaches infinity. If we let n_E equal the number of times the event E occurs out of n trials of an experiment, we may *estimate* $P(E)$ by

$$P(E) = \frac{n_E}{n}$$

This is the relative frequency of occurrence of event E.

This interpretation of probability depends on the idea of *statistical regularity*, which assumes that, although relative frequencies may fluctuate considerably after only a few repetitions of an experiment, they tend to "settle down" and approach some fixed value after a large number of repetitions. This behavior of relative frequencies has been demonstrated experimentally many times. The relative frequency of occurrence of defective items coming off an assembly line, for example, behaves in this manner, as does the proportion of male births in a sequence of total births. In 1974, in Georgia, 0.52 of all live births were male babies. It is unlikely that this proportion was constant throughout the year. For instance, of the first 100 births occurring during the year, the proportion of male births was probably lower or higher than 0.52. After several thousand births had taken place, however, the proportion of males would be expected to be close to 0.52.

Subjective probability

There are many events of interest whose probabilities of occurrence cannot be computed in accordance with either the *a priori* or the relative-frequency interpretation. For example, these interpretations of probability are of no help in assessing the probability that there is life on some distant planet, that a cure for cancer will be discovered in the next ten years, that a certain individual will excel in college, that a certain high school football team will win the state championship, or that there will be rain tomorrow. Few people are reluctant, however, to assign probabilities to such events, as evidenced by the frequency with which we hear it said that there is a 30% chance of rain tomorrow, that some team does not have a chance of winning, and so on. The interpretation of probability that allows us to assign probabilities to events such as these is called *subjective probability*.

The magnitude of the probability that a person subjectively assigns to an event depends on that person's *degree of belief* in the occurrence of the event. This is why it is possible to assign probabilities to events that occur only once, such as the winning of a particular athletic contest. Unlike the relative-frequency concept of probability, the subjective interpretation does not depend on the repeatability of an experiment.

2.2 SOME BASIC IDEAS FROM SET THEORY

The discussion of probability, and especially the calculation of the probability of an event, is greatly facilitated by the concept of a *set*. Set theory was introduced in the latter part of the nineteenth century by Georg Cantor (1845–1918). It is not only important in the area of probability and statistics, but is fundamental to all of modern mathematics. We shall present here only the basic concepts.

Actually, the concept of set is a familiar one. We encounter it frequently in everyday conversation. Such expressions as "a set of dishes," "a set of golf clubs," and "a set of books" are familiar to us all. In order to clarify and formalize the concept, we must define the word *set*, as well as some additional basic terminology.

DEFINITION *A* ***set*** *is a well-defined collection of objects.*

We call the objects of which a set is composed the *elements* of the set. The past presidents of the United States constitute a set, the elements of which include, among others, George Washington, Abraham Lincoln, and Franklin D. Roosevelt. Other examples of sets are the known planets of our solar system, graduate students currently enrolled in the colleges of the United States, and members of the U.S. Supreme Court. We shall use capital letters, such as A, B, C, or D, to designate sets, and lower-case letters, a, b, c, x, y, and so on, to designate the elements of sets.

One may describe a set in one of two ways: the *roster method* or the *rule method.* When one uses the *roster method*, one lists each element of the set. Suppose, for example, that the seventh-grade teachers at a certain elementary school are Mrs. Allen, Miss Brown, Mr. Smith, and Mr. Walker. We may designate this set by the letter A, and describe it by the roster method as follows.

$$A = \{\text{Mrs. Allen, Miss Brown, Mr. Smith, Mr. Walker}\}$$

To describe a set by the *rule method*, we state a rule which, when followed, permits the identification of each element of the set. We may describe this same set of seventh-grade teachers by the rule method as follows.

$$A = \{x \mid x \text{ is a seventh-grade teacher at the elementary school}\}$$

We read this as "A is the set of all x such that x is a seventh-grade teacher at the elementary school." The vertical line between the two x's inside the braces is read "such that."

DEFINITION *The* ***universal set*** *is that set consisting of the largest collection of elements for which there is interest in a given discussion.*

The universal set is usually designated by U. Note the similarity between the definition of universal set and the definition of population given in Chapter 1. This is not coincidence, but an example of expressing the same concept in two different ways. As in our discussion of population, we can say here that the universal set of a particular discussion changes as our sphere of interest changes. In one discussion, the universal set may consist of all college students in the United States. In another discussion, the universal set may consist of students enrolled at a particular college.

DEFINITION *Set A is a* ***subset*** *of set B if every element of A is also an element of B.*

Referring again to Chapter 1, we may say that a sample is a subset of the set (population) from which it was drawn. In the example given earlier, the set

$$B = \{\text{Mrs. Allen, Miss Brown, Mr. Smith}\}$$

is a subset of the set

$$A = \{\text{Mrs. Allen, Miss Brown, Mr. Smith, Mr. Walker}\}$$

DEFINITION *Two sets A and B are* ***equal*** *if they both contain exactly the same elements.*

When two sets A and B are equal, we may write $A = B$. If, however, either set A or set B contains at least one element that is not contained in the other, the two sets are unequal, and we write $A \neq B$. The following examples illustrate this point.

Example 2.1 Suppose that

$$A = \{2, 4, 6, 8, 10\}, \quad B = \{2, 4, 6, 8, 10\}$$

Then $A = B$. But suppose that

$$C = \{1, 3, 5, 7\}, \quad D = \{1, 3, 5, 7, 9\}$$

Then $C \neq D$.

DEFINITION *The set that contains no elements is called the* ***null set*** *or* ***empty set.***

The null or empty set is designated by the symbol $\emptyset$. The set

$$A = \{\text{All women who have been President of the United States}\}$$

is a null set. In symbols, $A = \emptyset$.

Set operations

We shall now show how it is possible to take a universal set U and some of its subsets and form new sets by executing one or more of several possible set operations. Each new set will also be a subset of the universal set U.

We shall illustrate the results of these set operations by means of Venn diagrams, named for the English logician John Venn (1834–1923).

DEFINITION *The* ***complement*** *of a subset A of the universal set U is the set that consists of all elements in U that are not members of the set A.*

We shall denote the complement of a given set by affixing the prime symbol (′) to the letter used to designate the given set. For example, we shall write the complement of the set A as A' (read "A prime").

Figure 2.1 Venn diagram illustrating the complement of a set

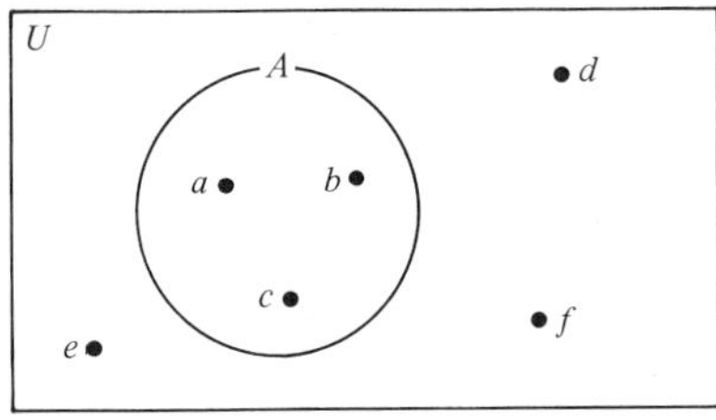

Example 2.2 Let the universal set consist of

$$U = \{a, b, c, d, e, f\}$$

and let

$$A = \{a, b, c\}$$

be a subset of U. The complement of A is

$$A' = \{d, e, f\}$$

This example is illustrated in Figure 2.1, where the universal set U is represented by the rectangle, the set A by the circle, and A' by the area within the rectangle that lies outside the circle.

DEFINITION *The **intersection** of two sets A and B is the set containing all elements that are elements of both A and B.*

The intersection of two sets A and B is designated by $A \cap B$ (read "A intersection B").

Example 2.3 Set A consists of all seniors at a certain high school who take geometry, and set B consists of all seniors who take French. The intersection of sets A and B, $A \cap B$, consists of all seniors who take geometry *and* French. Figure 2.2 illustrates the intersection of two sets.

DEFINITION *Two sets A and B are said to be **disjoint** or **mutually exclusive** if they have no elements in common.*

If A and B are disjoint, $A \cap B$ is the null set.

Figure 2.2 The intersection of two sets

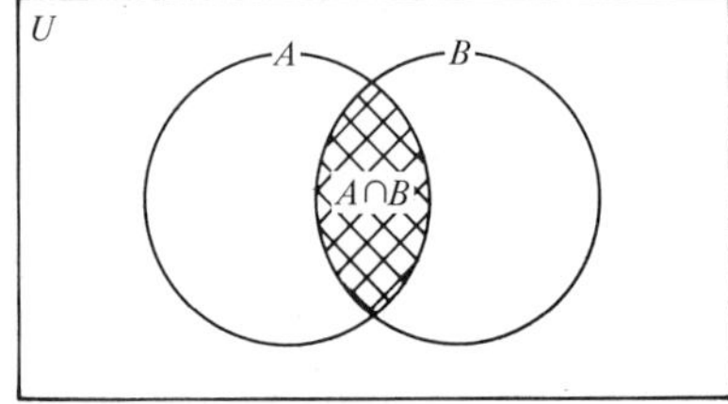

Figure 2.3 The union of two sets

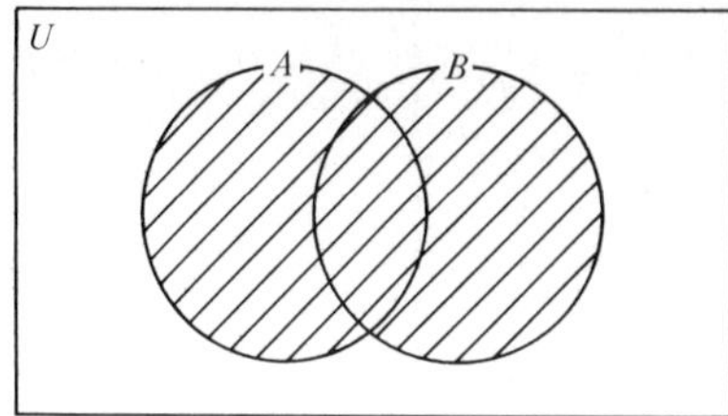

Example 2.4 Suppose that

$$A = \{\text{All seniors (at a certain high school) who take geometry}\}$$

and

$$B = \{\text{All seniors (at the same high school) who take French}\}$$

If no senior takes both geometry and French, A and B are disjoint, and $A \cap B = \emptyset$.

DEFINITION *The **union** of two sets A and B is the set consisting of all elements that are elements of A, or B, or both.*

The union of two sets is indicated by the symbol $\cup$. The union of A and B, for example, is written $A \cup B$ (read "A union B").

Example 2.5 Suppose that

$$A = \{\text{All teachers in a certain school system certified in mathematics}\}$$

and

$$B = \{\text{All teachers in the same school system certified in science}\}$$

The set of all teachers in the system certified in mathematics, or science, or both, is the union of sets A and B, written $A \cup B$.

Figure 2.3 illustrates the union of two sets.

EXERCISES

1 Given: $A = \{t, u, v, w, x, y, z\}$, $B = \{w, x, y, z\}$.
(a) Find $A \cup B$. (b) Find $A \cap B$.
(c) Are A and B equal? (d) Are A and B disjoint?

2 Set C consists of the citizens of a certain town who voted "yes" in a school bond referendum. Set D consists of the citizens of the same town who have children in school. Define:
(a) $C \cup D$ (b) $A \cap D$ (c) C' (d) D'

3 Express each of the following sets by a single symbol.
(a) $A \cap \emptyset$ (b) $A \cap A'$ (c) $A \cup \emptyset$ (d) $A \cup A'$

Table 2.1 Data for Exercise 7

Type of program	Educational level			Total
	HIGH SCHOOL (A)	COLLEGE (B)	GRADUATE SCHOOL (C)	
Sports (S)	15	8	7	30
News (T)	3	7	20	30
Drama (V)	5	5	15	25
Comedy (W)	10	3	2	15
Total	33	23	44	**100**

4 Express each of the following sets by a different symbol.
(a) U' (b) $(A')'$ (c) $\varnothing'$ (d) $(A \cap \varnothing)'$

5 Let $U = \{a, b, c, d, e, f, g, h, j, k, m\}$, $R = \{a, c, e, g\}$, $S = \{b, d, f, h\}$, $T = \{a, e, d, f, j, k\}$. Draw a Venn diagram to represent these sets, along with their respective elements. List the elements in each of the following sets.
(a) R' (b) S' (c) T' (d) $R \cap S$
(e) $R \cap T$ (f) $T \cap S$ (g) $R \cup S$ (h) $S \cup T$
(i) $T \cup R$

6 Refer to Exercise 5. List the elements in the following.
(a) $R \cap (S \cap T)$ (b) $(S \cap T) \cup R$
(c) $R \cap (S \cup T)$ (d) $R \cap S \cap T$

7 One hundred business people were asked to specify which type of television program they preferred. Table 2.1 shows the 100 responses, cross-classified by educational level and type of program preferred. Specify the number of members of each of the following sets.
(a) S (b) $V \cup C$ (c) A (d) W'
(e) U (f) B' (g) $T \cap B$ (h) $(T \cap C)'$

8 Refer to Exercise 7. Explain the following sets in words.
(a) $V \cap B$ (b) $S \cup C$ (c) T' (d) $(V \cap A)'$
(e) W (f) W' (g) $(S \cup B)'$ (h) $W \cap V$

9 Refer to the Venn diagram below, and list the elements in each of the following sets.
(a) $A \cup B$ (b) $A \cap B$
(c) $A \cup C$ (d) $C \cup B$
(e) $A \cap C$ (f) $C \cap B$
(g) $A \cap B \cap C$ (h) $A \cup B \cup C$
(i) $(A \cup B \cup C)'$ (j) B'
(k) $(A \cup B)'$

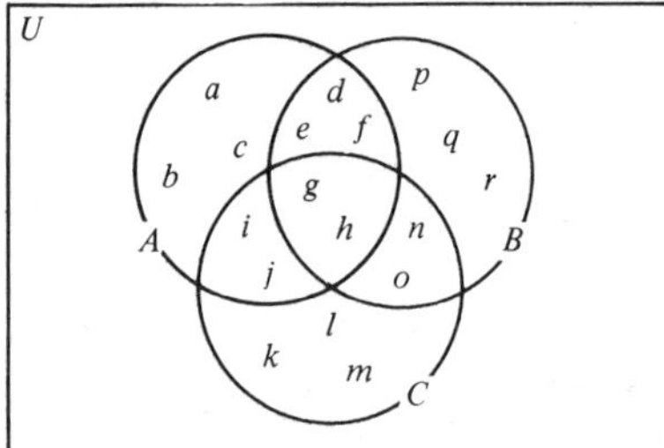

2.3 COUNTING TECHNIQUES

As we shall see in Section 2.4, when calculating probabilities one must be able to determine the number of times that a certain event occurs. In many situations of practical importance, it is virtually impossible to make a physical count of the number of occurrences of an event. When you are confronted with one of these situations, it is helpful to have available some rapid, efficient, short-cut method of counting. Let us look at some techniques that satisfy these requirements.

The multiplication principle

A basic technique for counting is described by what is known as the *multiplication principle*, which may be stated as follows.

If there are k operations to be performed, and if the first can be performed in n_1 *ways, and if, regardless of the way the first was performed, the second operation can be performed in* n_2 *ways, and if, regardless of the way the first two operations were performed, the third operation can be performed in* n_3 *ways, and so on for the k operations, then the sequence of k operations can be performed in* $(n_1)(n_2) \cdots (n_k)$ *ways, the product of the individual numbers.*

Example 2.6 One of the editors of a high school annual, in designing a page that was supposed to have one each of five different types of pictures, was interested in knowing how many different page layouts could be prepared from the available pictures of each type. The editor could choose from among four group pictures of teachers, ten pictures of athletic events, seven classroom shots, eight campus shots, and five pictures showing activities of the various clubs.

Using the multiplication principle, we find that there are $4 \cdot 10 \cdot 7 \cdot 8 \cdot 5 = 11{,}200$ possibilities. The editor decided not to try them all in order to decide which looked best.

Let us now consider a simpler example. Suppose a girl has three blouses, two skirts, and two pairs of shoes. Using these seven articles of clothing, how many different outfits can she wear? With each of the three blouses, she can wear either one of the skirts, for a total of six outfits, disregarding the shoes. With each of these six outfits, she can wear either pair of shoes, for a total of $3 \cdot 2 \cdot 2 = 12$ outfits.

If the number of possibilities is not too great, a graphic device known as a *tree diagram* is useful for displaying all possible sequences of such operations. A tree diagram consists of a series of "branches," with one branch for each way in which an operation can be performed.

Example 2.7 A bank official who drives to work gets from his home to the freeway on either one of three routes (A, B, C). He has three ways to get from the freeway to downtown (I, II, III); and from the downtown area to his parking lot

Figure 2.4 Tree diagram showing the possible routes from a bank official's home to his parking lot

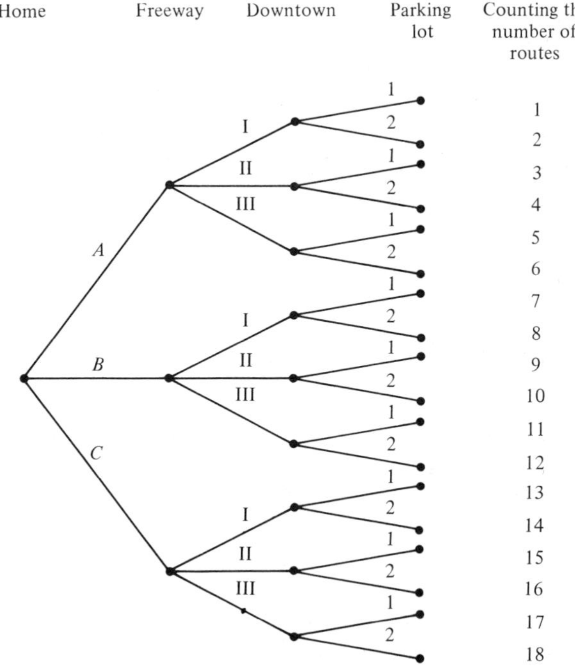

he has a choice of two routes (1, 2). By how many different routes can the official drive to work?

Using the multiplication principle, we find the answer to be $3 \cdot 3 \cdot 2 = 18$. The solution is illustrated by the tree diagram in Figure 2.4.

Permutations

Permutations are useful for efficiently counting the number of different arrangements or orderings of a set of objects. We may use the concept of permutations to determine the number of ways a group of children can be assigned to the seats in a classroom, the number of ways a group of speakers can be seated on a stage, the number of ways a set of books can be arranged on a shelf, and so on.

DEFINITION *A **permutation** is one of the possible different arrangements of all or part of a set of objects.*

Before giving a general rule for determining how many different arrangements (permutations) are possible in a given situation, let us see how this number can be found intuitively.

Example 2.8 There are to be five guest speakers at a PTA meeting. The program moderator wishes to know in how many ways the five speakers can be seated in a row on the stage. Each of these possible ways is a permutation, so the moderator is really asking for the number of permutations of five things taken five at a time.

Visualize five chairs in a row on the stage. The program moderator may choose any one of the five speakers to occupy the first chair. Once this choice has been made, the moderator may choose an occupant for the second chair from among the four remaining speakers. Each time a chair is filled, the number of available speakers to fill the next chair is decreased by one. This continues until the last chair is reached, at which time there is only one speaker left who can be assigned to it. Using the terminology employed in the statement of the multiplication principle, we may say that there are $k=5$ operations to be performed. The first can be performed in five ways, the second in four ways, and so on to the fifth operation, which can be performed in one way. By the multiplication principle, then, we find the number of ways the five speakers can be seated in a row on the stage to be $5 \cdot 4 \cdot 3 \cdot 2 \cdot 1 = 120$.

Let us designate the speakers by *A*, *B*, *C*, *D*, and *E*. Some of the possible seating arrangements (permutations) would be *A B C D E*, *A C B D E*, *A C D B E*, *A C D E B*, and so on, each arrangement differing from the other only with respect to who sits next to whom.

As we have noted, the 120 we have just calculated is called the number of permutations of five things taken five at a time. We may state the general rule for finding the number of permutations of n things taken n at a time as follows.

The number of permutations of n different objects taken n at a time is equal to n!.

The number of permutations of n objects taken n at a time is designated by ${}_nP_n$, so we may write

$${}_nP_n = n!$$

The symbol $n!$ is read "n factorial." We evaluate it by multiplying each of the n integers together, that is, $n! = n(n-1)(n-2) \cdots 2 \cdot 1$. As we have already seen, $5! = 5 \cdot 4 \cdot 3 \cdot 2 \cdot 1 = 120$. We define $0! = 1$. We also note that $n! = n(n-1)!$. For example, $10! = 10 \cdot 9!$.

Example 2.9 Someone wants to hang six pictures in a straight line on the wall of a library. How many arrangements are possible?

The answer is given by

$${}_6P_6 = 6! = 6 \cdot 5 \cdot 4 \cdot 3 \cdot 2 \cdot 1 = 720$$

On occasion, when arranging objects, one may not be able to use them all. In arranging pictures on a wall, for example, one may have more pictures than spaces available.

Example 2.10 An automobile dealer has seven models to display in a show window that has spaces for only five cars at a time. How many displays are possible? Note that we have seven objects available, but can use only five at a time. We are seeking, then, the number of permutations of seven things taken five at a time.

The first space can be filled in any one of seven ways, the second in six ways, and so on to the fifth space, which can be filled in three ways. By using the multiplication principle, then, we find that there are

$$7 \cdot 6 \cdot 5 \cdot 4 \cdot 3 = 2520$$

possible displays.

Alternatively, we can indicate the calculations as follows.

$$7 \cdot 6 \cdot 5 \cdot 4 \cdot 3 = \frac{7 \cdot 6 \cdot 5 \cdot 4 \cdot 3 \cdot 2 \cdot 1}{2 \cdot 1}$$

$$= \frac{7!}{2!} = \frac{7!}{(7-5)!}$$

Let us generalize the situation in the above example by letting n be the number of objects available and r the number of spaces to be filled. Using the multiplication principle again, we find that the number of ways r spaces can be filled when n objects are available is given by

$$n(n-1)(n-2) \cdots (n-r+1)$$

In other words, we may write the number of permutations of n objects taken r at a time, which we designate by ${}_nP_r$, as follows.

$${}_nP_r = n(n-1)(n-2) \cdots (n-r+1)$$

$$= \frac{n(n-1)(n-2) \cdots (n-r+1)(n-r)(n-r-1) \cdots 1}{(n-r)(n-r-1) \cdots 1}$$

Finally

$${}_nP_r = \frac{n!}{(n-r)!}$$

which leads to the following statement.

The number of permutations of n different objects taken r at a time, denoted by ${}_nP_r$, is given by ${}_nP_r = n!/(n-r)!$.

Example 2.11 A speaker has eight topics on which she can give a 30-minute lecture. She is asked to present a series of five 30-minute lectures to a certain group. How many sequences of lectures does she have to choose from?

The answer is given by

$$_8P_5 = \frac{8!}{(8-5)!} = \frac{8!}{3!} = \frac{8 \cdot 7 \cdot 6 \cdot 5 \cdot 4 \cdot 3!}{3!} = 6720$$

Combinations

A characteristic of permutations is that the *order* of arrangement of objects is important. Suppose, for example, that we have four textbooks, one each for history, mathematics, English, and science, which we can place between book ends that will hold only two books. When we use the method for permutations to calculate the number of ways the two spaces can be filled, we are implying that the order in which two books appear between the bookends is important. That is, we are implying that it is important, for example, to distinguish between the arrangement in which the history book is on the left and the mathematics book is on the right and the arrangement in which the positions of the books are reversed. If we use the letters H, M, E, and S to designate the history, mathematics, English, and science books, respectively, we may write the 12 possible permutations as follows.

HM	MH	EH	SH
HE	ME	EM	SM
HS	MS	ES	SE

If the order of arrangement does not matter, that is, if HM is considered the same as MH, HE the same as EH, and so on, then the number of arrangements reduces to the following six.

HM	ME
HE	MS
HS	ES

We call this the number of *combinations* of four things taken two at a time. We may define a combination, then, as follows.

DEFINITION *A* ***combination*** *is an arrangement of objects taken from a set of n objects in such a way that the order of selection is disregarded.*

To obtain again the 12 original permutations, we need only construct the permutations for each of the six combinations. Thus, in this case, for each combination there are two permutations.

In general, if we have n objects to be taken r at a time, we may construct $r!$ permutations from each of the possible combinations. Let us designate the number of combinations of n objects taken r at a time by $\binom{n}{r}$. Then we may designate the number of resulting permutations by $\binom{n}{r}r!$ It is true—and this was demonstrated by the textbook example above—that this product is equal to the total number of permutations of n things taken r at a time, so we may write

$$_nP_r = \binom{n}{r} r!$$

We obtain a formula for calculating the number of combinations of n things taken r at a time by solving this equation for $\binom{n}{r}$. This enables us to make the following statement.

The number of combinations of n objects taken r at a time is given by

$$\binom{n}{r} = \frac{_nP_r}{r!}$$

Recalling that $_nP_r = n!/(n-r)!$, we may write this equation in its most frequently observed form,

$$\binom{n}{r} = \frac{n!}{r!\,(n-r)!}$$

Example 2.12 A committee of three is to be appointed from among 15 teachers. How many different committees can be appointed?

The answer is given by

$$\binom{15}{3} = \frac{15!}{3!\,12!} = \frac{15 \cdot 14 \cdot 13 \cdot 12!}{3 \cdot 2 \cdot 1 \cdot 12!} = 455$$

Permutations of objects that are not all different

In the preceding discussion, the objects to be arranged have all been different. On occasion, we may wish to calculate the number of permutations of n objects of which n_1 are of type 1, n_2 are of type 2, . . . , and n_k are of type k. In such a situation, the formula for calculating the number of distinguishable arrangements, denoted by $_nP_{n_1,n_2,\cdots,n_k}$ is given by

$$_nP_{n_1,n_2,\cdots,n_k} = \frac{n!}{n_1!\,n_2! \cdots n_k!}$$

Example 2.13 In a kindergarten classroom there are eight plastic shapes: three squares, three triangles, and two rectangles. The shapes are otherwise indistinguishable. In how many ways can the students arrange the shapes in a row on a table?

From the above equation, we find the answer to be

$$_8P_{3,3,2} = \frac{8!}{3!\,3!\,2!} = 560$$

10 Evaluate the following.

(a) $6!$ (b) $4!$ (c) $\binom{5}{3}$ (d) ${}_5P_3$

(e) $\binom{13}{4}$ (f) ${}_{13}P_7$ (g) ${}_{10}P_{5,3,2}$ (h) ${}_9P_{4,3,2}$

11 A club that has 12 members plans to elect four officers—a president, a vice president, a secretary, and a treasurer—by secret write-in ballot. All 12 club members are eligible and willing to serve. How many possible sets of four members can serve if the office held is ignored?

12 Fifteen high school students have passed the semifinal qualifications for cheerleader. From these semifinalists, six will be chosen to be on the first-squad cheerleading team. How many different teams can be formed from the 15 semifinalists?

13 A contractor has 16 house plans from which to select nine for use in building nine houses on a cul-de-sac in a new subdivision. How many different arrangements of houses can be built if each house has a different floor plan?

14 A landscape architect has 11 shrubs from which to select five for use in landscaping the front of an office building. How many designs are possible on the basis of different shrubs?

15 The producer of a television variety show has ten acts for use on a one-hour show. How many different shows can be arranged if each act can perform only once?

16 In a secretarial pool, there are seven secretaries who must sit at desks placed one behind the other. In how many ways can the secretaries be assigned to the desks?

17 A disc jockey has ten records from which to select five to play on a 15-minute program. In how many ways can these five records be scheduled in the five time slots available between commercials, chitchat, etc?

18 A social worker has ten clients to visit in a week. The clients consist of three children, four adolescents, two middle-aged persons, and one elderly person. How many distinguishable visiting arrangements can the social worker prepare if he wishes to distinguish among the clients only on the basis of age category?

19 A psychologist has 14 clients from which to select nine for a group experiment. How many groups are possible?

20 A child is given a box containing 15 different toys from which he may select five. How many different sets of toys are possible?

21 For their vacation, a family living in Louisville wishes to visit, in succession, Nashville, Atlanta, and Miami. There are two routes from Louisville to Nashville, three from Nashville to Atlanta, and three from Atlanta to Miami. In how many different ways can the family travel from Louisville to Miami?

22 A teacher has eight students whom he would like to have sit in the front row. However, there are only five front-row seats. How many front-row seating arrangements are possible?

23 A teacher must appoint a committee of four boys and three girls from among her students. She has eight boys and six girls from whom to choose. How many committees are possible?

2.4 CALCULATING THE PROBABILITY OF AN EVENT

Up to now in this chapter we have been laying the foundation for the main objective of the chapter, which is to teach you how to calculate probabilities of events. Here we shall introduce some basic probability concepts and present some rules and techniques which greatly facilitate calculations of probability. To understand the topics that follow, you need to understand the contents of this section.

The addition rule

If $E_1, E_2, \ldots, E_k$ are mutually exclusive events, the probability of the occurrence of $E_1, E_2, \ldots,$ or E_k is equal to the sum of the probabilities of the individual events.

We may write the addition rule symbolically as

$$P(E_1 \text{ or } E_2 \text{ or } \ldots \text{ or } E_k) = P(E_1) + P(E_2) + \cdots + P(E_k)$$

Here is an example of the use of the addition rule.

Example 2.14 Suppose we have a group of 500 recent college graduates, of whom 175 majored in education, 150 in business, 100 in one of the humanities, and 75 in a health science. Suppose we select a person at random from this group. What is the probability that the person will have majored in either education or business?

From our discussion of probability in Section 2.1, we know that we may calculate the probability of selecting an education major as 175/500 and the probability of selecting a business major as 150/500. If we can assume that there were no multiple majors (the events are mutually exclusive), we may use the addition rule to find

$$P(\text{education major or business major}) = P(\text{education major}) + P(\text{business major}) = \frac{175}{500} + \frac{150}{500} = 0.65$$

If the events are not mutually exclusive, the addition rule has to be modified in order to handle the fact that events that are not mutually exclusive have elements in common. Let us consider the simple case in which

Figure 2.5 Venn diagram illustrating the calculation of $P(A \cap B)$

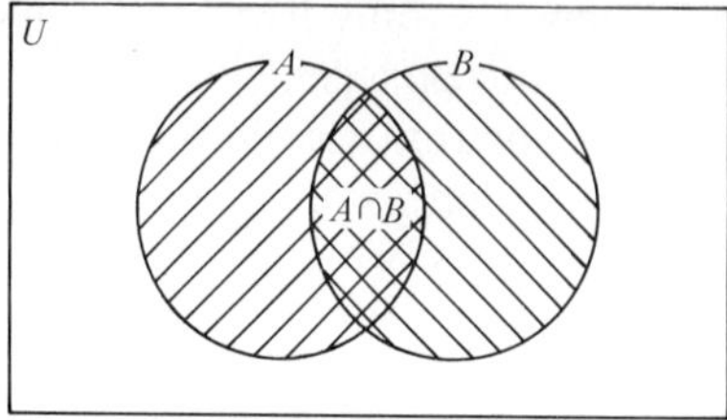

there are only two events of interest, A and B, which are not necessarily mutually exclusive. We may state the addition rule more generally as follows.

Given two events A and B, the probability that event A or event B or both occurs is equal to the probability that event A occurs plus the probability that event B occurs minus the probability that events A and B both occur.

Using set notation from Section 2.2, we may write this rule symbolically as

$$P(A \cup B) = P(A) + P(B) - P(A \cap B)$$

Figure 2.5 illustrates the addition rule. Events A and B are each represented by a hatched circle, and the total hatched area corresponds to $A \cup B$. The cross-hatched area, corresponding to $A \cap B$, is subtracted in the calculation of $P(A \cup B)$ because it is counted twice, once in $P(A)$ and once in $P(B)$.

In future discussions we shall designate the number of elements in a set by enclosing the set symbol in parentheses, preceded by a small n. If the set A has 25 elements, for example, we shall write $n(A) = 25$.

Example 2.15 A certain school system receives 25 applications for an opening that exists for an elementary school principal. Of these applicants, 10 are male and 15 are female. Seventeen hold master's degrees only, and eight have sixth-year certificates. A selection from among these 25 applicants is made at random. What is the probability that a female *or* a person with a sixth-year certificate will be selected?

Let us designate the event "male" by A_1, "female" by A_2, "a master's degree" by B_1, and "sixth-year certificate" by B_2, and display the information in Table 2.2.

From Table 2.2 we see that the number of females is $n(A_2) = 15$, and the number of persons with sixth-year certificates is $n(B_2) = 8$. The number of persons who both are female *and* have sixth-year certificates appears at the intersection of the row labeled "Female" and the column labeled "Sixth-year certificate." This number is equal to $n(A_2 \cap B_2) = 5$. By the addition rule, we find the desired probability to be

$$P(A_2 \cup B_2) = \frac{15}{25} + \frac{8}{25} - \frac{5}{25} = \frac{18}{25} = 0.72$$

Table 2.2 Twenty-five applicants for principal's position classified by sex and educational achievement

Sex	Educational achievement: MASTER'S DEGREE (B_1)	SIXTH-YEAR CERTIFICATE (B_2)	Total
Male (A_1)	7	3	10
Female (A_2)	10	5	15
Total	17	8	**25**

Conditional probability

In each of the probabilities computed up to this point, the denominator of the fraction representing the desired probability has been the universal set. This has been the case because the universal set has comprised the largest collection of elements for which there has been an interest. In Example 2.15 we were asked to compute a probability based on 25 possible selections, where these 25 selections exhausted our sphere of interest.

In many instances our sphere of interest may be only a subset of the universal set. In other words, we may have our sphere of interest reduced from an original population to a subpopulation.

Example 2.16 Let us look again at Example 2.15, which concerns the selection of a school principal. Suppose we are asked to find the probability that a female will be selected if it is known that the selection is to be made at random from among the persons with sixth-year certificates. This additional information has the effect of reducing our sphere of interest from the universal set of 25 applicants to the subset of the eight applicants who hold sixth-year certificates.

We are now being asked to find the probability that a female will be selected, *given* that the person selected holds a sixth-year certificate. In terms of the set notation of Table 2.2, we are seeking $P(A_2 \mid B_2)$, where the vertical line is read "given." Since our subset of interest consists of $n(B_2)=8$ applicants, and since of these, those with the characteristic of interest compose the set $A_2 \cap B_2$, we have

$$P(A_2 \mid B_2) = \frac{n(A_2 \cap B_2)}{n(B_2)} = \frac{5}{8} = 0.625$$

The probability $P(A_2 \mid B_2)$ is called a *conditional probability*, since its calculation depends on the condition specified by the additional information that is given; namely, that the selection is to be made from a subset of the universal set. A probability calculated with the universal set as the denominator is sometimes called an *unconditional probability*.

The concept of a conditional probability is illustrated in Figure 2.6. To obtain the unconditional probability of A, we would divide the number of

Figure 2.6 Venn diagram illustrating the calculation of $P(A \mid B)$

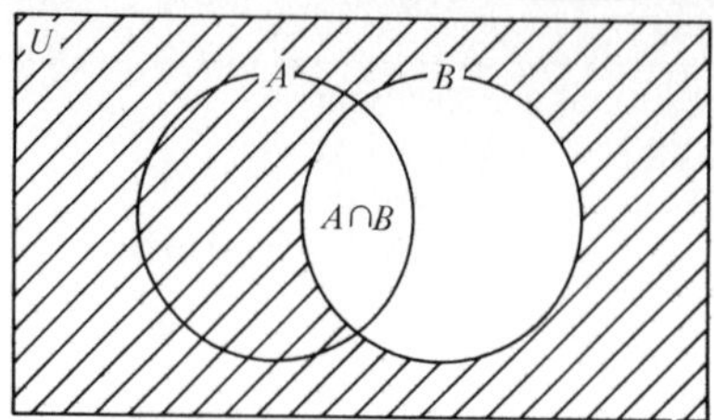

elements in A by the total number of elements in U. The conditional probability of $A \mid B$ eliminates all elements of A (and U) that are not also elements of B. Hence $P(A \mid B)$ is the ratio of the number of elements in the lens-shaped region $(A \cap B)$ to the number of elements in B.

The reasoning involved in the solution to Example 2.16 suggests the following definition of conditional probability.

DEFINITION *The probability of the occurrence of event A, given that event B has occurred, is called the* ***conditional probability*** *of the occurrence of event A. It is designated by* $P(A \mid B)$ *and is given by*

$$P(A \mid B) = \frac{P(A \cap B)}{P(B)}$$

If $P(B) = 0$, *then* $P(A \mid B)$ *is not defined.*

We call $P(A \cap B)$ the *joint probability* of events A and B, and we call the set $A \cap B$, which we have defined as the intersection of A and B, a *joint event*, since the elements of $A \cap B$ belong to both A and B. We may formally define a joint event as follows.

DEFINITION *Given two events A and B,* $A \cap B$ *is called a* ***joint event.***

The multiplication rule

Frequently, we need to determine the probability of the joint occurrence of two events A and B, that is, $P(A \cap B)$. This is often referred to as the probability that both event A *and* event B occur. In Example 2.15 we needed the probability of the joint occurrence of the events "female" and "sixth-year certificate" in order to apply the addition rule to find the probability that a female applicant or an applicant with a sixth-year certificate would be selected. In that example, we found the joint probability of the two events by dividing the number of elements in the intersection set by the number of elements in the universal set.

We may derive an alternative way of finding the probability of a joint event, which is useful when original data are not available, from the equation given earlier for finding a conditional probability. When we solve the conditional-probability equation for $P(A \cap B)$, we have

$$P(A \cap B) = P(B)P(A \mid B)$$

which is called the *multiplication rule*. It may be stated in words as follows.

DEFINITION *The probability of the joint occurrence of events A and B is equal to the product of P(B) and P(A | B).*

Note that in this definition A and B may be interchanged, so the right-hand term of the equation may also be written $P(A)P(B \mid A)$.

Example 2.17 Let us refer again to Example 2.15 and use the multiplication rule to compute the probability that a random selection will result in a female *and* a person with a sixth-year certificate.

By this rule, we have

$$P(A_2 \cap B_2) = P(B_2)P(A_2 \mid B_2)$$

$$= \frac{8}{25} \cdot \frac{5}{8} = \frac{1}{5} = 0.20$$

Example 2.18 A questionnaire was administered to a group of people to obtain their opinion on a certain proposal. Of those interviewed, 46% were against the proposal. Of those who were against it, 57% were under 30. A questionnaire is selected at random from those completed. What is the probability that it will be for a person who was against the proposal and under 30?

We interpret the given percentages as probabilities. If we let A stand for the set of persons against the proposal, we may designate these probabilities by $P(A) = 0.46$ and $P(<30 \mid A) = 0.57$. By the multiplication rule we find the probability we seek to be given by

$$P(A \cap <30) = (0.46)(0.57) = 0.26$$

Independent events

If, in the multiplication rule, $P(A \mid B) = P(A)$, the equation reduces to $P(A \cap B) = P(B)P(A)$, and we say that events A and B are *independent*. $P(A \mid B) = P(A)$ if the occurrence of event B in no way alters the probability of the occurrence of event A. Let us consider a concrete example.

Suppose you are asked to guess the probability that a house, picked at random from the households in a certain metropolitan area of the South, has central air conditioning. Let us call this probability $P(A)$, where event A is the set of all households in the area with central air conditioning. Suppose now you are given the additional information that the house is valued at $65,000 or more, and asked to guess again the probability that the house has central air conditioning. Let us call this probability $P(A \mid B)$, where the event B is the set of all households in the area valued at $65,000 or more. Do you think your guess of $P(A)$ and $P(A \mid B)$ would be the same? The answer is no, since we know that a more expensive house is more likely to have central air conditioning. That is, we know that the events "central air conditioning" and "valued at \$65,000 or more" are not independent, and thus we would guess $P(A \mid B)$ to be greater than $P(A)$. The two events under consideration here are said to be *associated* or *dependent*.

Consider now an example in which two events in fact probably are independent. Suppose you are asked to guess the probability that a person picked at random has blood type A. After being given the additional information that the person is over 30, you are asked to guess again. Would your guess change? No, since there is no reason to believe that the frequency of blood type A is different in the over-30 and 30-and-under age groups. We say, then, that age and blood group are independent, so if we let A be the event "blood type A" and B the event "over 30," we are willing to assume that $P(A \mid B) = P(A)$.

EXERCISES

24 Table 2.3 shows 1000 college students classified according to scores made on a college entrance examination. It also shows the quality of the high schools from which they graduated, as rated by a group of educators.

(a) Calculate the probability that a student picked at random from this group: (1) Made a low score on the examination. (2) Graduated from a superior high school. (3) Made a low score on the examination *and* graduated from a superior high school. (4) Made a low score on the examination, given that the student graduated from a superior high school. (5) Made a high score *or* graduated from a superior high school.

(b) Calculate the following probabilities.

(1) $P(A)$ (2) $P(H)$ (3) $P(M)$

(4) $P(A \mid H)$ (5) $P(M \cap P)$ (6) $P(H \mid S)$

25 In Exercise 24, is examination score independent of high school quality? Substantiate your answer.

26 A group of 50 adults is composed of 20 males and 30 females. Of the 35 persons in the group who are in favor of a certain candidate for

Table 2.3 Data for Exercise 24

	Quality of high school			
Score	POOR (P)	AVERAGE (A)	SUPERIOR (S)	Total
Low (L)	100	50	50	200
Medium (M)	75	175	150	400
High (H)	25	75	300	400
Total	200	300	500	**1000**

mayor, 15 are male. A person is selected at random from the group. What is the probability that this person will be a female who is opposed to the candidate?

27 Given: $P(A) = 0.6$, $P(A \cap B) = 0.25$, $P(B') = 0.7$.

(a) What is $P(B \mid A)$?

(b) Are A and B independent? Why?

(c) What is $P(A')$?

28 The probability that a public health nurse will find a client at home is 0.8. What is the probability (assuming independence) that, on two home visits made in a day, the nurse will find both clients at home?

29 A high school guidance counselor has assessed the probability of success in college of three graduating seniors to be 0.9, 0.8, and 0.7, respectively. If we assume independence, what is the probability that all three will succeed in college?

30 A psychological experiment cannot use subjects who are either colorblind or left-handed. In the population from which subjects are drawn, 7% are colorblind and 8% are left-handed. A subject is picked at random from the population. What is the probability that the person will not be able to participate in the experiment? What assumption must you make in order to answer the question from the information given?

31 Table 2.4 shows the outcome of 500 interviews attempted during a survey to study the opinions about UFOs of residents of a certain city. The data are also classified by the area of the city in which the questionnaire was attempted.

(a) A questionnaire is selected at random from the 500. What is the probability that: (1) The questionnaire was completed? (2) The potential respondent was not at home? Refused to answer? (3) The potential respondent lived in area A? B? D? E?(4) The questionnaire was completed, given that the potential respondent lived in area B? (5) The potential respondent refused to answer the questionnaire or lived in area D?

(b) Calculate the following probabilities.

(1) $P(A \cap R)$ (2) $P(B \cup C)$ (3) $P(D')$

(4) $P(N \mid D)$ (5) $P(B \mid R)$ (6) $P(C)$

32 Table 2.5 shows the results of a survey in which 100 college seniors indicated their political inclinations, and stated whether or not they belonged to a fraternity or sorority.

Table 2.4 Data for Exercise 31

Area of city	Outcome of interview			Total
	COMPLETED (C)	NOT AT HOME (N)	REFUSED (R)	
A	100	20	5	125
B	115	5	5	125
D	50	60	15	125
E	35	50	40	125
Total	300	135	65	**500**

Table 2.5 Data for Exercise 32

Fraternity-sorority membership	Political inclination CONSERVATIVE (A)	OTHER (A')	Total
Yes (B)	35	20	55
No (B')	5	40	45
Total	40	60	**100**

(a) Calculate the following probabilities.
(1) $P(A)$ (2) $P(A')$ (3) $P(B)$
(4) $P(B')$ (5) $P(A \cap B)$ (6) $P(A' \cap B)$
(7) $P(A \cap B')$ (8) $P(A' \cap B')$

(b) Express in words each of the probabilities calculated in (a) above.

(c) Display the probabilities calculated in (a) above in a table similar to Table 2.5, in which the frequencies are displayed.

33 In an urban area, A is the set of all Caucasians, and B is the set of all residents living within an incorporated city. Given that $P(A) = 0.35$ and $P(B) = 0.45$, complete the following table. (Assume that the two criteria of classification are independent.)

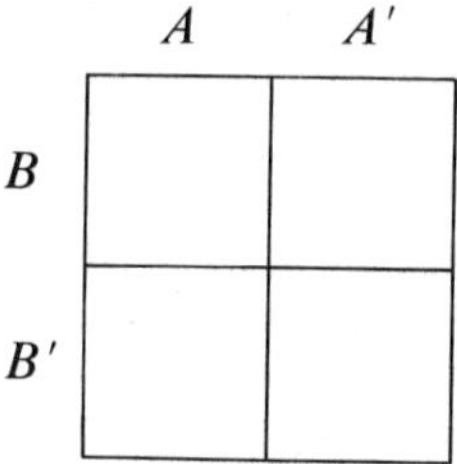

34 In a large city, 70% of the households receive a morning paper and 90% an evening paper. Suppose that these two events are independent. What is the probability that a randomly selected household will be one that receives both a morning and an evening paper?

2.5 BAYES'S THEOREM

The English clergyman, Thomas Bayes (1702–1761), who was also a mathematician of some stature, is generally given credit for the development of an interesting formula for the calculation of probabilities. The formula, known as *Bayes's theorem*, was published in 1763 after Bayes's death. This formula allows one to calculate the probability that some event which has occurred (the "effect") is the result of some "cause." It has been used extensively in medicine in the areas of so-called computer diagnosis and prediction of length of hospital stay. Bayes's theorem is also considered to be one of the foundations of modern *decision theory*, a body of knowledge devoted to the complex problem of decision-making under uncertainty, which has been developing since about 1950. The theorem may be stated as follows.

BAYES'S THEOREM *If $A_1, A_2, \ldots, A_n$ are n mutually exclusive events whose union is the universe, B is an arbitrary event such that $P(B) > 0$, and $P(B \mid A_k)$ and $P(A_k)$ are known for $1 \leq k \leq n$, then*

$$P(A_i \mid B) = \frac{P(A_i)P(B \mid A_i)}{\sum_{k=1}^{n} P(A_k)P(B \mid A_k)}$$

Let us now consider an application of Bayes's theorem.

Example 2.19 In a certain metropolitan area, low-, medium-, and upper-income groups constitute 20, 55, and 25% of the population, respectively. It is known that 80% of the low-income group is opposed to a certain bill now before Congress. The percentages for the medium- and high-income groups are 30 and 10, respectively. The available information is summarized in Table 2.6. An individual is selected at random from this population and found to be opposed to the bill. What is the probability that the person is from the low-income group? Medium-income group? High-income group?

We let B denote the event that a person opposing the bill has been selected, and A_1, A_2, and A_3 be the event that the selected person belongs in the low-, medium-, and high-income groups, respectively. The probabilities we seek are the conditional probabilities of belonging to the three income groups, given that the selected individual is opposed to the bill. That is, we wish to calculate $P(A_1 \mid B)$, $P(A_2 \mid B)$, and $P(A_3 \mid B)$.

We are given the unconditional probabilities $P(A_1) = 0.20$, $P(A_2) = 0.55$, and $P(A_3) = 0.25$. These probabilities are called *prior probabilities*, since they are available prior to the selection of the individual under discussion.

The conditional probabilities of selecting a person who opposes the bill, given that the individual belongs to one of the income groups, are also known. They are $P(B \mid A_1) = 0.80$, $P(B \mid A_2) = 0.30$, and $P(B \mid A_3) = 0.10$. These probabilities are referred to as *likelihoods*.

From the information given, we may calculate the following joint probabilities.

Table 2.6 Data summary for Example 2.19

Income group	**Proportion of population**	**Proportion of income group opposed**
Low	0.20	0.80
Medium	0.55	0.30
High	0.25	0.10

Table 2.7 Summary of calculations illustrating application of Bayes's theorem

Event	Prior probability, $P(A_i)$	Likelihood, $P(B \mid A_i)$	Joint probability, $P(B \cap A_i)$	Posterior probability, $P(A_i \mid B)$
A_1	0.20	0.80	0.160	0.46
A_2	0.55	0.30	0.165	0.47
A_3	0.25	0.10	0.025	0.07
	1.00		0.350	1.00

$$P(B \cap A_1) = P(B \mid A_1)P(A_1)$$

$$= (0.80)(0.20) = 0.16$$

$$P(B \cap A_2) = P(B \mid A_2)P(A_2)$$

$$= (0.30)(0.55) = 0.165$$

$$P(B \cap A_3) = P(B \mid A_3)P(A_3)$$

$$= (0.10)(0.25) = 0.025$$

From these calculations, we may use Bayes's theorem to calculate the desired probabilities.

$$P(A_1 \mid B) = \frac{0.16}{0.16 + 0.165 + 0.025} = \frac{0.16}{0.35} = 0.46$$

$$P(A_2 \mid B) = \frac{0.165}{0.35} = 0.47$$

$$P(A_3 \mid B) = \frac{0.025}{0.35} = 0.07$$

In words, we can now say that, given that a randomly selected individual is opposed to the bill, the probability that the person belongs to the low-income group is 0.46; to the medium-income group, 0.47; and to the high-income group, 0.07.

Since these conditional probabilities are computed after it is known that the person selected opposes the bill, we call them *posterior* probabilities.

Table 2.7 shows the calculations given above.

EXERCISES

35 In the medical records department of a hospital, three clerks are assigned the task of processing the records of patients. The first clerk, C_1, processes 45% of the records; the second clerk, C_2, 30% of the records;

and the third clerk, C_3, 25% of the records. The first clerk has an error rate of 0.03, the second of 0.05, and the third of 0.02. A record is selected at random from those processed during a week and is found to have an error. The medical records librarian wants to know the probability that the record was processed by each of the three clerks.

36 At a certain prison, 10% of the inmates have only a fourth-grade education or less. The educational status of the remainder is as follows: completed grades five, six, or seven, 50%; completed seventh grade or higher, 40%. Of the first group, 20% are under 25 years old; of the second group, 50%; and of the third group, 70%. An inmate is picked at random from this population and found to be under 25 years old. What is the probability that the inmate has a fourth-grade education or less? Find the probabilities for the other two groups.

37 A researcher in a specialty clinic has found that, over a period of several years, 20% of the patients referred to the clinic have had a particular disease, D_1; 30% have had a second disease, D_2; and 50% have had a third disease, D_3. The researcher has also found that a particular set of well-defined and easily recognized symptoms, S, is present in 25% of patients with disease D_1, 60% of patients with disease D_2, and 80% of patients with disease D_3. The researcher would like to use this information to quickly predict the diagnosis of newly admitted patients. Suppose a patient exhibiting the specified set of symptoms is admitted. What is the probability that this person has disease D_1? D_2? D_3?

38 In a certain population of adult males, 10% have an elementary school education only, 70% have a high school education, and 20% have a college education. Of those with an elementary school education, 5% are in the "high"-income group. The percentages in the "high"-income group for those with high school and college educations are 15 and 75, respectively. A man is randomly selected from this population and found to be in the "high"-income group. Find the probability that he has an elementary school education only; a high school education; a college education.

39 Of the persons seen in a county health department, 45% come from one area of the county, say area A; 30% from a second area, B; and 25% from a third area, C. Thirty percent of the patients from area A, 20% from area B, and 10% from area C are on the county welfare roll. A patient is picked at random from the department's files and is found to be on welfare. What is the probability that the patient is from area A? B? C?

40 A physician in a large industrial firm has found that 20% of the emergency cases she sees are from department A, 10% from department B, 45% from department C, and 25% from department D. She has also found that 10% of the emergency cases from department A, 5% from B, 15% from C, and 12% from D are accidents caused by apparent carelessness. An emergency case comes into the clinic suffering from an accident resulting from carelessness. What is the probability that the patient is from department A? B? C? D?

41 Table 2.8 shows some characteristics of a certain male population

Table 2.8 Data for Exercise 41

Age	Proportion of total population	Proportion of age group married
14–25	0.20	0.05
26–44	0.40	0.80
45 and over	0.40	0.95

expressed as proportions. A man is picked at random and found to be married. Calculate the probability that he is between 14 and 25 years old; between 26 and 44; 45 or over.

42 A researcher in a chronic-disease hospital has found that 15% of the patients admitted remain in the hospital less than 30 days, while 85% stay 30 days or more. The researcher has also found that 20% of the short-stay patients and 60% of the long-stay patients possess a certain set of characteristics. A patient who has the characteristics is admitted. What is the probability that the individual will remain in the hospital less than 30 days? 30 or more?

CHAPTER SUMMARY

In this chapter we covered the following topics: basic set theory, counting techniques, elementary probability concepts, and Bayes's theorem. The objective of this chapter was to give you sufficient background in probability to enable you to better understand the topics presented in later chapters.

SOME IMPORTANT CONCEPTS AND TERMINOLOGY

Experiment	Permutation
Event	Combination
Probability	Addition rule
Set	Conditional probability
Complement	Multiplication rule
Intersection	Independent events
Union	Bayes's theorem

REVIEW EXERCISES

43 Table 2.9 shows the number of patients admitted to a psychiatric outpatient clinic over a period of time. The patients have been cross-classified by diagnosis and age. From these data, find the following.

(a) $P(A_3 \cap B_4)$ (b) $P(A_4 \cup B_3)$ (c) $P(A_1')$
(d) $P(B_1 \mid A_7)$ (e) $P(B_6)$ (f) $P(B_6 \mid A_1)$
(g) The probability that a patient picked at random is schizophrenic

Table 2.9 Data for Exercise 43

Diagnosis	Age (years) <15 (A_1)	15–24 (A_2)	25–34 (A_3)	35–44 (A_4)	45–54 (A_5)	55–64 (A_6)	65 AND OLDER (A_7)	Total
Involutional psychotic reaction (B_1)	0	0	0	7	27	20	14	68
Manic-depressive reaction (B_2)	3	1	4	5	9	10	5	37
Schizophrenia (B_3)	5	95	140	160	103	44	7	554
Pschoneurotic reactions (B_4)	8	26	50	48	30	13	3	178
Alcohol addiction (B_5)	2	10	40	85	68	26	5	236
Drug addiction (B_6)	5	12	25	24	5	2	1	74
Total	23	144	259	329	242	115	35	1147

(h) The probability that a patient picked at random is schizophrenic, given that he/she is 25–34 years old

(i) The probability that a patient picked at random is schizophrenic and 25–34 years old

44 Evaluate the following.

(a) ${}_6P_2$ (b) ${}_7P_3$ (c) ${}_{10}P_5$

(d) $\binom{6}{2}$ (e) $\binom{7}{3}$ (f) $\binom{10}{5}$

(g) $\binom{8}{5}$ (h) $\binom{9}{5}$ (i) $\binom{5}{2}$

45 A saleswoman has 10 products she wishes to display at a national convention. She can display only four. How many displays does she have from which to choose, given that the order in which she displays the products is immaterial?

46 A foreman has eight men available from which to form a production team of four men. How many different teams can he form?

47 A warehouse is stocked with 5000 television sets. Table 2.10 shows the sets classified by brand and style. From these data, find the following.

(a) $P(B_1)$ (b) $P(S_4 \cap B_2)$

(c) $P(S_3 \mid B_3)$ (d) $P(S_1 \cup B_1)$

Table 2.10 Data for Exercise 47

Style	Brand B_1	B_2	B_3	Total
S_1	700	225	500	1425
S_2	650	175	400	1225
S_3	450	350	325	1125
S_4	500	175	600	1225
Total	2300	925	1825	**5000**

Table 2.11 Data for Exercise 49

Occupational status	Finished high school NO	YES
Unskilled labor (A)	250	100
Semiskilled and skilled labor (B)	150	100
Clerical and sales (C)	115	110
Semiprofessional and low-level management (D)	70	105

(e) The probability that a randomly selected set will be brand B_1 given that it is style S_4

(f) The probability that a randomly selected set is style S_2 and brand B_3

(g) The probability that a randomly selected set is either B_1 or B_3

48 In a certain organization, the probability that an employee picked at random will be over 30 is 0.55. What is the probability that an employee picked at random will be 30 or younger?

49 In a survey, 1000 adult male respondents were cross-classified by occupational status and by whether or not they had finished high school. Table 2.11 shows the results of the survey. A respondent, selected at random from this group for further interviews, was found not to have finished high school. What is the probability that he:

(a) Was an unskilled laborer?

(b) Was a semiskilled or skilled laborer?

(c) Held a clerical or sales job?

(d) Held a semiprofessional or low-level-management job?

50 Six high school seniors enter an essay contest. A prize of decreasing value is to be offered to each contestant. There can be no ties. How many different results are possible? How many different sets of first-, second-, and third-place winners are possible?

REFERENCES

1 John R. Dixon, *A Programmed Introduction to Probability*, New York: Wiley, 1964

2 Boyd Earl, J. William Moore, and Wendell I. Smith, *Introduction to Probability*, New York: McGraw-Hill, 1963

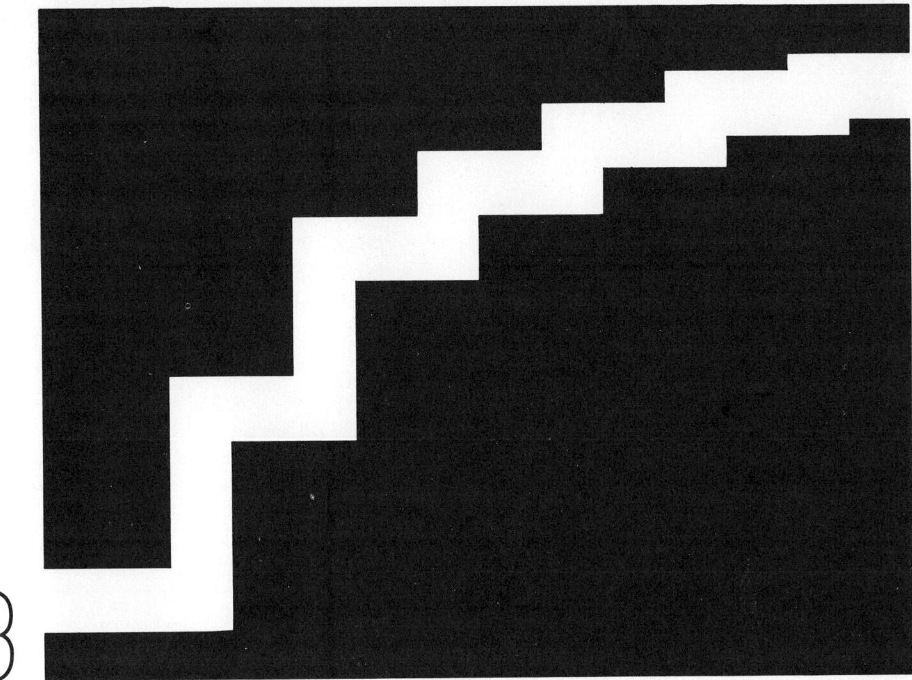

3 Probability Distributions

Table 3.1 shows, for the state of Georgia, the proportion in various age groups of all felons convicted during fiscal year 1971 and committed to custody of the State Board of Corrections. In Chapter 1 we learned to call such a table a relative frequency distribution, and in Chapter 2 we learned that the probability of an event may be interpreted as its relative frequency of occurrence. Hence we may refer to Table 3.1 as a *probability distribution*, which is the subject of this chapter. Tables like Table 3.1 are frequently found in the annual reports of government agencies. The information is of interest to a variety of people. The data in Table 3.1, for example, would be of interest to sociologists, legislators, social workers, and the officials and professional staff of the agency preparing the annual report.

In this chapter we build on what we learned in Chapters 1 and 2 to bring us closer to understanding how it is possible to draw inferences about a population by examining the data of a sample drawn from that population. We shall first discuss the construction of empirical probability distributions, and then examine some special probability distributions of both discrete and continuous random variables.

Table 3.1 Age of felons convicted during fiscal year 1971 who were committed to custody of the Georgia State Board of Corrections

Age	Proportion
17–18	0.151
19–24	0.408
25–29	0.173
30–34	0.098
35–39	0.067
40–44	0.044
45–49	0.020
50–54	0.020
55–59	0.009
60–64	0.004
65–69	0.003
Over 70	0.001
	1.000 (approx.)

Source: Adapted from Table 9 of *Annual Report 1970–1971*, Georgia Board of Corrections

3.1 DISTRIBUTIONS OF DISCRETE VARIABLES

In Chapter 1 we defined a discrete variable as a variable whose values are separated by some amount. An example of a discrete variable is one whose values we obtain by counting, as when we count the number of admissions to a hospital. In this example, the variable can assume only the values 0, 1, 2, ... It cannot assume such values as $\frac{1}{2}$, 1.33, and 2.1605.

It is frequently convenient to be able to construct some device, or rule, that will allow us to determine, or estimate, the probability that the discrete variable X assumes some particular value x.

DEFINITION *Any rule, or device, for determining $P(X = x)$, the probability that the random variable X assumes each of the possible values x, is called a* ***probability distribution.***

The rule, or device, may be a table, a graph, or a formula. We call a formula used to calculate $P(X = x)$ a *probability function*, and usually designate it by $f(x)$. That is, $f(x) = P(X = x)$.

Example 3.1 The first two columns of Table 3.2 show the distribution of the number of previous arrests of 500 adolescents seen for the first time during a year in a juvenile court.

Let us designate by X the discrete random variable "number of previous arrests" and by x the values that X can assume. Using the relative-frequency concept of probability, we may compute, for each x, the probability that X assumes that value. These probabilities, designated by $P(X = x)$, are shown in the last column of Table 3.2. This table illustrates the manner in which a probability distribution may be represented by a table.

Table 3.2 Number of previous arrests of 500 adolescents seen for the first time during a year in a juvenile court

Number of previous arrests	Number of adolescents	$P(X = x)$
0	300	$300/500 = 0.60$
1	100	$100/500 = 0.20$
2	60	$60/500 = 0.12$
3	20	$20/500 = 0.04$
4	10	$10/500 = 0.02$
5	5	$5/500 = 0.01$
6	5	$5/500 = 0.01$
	500	$500/500 = 1.00$

We may use Table 3.2 to answer questions about the probabilities of observing adolescents from the group who have had various numbers of previous arrests. For example, the probability that an adolescent picked at random from this group has had two previous arrests is 0.12. We may express this more compactly as $P(X=2)=0.12$. The two necessary characteristics of a probability distribution are illustrated by Table 3.2. They are as follows.

1 $P(X = x) \geq 0$ for all values of x.
2 $\Sigma_{\text{over all } x} P(X = x) = 1$.

In other words, the probability of occurrence of each x must be some number equal to or greater than 0, and the sum of the probabilities for all possible values of X must equal 1. Taken together, these characteristics imply what we have already learned, namely, that the probability of an event is some number between 0 and 1.

Remember that, for a probability distribution to exist, the two characteristics stated above must be present.

Frequently we are interested in knowing the probability that X assumes any value less than or equal to x. In Example 3.1, for example, we may wish to know $P(X \leq 3)$, the probability that an adolescent from the group has had three or fewer previous arrests. Information of this type is useful in planning education and rehabilitation programs for juvenile offenders.

DEFINITION *The probability that the random variable X assumes values less than or equal to x is called the* ***cumulative distribution function*** *of X, and is denoted by F(x).*

Thus we may write

$$F(x) = P(X \leq x)$$

Then, to evaluate the probability that X is less than or equal to any value, say a, we evaluate

$$F(a) = P(X \leq a) = \sum_{x \leq a} f(x) = \sum_{x \leq a} P(X = x)$$

Figure 3.1 Graphic representations of the probability distribution and cumulative distribution function of Example 3.1

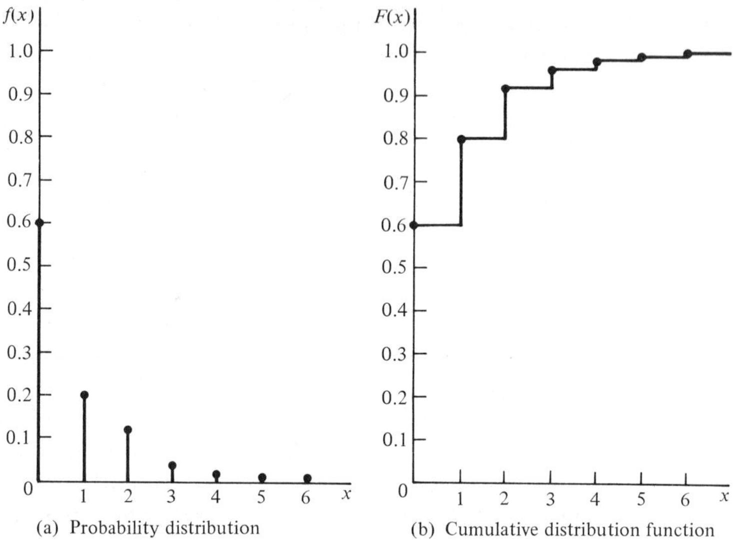

(a) Probability distribution (b) Cumulative distribution function

We may facilitate the calculation of $P(X \leq x)$ by cumulating the $P(X = x)$ values given in the table of the probability distribution. Table 3.3 shows the results obtained by cumulating the $P(X = x)$ values in Table 3.2. By examining this table, we see that $P(X \leq 3) = 0.96$.

We may represent probability distributions and cumulative distribution functions graphically. Figure 3.1(a) shows the probability distribution for Example 3.1, and Figure 3.1(b) is a graphic representation of the corresponding cumulative distribution function. The physical appearance of these graphs is typical of such graphs when the variable under study is discrete.

In Figure 3.1(b) the probabilities are represented by the horizontal lines only. The vertical lines merely serve to tie the lines representing the probabilities together, so they don't appear to be floating off into space. Bear in mind that $f(x)$ is 0 for any value of x that X cannot assume, and consequently $F(x)$ is 0 up to $x = 0$. From Figure 3.1(a) we can determine that $P(X = 3) = f(3) = 0.04$. From Figure 3.1(b) we see that $P(X \leq 3) = F(3) = 0.96$.

We may call the distributions discussed in this section so far *empirical distributions*, because they have been constructed directly from data generated by an experiment at hand. In the remainder of this section we discuss a special discrete distribution, the binomial distribution. This distribution is based on a certain process, which, when operative, yields data that are dis-

Table 3.3 Cumulative distribution function for Example 3.1

Values, x, of number of previous arrests	0	1	2	3	4	5	6
$F(x) = P(X \leq x)$	0.60	0.80	0.92	0.96	0.98	0.99	1.00

tributed according to a well-defined rule. Generally we are unable to say with certainty that the data at hand are distributed exactly as specified by the particular rule. Frequently, however, we are willing to say that a random variable in which we are interested is so distributed that its probability distribution can be *approximated* by the rule of this special distribution. Consequently, we shall use the rule (probability function) to calculate probabilities of interest.

The binomial distribution

Consider a random experiment which, on a single trial, can yield only one of two different mutually exclusive outcomes. Examples of such experiments include the following.

1 The birth of a child: The possible outcomes are "male" and "female."

2 A child is picked at random from a group of children in which m have above-average intelligence and n have average or below-average intelligence. The possible outcomes are "a child with above-average intelligence" and "a child with average or below-average intelligence."

3 A package of cereal is selected at random from a grocery shelf containing a packages of brand A and b packages of brand B. The possible outcomes are "a package of brand A cereal" and "a package of brand B cereal."

4 A student takes a final examination. The possible outcomes are "pass" and "fail."

Experiments such as these, in which a single trial can yield only one of two mutually exclusive outcomes, are called *Bernoulli trials*, after the Swiss mathematician Jakob Bernoulli (1654–1705), who is considered to be one of the founders of probability theory. For convenience, we may arbitrarily call one of the outcomes of a Bernoulli trial a success (S), and the other a failure (F). We may designate the probability of a success on a single trial by p, and the probability of a failure by $(1 - p) = q$.

Now consider a series of repeated Bernoulli trials which satisfy the following conditions.

1 The probability of a success p remains constant from trial to trial.
2 The trials are independent.

It is usually of interest to be able to determine the probability that x successes will be observed when an experiment is repeated n times under the conditions stated above.

Suppose, for example, that 20% of the patients discharged from a community hospital over the past ten years have been adolescents. A medical records librarian, asked by a physician to retrieve the records of any three adolescents, may wonder what the chances of obtaining $x = 3$ records of adolescents (successes) are if $n = 5$ records are retrieved at random. It will be both convenient and efficient if we can find a formula that will work for all values of n and x.

Suppose that n Bernoulli trials resulted in $n - x$ failures, followed by x successes. We may represent this sequence as follows.

$$\underbrace{FFF \cdots F}_{\substack{n-x \\ \textit{Failures}}} \quad \underbrace{SSS \cdots S}_{\substack{x \\ \textit{Successes}}}$$

By the multiplication rule, the probability of this particular sequence of outcomes is $qqq \cdots q\, ppp \cdots p = q^{n-x}p^x$. The probability would be the same regardless of the order in which the x successes and $n - x$ failures occurred.

In determining the probability that x successes will be observed in n trials, the order in which the successes occur is immaterial. Therefore, we must know the number of possible sequences of successes and failures that will yield exactly x successes. We may calculate this number from the formula, given in Chapter 2, for finding the number of permutations of n things, of which x are of one type and $n - x$ are of another type. By this formula, we find the desired number of sequences to be

$$_nP_{x,n-x} = \frac{n!}{x!\,(n-x)!} = \binom{n}{x}$$

Given a small value of n, it would be easy enough to list each of the possible sequences and number them 1, 2, 3, . . . , $\binom{n}{x}$.

Consider the medical-records example. The number of sequences of five records of which three are for adolescents, designated S for successes, and the remainder are designated F for failures, are given by

$$\binom{5}{3} = \frac{5!}{3!\,2!} = 10$$

We may list these sequences as follows.

1	*SSSFF*	2	*SFSSF*	3	*SFFSS*	4	*SSFSF*	5	*SSFFS*
6	*SFSFS*	7	*FSSSF*	8	*FFSSS*	9	*FSFSS*	10	*FSSFS*

Exactly x successes out of n trials would result if we observed either sequence number 1, or sequence number 2, or sequence number 3, . . . , or sequence number $\binom{n}{x}$. By the addition rule, then, the probability of observing exactly x successes out of n Bernoulli trials is equal to the probability of sequence number 1, plus the probability of sequence number 2, plus the probability of sequence number 3, . . . , plus the probability of sequence number $\binom{n}{x}$. Since there are a total of $\binom{n}{x}$ sequences, and since the probability of occurrence of each is $q^{n-x}p^x$, we may express the probability of observing exactly x successes in n Bernoulli trials by

$$P(X = x \mid n, p) = \binom{n}{x} q^{n-x} p^{x}$$

where X is the random variable "number of successes," and the possible values X can assume are $x = 0, 1, 2, 3, \ldots, n$. The left-hand side of the equation is read, "the probability that X is equal to x, given n and p." The equation may also be written

$$P(X = x \mid n, p) = \binom{n}{x} p^{x} q^{n-x}$$

Returning again to the medical-records example, we find, by using the above equation, that the probability of retrieving the records of three adolescents when $n = 5$, and when the probability of obtaining an adolescent's record is 0.20, is

$$P(X = 3 \mid 5, 0.20) = \binom{5}{3}(0.20)^{3}(0.80)^{2} = 10(0.008)(0.64) = 0.0512$$

It is now possible for us to list all values that X can assume, along with the probability of occurrence of each value. In other words, by means of the equation, we can display the probability distribution of X in tabular form, as shown in Table 3.4.

We have learned that two necessary characteristics of a probability distribution are that each value of $P(X = x)$ must be greater than or equal to 0, and that the sum of all $P(X = x)$ must equal 1. To verify that the first re-

Table 3.4 Probability distribution of X, the number of successes in n Bernoulli trials

Number of successes, x	$P(X = x)$
0	$\binom{n}{0} q^{n-0} p^{0}$
1	$\binom{n}{1} q^{n-1} p^{1}$
2	$\binom{n}{2} q^{n-2} p^{2}$
.	.
.	.
.	.
x	$\binom{n}{x} q^{n-x} p^{x}$
.	.
.	.
.	.
n	$\binom{n}{n} q^{n-x} p^{n}$
Total:	1

quirement is satisfied in Table 3.4, we note that p and n are both nonnegative, and therefore $P(X=x)$ can never be negative. To verify that the second condition is satisfied, we note that, term for term, the probabilities $P(X=x)$ are identical to the terms in the binomial expansion of $(q+p)^n$, which is expanded as follows:

$$(q+p)^n = q^n + nq^{n-1}p^1 + \frac{n(n-1)}{2} q^{n-2}p^2 + \cdots + nq^1p^{n-1} + p^n$$

Since $q+p=1$, we have

$$\sum \binom{n}{x} q^{n-x}p^x = (q+p)^n = 1^n = 1$$

As a result of this relationship, we usually refer to the probability distribution given in Table 3.4 as the *binomial probability distribution*, or more simply, the *binomial distribution*.

The binomial distribution is really a family of distributions, since a different distribution is defined for each different value of n and p—which are called the *parameters* of the binomial distribution. Figure 3.2 illustrates the manner in which the binomial distribution varies for different values of p and n. Regardless of the value of n, the distribution is symmetric when $p=0.5$. When p is greater than 0.5, the distribution is asymmetric, with the peak occurring to the right of center. When p is less than 0.5, the distribution is asymmetric, with the peak occurring to the left of center.

When the conditions of a series of Bernoulli trials are met, the binomial distribution is a convenient device for determining useful probabilities. To determine the probability that x successes will be observed in n Bernoulli trials, we simply evaluate our equation for specified values of x and n. We should realize, however, that the valid use of the binomial distribution depends on the assumption that the n trials of an experiment are n trials from an infinite number of possible trials. In practical applications of the binomial distribution, we say that it is applicable when sampling is from an infinite population, or when sampling is from a finite population with replacement. Since, in practice, sampling is carried out without replacement, care must be exercised in applying the binomial distribution. It is generally agreed that, when the size of the population is large relative to the size of the sample, and when p is not too close to 0 or 1, the binomial distribution may be used when sampling is without replacement from a finite population. This is not a very specific guideline, but better ones are not readily available. Some practitioners recommend that the binomial distribution be used only if the population is at least ten times the size of the sample.

Another example will illustrate the use of the binomial distribution.

Example 3.2 In a certain geographic area, 40% of the adult population are Democrats. A random sample of ten adults is selected from this population. What is the probability that three of them will be Democrats?

We assume that the size of the population is large relative to the size of

Figure 3.2 The binomial distribution for selected values of p and n

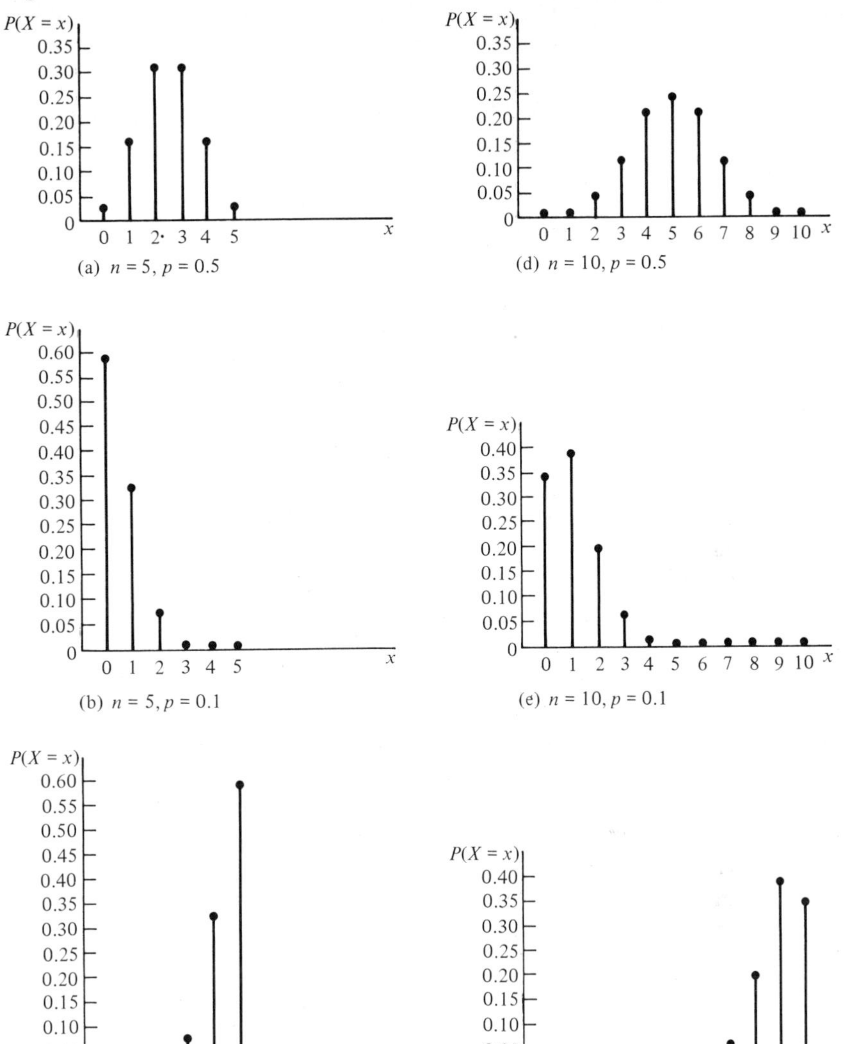

(a) $n = 5, p = 0.5$

(b) $n = 5, p = 0.1$

(c) $n = 5, p = 0.9$

(d) $n = 10, p = 0.5$

(e) $n = 10, p = 0.1$

(f) $n = 10, p = 0.9$

the sample. We then have a question that can be answered by means of the binomial distribution, where $n = 10$, $p = 0.4$, and $x = 3$. By means of our equation, we find

$$P(X = 3 \mid 10, 0.4) = \binom{10}{3}(0.4)^3(1 - 0.4)^{10-3}$$

$$= \frac{10!}{3!\,7!}(0.4)^3(0.6)^7 = 0.2150$$

Therefore, the probability that three Democrats will be found in ten adults selected at random from a large population of adults, of whom 40% are Democrats, is 0.2150.

Tables of the binomial distribution Calculating binomial probabilities by evaluating the equation can be unbelievably laborious when n is large. Fortunately, tables of binomial probabilities have been prepared, so direct use of the equation is not necessary. We merely enter a table with given values of n, p, and x, to obtain the desired probability. Table D of the Appendix is a portion of one such table.

To illustrate the use of Table D, let us again consider Example 3.2, in which we wished to determine the probability that there would be three Democrats out of ten adults randomly selected from a population of adults of whom 40% are Democrats. To find this probability in Table D, we first locate $n = 10$, then the column labeled $p = 0.40$, and finally, for $n = 10$, the row labeled 3. The tabular entry at the intersection of this row and column is the desired probability. We find that it is 0.2150, the value we obtain by evaluating the equation.

Finding cumulative binomial probabilities Frequently it is of interest to know with what probability X assumes values greater than some x, less than some x, or within some interval bounded by two values of X, say x_1 and x_2.

Let us refer again to the geographic area in which 40% of the population are Democrats. Suppose we wish to know the probability of finding between three and five Democrats in a random selection of ten people from the area.

Given some n, p, and x, we find the probability that X is greater than x by summing the individual probabilities.

$$P(X = x + 1) + P(X = x + 2) + \cdots + P(X = n)$$

That is,

$$P(X > x \mid n, p) = \sum_{X=x+1}^{n} \binom{n}{X} p^X q^{n-X}$$

Similarly, we obtain the probability that X assumes values less than x by summing the probabilities for individual values of X from 0 up to $x - 1$. That is,

$$P(X < x \mid n, p) = \sum_{X=0}^{x-1} \binom{n}{X} p^X q^{n-X}$$

Finally we find the probability that X assumes values between x_1 and x_2 inclusive by adding the probabilities for individual values of X between x_1 and x_2 inclusive. We express this in more compact form as

$$P(x_1 \le X \le x_2 \mid n, p) = \sum_{X=x_1}^{x_2} \binom{n}{X} p^X q^{n-X}$$

We may obtain all these probabilities by adding appropriate entries from Table D of the Appendix. We illustrate the procedure with an example.

Example 3.3 Thirty-five percent of the inmates in a correctional institution are repeaters. A random sample of 15 inmates is selected for evaluation.

(a) Find the probability that the number of repeaters in the group will be greater than 10.

Solution From Table D in the Appendix, we find

$$P(X > 10 \mid 15, 0.35) = P(X = 11) + P(X = 12) + P(X = 13)$$
$$+ P(X = 14) + P(X = 15) = 0.0024 + 0.0004$$
$$+ 0.0001 + 0.0000 + 0.0000 = 0.0029$$

(b) Find the probability that 5 or more will be repeaters.

Solution From Appendix Table D we obtain

$$P(X \geq 5 \mid 15, 0.35) = P(X = 5) + P(X = 6) + \cdots + P(X = 15)$$
$$= 0.2123 + 0.1906 + \cdots + 0.0000 = 0.6481$$

(c) Find the probability that fewer than 8 will be repeaters.

Solution From Appendix Table D we find

$$P(X < 8 \mid 15, 0.35) = P(X = 0) + P(X = 1) + \cdots + P(X = 7)$$
$$= 0.0016 + 0.0126 + \cdots + 0.1319 = 0.8868$$

(d) Find the probability that 9 or fewer will be repeaters.

Solution From Appendix Table D we have

$$P(X \leq 9 \mid 15, 0.35) = P(X = 0) + P(X = 1) + \cdots + P(X = 9)$$
$$= 0.0016 + 0.0126 + \cdots + 0.0298 = 0.9876$$

(e) Find the probability that the number of repeaters will be between 5 and 12 inclusive.

Solution From Appendix Table D we obtain

$$P(5 \leq X \leq 12 \mid 15, 0.35) = P(X = 5) + P(X = 6) + \cdots + P(X = 12)$$
$$= 0.2123 + 0.1906 + \cdots + 0.0004$$
$$= 0.6480$$

(f) Find the probability that the number of repeaters will be greater than 6, but less than 12.

Solution From Appendix Table D we find

$$P(6 < X < 12) = P(X = 7) + P(X = 8) + \cdots + P(X = 11)$$

$$= 0.1319 + 0.0710 + \cdots + 0.0024 = 0.2447$$

The use of Appendix Table D for $p > 0.5$ Note that Table D of the Appendix does not contain values of p beyond 0.50. To find binomial probabilities in Appendix Table D, when p is greater than 0.50, we enter the table with n as usual, but we locate the column labeled $1 - p$, rather than p, and the row labeled $n - x$, rather than x. In other words, we find the desired probability at the intersection of the column labeled $1 - p$ and the row labeled $n - x$, for a given n.

Utilizing our usual notation, we may write

$$P(X = x \mid n, p > 0.5) = P(X = n - x \mid n, 1 - p) = \binom{n}{n-x} q^{n-x}p^{x}$$

Let us illustrate by means of an example.

Example 3.4 Seventy percent of the adults living in a certain community are over 25 years old. Five adults are selected at random from this community. What is the probability that three will be over 25?

To find the desired probability, we locate $n = 5$ in Appendix Table D, and then find the intersection of the column labeled $1 - 0.7 = 0.3$ and the row labeled $5 - 3 = 2$. We find the probability to be 0.3087.

We may verify that this is correct by evaluating our equation. We have

$$\binom{5}{2}(0.3)^2(0.7)^3 = \frac{5!}{2!\,3!}(0.09)(0.343) = 0.3087$$

EXERCISES

1 A student, totally unprepared for an examination, finds that it contains 20 true-false questions. She decides to flip a coin (assumed to be fair) for each question and answer "true" if the coin comes up heads and "false" if it comes up tails.

(a) What is the probability that she will pass the examination if she must answer 70% of the questions correctly in order to pass?

(b) What is the probability that she will answer at least half of the questions correctly?

2 The probability that a particular man will hit the bull's eye of a target on a single throw of a dart is 0.40. He throws ten darts in succession. Find the probability that he will hit the bull's eye:

(a) all ten times

(b) five times
(c) between three and seven times, inclusive
(d) fewer than three times

3 The probability of a male birth is 0.52. What is the probability that a couple who have three children will have:

(a) all boys?
(b) no boys?
(c) at least one boy?

4 Suppose that 30% of the students at a large university are opposed to paying a student activities fee. Find the probability that, in a random sample of 25 students, the number who oppose the fee will be:

(a) exactly five
(b) greater than five
(c) five or fewer
(d) between six and ten inclusive

5 A secretary who is supposed to be at work at 8:00 every morning is 15 or more minutes late 20% of the time. The company president, who doesn't come in until 10:00, occasionally calls the office between 8:00 and 8:15 to dictate a letter. What is the probability that the secretary will not be in the office three mornings out of the next six on which the president calls?

6 In a certain metropolitan area, in 15% of the households, no one is at home each night between 7:00 and midnight. Someone conducting a telephone survey randomly selects ten households from this area to call. What is the probability that the person doing the survey will not get an answer to:

(a) all ten calls?
(b) exactly five calls?
(c) three or fewer calls?
(d) What is the probability that the person will get an answer to all ten calls?

7 The standard method for teaching a particular skill to certain handicapped persons is effective in 50% of the cases. A new method is tried with 15 persons. Given that if the new method is no better than the standard method, what is the probability that 11 or more will learn the skill?

8 Personnel records of a large manufacturer show that 10% of assembly-line employees quit within one year after being hired. Ten new employees have just been hired.

(a) What is the probability that exactly half of them will still be working after one year?
(b) What is the probability that all will be working after one year?
(c) What is the probability that three of the ten will quit before the year is up?

9 In a certain geographic area, 15% of the adults are illiterate. In a random sample of 25 adults from this area, what is the probability that the number of illiterates will be:

(a) exactly ten?
(b) less than five?

(c) five or more?
(d) between three and five inclusive?
(e) less than seven, but more than four?

3.2 DISTRIBUTIONS OF CONTINUOUS VARIABLES

In Chapter 1 we defined a continuous variable as a variable that can assume any value within some interval of values, and gave as examples variables that are measured in units of height, weight, time, or temperature.

In Section 3.1, which discussed discrete variables, we learned how it is possible to determine the probability that the discrete random variable X assumes a particular value, x. We examined tabular, graphic, and mathematical (by means of a formula) methods. When we focus on a continuous variable, we usually want to be able to determine the probability that X assumes values within some interval. For example, we may wish to know the probability that X assumes values between x_a and x_b, the probability that X is greater than x_b, or the probability that X is less than x_a. In Chapter 1 we learned how to construct frequency distributions and relative frequency distributions of continuous variables by defining class intervals and determining the frequency or relative frequency of occurrence of observations falling in the various class intervals. Example 1.1 illustrates the procedure. We also learned that frequency distributions and relative frequency distributions may be represented graphically by means of histograms and frequency polygons. The area under a histogram or frequency polygon between any two values, say x_a and x_b, of the random variable X under study is equal to the relative frequency of occurrence of values of X between x_a and x_b. From our discussion of probability in Chapter 2, we know that if the data at hand constitute a sample of observations from some population, we may interpret these relative frequencies as estimates of the corresponding true probabilities. Thus we may interpret the relative frequency of occurrence of sample values of X between x_a and x_b inclusive as an estimate of $P(x_a \leq X \leq x_b)$, the probability that X assumes values between x_a and x_b (inclusive).

Suppose we have available a sample of measurements of some continu-

Figure 3.3 Histogram of sample observations

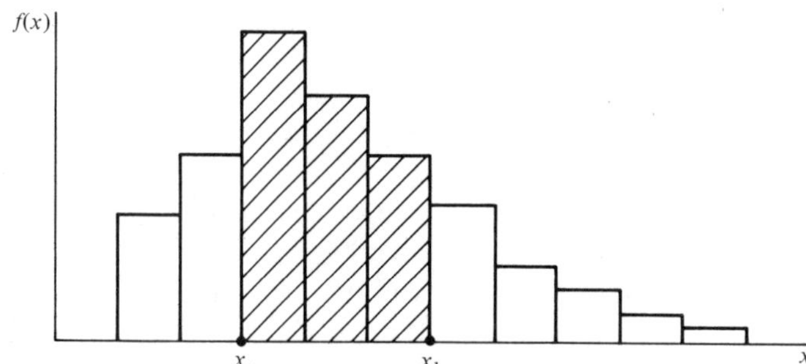

Figure 3.4 Histogram with small class intervals constructed from a large sample

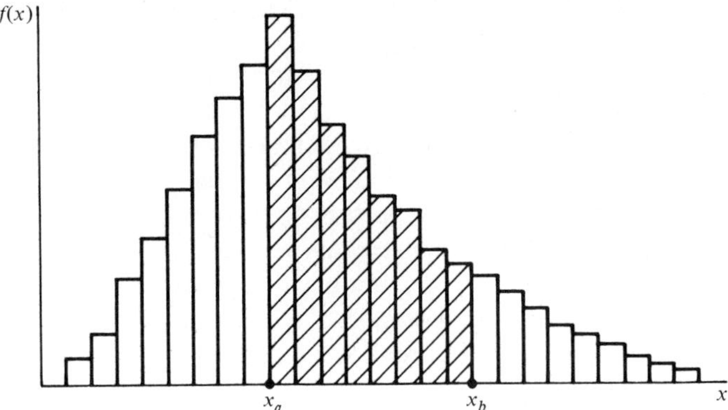

ous random variable X. A relative frequency histogram for the sample data might look like Figure 3.3. We interpret the hatched area in this figure as an estimate of the probability that X assumes values between x_a and x_b.

Now suppose we have a much larger sample of X values. Since our random variable is continuous, let us assume that we can make measurements to any number of decimal places. Therefore, in constructing a histogram for this large sample of observations, we may make our class intervals as small as we wish. A relative frequency histogram, constructed from a very large sample of values of X, might look like Figure 3.4 when very small class intervals are used.

Thus we see that as the number of observations gets larger and the class interval widths get smaller, the histogram approaches a smooth curve, such as that in Figure 3.5. It seems reasonable to assume, then, that the area between x_a and x_b in Figure 3.4 is very nearly equal to the area between x_a and x_b in Figure 3.5.

A continuous probability distribution is always represented graphically as a smooth curve. The area under the curve, above the horizontal axis, and

Figure 3.5 Distribution of a continuous variable

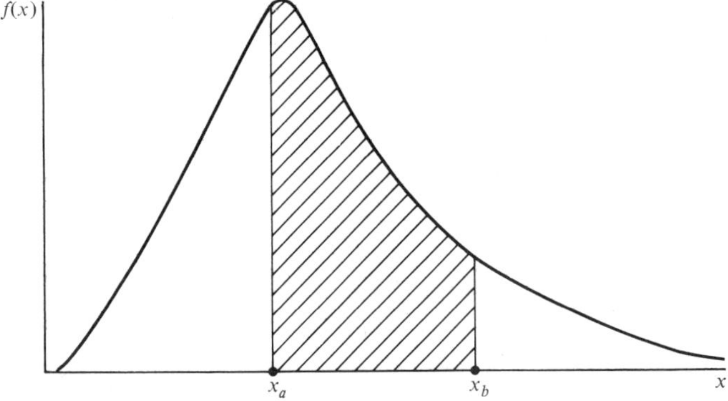

between perpendiculars erected at two points a and b is equal to the probability that the random variable assumes values between the two points.

Note that the discussion here has been in terms of determining the probability associated with an interval. The reason for this is that, for a continuous variable, $P(X = x) = 0$. That is, the probability that X assumes a specific value is 0. Consider a smooth curve drawn to represent a continuous probability distribution. You can see that the area above a point is equal to 0.

When dealing with histograms computed from sample data, we obtain the area under a histogram between two points, each of which falls at some class boundary, by adding the areas represented by adjacent cells of the histogram. For example, if we want to find the area between x_a and x_b of the histogram in Figure 3.3, we add the individual areas of the three cells lying between x_a and x_b.

Similarly, to obtain the area between x_a and x_b of the histogram in Figure 3.4, we again obtain the sum of the areas of the individual cells between x_a and x_b. Here the task is less appealing than that required for the first histogram.

Determining the area under a smooth curve, such as the area between x_a and x_b in Figure 3.5, is more of a problem because we have no cells, and therefore no cell areas to add. To find such an area, we need to use integral calculus. Since you are not expected to be familiar with calculus, we shall deal only briefly with the subject at this point.

Through the process of integration, integral calculus provides a mathematical technique that is the limit of summation. So when one uses integral calculus to find the area under a smooth curve, one in effect adds areas of infinitesimally small rectangles (cells). In order to perform the calculations required in integration, one must know the formula for the specific distribution under study.

Fortunately, we shall not have to use integral calculus with any of the continuous probability distributions in this book. Areas under curves that will be of interest have been determined, tabulated, and made available in Appendix tables that we shall cite when necessary.

The normal distribution

Of all the continuous distributions known, one of the most important in statistics is the *normal distribution.* The formula for this distribution was first published by Abraham Demoivre (1667–1754) in 1733. Other mathematicians who figure prominently in the early history of the normal distribution are Pierre Simon, Marquis de Laplace (1749–1827), and Carl Friedrich Gauss (1777–1855), in whose honor it is sometimes called the *Gaussian distribution.*

The formula for the normal distribution is

$$f(x) = \frac{1}{\sigma\sqrt{2\pi}} \exp\left[-\frac{1}{2}\left(\frac{x-\mu}{\sigma}\right)^2\right]$$

Figure 3.6 A normal distribution with mean μ and variance σ^2

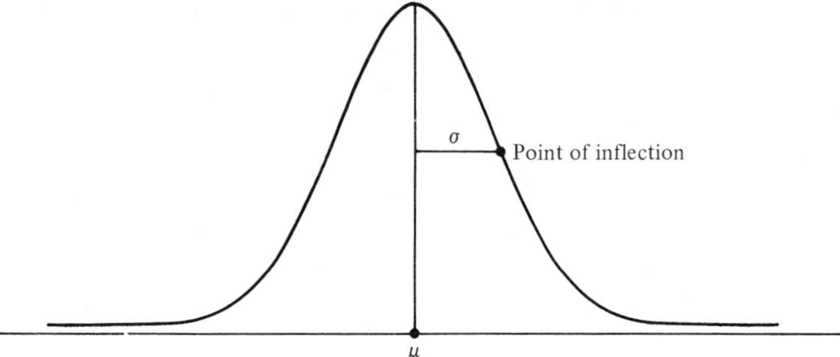

where

$\mu =$ the mean of the distribution
$\sigma =$ the standard deviation of the distribution
$\pi =$ the constant 3.14159...

and

$$\exp\left[-\frac{1}{2}\left(\frac{x-\mu}{\sigma}\right)^2\right]$$

indicates that the term in brackets is the exponent of e, where

$e =$ the constant 2.71828...

Figure 3.6 shows the graph of a normal distribution with mean μ and variance σ^2.

Some important characteristics of the normal distribution are the following.

1 The total area under the curve and above the horizontal axis is equal to 1 (square unit).

2 The distribution is symmetric about its mean. That is, 50% of the area lies to the right of the mean and 50% to the left.

3 The mean, the median, and mode are all equal.

4 The horizontal distance from the point of inflection on the curve (the point where the curve stops being concave downward and starts being concave upward) to a perpendicular erected at the mean is equal to the standard deviation σ, as shown in Figure 3.6.

5 The normal distribution is really a "family" of distributions, since there is a different distribution for each value of μ and σ. Figure 3.7 shows three normal distributions, with the same standard deviation but different means. As we can see, the value of μ locates a distribution on the horizontal axis. Distributions with different means are located at different positions on the horizontal axis.

Figure 3.7 Three normal distributions with equal standard deviations but different means

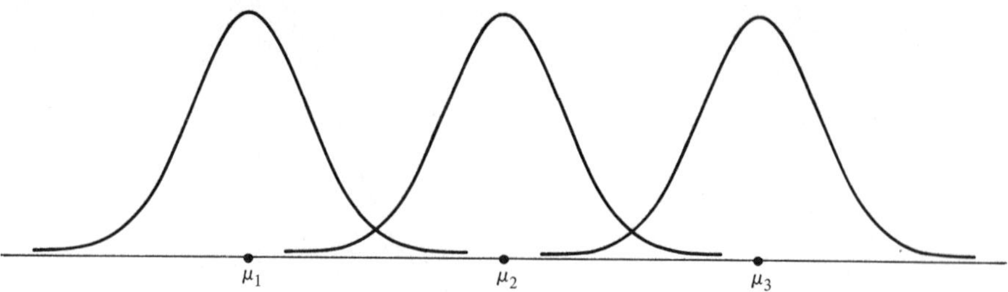

Figure 3.8 shows three normal distributions with the same mean, but different standard deviations. This figure shows that the greater the standard deviation, the flatter and more spread out the graph of the distribution.

6 The curve of a normal distribution extends from $-\infty$ to $+\infty$.

7 If we erect perpendiculars from the horizontal axis a distance of one standard deviation from the mean on either side, the area bounded by these perpendiculars, the curve, and the horizontal axis is equal to approximately 0.68, that is, about 68% of the total area. In a similar manner, we can enclose approximately 95% of the total area under the curve of a normal distribution by erecting perpendiculars a distance of two standard deviations from the

Figure 3.8 Three normal distributions with equal means but different standard deviations

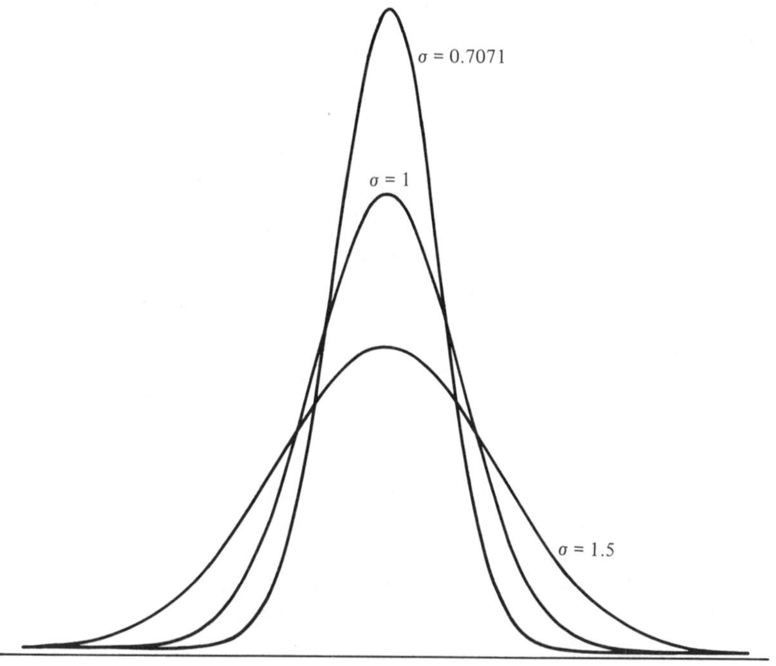

Figure 3.9 Curves of normal distributions, showing approximate proportion of total area enclosed by perpendiculars erected at various distances from the mean

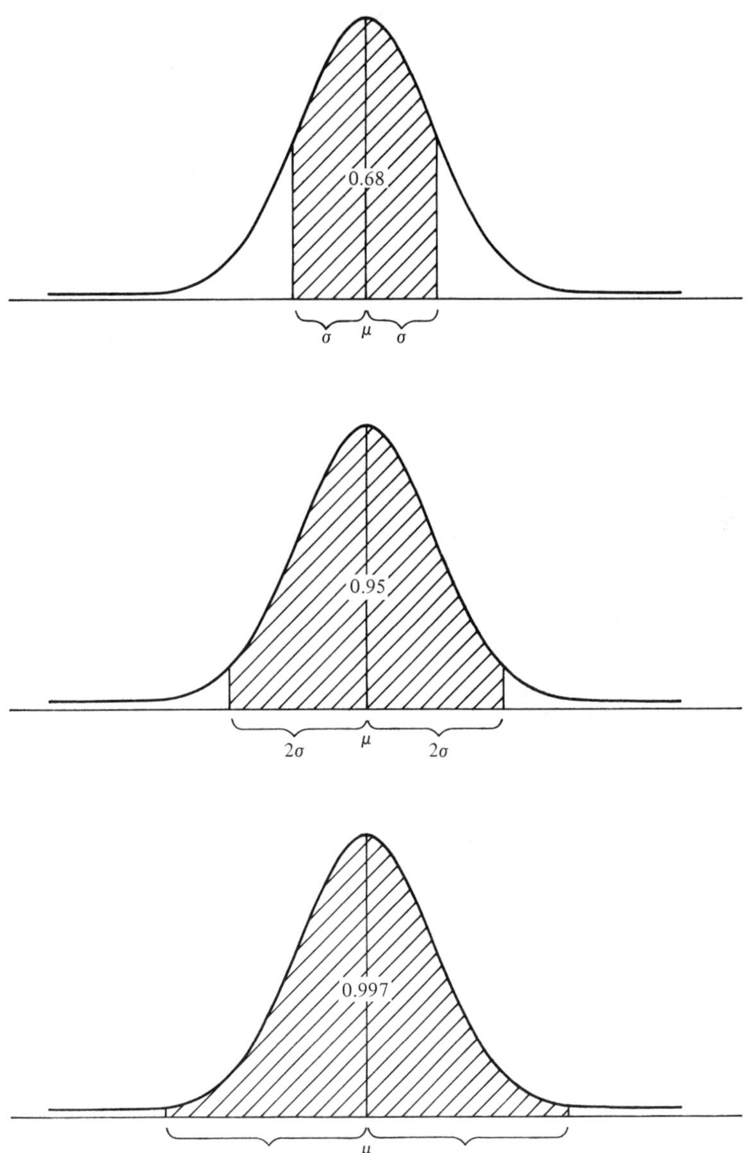

mean on either side. And we can enclose approximately 99.7% of the area by erecting perpendiculars a distance of three standard deviations from the mean on either side. Figure 3.9 illustrates these facts.

Given any random variable X that is normally distributed, one can find $P(x_a \leq X \leq x_b)$, the probability that X assumes values between x_a and x_b, by

Figure 3.10 Area under the curve of the normal distribution, showing $P(x_a \leq X \leq x_b)$

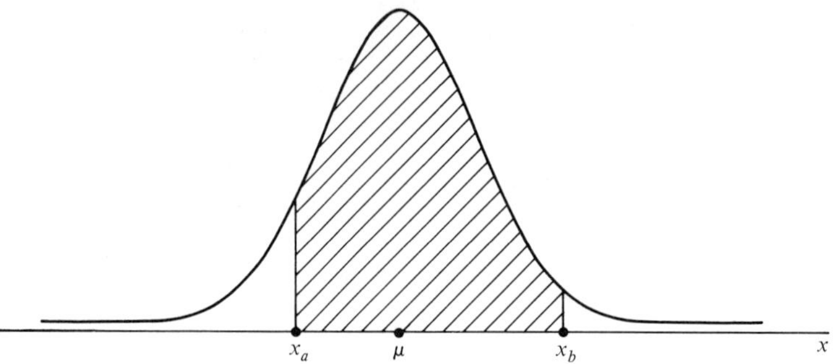

integrating the equation for the normal distribution from x_a up to x_b. Figure 3.10 shows the area corresponding to the desired probability.

Fortunately, as explained in the discussion below, it won't be necessary for you to perform the actual integration in order to find $P(x_a \leq X \leq x_b)$ or any other probability being sought.

Note that, in the case of continuous distributions, $P(x_a \leq X \leq x_b)$ yields the same result as $P(x_a < X < x_b)$. Since the area above the points x_a and x_b is 0, these end points make no contribution to the probability associated with the interval.

The standard normal distribution

We have already noted that there is a different normal distribution for each different value that μ and σ can assume. A normal distribution of particular importance in statistics is the *standard normal distribution*, which has a mean $\mu = 0$ and a variance $\sigma^2 = 1$.

Areas for the standard normal distribution corresponding to various probabilities have been tabulated. Appendix Table E is one such table. The body of this table contains areas under the curve between 0 and z_0, a specified value of Z, the random variable that is normally distributed with mean 0 and variance 1. The hatched area of Figure 3.11 corresponds to the entry in Appendix Table E for $P(0 \leq Z \leq z_0)$.

Figure 3.11 Area under the standard normal curve between 0 and z_0

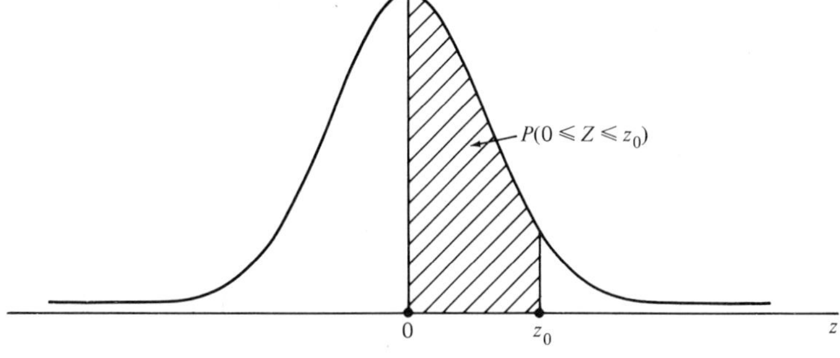

Figure 3.12 Area under the standard normal curve between 0 and 2.05

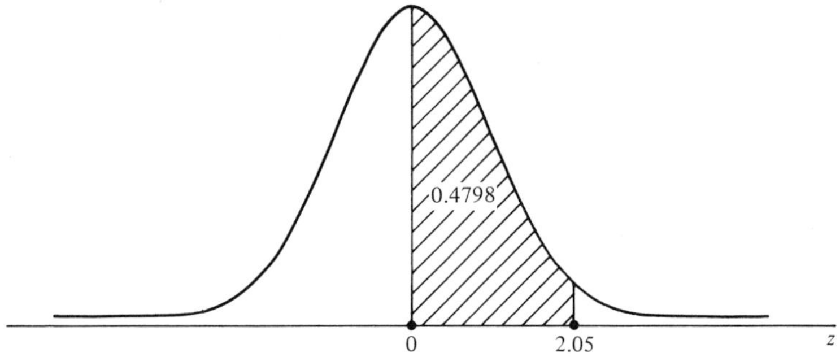

Let us illustrate the use of Appendix Table E by some examples.

Example 3.5 Find the area under the standard normal curve between 0 and $z = 2.05$.

Solution The desired area is the hatched area in Figure 3.12.

To find the numerical value of the area, we locate 2.0 in the leftmost column and 0.05 in the top row of Appendix Table E. The entry at the intersection of this row and column, 0.4798, is the area we seek.

Example 3.6 What is the probability that a value of z picked at random will be between 0 and −2.05?

Solution Because of the symmetry of the normal distribution, the area between 0 and −2.05 is exactly equal to the area between 0 and +2.05, as shown in Figure 3.13.

To find the numerical value of the area, we enter Appendix Table E, as explained in Example 3.5. Since our value of z is negative, however, the area is located to the left of 0 in the graphic representation.

Example 3.7 Find $P(-1.78 \leq Z \leq 1.52)$.

Solution We may think of the area sought in this example as consisting of two parts. One part is to the left of 0, and the other part to the right, as shown in Figure 3.14.

Figure 3.13 Area under the standard normal curve between 0 and −2.05

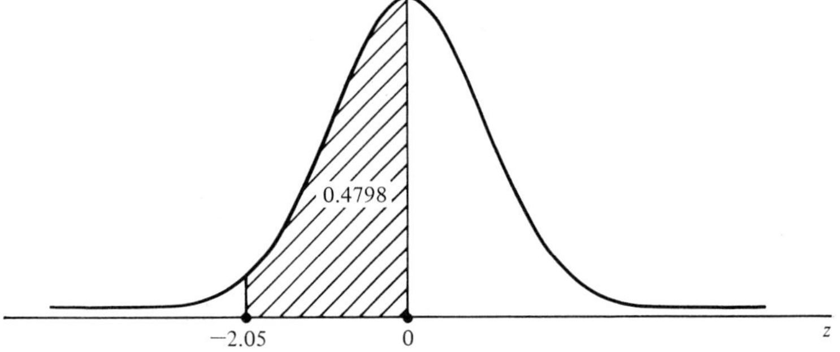

Figure 3.14 Area under the standard normal curve between −1.78 and 1.52

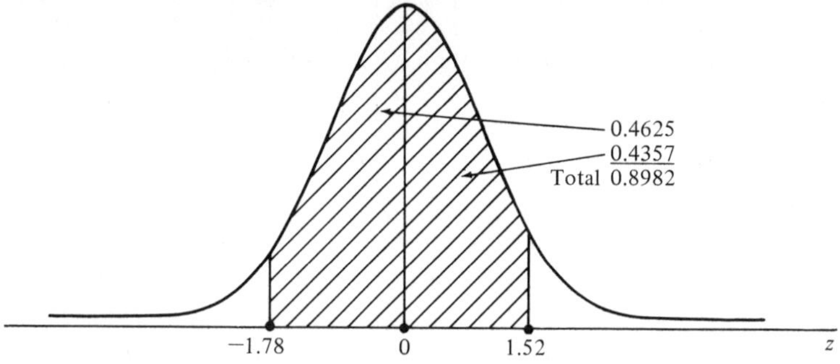

To find the area in Appendix Table E, we first find the area between 0 and 1.52, namely 0.4357. We then find the area between 0 and −1.78, namely 0.4625. We find the total area by adding 0.4357 and 0.4625. Our answer is 0.8982.

Example 3.8 Find $P(Z > 1.67)$.

Solution The area we are asked to find is shown in Figure 3.15. This area is not given directly in Appendix Table E. We recall that the total area to the right of 0 is 0.5000. To find the area to the right of 1.67, then, we find the area between 0 and 1.67 and subtract it from 0.5000. We obtain $0.5000 - 0.4525 = 0.0475$.

Applications of the normal distribution

We are not likely to encounter in nature random variables that are exactly normally distributed, since the normal distribution is a mathematical ideal. Many continuous random variables, however, may be adequately characterized by a normal distribution. Such variables include heights of individuals belonging to certain well-defined populations, dimensions of certain manufactured articles such as the diameters of washers, IQs of individuals, and various test scores.

Figure 3.15 Area under the standard normal curve to the right of 1.67

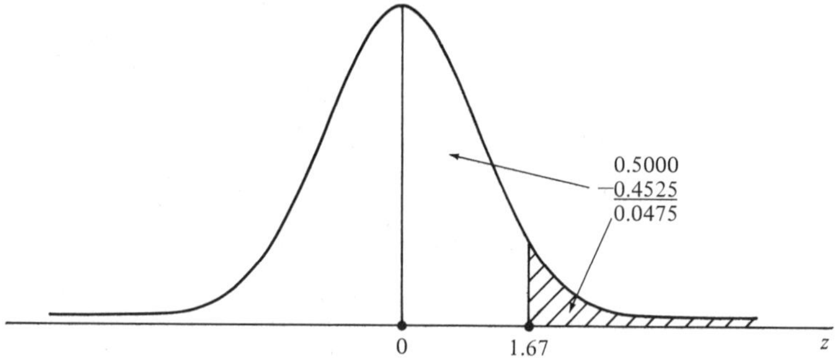

Figure 3.16 Normal distribution, original units and corresponding standard normal units

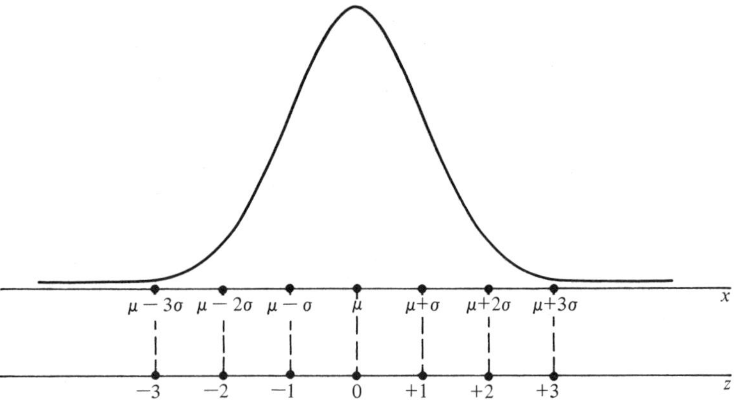

When the variable of interest in an investigation is at least approximately normally distributed, we frequently use our knowledge of the normal distribution in our analysis. As with the standard normal distribution, we may answer probability questions about any random variable X that is at least approximately normally distributed. For example, we may wish to know the probability that some approximately normally distributed random variable X with mean μ and standard deviation σ assumes values between, say, x_a and x_b. To obtain such probabilities, we transform the variable X with mean μ and variance σ^2 to the standard normal Z, with mean 0 and variance 1. We accomplish this by means of the formula

$$z = \frac{x - \mu}{\sigma}$$

by which any value x of the random variable X is transformed into a value z of the standard normal variable Z. Figure 3.16 shows the relationship between the original scale in units of X and the transformed scale in units of Z.

Once we have made the transformation, we can use Appendix Table E to find probabilities of interest.

Let us illustrate with some examples.

Example 3.9 The IQs of individuals composing a certain population are approximately normally distributed, with a mean of 100 and a standard deviation of 10.

(a) Find the proportion of individuals with IQs greater than 125.

Solution We first convert 125 to a z value by means of the formula.

$$z = \frac{125 - 100}{10} = 2.5$$

The problem now reduces to that of finding the area under the standard

Figure 3.17 Proportion of individuals with IQs greater than 125, Example 3.9(a)

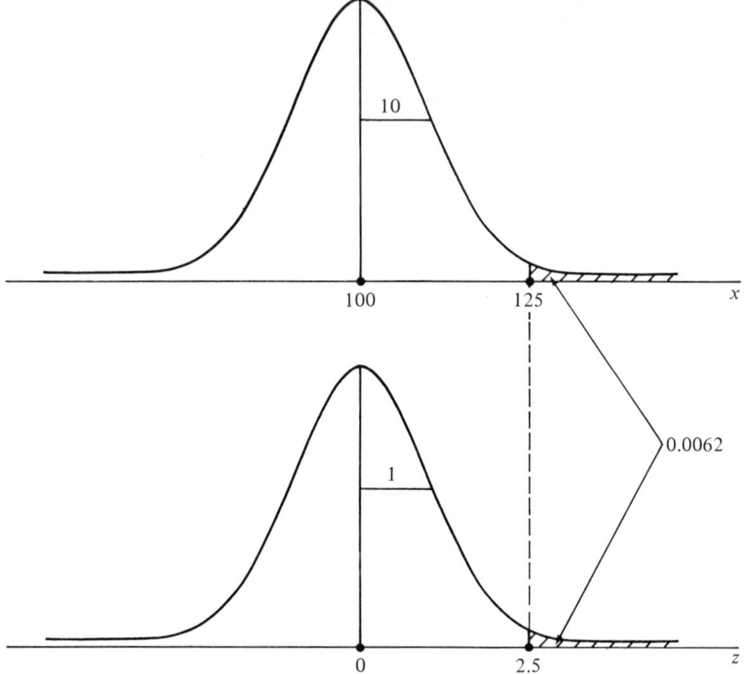

normal curve that is to the right of 2.5. From Appendix Table E, we find this area to be 0.0062. Figure 3.17 illustrates the procedure.

(b) What is the probability that an individual picked at random will have an IQ between 105 and 115?

Solution We let $x_a = 105$ and $x_b = 115$, and find the corresponding values of z_a and z_b as follows.

$$z_a = \frac{105 - 100}{10} = 0.5, \qquad z_b = \frac{115 - 100}{10} = 1.5$$

From Appendix Table E, we find the desired area to be 0.2417. Figure 3.18 shows the probability graphically.

(c) Find $P(80 \leq X \leq 95)$

Solution We have

$$P(80 \leq X \leq 95) = P\left(\frac{80 - 100}{10} \leq z \leq \frac{95 - 100}{10}\right)$$

$$= P(-2.0 \leq z \leq -0.5)$$

$$= P(-2.0 \leq z \leq 0) - P(-0.5 \leq z \leq 0)$$

$$= 0.4772 - 0.1915 = 0.2857$$

Figure 3.18 Probability that an individual picked at random will have an IQ between 105 and 115, Example 3.9(b)

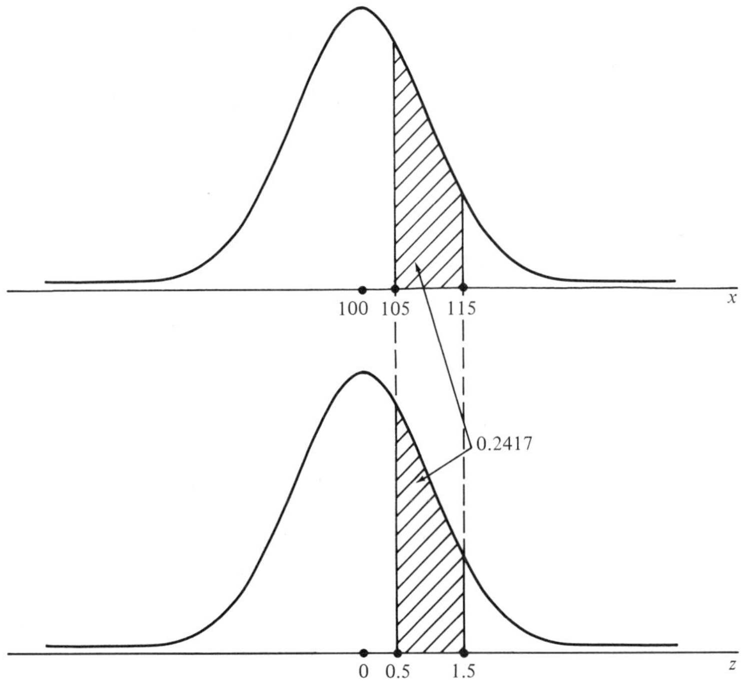

Figure 3.19 shows this probability graphically.

The normal approximation to the binomial

The usefulness of the normal distribution in statistics is further illustrated by the fact that this distribution provides a good approximation to the binomial distribution when n is large and p is not too close to 0 or 1. A rule of thumb that is frequently followed states that the normal approximation to the binomial is appropriate when np and $n(1-p)$ are both greater than five.

To use the normal approximation, we let $\mu = np$, $\sigma = \sqrt{np(1-p)}$, and convert values of the original variable to values of z in order to find probabilities of interest. When the size of the sample to be analyzed is not one of the values of n given in available binomial tables, the normal approximation to the binomial provides a convenient alternative.

Since the normal distribution is continuous, and the binomial is a discrete distribution, we can obtain better results if we make an adjustment to account for this fact when we use the approximation. We can best understand the adjustment, called the *continuity correction*, by observing a histogram, constructed from binomial data, with a superimposed smooth curve. Figure 3.20 shows this for $n = 20$ and $p = 0.3$.

In Figure 3.20, the probability that $X = x$ is equal to the area of the

Figure 3.19 $P(85 \leq X \leq 95)$, Example 3.9(c)

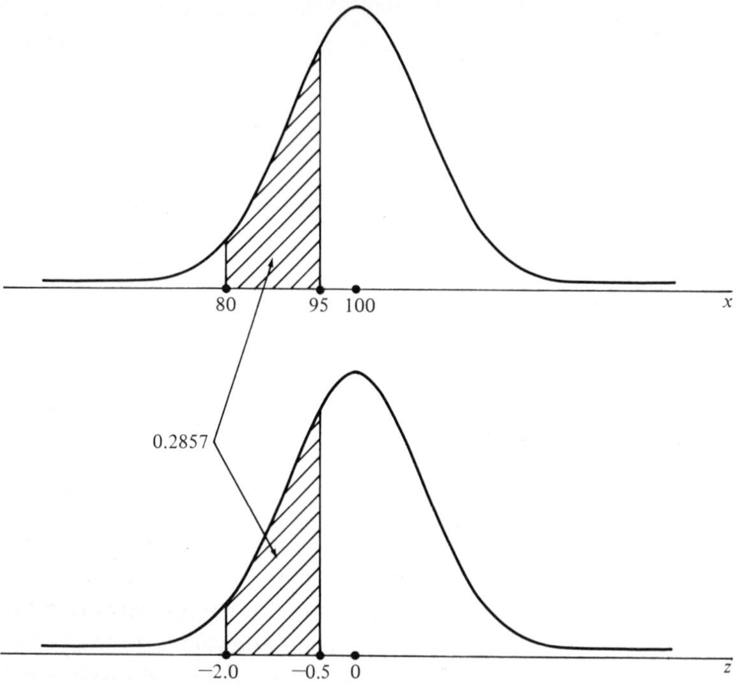

rectangle centered at x. For example, the probability that $X = 8$ is equal to the area of the rectangle centered at 8. We can see that this rectangle extends from 7.5 to 8.5. Consulting Appendix Table D, we find that this area is equal to 0.1144. The corresponding area is hatched in Figure 3.20(a).

When we use the normal approximation to the binomial, we must take account of the fact that, for the binomial, $P(X = x)$ is the area of a rectangle centered at x. When we convert values of x to values of z, the continuity correction consists of adding 0.5 to, and/or subtracting 0.5 from, x, as appropriate. To illustrate, let us use the continuity correction and normal approximation to find the probability that X assumes a value between $x_a = 7.5$ and $x_b = 8.5$. Converting to z values, we have

$$z_a = \frac{7.5 - 6}{\sqrt{(20)(0.3)(0.7)}} = \frac{1.5}{2.05} = 0.73$$

$$z_b = \frac{8.5 - 6}{\sqrt{(20)(0.3)(0.7)}} = \frac{2.5}{2.05} = 1.22$$

From Appendix Table E, we find that the probability we seek is 0.1215, which is reasonably close to the exact probability of 0.1144. The area under the normal curve corresponding to $P(7.5 \leq X \leq 8.5)$ is hatched in Figure 3.20(b).

Figure 3.20 Normal approximation to the binomial with $n = 20$, $p = 0.3$, and $\mu = np = 6$, showing $P(X = 8)$

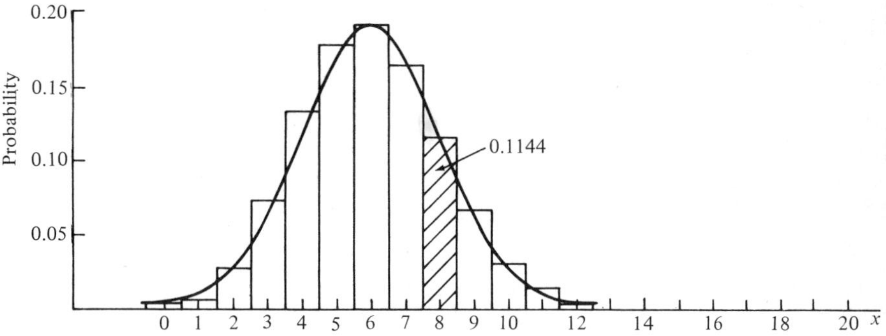

(a) $P(X = 8)$ using binomial probabilities

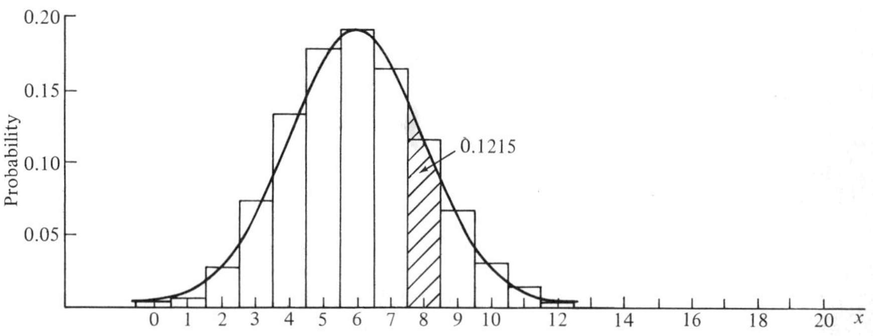

(b) $P(7.5 \leqslant X \leqslant 8.5)$ using normal approximation

Let us use this same example to find $P(5 \leq X \leq 10)$. Using binomial probabilities from Appendix Table D, we find the answer to be 0.7454. The corresponding area is hatched in Figure 3.21(a).

To use the normal approximation, we find

$$P(4.5 \leq X \leq 10.5) = P\left(\frac{4.5 - 6}{2.05} \leq z \leq \frac{10.5 - 6}{2.05}\right)$$

$$= P(-0.73 \leq z \leq 2.20)$$

$$= 0.2673 + 0.4861 = 0.7534$$

The corresponding area is hatched in Figure 3.21(b).

Again we see that the normal approximation gives a result that is quite close to the exact probability. If we had not used the continuity correction, the normal approximation would have given

Figure 3.21 Normal approximation to the binomial with $n = 20$, $p = 0.3$, and $\mu = np = 6$, showing $P(5 \leq X \leq 10)$

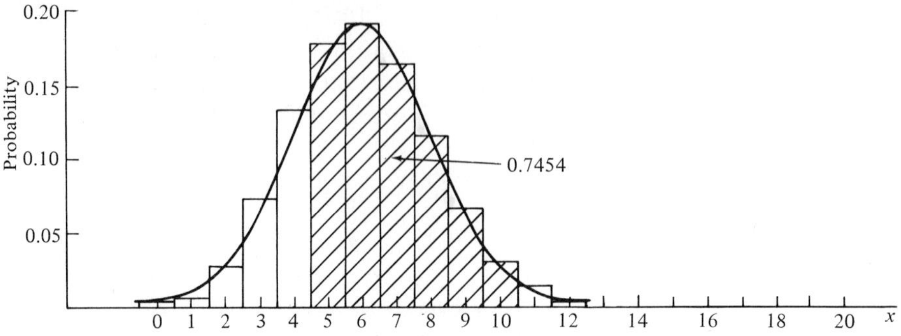

(a) $P(5 \leqslant X \leqslant 10)$ using binomial probabilities

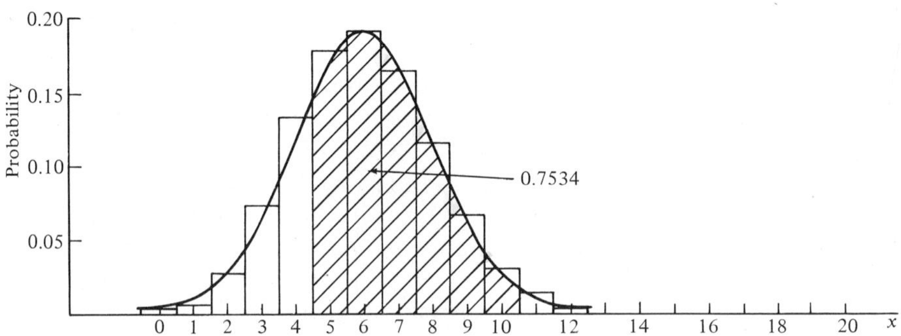

(b) $P(4.5 \leqslant X \leqslant 10.5)$ using normal approximation

$$P(5 \leq X \leq 10) = P\left(\frac{5-6}{2.05} \leq z \leq \frac{10-6}{2.05}\right)$$

$$= P(-0.49 \leq z \leq 1.95)$$

$$= 0.1879 + 0.4744 = 0.6623$$

We see that this approximation is not nearly as close to the true probability of 0.7454 as the approximation obtained with the continuity correction. When n is large and p is not too close to 0 or 1, one usually omits the continuity correction when finding probabilities associated with intervals such as $P(x_a \leq X \leq x_b)$, $P(X \leq x)$, or $P(X \geq x)$.

EXERCISES

10 Given the random variable Z which is distributed as the standard normal distribution, find the following.

(a) $P(z \leq 1.77)$
(b) $P(z < 2.46)$

(c) $P(0 \leq z \leq 0.95)$
(d) $P(-2.00 \leq z \leq 1.80)$
(e) $P(-2.10 \leq z \leq -1.65)$
(f) $P(1.45 < z < 2.15)$
(g) The proportion of z values greater than 1.96
(h) The probability that Z assumes values between -1.96 and $+1.96$
(i) The relative frequency of occurrence of values greater than -1.65
(j) The probability that a z picked at random will be between -2.58 and $+2.58$.

11 A certain firm has found that the lengths of its outgoing long-distance telephone calls are approximately normally distributed with a mean of three minutes and a standard deviation of one minute.

(a) What proportion of outgoing long-distance calls are longer than two minutes, but shorter than three and a half minutes?
(b) What proportion of calls are completed within one minute or less?
(c) A secretary is about to place a long-distance call. What is the probability that the call will be longer than five minutes?

12 A psychologist has found that "normal" subjects on the average complete a certain task in ten minutes. The times required to complete the task are approximately normally distributed with a standard deviation of three minutes. Find the following.

(a) The proportion of normal subjects completing the task in less than four minutes
(b) The proportion of subjects requiring more than five minutes to complete the task
(c) The probability that a normal subject who has just been assigned the task will complete it within three minutes

13 Agricultural experts have found that, for one type of grain, yields per acre are approximately normally distributed, with a mean and standard deviation of 40 and 10 bushels per acre, respectively.

(a) What proportion of the acreage planted in this grain yields more than 50 bushels per acre?
(b) If an acre of farmland planted in this grain is picked at random, what is the probability that it will yield less than 15 bushels?

14 Scores made on a scholastic aptitude test are normally distributed with a mean of 600 and a variance of 10,000.

(a) What proportion of those taking the test score below 300?
(b) A person is about to take the test. What is the probability that the individual will make a score of 850 or more?
(c) What proportion of scores fall between 450 and 700?

15 In a certain population, the heights of adult males are approximately normally distributed, with a mean and standard deviation of 70 and 3 inches, respectively.

(a) What proportion of adult males are between 65 and 73 inches tall?

(b) An adult male is picked at random from this population. What is the probability that he will be over six feet tall?

(c) What proportion of adult males in this population are shorter than 5 feet 8 inches?

16 The petal lengths of a certain species of flower are normally distributed, with a mean and standard deviation of 4 and 2 cm (centimeters), respectively.

(a) What proportion of petals are longer than 5 cm?

(b) What proportion are shorter than $2\frac{1}{2}$ cm?

17 The speeds of automobiles passing a certain freeway checkpoint are approximately normally distributed, with a mean of 45 mph and a variance of 25.

(a) What proportion of automobiles passing the checkpoint are traveling faster than 50 mph?

(b) What proportion pass the checkpoint at a rate of speed of less than 40 mph?

(c) Suppose the speed limit at the checkpoint is 55 mph. What proportion of automobiles are exceeding the speed limit when they pass the checkpoint?

18 In a certain section of the country, weekly expenditures for food per family are approximately normally distributed, with a mean and standard deviation of $60 and $15, respectively.

(a) Find the proportion of families who spend more than $80 per week on food.

(b) Find the proportion who spend less than $50 per week.

19 In a certain prison population, 30% of the inmates are there on burglary charges. A random sample of 50 is selected from this population. What is the probability that the number imprisoned for burglary will be between 20 and 24?

20 Suppose that 40% of the individuals composing a particular population favor a certain bill before Congress. What is the probability that a random sample of 100 individuals will contain between 30 and 50 who favor the bill?

CHAPTER SUMMARY

In this chapter we have introduced the concept of a probability distribution and have discussed distributions of both discrete and continuous variables.

We have covered in considerable detail a special discrete distribution, the binomial distribution. We have also introduced one of the most important distributions in statistics, the normal distribution, which is a continuous distribution.

In addition, we have dealt with the normal approximation to the binomial.

SOME IMPORTANT CONCEPTS AND TERMINOLOGY

Probability distribution
Cumulative distribution function
Binomial distribution
Normal distribution
Standard normal distribution

REVIEW EXERCISES

21 In a suburban community, on a given evening, someone is at home in 65% of the households. A researcher conducting a telephone survey randomly selects 15 households to call that evening. What is the probability that the researcher will find someone at home in exactly 8 households?

22 It is known that 80% of the seeds of a rare plant will germinate when planted. A random sample of five seeds is selected from current stock. What is the probability that exactly three will germinate?

23 The ages at time of onset of a disease are approximately normally distributed, with a mean of 10 years and a standard deviation of 2 years. A child has just come down with the disease. What is the probability that the child is: (a) between the ages of 8 and 12 years? (b) over 11 years of age? (c) under 12?

24 The weights of a certain small species of animal are normally distributed, with a mean of 14 ounces and a standard deviation of 1.5 ounces. What is the probability that an animal drawn at random from this population will weigh less than 12 ounces?

25 The total cholesterol values for a certain population are approximately normally distributed, with a mean of 180 mg/100 ml and a standard deviation of 15 mg/100 ml. What is the probability that an individual picked at random from this population will have a cholesterol value: (a) between 165 and 195 mg/ml? (b) greater than 210 mg/ml? (c) less than 150 mg/ml? (d) between 160 and 200 mg/ml?

26 It is estimated that, for a certain group of motorists, 30% use a particular brand of gasoline. Given that the estimate is correct, and that 25 persons picked at random from this group are questioned on the matter, what is the probability that the number using the brand of gasoline will be: (a) 5 or fewer? (b) between 10 and 15? (c) 15 or more?

27 In a certain population of adolescents, the proportion who smoke is 0.40. A random sample of 20 is selected from the population. What is the probability that the number who smoke will be: (a) greater than 10? (b) fewer than 5? (c) between 5 and 15 inclusive?

28 The weights of a population of children are approximately normally distributed, with a mean of 50 pounds and a standard deviation of 5 pounds. What is the probability that a child picked at random from this population will weigh: (a) more than 58 pounds? (b) less than 45 pounds? (c) between 45 and 54 pounds? (d) Between what two values does the middle 90% of

weights lie? (e) A particular child from the population weighs 65 pounds. What proportion of the children weigh more than this?

29 In a population of disabled workers, 40% are under 55 years old. What is the probability that in a random sample of 15 persons from this population, 8 or more will be under 55?

30 Suppose it is known that 20% of convicted felons were previously convicted in a state other than the state of current conviction. In a random sample of 20 currently convicted felons, what is the probability that between 10 and 15 will have had a previous conviction in another state?

4

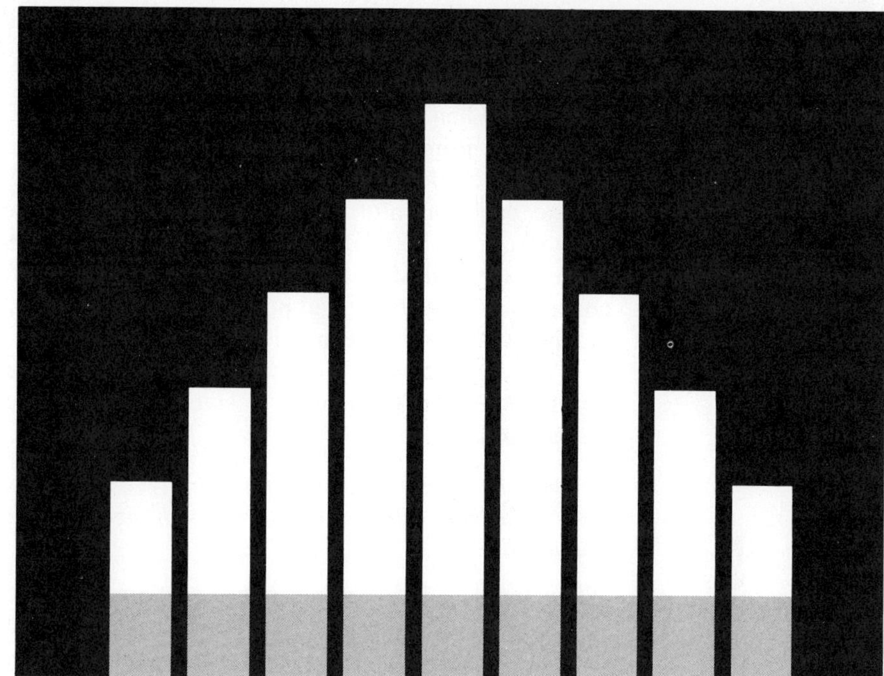

Sampling Distributions

Much of the knowledge we have of populations is based on the information contained in samples drawn from those populations. When the sponsors of a television series want to know how many people watched a given program, they don't question all potential viewers in the country; they interview only a sample. A drug manufacturer who wants to know how well a drug for lowering high blood pressure compares with a competitor's drug does not conduct an experiment involving all known patients with high blood pressure. Instead a sample of patients participate in the experiment. The sociologist who wants to know the attitudes of young adults toward abortion does not interview all the young adults in the country, but talks to a sample of young adults. We could give many similar examples.

This book is concerned primarily with the proper inferential procedures for reaching valid conclusions about populations on the basis of the information contained in a sample. We discuss these procedures in Chapters 5 through 11. In Chapters 1 through 3 we presented the underlying concepts and principles on which inferential procedures are based. This chapter serves as a bridge connecting Chapters 1 through 3 with Chapters 5 through 11. Before proceeding, let us briefly review what we have covered so far.

In Chapter 1 we introduced the idea of describing a set of data by means of various descriptive measures, such as the mean and the variance. These descriptive measures, when computed from the data of a sample, are called statistics. When computed from population data, they are called parameters. One of the primary concerns of researchers and decision makers is to be able to reach decisions about parameters (which are generally unknown) on the basis of information regarding statistics computed from samples drawn from the populations of interest. Such a procedure, as we pointed out earlier, is called statistical inference. Statistical inference is based on the concepts of probability and probability distributions, topics that are covered in Chapters 2 and 3, respectively.

We come now to a discussion of *sampling distributions*, a topic that ties together the concept of a statistic and the concept of a probability distribution.

DEFINITION *A **sampling distribution** is a probability distribution of a sample statistic computed from all possible samples of size n randomly drawn from some population.*

When the population under consideration is infinite, we must think of the sampling distribution as a *theoretical sampling distribution*, since it is impossible to draw all possible random samples from an infinite population. When the population is finite and of moderate size, we may construct an *experimental sampling distribution* by actually drawing all possible samples of a given size, computing the statistic of interest for each sample, and listing the different computed values of the statistic along with their probabilities of occurrence. We may experimentally approximate the true sampling distributions based on sampling from infinite or large finite populations by drawing a large number of random samples and proceeding as described above. In the next two sections, we shall construct an experimental sampling distribution for a sample mean. Succeeding sections are devoted to various theoretical sampling distributions.

We are generally interested in knowing one or more of the following characteristics of a sampling distribution.

1 Its functional form (how it looks when graphed)

2 Its mean

3 Its standard deviation

Before we can understand statistical inference, we must understand the concept of sampling distributions. This fact cannot be overemphasized, and you are urged to make sure you understand the material in this chapter before proceeding to the next.

Note that the purpose of describing in detail the methods of constructing experimental sampling distributions is to give you an understanding of the nature of these distributions and their role in statistical inference, not to teach a skill that you will use in your job. As a rule, in statistical applications, one does not construct sampling distributions. In order to employ the methods of statistical inference, one needs to know only the characteristics of the sampling distribution of the statistic appropriate to the problem at hand. The purpose of this chapter is to acquaint you with the characteristics of some of the more frequently used statistics and to show how a knowledge of sampling distributions enables one to make valid inferences.

4.1 THE MEAN AND VARIANCE OF SAMPLE MEANS

In order to understand inferential procedures relative to population means, you must understand the nature of the sampling distribution of the sample mean. To help you achieve this goal, let us construct an experimental sampling distribution for sample means computed from all possible samples drawn from a small population.

Example 4.1 An elementary school employs $N = 10$ teachers. The variable of interest, X, is the number of years of teaching experience of the faculty. Table 4.1 shows the data.

The mean and variance for this population are

$$\mu = \frac{\Sigma x_i}{N} = \frac{55}{10} = 5.5, \qquad \sigma^2 = \frac{\Sigma (x_i - \mu)^2}{N} = 8.25$$

We wish now to construct the sampling distribution of the sample mean $\bar{x}$, based on samples of size $n = 2$ from this population. We can accomplish this in three steps.

1 First we draw, with replacement, all possible samples of size $n = 2$. Table 4.2 shows the resulting samples. When we are sampling with replacement from a population of size N, the number of possible samples of size n is equal to N^n. Here we have $10^2 = 100$ samples.

2 Second, we calculate the mean $\bar{x}$ for each of the samples. Table 4.2 shows the sample means given in parentheses.

3 Finally, we make a list of the different distinct values of $\bar{x}$ that were observed, along with the probability of occurrence of each. This table, which is shown as Table 4.3, constitutes the sampling distribution of the sample mean for samples of size $n = 2$ drawn from the specified population.

Table 4.1 Number of years of teaching experience for a population of ten elementary school teachers

Teacher number	1	2	3	4	5	6	7	8	9	10
Years of teaching experience, X	6	1	2	9	5	8	4	3	10	7

Table 4.2 All possible samples of size $n = 2$ drawn from a population of size $N = 10$. Samples above or below the principal diagonal result when sampling is without replacement. Sample means are in parentheses.

First draw	Second draw 1	2	3	4	5	6	7	8	9	10
1	**1, 1 (1)**	1, 2 (1.5)	1, 3 (2)	1, 4 (2.5)	1, 5 (3)	1, 6 (3.5)	1, 7 (4)	1, 8 (4.5)	1, 9 (5)	1, 10 (5.5)
2	2, 1 (1.5)	**2, 2 (2)**	2, 3 (2.5)	2, 4 (3)	2, 5 (3.5)	2, 6 (4)	2, 7 (4.5)	2, 8 (5)	2, 9 (5.5)	2, 10 (6)
3	3, 1 (2)	3, 2 (2.5)	**3, 3 (3)**	3, 4 (3.5)	3, 5 (4)	3, 6 (4.5)	3, 7 (5)	3, 8 (5.5)	3, 9 (6)	3, 10 (6.5)
4	4, 1 (2.5)	4, 2 (3)	4, 3 (3.5)	**4, 4 (4)**	4, 5 (4.5)	4, 6 (5)	4, 7 (5.5)	4, 8 (6)	4, 9 (6.5)	4, 10 (7)
5	5, 1 (3)	5, 2 (3.5)	5, 3 (4)	5, 4 (4.5)	**5, 5 (5)**	5, 6 (5.5)	5, 7 (6)	5, 8 (6.5)	5, 9 (7)	5, 10 (7.5)
6	6, 1 (3.5)	6, 2 (4)	6, 3 (4.5)	6, 4 (5)	6, 5 (5.5)	**6, 6 (6)**	6, 7 (6.5)	6, 8 (7)	6, 9 (7.5)	6, 10 (8)
7	7, 1 (4)	7, 2 (4.5)	7, 3 (5)	7, 4 (5.5)	7, 5 (6)	7, 6 (6.5)	**7, 7 (7)**	7, 8 (7.5)	7, 9 (8)	7, 10 (8.5)
8	8, 1 (4.5)	8, 2 (5)	8, 3 (5.5)	8, 4 (6)	8, 5 (6.5)	8, 6 (7)	8, 7 (7.5)	**8, 8 (8)**	8, 9 (8.5)	8, 10 (9)
9	9, 1 (5)	9, 2 (5.5)	9, 3 (6)	9, 4 (6.5)	9, 5 (7)	9, 6 (7.5)	9, 7 (8)	9, 8 (8.5)	**9, 9 (9)**	9, 10 (9.5)
10	10, 1 (5.5)	10, 2 (6)	10, 3 (6.5)	10, 4 (7)	10, 5 (7.5)	10, 6 (8)	10, 7 (8.5)	10, 8 (9)	10, 9 (9.5)	**10, 10 (10)**

We may verify that Table 4.3 is a probability distribution by noting that each of the probabilities is greater than 0 and the sum of all probabilities is equal to 1.

Let us now compute the mean and variance for the population of 100 sample means displayed in Table 4.2.

We obtain the mean, denoted by $\mu_{\bar{x}}$, by adding the 100 sample means and dividing by 100 as follows.

$$\mu_{\bar{x}} = \frac{\Sigma \bar{x}_i}{N^n} = \frac{550}{100} = 5.5$$

Note that:

The mean of all the sample means, $\mu_{\bar{x}}$, is exactly equal to the population mean.

Now let us compute the variance of our 100 sample means. We call this variance $\sigma^2_{\bar{x}}$, and compute it as follows.

$$\sigma_{\bar{x}}^2 = \frac{\Sigma\,(\bar{x}_i - \mu_{\bar{x}})^2}{N^n} = \frac{(1-5.5)^2 + (1.5-5.5)^2 + \cdots + (10-5.5)^2}{100}$$

$$= 4.125$$

Note that the variance of the sample means is not equal to the population variance. A little investigation, however, reveals the following.

The variance of the sample means is equal to the population variance divided by the sample size n.

That is,

$$\sigma_{\bar{x}}^2 = \frac{\sigma^2}{n} = \frac{8.25}{2} = 4.125$$

The square root of $\sigma_{\bar{x}}^2$, which is usually written $\sigma_{\bar{x}} = \sigma/\sqrt{n}$, is called the *standard error* of the mean.

The results obtained in this example are not coincidental, but a demonstration of the following general facts:

When sampling is with replacement from a finite population, the mean of the sampling distribution of $\bar{x}$ is equal to the mean μ of the original population, and the variance of the sampling distribution is equal to the population variance divided by the sample size.

Table 4.3 Sampling distribution of the sample mean computed from samples of size $n = 2$ drawn from the population of Example 4.1

$\bar{x}$	*Probability*
1.0	1/100
1.5	2/100
2.0	3/100
2.5	4/100
3.0	5/100
3.5	6/100
4.0	7/100
4.5	8/100
5.0	9/100
5.5	10/100
6.0	9/100
6.5	8/100
7.0	7/100
7.5	6/100
8.0	5/100
8.5	4/100
9.0	3/100
9.5	2/100
10.0	1/100
	100/100

Let us now consider the case in which sampling is without replacement. The possible samples are shown both above and below the principal diagonal in Table 4.2. The samples appearing on the diagonal are not included, since they are not possible when sampling is without replacement. The 45 samples above the principal diagonal are identical to the 45 samples below the diagonal, except for the order in which the observations composing the individual samples were drawn. Since the order of drawing the observations has no effect on sample results, we may limit our discussion to the 45 samples above the diagonal. In general, when sampling is without replacement from a finite population, and when order is ignored, the number of possible samples is given by $\binom{N}{n}$. In our present example, then, we have

$$\binom{10}{2} = \frac{10!}{8!\,2!} = 45$$

The mean of the 45 sample means, which we will also call $\mu_{\bar{x}}$, is equal to

$$\mu_{\bar{x}} = \frac{1.5 + 2.0 + \cdot \cdot \cdot + 9.5}{45} = \frac{247.5}{45} = 5.5$$

Again we see that the mean of the sample means is equal to the population mean.

We may obtain the variance $\sigma_{\bar{x}}^2$ of the 45 possible sample means that result when sampling is without replacement as follows.

$$\sigma_{\bar{x}}^2 = \frac{(1.5 - 5.5)^2 + (2.0 - 5.5)^2 + \cdot \cdot \cdot + (9.5 - 5.5)^2}{45} = 3.67$$

This time $\sigma_{\bar{x}}^2$ is not equal to σ^2/n. We should note, however, that multiplication of σ^2/n by the factor $(N - n)/(N - 1)$ gives 3.67. That is,

$$\left(\frac{8.25}{2}\right)\left(\frac{10 - 2}{10 - 1}\right) = \left(\frac{8.25}{2}\right)\left(\frac{8}{9}\right) = 3.67$$

We can summarize the results obtained with the 45 sample means as follows.

In general, when sampling is without replacement from a finite population, the mean $\mu_{\bar{x}}$ of the sampling distribution of sample means is equal to the population mean, and the variance $\sigma_{\bar{x}}^2$ is equal to

$$\left(\frac{\sigma^2}{n}\right)\left(\frac{N - n}{N - 1}\right)$$

The factor $(N - n)/(N - 1)$ is called the *finite-population correction* (fpc). We may ignore it when the size of the sample is small relative to the size of the population. If N is a great deal larger than n, the difference between σ^2/n and $(\sigma^2/n)[(N - n)/(N - 1)]$ will be negligible. A frequently used rule of thumb states that the finite-population correction may be ignored when n/N is less than or equal to 0.05, that is, when the sample contains 5% or less of the population.

Since this section has centered on the case in which sampling is from a finite population, you may wonder about the results of sampling from an infinite population. Sampling from a finite population with replacement is equivalent to sampling from an infinite population, and the results discussed in this section that apply in the former case also apply in the latter. That is:

If sampling is from an infinite population, the mean $\mu_{\bar{x}}$ *of the sampling distribution of the sample mean is equal to the mean of the sampled population, and the variance* $\sigma^2_{\bar{x}}$ *is equal to* σ^2/n*, provided the sampled population has a finite variance.*

A consequence of the results demonstrated in this section is that we can determine the mean and variance of the sampling distribution of sample means without actually calculating them. As we shall see in succeeding sections, this is indeed a very useful consequence.

4.2 THE FUNCTIONAL FORM OF THE DISTRIBUTION OF SAMPLE MEANS

We have seen how the mean and variance of the distribution of sample means may be determined. We now wish to explore the nature of the functional form of the distribution of sample means. We shall distinguish between two situations: the case in which sampling is from a normally distributed population and the case in which sampling is from a nonnormally distributed population.

Sampling from a normally distributed population

The following statement is given as a theorem and proved in books on mathematical statistics. No proof will be given here.

If $\bar{x}$ *is the mean of a random sample of size n drawn from a normally distributed population with mean* μ *and finite variance* σ^2*, then the sampling distribution of* $\bar{x}$ *is normally distributed with mean* μ *and variance* σ^2/n*.*

As a consequence of this fact, the quantity

$$z = \frac{\bar{x} - \mu}{\sigma/\sqrt{n}}$$

where $\bar{x}$ is the mean of a random sample of size n drawn from a normally distributed population, is normally distributed with mean 0 and variance 1.

When sampling from normally distributed populations, with these results we can make probability statements about sample means in exactly the same way in which we were able to make probability statements about individual values of normally distributed random variables in Section 3.2. As we shall demonstrate in Chapters 5 and 6, the ability to make probability statements about statistics such as the sample mean leads directly to inferential statements about the corresponding population parameters.

Let us now illustrate, with an example, the usefulness of the results given above.

Example 4.2 It is known that the response time of a certain experimental animal to a particular stimulus is a normally distributed random variable with a mean of 10 seconds and a variance of 9. What is the probability that a random sample of 16 animals will have a mean response time of 11 seconds or more?

Since the population of response times is normally distributed, with $\mu = 10$ and $\sigma^2 = 9$, we know that the sampling distribution of $\bar{x}$ based on samples of size 16 is normally distributed with $\mu_{\bar{x}} = 10$ and variance $\sigma^2_{\bar{x}} = 9/16$. To find the probability that $\bar{x}$ is equal to or greater than 11, we transform the normal distribution of $\bar{x}$'s to the standard normal distribution and use Appendix Table E to find the desired probability. Thus we have

$$z = \frac{11 - 10}{\sqrt{9/16}} = \frac{1}{0.75} = 1.33$$

From Appendix Table E we find that the area to the right of 1.33 is 0.0918. Thus we see that, in this example, the probability of obtaining a mean as large as 11 from a sample of size 16 is 0.0918. Figure 4.1 illustrates the procedure.

This example shows that the probability that a sample of size 16, from the specified population, will yield a mean of 11 seconds or more is equivalent to the probability that the sample data will yield a z value of 1.33 or greater. In fact, we may at times want to determine the probability that a sample of size n will yield a value of z falling within some specified interval. We might, for example, want to know the probability that, from a sample of size n from a specified population, we will compute a value of z equal to or greater than 1.96, or 2.00, or 3.00, or any value we care to specify.

Sampling from nonnormally distributed populations

Frequently a researcher either will know that the population from which a sample is drawn is not normally distributed or, in the absence of knowledge of its functional form, will be unwilling to assume that it is normal. In such a case, the results stated earlier for sampling from normally distributed populations do not apply.

The situation is by no means hopeless, since there are alternative approaches to use when one must make an inference about the mean of such a population. One alternative that is widely used requires that a large sample be drawn from the population of interest. Armed with this large sample, the investigator may use the *central-limit theorem.* We may summarize the practical importance of this theorem as follows.

Figure 4.1 Probability distribution and distribution of sample means, Example 4.2

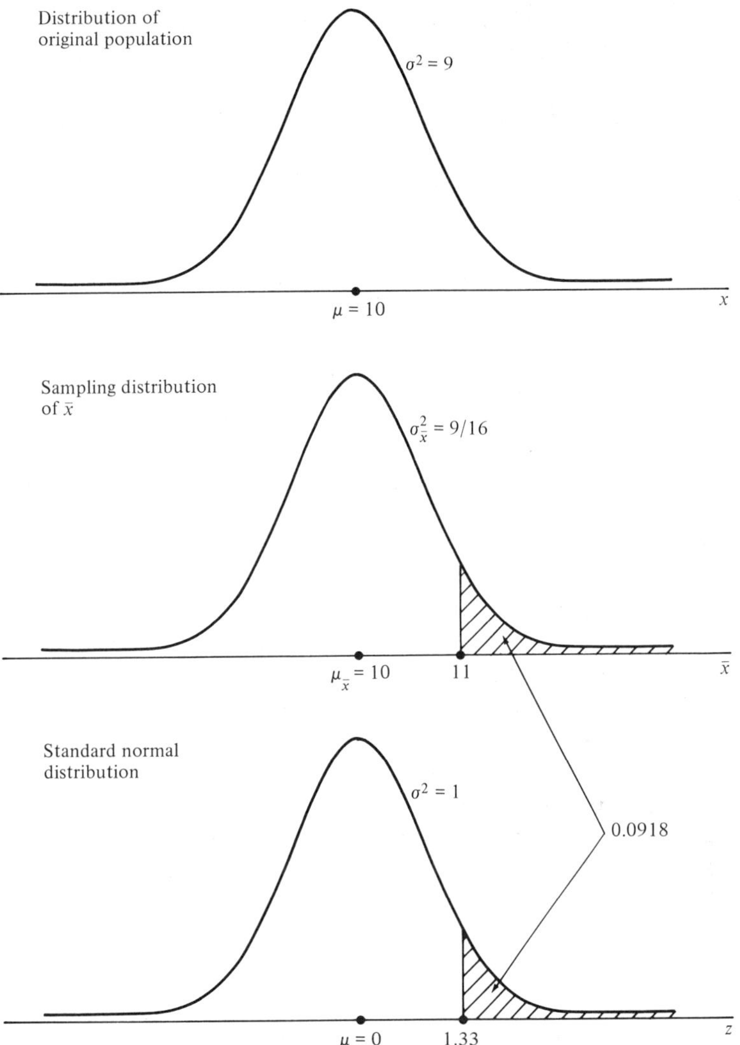

Regardless of the functional form of the sampled population, the distribution of sample means computed from samples of size n drawn from a population with mean μ and finite variance σ^2 approaches a normal distribution with mean μ and variance σ^2/n, as n increases. If n is large, the sampling distribution of the sample mean may be approximated quite closely with a normal distribution.

Note that the central-limit theorem does not depend on the functional form of the population from which the samples are drawn. This feature makes the central-limit theorem perhaps the most important theorem in all of statistics.

This theorem tells us that, regardless of the form of the population under study, we can still use normal theory in drawing inferences about the population mean, provided we draw a large sample, because the sampling distribution of $\bar{x}$ will be *approximately normal* when n is large. In other words, we will be able to use the fact that, when n is large,

$$z = \frac{\bar{x} - \mu}{\sigma/\sqrt{n}}$$

is approximately normally distributed with mean 0 and variance 1.

You might ask how large n must be in order for the use of the central-limit theorem to be valid. Many practitioners follow the suggestion that a sample size of 30 is usually large enough to warrant use of the theorem. We shall follow this somewhat arbitrary rule in this text.

To get a feel for how the central-limit theorem works, let us consider the population of years of teaching experience of the ten elementary school teachers described in Example 4.1. This population is certainly not normally distributed, as its histogram, shown in Figure 4.2, clearly demonstrates. When we graph the sampling distribution of means shown in Table 4.3, we get quite a different picture, as shown in Figure 4.3.

A striking feature of the histogram in Figure 4.3 is that, although the parent population is uniformly distributed, the sampling distribution of means computed from samples of size $n = 2$ rises to a peak at the mean and drops off almost symmetrically. In light of these results, you should have no trouble believing that, for large n, the sampling distribution of $\bar{x}$ approaches a normal distribution.

Let us now illustrate the application of the central-limit theorem with an example.

Example 4.3 The mean daily water intake for a certain laboratory animal is 16 grams. The standard deviation is 2 grams. What is the probability that the mean daily

Figure 4.2 Histogram of years of teaching experience of ten elementary school teachers, Example 4.1

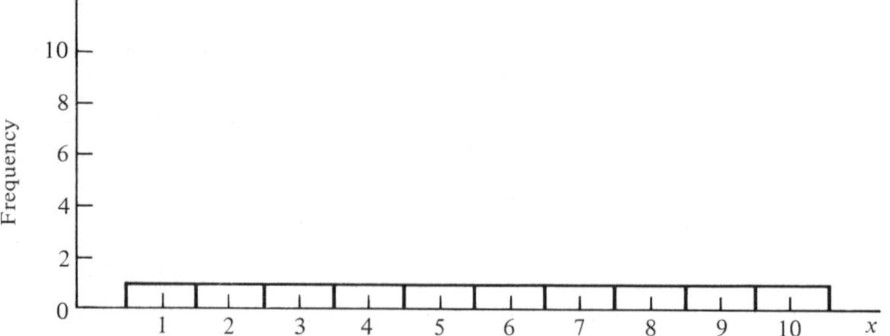

Figure 4.3 Histogram for sampling distribution of means, Table 4.3

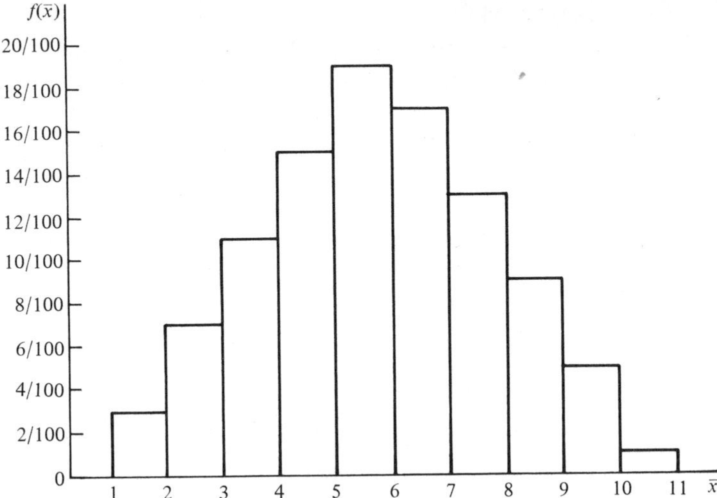

water intake for a random sample of 65 animals will be between 15.50 and 16.25 grams?

No mention is made of the functional form of the population of values of daily water intake, but this lack of information poses no problem. Since the sample size is large, the central-limit theorem applies, and we know that the distribution of sample means computed from samples of size 65 is approximately normally distributed, with a mean of 16 grams and a standard error of $2/\sqrt{65} = 0.25$ gram.

To find the probability that $\bar{x}$ will be between $\bar{x}_1 = 15.50$ and $\bar{x}_2 = 16.25$, we compute

$$z_1 = \frac{15.5 - 16}{0.25} = -2.00, \qquad z_2 = \frac{16.25 - 16}{0.25} = 1.00$$

From Table E of the Appendix we find that the area under the standard normal curve between $z = -2.00$ and $z = 1.00$ is 0.8185. The probability that our sample will yield a mean between 15.50 and 16.25, then, is 0.8185.

This example, as well as Example 4.2, illustrates those situations in which the population of interest is hypothetical rather than real. There is no existing definitive population of response times or water-intake values for the animals referred to in these two examples. It is implied that the populations of interest consist of response-time values and water-intake values for all such animals that have ever lived, are now living, or will live in the future. It would be impossible to obtain a population of measurements on the animals under those conditions. Consequently, we say that the populations defined in Examples 4.2 and 4.3 are theoretical. At this point, you may well ask how it is possible to make statements about the values of the means and standard deviations of such populations. An investigator may merely postulate values for the mean and standard deviation of a hypothetical population. In other words, he or she may ask us to suppose that the mean and standard deviation are equal to some particular values, and then find the probability that a random sample of specified size will yield a mean within some specified range. In other situations, an investigator's statements with regard to the magnitude of the mean and standard deviation of some population may be based on the accumulation of very large amounts of data over several years.

The examples and exercises in this section are presented primarily to lay a conceptual foundation for the statistical inference procedures of estimation and hypothesis testing, to be introduced in Chapters 5 and 6. As we have noted, our ability to draw inferences about populations will depend on our ability to calculate the probabilities associated with specified intervals of relevant sample statistics.

Sampling from finite populations

As pointed out in Section 4.1, the variance of the sampling distribution of $\bar{x}$ is given by

$$\sigma_{\bar{x}}^2 = \left(\frac{\sigma^2}{n}\right)\left(\frac{N-n}{N-1}\right)$$

when sampling is without replacement (the usual procedure in practice) from a finite population. We also noted that we may ignore the fpc, $(N-n)/(N-1)$, when n/N is less than or equal to 0.05; that is, when the sample contains fewer than 5% of the items in the population.

The following example illustrates a situation in which we would use the fpc.

Example 4.4 A firm employs 1500 people. During a particular year, the mean amount spent on personal medical expenses per employee was \$25.75, and the standard deviation was \$5.25. What is the probability that a simple random sample of 100 employees will yield a mean between \$25.00 and \$27.00?

The functional form of the population of expenditure values is not specified, but this presents no problem, since the sample size is large enough for us to use the central-limit theorem.

The sampling distribution of sample means, then, is approximately normally distributed, with a mean of \$25.75. Since the sample size of 100 constitutes approximately 7% of the total population, we should use the fpc in calculating the variance of the sampling distribution. Thus

$$\sigma_{\bar{x}}^2 = \left(\frac{(5.25)^2}{100}\right)\left(\frac{1500-100}{1499}\right) = 0.2574$$

To find the probability that $\bar{x}$ will be between 20 and 30, we compute

$$z_1 = \frac{25 - 25.75}{\sqrt{0.2574}} = -1.48, \qquad z_2 = \frac{27 - 25.75}{\sqrt{0.2574}} = 2.46$$

From Appendix Table E we find the desired probability to be 0.9237.

EXERCISES

Unless indicated otherwise, assume that the fpc can be ignored.

1 Reading-readiness scores of certain kindergarten children are normally distributed with a mean and standard deviation of 75 and 10, respectively. What is the probability that a random sample of 25 such kindergarten children will yield a mean score between 70 and 78?

2 The mean number of years of driving experience of a certain group of truck drivers is 10 years, and the standard deviation is 3 years. What is the probability that a random sample of 81 of these truck drivers will yield a mean greater than 10 years and 8 months?

3 The placenta weights for a certain species of experimental animal are normally distributed with a mean of 7 grams and a standard deviation of 1.5 grams. What is the probability that the mean weight of placentas from a random sample of 9 animals will be less than 6 grams?

4 The net weights of the contents of packages of a certain breakfast cereal have a mean of 16 ounces and a standard deviation of 0.5 ounce. The weights are normally distributed. What is the probability that a random sample of 25 packages will have a mean net weight between 15.8 and 16.2 ounces?

5 In a certain population of alcoholics, the mean duration of abuse is 12 years and the standard deviation is 6 years. What is the probability that a random sample of 36 subjects from this population will yield a mean duration of abuse between 10 and 11 years?

6 In a population of 1200 adolescents, the mean amount of money spent on recreation per week is \$6.50, and the standard deviation is \$6.00. What is the probability that a simple random sample of 36 adolescents from this population will yield a mean between \$5.00 and \$10.00?

4.3 THE DISTRIBUTION OF THE DIFFERENCE BETWEEN TWO SAMPLE MEANS

Sometimes interest in an investigation is focused on two populations. For example, we often want to draw inferences about the difference between two population means. We may wish to know whether it is reasonable to conclude that two population means are not equal. Or we may be interested in assessing the magnitude and direction of any difference which may exist. A medical researcher may wish to know whether the mean serum cholesterol level is higher in sedentary office workers than in laborers. A manufacturer of fertilizer may want to know whether the mean yield per acre of some grain is different when different formulas are used in the manufacture of the product in question. A teacher may wish to know whether the mean achievement score in arithmetic is different in two groups, each of which has been taught by a different method. The relevant statistic in such inferential procedures is the difference between two sample means, which we may designate by $\bar{x}_1 - \bar{x}_2$. It is necessary, then, for us to know the nature of the sampling distribution of this statistic.

If the two populations of interest were very small, one could actually construct the sampling distribution of $\bar{x}_1 - \bar{x}_2$ by the following procedure.

1 Draw all possible samples of size n_1 from population 1, and compute the mean $\bar{x}_1$ for each of the samples.

2 Draw all possible *independent* samples of size n_2 from population 2, and compute the mean $\bar{x}_2$ for each of these samples.

3 Take all possible differences between the means computed from samples from population 1 and those computed from samples from population 2.

4 The sampling distribution of $\bar{x}_1 - \bar{x}_2$ consists of the differences computed in step 3 and their frequencies of occurrence.

One can show mathematically that the mean $\mu_{\bar{x}_1 - \bar{x}_2}$ and variance $\sigma^2_{\bar{x}_1 - \bar{x}_2}$ of the sampling distribution of $\bar{x}_1 - \bar{x}_2$ are $\mu_1 - \mu_2$ and $(\sigma_1^2/n_1) + (\sigma_2^2/n_2)$, respectively, when the samples are independent. When the populations of interest are infinite or finite, but large, one may approximate the true sampling distribution of $\bar{x}_1 - \bar{x}_2$ by taking a large number of samples from each population and proceeding as outlined above.

Remember that, in a practical situation, sampling distributions are not actually constructed. Investigators wishing to make inferences on the basis of a single sample from each of two populations are able to do so because of the results they would obtain if they were to construct the relevant sampling distribution.

The functional form of the sampling distribution of $\bar{x}_1 - \bar{x}_2$ depends on the form of the sampled populations. If both populations of interest are normally distributed, the sampling distribution of $\bar{x}_1 - \bar{x}_2$ will be normal. If one or both of the original populations are not normally distributed, the sampling distribution of $\bar{x}_1 - \bar{x}_2$ will be approximately normally distributed when n_1 and n_2 are large. An extension of the central-limit theorem is responsible for this last result.

We may summarize an important characteristic of the sampling distribution of the difference between two sample means as follows.

The sampling distribution of the difference between two sample means computed from independent random samples of size n_1 and n_2 from two normally *distributed populations is normally distributed with mean $\mu_1 - \mu_2$ and variance $(\sigma_1^2/n_1) + (\sigma_2^2/n_2)$.*

If n_1 and n_2 are both large, the sampling distribution of the difference between two sample means will be approximately normally distributed, with mean $\mu_1 - \mu_2$ and variance $(\sigma_1^2/n_1) + (\sigma_2^2/n_2)$, regardless of the form of the original populations.

Note that the statement about the sampling distribution of $\bar{x}_1 - \bar{x}_2$ specifies that the samples be independent. This condition will be met if the observations appearing in one sample are in no way dependent on the observations appearing in the other sample. Suppose, for example, that a public health official is interested in knowing whether teen-age boys and girls differ with respect to mean protein intake. If a sample of teen-age boys were selected for interview, and whenever possible sisters of these boys were included in the sample of teen-age girls, the two samples would not be independent. The presence of these girls in the sample would depend on their having a teen-age brother in the other sample. Since members of the same family might tend to have similar dietary habits, the results of this type of sampling might be misleading.

To find probabilities associated with intervals on the $\bar{x}_1 - \bar{x}_2$ axis, one may transform values of $\bar{x}_1 - \bar{x}_2$ to the standard normal distribution by the formula

$$z = \frac{(\bar{x}_1 - \bar{x}_2) - (\mu_1 - \mu_2)}{\sqrt{(\sigma_1^2/n_2) + (\sigma_2^2/n_2)}}$$

which, you will note, is an adaptation of the formula given previously for transforming $\bar{x}$ into z.

Example 4.5 Two populations of high school seniors were identified. The variable of interest in an investigation was the mathematics achievement test scores made by the students in these two populations. Researchers felt that the scores in the two populations were normally distributed with means and variances as follows: $\mu_1 = 50$, $\sigma_1^2 = 40$, $\mu_2 = 40$, $\sigma_2^2 = 60$. A random sample of size $n_1 = 10$ is drawn from population 1, and an independent random sample of size $n_2 = 12$ is drawn from population 2. What is the probability that the difference between sample means will be between 5 and 15?

Since the populations are normally distributed, we know that the sampling distribution of $\bar{x}_1 - \bar{x}_2$ is normally distributed with mean $\mu_{\bar{x}_1 - \bar{x}_2} = 50 - 40 = 10$ and variance $\sigma^2_{\bar{x}_1 - \bar{x}_2} = \frac{40}{10} + \frac{60}{12} = 9$. To find the desired probability, we transform the values of $\bar{x}_1 - \bar{x}_2$ into the standard normal as follows.

$$z_1 = \frac{5 - (50 - 40)}{\sqrt{\frac{40}{10} + \frac{60}{12}}} = \frac{-5}{3} = -1.67$$

$$z_2 = \frac{15 - (50 - 40)}{\sqrt{\frac{40}{10} + \frac{60}{12}}} = \frac{5}{3} = 1.67$$

Consulting Appendix Table E, we find $P(-1.67 \le z \le 1.67) = 2(0.4525) = 0.9050$. Thus the probability that $\bar{x}_1 - \bar{x}_2$ will be between 5 and 15, when samples of size $n_1 = 10$ and $n_2 = 12$ are drawn, is equal to 0.9050.

Example 4.6 At a certain high school, tenth-grade chemistry students were randomly assigned to one of two groups. Group 1, called the experimental group, consisted of 60 students, and was taught by a programmed instruction method. Group 2, the control group, consisted of 40 students, and was taught by traditional methods. At the end of the year, the mean grade for the experimental group was 83, and the mean score for the control group was 77. Previous experience shows that the variance for both groups is about 100. Suppose both teaching methods are equally effective. What is the probability of getting a difference between sample means as large as that observed in this study?

No mention is made of the forms of the two populations, but this presents no problem, since the sample sizes are large. We know that, in this case, the sampling distribution of the difference between sample means will be approximately normally distributed, and we can use the standard normal distribution to determine the desired probability. If there is no difference in the population means, we know that the mean and variance of the relevant sampling distribution will be

$$\mu_{\bar{x}_1 - \bar{x}_2} = \mu_1 - \mu_2 = 0, \qquad \sigma^2_{\bar{x}_1 - \bar{x}_2} = \frac{100}{60} + \frac{100}{40}$$

To find the probability of obtaining a value of $\bar{x}_1 - \bar{x}_2$ greater than or equal to what was observed in the present study, we first find

$$z = \frac{(83 - 77) - 0}{\sqrt{\frac{100}{60} + \frac{100}{40}}} = \frac{6}{2.04} = 2.94$$

From Appendix Table E we find that the area to the right of $z = 2.94$ is 0.0016. If, in fact, the two methods are equally effective (so that, on the average, students would make the same final grade, regardless of the method used), the probability of getting a difference between sample means as large as or larger than 6 is only 0.0016, a very small probability.

EXERCISES

7 Scores on a certain motor-performance test for high school boys who participate in varsity sports (group 1) are normally distributed, with a mean and variance of 60 and 100, respectively. Scores for high school boys who do not participate in varsity sports (group 2) are normally distributed with a mean of 50 and a variance of 121. A random sample of 10 boys is selected from group 1, and an independent random sample of size 11 is selected from group 2. What is the probability that the difference between sample means will be between 8 and 14?

8 An anthropologist believes that the inhabitants of a certain region (region 1) have a mean cephalic index of 80 with a standard deviation of 3, and that for the inhabitants of another region (region 2), the mean and standard deviation are 75 and 2, respectively. Suppose the anthropologist is correct. What is the probability that a random sample of 40 inhabitants from region 1 and an independent sample of 50 inhabitants from region 2 will yield a difference between sample means of 6 or more?

9 A random sample of 50 two-bedroom apartments in a certain metropolitan area yielded a mean monthly rent of \$175. From an independent random sample of 45 two-bedroom apartments in another metropolitan area, a mean monthly rental of \$180 was computed. Suppose there is no difference between the two areas with respect to the true mean monthly rental charges for two-bedroom apartments. What is the probability of observing a difference between sample means as large as or larger than that reported above? Assume that $\sigma^2 = 225$ for both areas.

10 The mean daily protein intake for a certain population is 100 grams, and that of another population is 75 grams. Suppose daily protein-intake values in the two populations are normally distributed, with a standard deviation of 20 grams. What is the probability that random and independent samples of size 20 from each population will yield a difference between sample means of 10 or less?

4.4 THE *t* DISTRIBUTION

We saw in Section 4.2 that, when sampling is from a normally distributed population, the variable

$$z = \frac{\bar{x} - \mu}{\sigma/\sqrt{n}}$$

is normally distributed with mean 0 and variance 1. In addition, we learned that, even when the sampled population is not normally distributed, z, as computed above, is approximately normally distributed when n is large. We also illustrated the usefulness of these results in computing probabilities associated with intervals of $\bar{x}$.

In actual practice, it is the rule, rather than the exception, that the population standard deviation σ is unknown. This raises the obvious question of whether our efforts to make statistical inferences will be thwarted by a lack of knowledge of σ. The answer, fortunately, is no.

In 1908 W. S. Gosset, writing under the pseudonym of Student, described the distribution of the variable

$$t = \frac{\bar{x} - \mu}{s/\sqrt{n}}$$

when sampling is from a normally distributed population. This distribution, known as *Student's t distribution*, or simply the *t distribution*, enables us to

Figure 4.4 The normal distribution and a t distribution

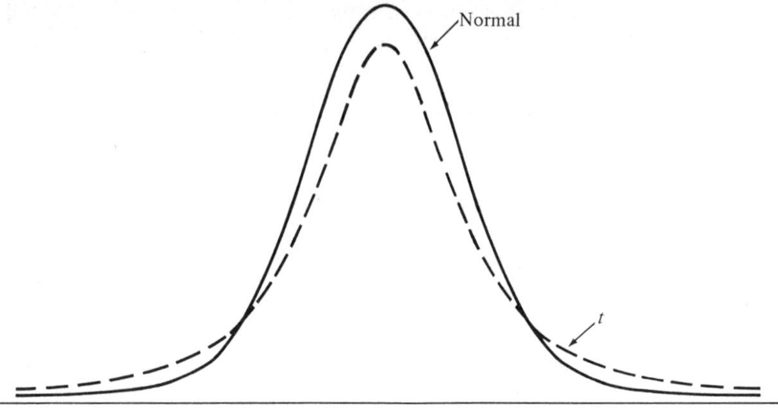

make inferences about population means when the population standard deviation is unknown. Note that the variable t contains the sample standard deviation s, rather than σ, in the denominator.

The t distribution, like the standard normal, is bell-shaped and has a mean of 0, about which it is symmetric. The variance, however, is greater than 1, a fact that causes the typical t distribution to be less peaked in the center and "higher" in the tails than the standard normal. Figure 4.4 illustrates the general relationship between the normal distribution and a t distribution.

The total area under the t distribution is equal to 1. There is a different t distribution for every value of $n-1$, the degrees of freedom used in calculating s. Figure 4.5 shows the t-distribution curves for several values of $n-1$.

Tables are available for use in applications requiring the use of the t distribution. One such table is given as Appendix Table F. The leftmost column of this table contains various values of $n-1$, the degrees of freedom. The column headings tell what proportion of the total area under the curve

Figure 4.5 The t distribution curves for selected values of degrees of freedom $n-1$

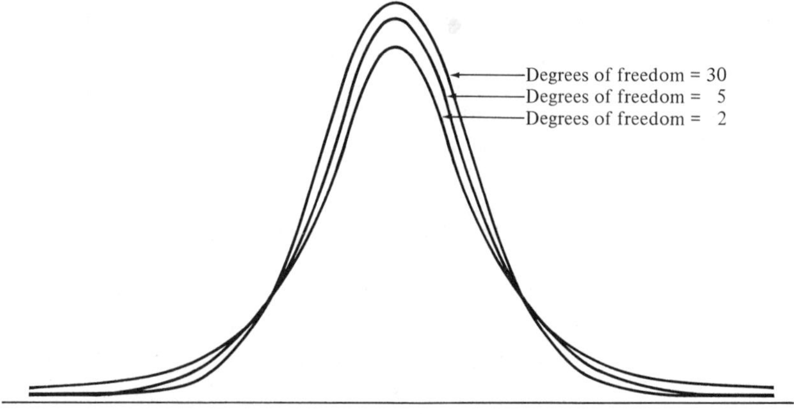

Figure 4.6 The t distribution for ten degrees of freedom, showing area to right and left of $t = 2.2281$

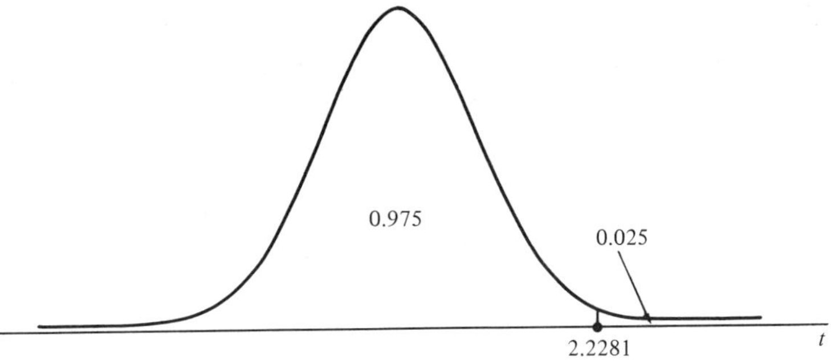

of the t distribution, for specified degrees of freedom, lies to the left of the corresponding value of t given in the body of the table. For example, if we are interested in the t distribution for 10 degrees of freedom, we can determine from Appendix Table F that 0.975 of the area under the curve lies to the left of $t = 2.2281$. The proportion of the total area to the right of 2.2281, then, is $1 - 0.975 = 0.025$. Using t_{10} to indicate that we are referring to the t with 10 degrees of freedom, we may express the above ideas by writing

$$P(t_{10} \leq 2.2281) = 0.975 \qquad \text{or} \qquad P(t_{10} \geq 2.2281) = 0.025$$

to show that the area under the curve of the t distribution represents probability. Figure 4.6 shows these areas.

The t distribution approaches the normal distribution as the degrees of freedom increase. For infinite degrees of freedom, the two distributions are identical. We may verify this fact by noting that the values of t in the last row of Appendix Table F are equal to the values of z corresponding to given probability values such as 0.90, 0.95, and so on.

Finally, we note that, although the distribution of t is based on the assumption that sampling is from a normally distributed population, moderate departures from this assumption can be tolerated. In other words, one may use the t distribution in practice even though the sampled population is not normally distributed, provided the departure from normality is not severe. Some practitioners recommend the use of the t distribution so long as the sampled population is at least mound-shaped.

We emphasize here that the statistic

$$\frac{\bar{x} - \mu}{s/\sqrt{n}}$$

is distributed approximately like the standard normal distribution when the sample size is large. As a consequence, many people prefer to use this approximate z statistic rather than the t statistic when the sample size is large, even though the population variance is unknown. This practice can be justified in part by the fact that when the sample size is large, s^2 provides a reliable estimate of the population variance.

Let us illustrate, by means of examples, ways in which the t distribution may be used. The two examples that follow illustrate uses of the t distribution when a single population is sampled. Later we shall discuss and illustrate the case in which two populations are sampled.

Example 4.7 Suppose it is known that the consumption of oxygen (in liters per hour) of infants of a certain age, weight, and sex is a normally distributed variable with a mean of 9. What is the probability that a random sample of 15 of these infants will yield a t value of 2.00 or more?

Since the sample is of size 15, we are interested in the t distribution for $15 - 1 = 14$ degrees of freedom. By consulting Appendix Table F, we find that, for 14 degrees of freedom, the probability of obtaining a t value greater than or equal to 1.7613 is $1 - 0.95 = 0.05$, and the probability of getting a t greater than or equal to 2.1448 is $1 - 0.975 = 0.025$. The probability of getting a t value greater than or equal to 2.00, then, is between 0.025 and 0.05.

Example 4.8 When a certain manufacturing process is under control, the mean breaking strength μ of a certain synthetic fiber is 100 pounds. To maintain control, the quality-control department periodically draws a random sample of 25 specimens of fiber for testing. So long as the t value computed from a sample is between -2.0639 and $+2.0639$, the process is considered to be under control. Suppose a random sample of 25 yields a mean $\bar{x}$ of 94 pounds and a standard deviation s of 10 pounds. What conclusions should be drawn?

From the sample data, we compute

$$t = \frac{94 - 100}{10/\sqrt{25}} = -3.00$$

Since -3.00 is less than -2.0639, one may draw one or the other of the following conclusions:

1 The manufacturing process is out of control.

2 The manufacturing process is still under control, and a rare event has occurred.

We say that the occurrence of sample data leading to a t value of -3.00 is a rare event because reference to Appendix Table F reveals that the probability of obtaining a t value as small as -3.00, when $\mu = 100$ and df (degrees of freedom) $= 24$, is less than $1 - 0.995 = 0.005$. The outcome observed here of a $t = -3.00$ would be more compatible with a mean of less than 100, which would indicate a weaker fiber, on the average, than is desired.

The typical quality-control department would operate on the basis of conclusion 1 and set about seeking the cause and remedy for the apparent lack of control.

The t distribution and sampling from two populations

In the preceding section we learned that when the statistic of interest is the difference between two sample means, we make use of

$$z = \frac{(\bar{x}_1 - \bar{x}_2) - (\mu_1 - \mu_2)}{\sqrt{(\sigma_1^2/n_1) + (\sigma_2^2/n_2)}}$$

which is normally distributed when the sampled populations are normally distributed and the samples are independent. We also learned that this z value will be approximately normally distributed, regardless of the forms of the sampled populations, provided the samples are independent and n_1 and n_2 are large. We now consider the typical situation, in which the population variances σ_1^2 and σ_2^2 are unknown.

When two population variances are unknown, two situations are possible: the population variances, though unknown, may be equal; or the unknown population variances may be unequal. We consider here only the case in which the two population variances are the same (equal).

If $\bar{x}_1$ and s_1^2 are the mean and variance, respectively, of a sample of size n_1 drawn from a normally distributed population with mean μ_1 and variance σ_1^2, and if $\bar{x}_2$ and s_2^2 are the mean and variance, respectively, of an independent random sample of size n_2 drawn from another normally distributed population with mean μ_2 and a variance σ_2^2 which is equal to σ_1^2, then

$$t = \frac{(\bar{x}_1 - \bar{x}_2) - (\mu_1 - \mu_2)}{\sqrt{(s_p^2/n_1) + (s_p^2/n_2)}}$$

is distributed as Student's t distribution, with $(n_1 - 1) + (n_2 - 1) = n_1 + n_2 - 2$ degrees of freedom.

In this equation, s_p^2, which is called the *pooled sample variance*, is the weighted average of the two sample variances obtained by weighting each by its respective degrees of freedom. That is,

$$s_p^2 = \frac{(n_1 - 1)s_1^2 + (n_2 - 1)s_2^2}{n_1 + n_2 - 2}$$

The rationale behind pooling the sample variances rests on the fact that s_1^2 serves as an estimate of σ_1^2 and s_2^2 provides us with an estimate of σ_2^2. Since the equation for t depends on the assumption that σ_1^2 and σ_2^2 are equal, s_1^2 provides an estimate of σ_2^2 and s_2^2 provides an estimate of σ_1^2. In other words, both s_1^2 and s_2^2 are estimates of that single variance that is represented by the symbols σ_1^2 and σ_2^2.

Since, in the typical practical situation, s_1^2 and s_2^2 will be different, the motivation for pooling is the desire to take the most efficient advantage of all the information provided by the sample data. This is a particularly important consideration when the two samples are of unequal size, since we look with greater favor on sample variances that are computed from larger samples.

Example 4.9 Suppose that two drugs, purported to reduce the response time of rats to a certain stimulus, are compared in a laboratory experiment. The experimenter, who is willing to assume that response times, following administration of the two drugs, are normally distributed with equal variances, wants to use the t distribution in the statistical analysis.

Drug A is to be administered to 12 rats, and drug B to 13 rats. The experimenter would like to know between what two values the central 95% of all t values would lie if the experiment were repeated many times using these sample sizes. The relevant t distribution is the one with $12 + 13 - 2 = 23$ degrees of freedom. Appendix Table F reveals that for 23 degrees of freedom, the central 95% of all t values lies between -2.0687 and $+2.0687$.

The experimenter's ultimate objective is to reach a decision about any difference in effectiveness of the two drugs with respect to reducing response time in rats. When the experiment is conducted, the mean reduction in response time for those rats receiving drug A is $\bar{x}_A = 30$ msec (milliseconds), with a standard deviation of $s_A = 5$ msec. The corresponding statistics for drug B are $\bar{x}_B = 23$ msec and $s_b = 6$ msec. What can the experimenter conclude?

If there is no difference between the two drugs with respect to the mean reduction in response times, the difference between population means, $\mu_A - \mu_B$, will equal 0. On the assumption that the two drugs are equally effective, we may compute the following t value from the sample data.

$$t = \frac{(30 - 23) - 0}{\sqrt{(30.74/12) + (30.74/13)}} = 3.15$$

We obtain the value of s_p^2 as follows.

$$s_p^2 = \frac{(12 - 1)(5)^2 + (13 - 1)(6)^2}{12 + 13 - 2} = 30.74$$

Since the computed t of 3.15 is greater than 2.0687, the experimenter must choose between the following alternatives.

1 There is, indeed, no difference in the mean effectiveness of the two drugs. The reason for the large t value is that chance alone has caused the experimenter to obtain a rare result.

2 The reason for the large t value cannot be attributed to chance alone. It is believed that the operating factor, other than chance, is that drug A is, in fact, more effective, on the average, than drug B in reducing rats' mean reaction time to the stimulus under investigation.

As we shall see in Chapter 6, the typical experimenter will select alternative 2.

11 The serum albumin concentration in a certain population of individuals is normally distributed with a mean of 4.1 grams/100 ml. A random sample of 16 of these individuals placed on a daily dosage of a certain oral steroid yielded a mean serum albumin concentration value of 3.8 grams/100 ml and a standard deviation of 0.4 gram. Does it appear likely from these results that the oral steroid reduces the level of serum albumin?

12 A manufacturer of frozen food is interested in the quality of frozen biscuits prepared by two different recipes. Shear-force measurements are made on samples of biscuits prepared by the two recipes, with the following results.

	Recipe *A*	**Recipe *B***
n	20	25
$\bar{x}$	135	165
s	10	11

Do the recipes appear to be comparable with respect to the shear force of the resulting biscuits?

4.5 THE DISTRIBUTION OF A SAMPLE PROPORTION

In practice, it is frequently desirable to be able to make inferences about population proportions. Therefore the sampling distribution of the sample proportion is of considerable interest. A marketing team may want to know what proportion of the consumers in some area prefer their firm's product to that of the competitor. A physician may wonder what proportion of arthritic patients will respond to a new drug. A candidate for elective office may wish to know what proportion of the voters are likely to vote for him. We could cite many other examples.

Suppose one draws a sample of entities from a population in order to count the number in the sample possessing some characteristic of interest. Suppose also that one wishes to determine in advance the probability that the sample will yield some specified number of entities that possess the characteristic under study. In Chapter 3 we saw that if certain assumptions are met, the binomial distribution is useful for calculating such probabilities. We also learned in Chapter 3 that the direct use of the binomial formula can be a computational burden even when the sample size is only moderately large. Tables of the binomial distribution, as we have seen, provide some relief in this respect, but they are also limited. Finally, we learned that under certain conditions, we may obtain approximate results by using the normal approximation to the binomial.

Partly as a result of these limitations, investigations into binomial populations are usually couched in slightly different terms. Typically, in practice, the investigator anticipates the selection of a sample of entities from some large finite population (it is, however, not infrequent for the population of interest to be infinite) for the purpose of drawing an inference about the proportion of entities in the population possessing some characteristic of interest.

The relevant statistic in such cases is the sample proportion, which is given by

$$\hat{p} = \frac{\text{number in sample with characteristic of interest}}{\text{total number in sample}}$$

For example, if, in a sample of 500 voters, 300 prefer candidate A, the sample proportion preferring candidate A is

$$\hat{p} = \frac{300}{500} = 0.60$$

Consequently, the inferential procedures depend on the sampling distribution of $\hat{p}$.

Let us visualize how we could construct the sampling distribution of $\hat{p}$ empirically. From the population of interest, we would select a large number of samples of size n, and for each we would compute the sample proportion $\hat{p}$. If the population were finite and reasonably small, we could select all possible samples of size n. For infinite populations, we can speak meaningfully only of taking a large number of samples. The values of $\hat{p}$, along with their frequencies of occurrence, would constitute the sampling distribution of $\hat{p}$. At this point we are interested in knowing the mean, standard deviation, and functional form of this distribution.

We may summarize the characteristics of the sampling distribution of $\hat{p}$, the sample population proportion, as follows.

The sampling distribution of $\hat{p}$, the sample proportion, computed from simple random samples of size n drawn from a population in which the population proportion is p, is approximately normally distributed if n is large. If the sampled population is finite, and of size N, the mean $\mu_{\hat{p}}$ of the distribution of $\hat{p}$ will be equal to p, and the standard deviation $\sigma_{\hat{p}}$ will be equal to $\sqrt{p(1-p)/n}\sqrt{(N-n)/(N-1)}$. If the sampled population is infinite, the mean and standard deviation of the distribution of $\hat{p}$ will be equal to p and $\sqrt{p(1-p)/n}$, respectively.

As we pointed out in Chapter 3, the sampling distribution of $\hat{p}$ will be approximately normal if both np and $n(1-p)$ are greater than five.

You will recognize the presence of the finite-population correction in the above statement about the sampling distribution of $\hat{p}$. As was the case with the finite-population correction factor appearing in the standard deviation of the sample mean given in Section 4.1, the fpc may be ignored in practical applications involving proportions when n/N is less than or equal to 0.05.

Figure 4.7 Sampling distribution of $\hat{p}$ and corresponding standard normal scale for Example 4.10

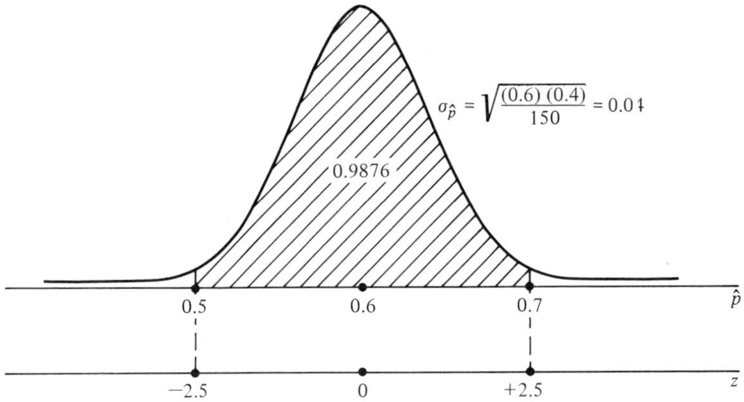

Example 4.10 It is known that 60% of the adults in a certain geographic area attend church regularly. A random sample of 150 adults is selected from this area. What is the probability that the sample proportion will be between 0.50 and 0.70? (The sample contains less than 5% of the population.)

We know that the sampling distribution of $\hat{p}$ will be approximately normally distributed with

$$\mu_{\hat{p}} = 0.60 \qquad \text{and} \qquad \sigma_{\hat{p}} = \sqrt{\frac{(0.6)(0.4)}{150}} = 0.04$$

We need now to convert 0.50 and 0.70 to z values, and use Appendix Table E to find the desired probability.

$$z_1 = \frac{0.50 - 0.60}{0.04} = \frac{-0.10}{0.04} = -2.5$$

By symmetry, z_2, the z value corresponding to $\hat{p} = 0.70$, is +2.5.

Appendix Table E reveals that the proportion of area under the standard normal curve between −2.5 and +2.5 is 2(0.4938) = 0.9876. The probability that a sample of 150 adults from the area will yield between 50 and 70% regular church attenders, therefore, is 0.9876.

Figure 4.7 shows a graph of the sampling distribution of $\hat{p}$ for this example, along with the corresponding scale of the standard normal distribution.

EXERCISES

In each of the following exercises, assume that the population is large enough for the fpc to be ignored, unless indicated otherwise.

13 In a certain metropolitan area, 18% of the teen-age boys have had some contact (ranging from warnings to arrests) with the police for delinquent behavior. A random sample of 100 teen-age boys is selected from this

area. What is the probability that between 15 and 25% will have had contact with the police as defined above?

14 It has been reported that, in a certain population of students, 40% have, at one time or another, smoked marijuana. In a random sample of 150 of these students, only 42 admit to having ever smoked it. What do you conclude?

15 A study conducted in a certain school district revealed that 70% of the elementary school children had moved at least once during their lifetime. A simple random sample of 200 elementary school children is drawn from this district. What is the probability that the proportion who have moved at least once will be between 0.65 and 0.75? What is the probability that the proportion will be greater than 0.75?

16 Of the 1150 teachers employed by a certain public school system, 60% hold master's degrees. A simple random sample of 150 of these teachers is selected. What is the probability that the proportion in the sample with master's degrees will be between 0.50 and 0.65?

4.6 THE DISTRIBUTION OF THE DIFFERENCE BETWEEN TWO SAMPLE PROPORTIONS

In many practical situations the researcher needs to make inferences about the difference between two population proportions. A store manager would like to know whether charge-card customers who live in two different areas of the city differ with respect to the proportion who have delinquent accounts. A sociologist wants to know whether urban and rural populations differ with respect to the proportion who favor capital punishment. An educator wants to know whether a population of high school dropouts and a population of those who remain in school differ on the basis of the proportion who come from broken homes. A medical researcher would like to know whether the proportion of lung cancer cases is higher in a population that smokes than in a population that doesn't smoke. The relevant sampling distribution in such cases is that of the difference between two sample proportions. To construct this distribution empirically from two sufficiently small finite populations, one would draw, from population 1, all possible simple random samples of size n_1 and compute, from each set of sample data, the sample proportion $\hat{p}_1$. From population 2, one would draw independently all possible simple random samples of size n_2 and compute, for each, the sample proportion $\hat{p}_2$. One would then compute the differences between all possible pairs of sample proportions, where one member of each pair was a value of $\hat{p}_1$ and the other a value of $\hat{p}_2$. The sampling distribution of $\hat{p}_1 - \hat{p}_2$, then, would consist of all such distinct differences, accompanied by their relative frequencies of occurrence. For large finite or infinite populations, one could approximate the sampling distribution of $\hat{p}_1 - \hat{p}_2$ by drawing a large number of independent simple random samples and proceeding as for small finite populations.

We are interested in the form and the mean and variance of the sampling distribution of $\hat{p}_1 - \hat{p}_2$ when the sample sizes are large. These characteristics are summarized in the following statements.

The sampling distribution of $\hat{p}_1 - \hat{p}_2$, the difference between two sample proportions, where $\hat{p}_1$ is computed from simple random samples of size n_1 from a population with parameter p_1, and $\hat{p}_2$ is computed from independent simple random samples of size n_2 from a population with parameter p_2, has mean

$$\mu_{\hat{p}_1 - \hat{p}_2} = p_1 - p_2$$

and standard deviation

$$\sigma_{\hat{p}_1 - \hat{p}_2} = \sqrt{\frac{p_1(1-p_1)}{n_1} + \frac{p_2(1-p_2)}{n_2}}$$

If n_1 and n_2 are large, the sampling distribution of $\hat{p}_1 - \hat{p}_2$ is approximately normally distributed.

Example 4.11 A psychiatric social worker believes that 15% of the adolescents in community A and 10% of those in community B suffer from some emotional or mental problem. In a simple random sample of 150 adolescents from community A, the social worker found 30 who had such a problem. An independent simple random sample of 100 adolescents from community B revealed that 7 were suffering from some emotional or mental problem.

Suppose the social worker's belief about the adolescents in these two communities is correct. What is the probability of observing a difference in sample proportions as large as or larger than that actually observed?

Since the sample sizes are large, we may proceed on the basis of a sampling distribution that is approximately normally distributed with a mean of

$$\mu_{\hat{p}A - \hat{p}B} = 0.15 - 0.10 = 0.05$$

and a standard deviation of

$$\sigma_{\hat{p}A - \hat{p}B} = \sqrt{\frac{(0.15)(0.85)}{150} + \frac{(0.10)(0.90)}{100}} = 0.04$$

if the social worker's belief is correct. The observed sample proportions are

$$\hat{p}_A = \frac{30}{150} = 0.20 \qquad \text{and} \qquad \hat{p}_B = \frac{7}{100} = 0.07$$

We wish to know the probability of observing a difference as large as or larger than

$$\hat{p}_A - \hat{p}_B = 0.20 - 0.07 = 0.13$$

To find this probability, we compute

$$z = \frac{0.13 - 0.05}{0.04} = \frac{0.08}{0.04} = 2.00$$

From Table E of the Appendix, we find that $P(z \geq 2.00) = 0.5 - 0.4772 = 0.0228$. In other words, if the social worker's belief is correct, the probability of observing a difference in sample proportions as large as or larger than 0.13 is 0.0228. In view of this small probability, the social worker may wish to revise his or her beliefs about the two populations.

EXERCISES

17 It is believed that 0.16 of the households in metropolitan area I have total incomes that fall below the "low"-income level. The proportion in metropolitan area II is believed to be 0.11. If these figures are accurate, what is the probability that a simple random sample of 200 households from area I and an independent simple random sample of 225 households from area II will yield a difference in sample proportions as large as or larger than 0.10?

18 In a simple random sample of 150 children who had attended kindergarten, 45 were reading at the time they entered first grade. In an independent simple random sample of 200 children who had not attended kindergarten, 20 were reading at the time of entering first grade. Suppose the proportion who can read is 0.15 in each group. What is the probability of observing sample results this extreme or more extreme?

19 It is believed that two drugs, A and B, are equally effective in reducing the level of anxiety in certain emotionally disturbed persons. The proportion of persons with whom the drugs are effective is believed to be 0.70. In a random sample of 100 emotionally disturbed persons who were given drug A, 75 experienced a reduction in anxiety level. Drug B was effective with 105 of an independent random sample of 150 subjects. If the two drugs are, in fact, equally effective, as believed, what is the probability of observing a value of $\hat{p}_A - \hat{p}_B$ as large as or larger than that reported here?

20 It is believed that 15% of the members of population A have one or more tattoos on their bodies, and that only 8% of the individuals in population B are tattooed. If these figures are accurate, what is the probability that a random sample of 120 individuals from population A and an independent random sample of 130 individuals from population B will yield a value of $\hat{p}_A - \hat{p}_B$ equal to or greater than 0.16?

4.7 THE CHI-SQUARE DISTRIBUTION AND THE DISTRIBUTION OF s^2

In the previous sections of this chapter, we have discussed the nature of the sampling distributions of the sample mean, the difference between two sample means, the sample proportion, and the difference between two sample proportions. In this section we shall comment on the distribution of the sample variance, $s^2 = \Sigma_{i=1}^n (x_i - \bar{x})^2/(n-1)$. Actually, the distribution of this statistic alone is not of particular interest in applied statistics. However, if sampling is from a normally distributed population, the distribution of a modification of s^2 is of tremendous importance. The nature of this modification and its distribution are given in the following statement.

If

$$s^2 = \frac{\Sigma_{i=1}^n (x_i - \bar{x})^2}{n-1}$$

is the variance of a random sample of size n from a normally distributed population with mean μ *and variance* σ^2, *then* $(n-1)s^2/\sigma^2$ *has a distribution known as the chi-square distribution.*

Note that

$$(n-1)s^2 = \sum_{i=1}^{n} (x_i - \bar{x})^2$$

which is the sum of squared deviations of sample values from their mean. So, alternatively, we might discuss our distribution of interest in terms of

$$\frac{\Sigma_{i=1}^n (x_i - \bar{x})^2}{\sigma^2}$$

Empirically, we could approximate the distribution of $\Sigma_{i=1}^n (x_i - \bar{x})^2/\sigma^2$ by drawing, from a normally distributed population, a large number of samples of size n, computing for each sample the sum of squared deviations of values from their mean, and dividing each sum by the population variance. A list of the different numerical values resulting from this procedure, along with their relative frequencies of occurrence, would constitute an approximation of the sampling distribution of $(n-1)s^2/\sigma^2$.

According to the statement above, this distribution follows a distribution known as the *chi-square distribution.* The variate that is so distributed is designated by the symbol χ^2, the Greek letter chi with the exponent 2. Thus we may say that

$$\chi^2 = \frac{(n-1)s^2}{\sigma^2}$$

follows a chi-square distribution.

The chi-square distribution, like the t distribution, is a family of distributions, since there is a different distribution for each of the possible values of a quantity, known as the *degrees of freedom.* The term degrees of freedom, as used here, is an expression of the same general concept that it was used for earlier. The chi-square distribution that is followed by $(n-1)s^2/\sigma^2$ is the one with $n-1$ degrees of freedom.

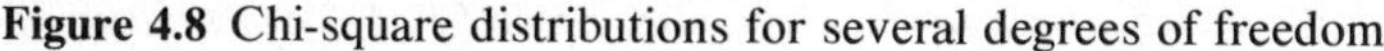

Figure 4.8 Chi-square distributions for several degrees of freedom

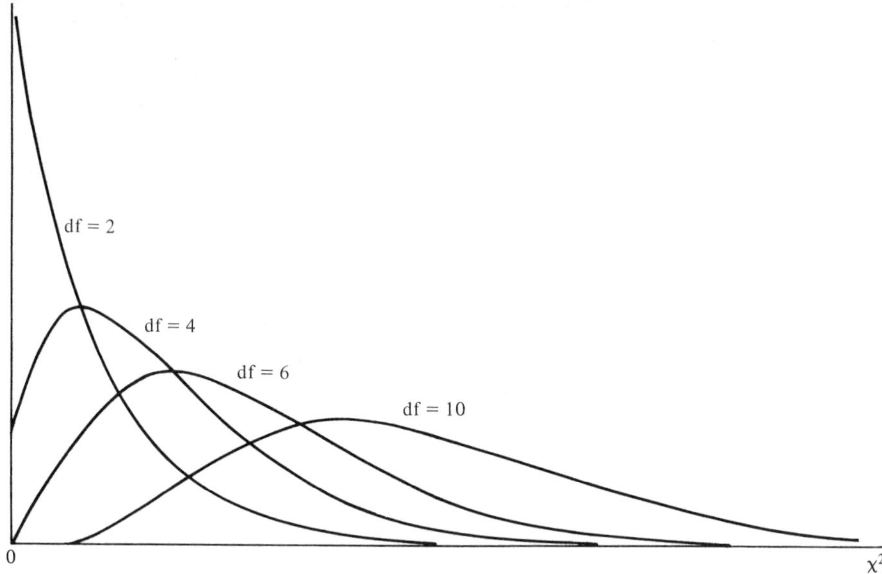

Figure 4.8 shows chi-square distributions for several degrees of freedom. Note that the curves tend to be skewed to the right and are not symmetric.

The total area bounded by the curve of a chi-square distribution and the axes is equal to 1. The variate χ^2 assumes only nonnegative values. The mean of any chi-square distribution is equal to its degrees of freedom, and its variance is equal to two times the degrees of freedom.

The chi-square distribution is one of the most widely used distributions in applied statistics. To facilitate its use, tables are available for finding areas, which are probabilities, associated with intervals bounded by designated values of χ^2. One such table is Table G of the Appendix. In this table the leftmost column shows the degrees of freedom needed to enter the table, and the column headings indicate the proportion of area to the left of the values of χ^2 in the body of the table. Suppose, for example, that we want to know, for the chi-square distribution with 10 degrees of freedom, what value of χ^2 has 0.95 of the area under the curve to its left. We locate 10 in the degrees-of-freedom column and the column headed $\chi^2_{0.95}$. The value of χ^2 at the intersection of the row labeled 10 and the column headed $\chi^2_{0.95}$, then, is the one we seek, and we see that it is 18.307. This tells us that under the curve of the chi-square distribution with 10 degrees of freedom, 95% of the area is to the left of 18.307. Since the total area under the curve is equal to 1, we know that 5% of the area is to the right of 18.307. We interpret area under the curve as probability, and therefore may say that if a value of χ^2 is picked at random from the chi-square distribution with 10 degrees of freedom, the probability that it will be less than 18.307 is 0.95. Or we may say that the probability that a value of χ^2 selected at random from the chi-square

Figure 4.9 Chi-square distribution with 10 degrees of freedom, showing area to left and right of $\chi^2 = 18.307$

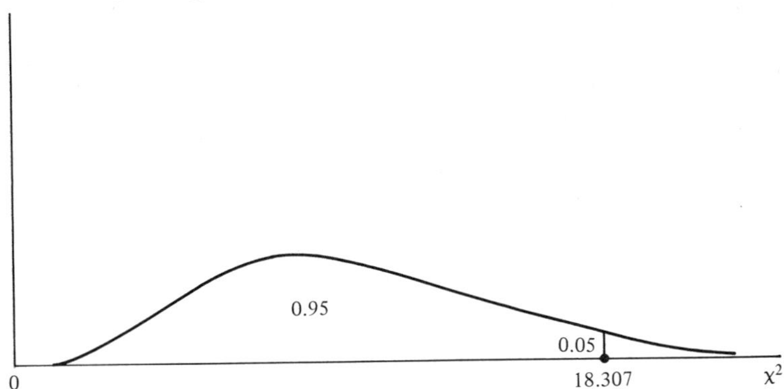

distribution will be greater than or equal to 18.307 is 0.05. Figure 4.9 shows these probabilities.

Now let us illustrate, by means of an example, one of the many ways to use the chi-square distribution.

Example 4.12 Suppose the variance of the weights of 12-year-old males is 39 kg² (kilograms squared), and weights are normally distributed. What is the probability that a random sample of 25 12-year-old males will yield a variance equal to or greater than 57?

From the sample data we compute

$$\chi^2 = \frac{(24)(57)}{39} = 35.077$$

To find the probability of observing a value of χ^2 as large as or larger than 35.077, we enter Appendix Table G with 24 degrees of freedom. Moving along the row labeled 24, we find that our computed value of $\chi^2 = 35.077$ lies between χ^2 values of 33.196 and 36.415. These values are in the columns headed $\chi^2_{0.90}$ and $\chi^2_{0.95}$, respectively. Thus we conclude that the probability of observing a value of χ^2 equal to or greater than 35.077 is something between 0.05 and 0.10. Consequently, we say that, under the given conditions, the probability of observing a value of s^2 equal to or greater than 57 is also something between 0.05 and 0.10.

We shall use our knowledge of the chi-square distribution again in Chapter 10.

EXERCISES

21 Given a chi-square distribution with 15 degrees of freedom, find the value of χ^2 that has 0.01 of the area under the curve to its right.

22 Find the value of χ^2 that divides the area under the curve of the

chi-square distribution with 20 degrees of freedom into two parts such that the smaller area is equal to 0.025.

23 For the chi-square distribution with 25 degrees of freedom, find two values of χ^2 between which 95% of the area under the curve is located. (Do not use zero or infinity.)

24 Consider the chi-square distribution with 30 degrees of freedom. What is the probability that a value of χ^2 picked at random from this distribution will be equal to or greater than 50.892?

25 The quality-control department of a manufacturing firm buys certain electrical components from an outside vendor. The firm specifies that the variance of the resistances of the components must not exceed 0.20 ohms squared. To guard against accepting incoming shipments that do not meet this specification, the quality-control department takes a random sample of 25 components from each shipment, and measures the resistance of each. If the sample variance is too large, the department refuses the shipment. A sample variance is considered to be too large if the probability of obtaining such a large value is equal to or less than 0.01. A shipment has just been sampled, and a variance of $s^2 = 0.35$ is computed. Should the shipment be accepted? Assume that resistances are normally distributed.

26 The variance of the systolic blood-pressure readings in a certain population of adults is known to be 196. A random sample of 16 individuals from this population had their blood pressures read after having taken a certain experimental drug. The sample variance was found to be $s^2 = 300$. Does it seem likely that the drug increases the variability of systolic blood-pressure readings? Assume that systolic blood-pressure readings in this population are normally distributed.

27 The weights of adult insects of a certain species are assumed to be normally distributed with a standard deviation of 0.7 gram. If these assumptions are correct, what is the probability that a random sample of ten insects will yield a variance equal to or greater than 0.85?

4.8 THE *F* DISTRIBUTION

As we have seen in previous sections of this chapter, we are frequently interested in the difference between two population means or the difference between two population proportions. In contrast to this, we find that when we are interested in inferences about two population variances, it is the ratio between them that provides the appropriate avenue of approach. Consequently, the nature of the sampling distribution of the ratio of two sample variances interests the researcher.

In 1924, R. A. Fisher (1890–1962) introduced a distribution that enables one to carry out useful inferential procedures using the ratio of two sample variances. We may summarize the nature of this distribution, called the *F distribution*, as follows.

Given s_1^2 and s_2^2, the variances computed from independent random samples of size n_1 and n_2 drawn from normally distributed populations with variances of σ_1^2 and σ_2^2, respectively, the random variable

$$F = \frac{s_1^2/\sigma_1^2}{s_2^2/\sigma_2^2}$$

follows the F distribution with $n_1 - 1$ and $n_2 - 1$ degrees of freedom.

An alternative expression for the variable F is

$$F = \frac{\dfrac{(n_1 - 1)s_1^2/\sigma_1^2}{n_1 - 1}}{\dfrac{(n_2 - 1)s_2^2/\sigma_2^2}{n_2 - 1}}$$

Note that the numerators in both the numerator and denominator are χ^2 variables, so we may express F in still another way:

$$F = \frac{\chi_1^2/(n_1 - 1)}{\chi_2^2/(n_2 - 1)}$$

Thus we see that F is the ratio of two independent χ^2 variables, each divided by its degrees of freedom.

It takes two degrees-of-freedom values to specify an F distribution, one associated with the numerator and one associated with the denominator. Figure 4.10 shows several F distributions.

The total area bounded by the curve of an F distribution and the axes is

Figure 4.10 The F distribution for various degrees of freedom (ν_1, ν_2). (Reproduced from *Documenta Geigy, Scientific Tables,* seventh edition. Courtesy of Ciba-Geigy Ltd., Basel, Switzerland, 1970)

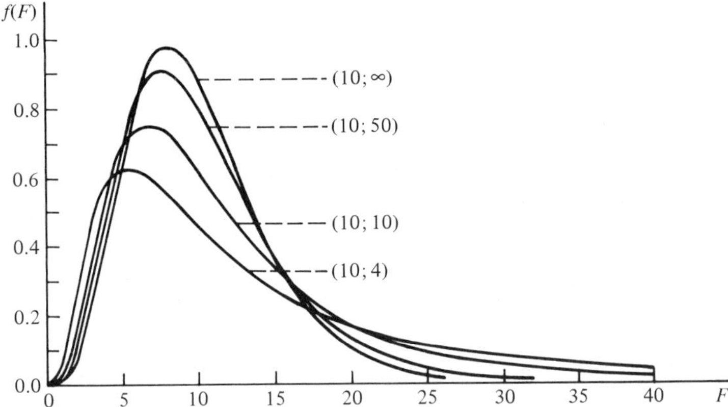

equal to 1, and, like the χ^2 variable, F assumes only nonnegative values. We interpret the area under the curve of an F distribution as the probability associated with the interval between specified values of F. To facilitate the determination of probabilities of interest, special tables, such as Table H of the Appendix, are available. The format of Appendix Table H is similar to that of Appendix Table G. The entries in the body of the table are values of F from the F distribution (specified by the degrees of freedom for the numerator and denominator) which have to their left, under the curve of the distribution, a proportion of area equal to the subscript on F shown at the top of the table.

Suppose, for example, we are interested in the F distribution with 15 numerator degrees of freedom and 20 denominator degrees of freedom. Now suppose we want to know the value of F that has to its left 0.95 of the area under the curve. To find this value of F, we locate that portion of the table labeled $F_{0.95}$. The value of F we seek is at the intersection of the column headed 15 (numerator degrees of freedom) and the row labeled 20 (denominator degrees of freedom). We see that it is 2.20. Thus 95% of the area under the curve under consideration is to the left of $F = 2.20$, and since the total area under the curve is equal to 1, we know that 5% of the area is to the right of $F = 2.20$. Since we interpret area as probability, we may say that the probability that an F picked at random from the F distribution with 15 and 20 degrees of freedom will be less than 2.20 is 0.95. Alternatively, we say that the probability that an F so chosen will be equal to or greater than 2.20 is 0.05. Figure 4.11 shows these areas (probabilities). In this figure, we have

Figure 4.11 Representation of the F distribution for 15 numerator degrees of freedom and 20 denominator degrees of freedom, showing area to left and right of $F = 2.20$. (Figure not drawn to scale.)

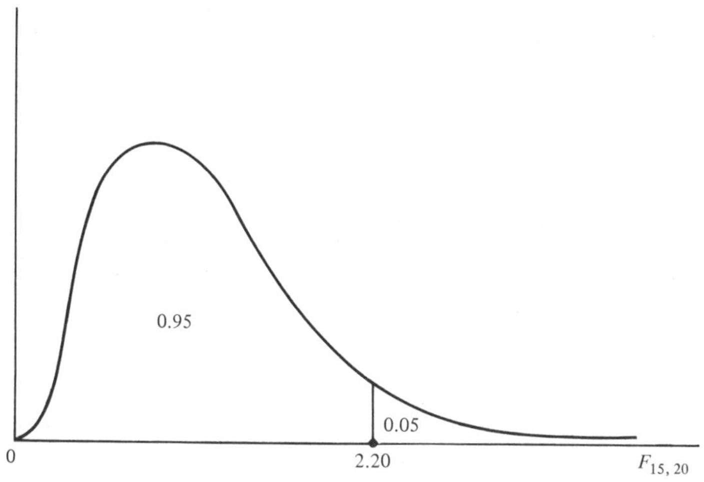

identified the particular F under discussion by giving the numerator and denominator degrees of freedom (in that order) as a subscript on F.

Suppose that independent random samples of size n_1 and n_2 are drawn from population 1 and population 2, respectively, and suppose that these populations are normally distributed with equal variances (that is, $\sigma_1^2 = \sigma_2^2$). Then we may rewrite the equation for F as

$$F = \frac{s_1^2/\sigma^2}{s_2^2/\sigma^2} = \frac{s_1^2}{s_2^2}$$

This also follows the F distribution with $n_1 - 1$ and $n_2 - 1$ degrees of freedom. In Chapters 5, 6, and 8 we shall use the F distribution in a variety of inferential procedures. The following example serves as a preview of our future uses of F.

Example 4.13 It is conjectured that the variability of the daily protein intake is the same in 15-year-old boys as in 15-year-old girls. A random sample of 16 girls and an independent random sample of 20 boys yield variances of 608 grams squared and 320 grams squared, respectively. Are these results consistent with the conjecture?

Let us assume that daily protein-intake values in the two populations are normally distributed. Then we may compute the ratio of the two sample variances as follows.

$$F = \frac{608}{320} = 1.90$$

To find the probability of getting an F value this large or larger under the stated conditions, we look at the intersection of 15 and 19 degrees of freedom in Appendix Table H. In the table, for $F_{0.90}$, we find, at the intersection of 15 numerator degrees of freedom and 19 denominator degrees of freedom, a value of $F = 1.86$. At the intersection of 15 and 19 under $F_{0.95}$, we find $F = 2.23$. Thus the probability of observing an F as large as or larger than 1.90 when the two population variances are equal is between 0.05 and 0.10.

EXERCISES

In the following exercises, compute F by placing the larger sample variance in the numerator.

28 Consider the F distribution with 9 and 16 degrees of freedom. What value of F has 0.05 of the area to its right?

29 Consider the F distribution with 12 and 21 degrees of freedom. What is the probability that an F picked at random from this distribution will be greater than or equal to 3.17?

30 Consider the F distribution with 6 and 17 degrees of freedom. Find the F_0 such that the probability that an F picked at random from the distribution will exceed F_0 is 0.025.

31 It is conjectured that the variance of the scores on a certain aptitude test is the same for men and women. A random sample of 21 men and an independent random sample of 19 women yield variances of 876 and 400, respectively. If the scores for both men and women are normally distributed with equal variances, what is the probability of observing sample results as extreme as or more extreme than these?

32 The directors of a firm that manufactures plastics want to know whether two manufacturing processes are comparable with respect to the variability of the tensile strength of the product. They say they will believe that the two processes yield plastics with equal tensile-strength variability if the ratio of variances based on samples of specimens from the two processes is not too large. They will consider the ratio too large if the probability of getting a value as large or larger is less than 0.05. Random (independent) samples of 25 specimens made by each process give variances of 540 and 256. Will the directors believe that the two processes yield plastics with the same tensile-strength variability? Assume that the tensile strengths of products made by both processes are normally distributed.

CHAPTER SUMMARY

This chapter has been concerned with sampling distributions. We again remind you that an understanding of the concept of a sampling distribution is necessary for a true understanding of the inferential procedures that are dealt with in the remaining chapters. Before proceeding further, carefully review this chapter, paying particular attention to any areas that are not clear.

This chapter demonstrated the empirical construction of a sampling distribution for a sample mean. In addition, it pointed out that the three characteristics of a sampling distribution that are of greatest interest are its functional form, its mean, and its standard deviation.

In this chapter we also covered sampling distributions of the following variables.

The sample mean
The sample proportion
The difference between two sample means
The difference between two sample proportions
$(n - 1)s^2/\sigma^2$
s_1^2/s_2^2

In addition, we also discussed Student's t distribution.

SOME IMPORTANT CONCEPTS AND TERMINOLOGY

Sampling distribution	Central-limit theorem
Degrees of freedom	*t* distribution
Finite population correction	Pooled estimate of population variance
Sampling with replacement	Chi-square distribution
Sampling without replacement	*F* distribution

REVIEW EXERCISES

33 The force, in pounds per square inch (psi), required to break a certain type of plastic sheet is an approximately normally distributed random variable with a mean of 2800 psi and a variance of 9000 psi^2. Suppose that a simple random sample of size 10 is to be selected from this population and tested until each sheet breaks. What is the probability that the mean force required to tear the sheets in the sample will be 2750 psi or less?

34 Suppose it is known that the mean and standard deviation of serum iron values for normal adult males are 120 and 15 μg (micrograms) per 100 ml, respectively. What is the probability that a random sample of 60 normal men will yield a mean between 115 and 125 μg/100 ml?

35 A psychologist has found the mean time required for mice to run a maze to be 24.5 seconds, with a standard deviation of 4.5 seconds.

(a) What are the mean and standard deviation of the sampling distribution of the sample mean, based on simple random samples of size 81 from this population?

(b) A simple random sample of 81 mice is drawn for special study. What is the probability that the mean time for running the maze for mice in this sample will be greater than 25 seconds?

36 In a study of annual family expenditures for recreation, two populations were surveyed, with the following results.

Population 1: $n_1 = 40, \bar{x}_1 = \$640$
Population 2: $n_2 = 35, \bar{x}_2 = \$600$

It is known that the population variances are $\sigma_1^2 = 2800$ and $\sigma_2^2 = 3150$. What is the probability of obtaining sample results as extreme as those shown above if there is no difference in the means of the two populations?

37 A recreation specialist studying the length of time spent by tourists in two types of tourist attractions observed a sample of 75 tourists in each type. The mean time spent by the sample of tourists in type *A* was 55 minutes, and the mean time for the sample of tourists in type *B* was 49 minutes. What is the probability of observing a sample difference as large as this if there is no difference in the true mean time spent by tourists in the two types of attractions, and if the standard deviation is 15 minutes for both populations? What assumptions are made regarding the samples?

38 In a random sample of 75 adults, 35 said they favored candidate A for mayor. In the population from which the sample was drawn, the true proportion who favor candidate A is 0.55. What is the probability of obtaining a sample proportion as small as or smaller than that obtained in this sample?

39 The owners of a firm that manufactures flashlight batteries have determined from experience that 3% of the batteries they produce are defective. A random sample of 300 batteries is examined. What is the probability that the proportion defective is between 0.02 and 0.035?

40 In a certain community, it is believed that 40% of the homemakers prefer brand A coffee. In another community, it is believed that only 15% of the homemakers prefer this brand. If these figures are correct, what is the probability that simple random samples of 100 from each community would yield a difference in the proportion of homemakers preferring this brand of coffee of 0.40 or more?

5

Estimating Population Parameters

Many situations arise in which someone—a decision maker, a program planner, or a researcher—wishes to know the values of such parameters as population variances, population means, population proportions, the difference between population means, and the difference between population proportions. A criminologist may want to know what proportion of persons convicted of a crime suffer from some mental disability. A public health official may want to know the mean age at which some population of smokers first began the habit. A sociologist may be interested in knowing whether the proportion reared in a home with only one parent present differs in two populations of juvenile delinquent first offenders. A psychologist may want to know whether the mean time required for mice to run a maze differs in two populations distinguishable on the basis of the stimulus received. We saw in previous chapters that such information is usually obtained from samples rather than the populations themselves, by means of statistical inference. In this chapter our discussion becomes more specific, as we show how to assess the magnitudes of such parameters by means of the statistical inference procedure called *estimation*. Before proceeding, however, let us briefly review the highlights of the discussion up to this point.

The concepts and techniques learned in Chapters 1 through 4 provide the theoretical, as well as the technical, foundation for making statistical inferences. In this chapter we shall, for the first time, tie together the ideas of descriptive measures, probability, probability distributions, and sampling distributions.

In Chapter 1 we defined statistical inference as the procedure whereby conclusions about a population are formed on the basis of information contained in a sample drawn from the population. We also pointed out that the need for statistical inference arises from the fact that, for many reasons, it may be impractical or impossible to examine a population in its entirety. The population may be too large, or the act of taking measurements may be destructive. In such cases, only a sample from the population may be conveniently examined. When one seeks information about a population, but only information about a sample is available, one needs some means for using the sample data to draw conclusions about the population. The concepts and techniques for meeting this need constitute what is known as *statistical inference*.

The two types of statistical inference are *estimation* and *hypothesis testing*. We shall cover the basic concepts and techniques of estimation in this chapter. In Chapter 6 we shall discuss the fundamentals of hypothesis testing.

5.1 ESTIMATES AND ESTIMATORS

Our objective in this chapter is to learn how we may use sample data to obtain estimates of population parameters. Here the parameters for which we shall wish to provide estimates are the population mean, μ; the difference between two population means, $\mu_1 - \mu_2$; a population proportion, p; the difference between two population proportions, $p_1 - p_2$; the population variance, σ^2; and the ratio of two population variances, σ_1^2/σ_2^2. Later we shall be concerned with estimating additional population parameters. Although the general procedures for estimating parameters are the same for all parameters, differences emerge when we apply the procedures in specific cases.

Two types of estimation are of interest: point estimation and interval estimation. As its name implies, *point estimation* consists of using a single sample statistic—for example, the sample mean $\bar{x}$—to estimate the corresponding population parameter—in this case, μ. For instance, we may draw a sample of households from a certain geographic area, compute a sample mean family income of \$10,000, and use it as a point estimate of the mean family income for all households in the area. Although this may be the best "guess" for μ available, we suspect that it is not exactly equal to μ.

An *interval estimate* consists of two points that define an interval which we estimate contains the parameter of interest. Based on sample information regarding family income for households of a certain area, we may, for example, estimate that the mean family income of all households in the area is between \$7,000 and \$15,000. The mechanics of constructing such an interval estimate are described in a later section.

At this point, let us distinguish between an *estimator* and an *estimate*. An *estimator* is a procedure, expressed as a rule or formula, whereby a numerical value, called an *estimate*, is obtained. Thus $\bar{x} = \Sigma x_i/n$, representing the method whereby a sample mean is computed, is an estimator. But the numerical result obtained by carrying out the indicated operation is an *estimate*.

5.2 PROPERTIES OF GOOD ESTIMATORS

In selecting an estimator, $\hat{\theta}$, of a parameter, θ (Greek letter theta), it seems logical that we would want to select the "best" estimator. In practice, we may be unable to determine what estimator is "best" in a given situation. What we can do, however, is select a "good" estimator. Several criteria for evaluating the "goodness" of estimators have been proposed. Four of these criteria are (1) unbiasedness, (2) consistency, (3) efficiency, and (4) sufficiency. We shall limit our discussion to the criterion of unbiasedness. The other criteria, though important, cannot be discussed adequately without recourse to mathematical concepts beyond the level of this book.

Unbiasedness

An estimator $\hat{\theta}$ is said to be an *unbiased* estimator of the population parameter θ if the mean value of $\hat{\theta}$, computed from all possible random samples of a given size drawn from the population of interest, is equal to θ. This property is usually expressed symbolically as

$$E(\hat{\theta}) = \theta$$

where the expression on the left-hand side is read, "expected value of theta hat." In other words, θ is the mean of the sampling distribution of $\hat{\theta}$. In Chapter 4 we stated that $\mu_{\bar{x}}$, the mean value of $\bar{x}$ computed from all possible random samples of a given size from some population, is equal to μ. We demonstrated this fact in Example 4.1 for a small population and samples of size $n = 2$. Thus we say that the expected value of $\bar{x}$ is μ, and therefore that $\bar{x}$ is an unbiased estimator of μ. Other unbiased estimators, of corresponding population parameters, discussed in Chapter 4 are $\bar{x}_1 - \bar{x}_2$, $\hat{p}$, and $\hat{p}_1 - \hat{p}_2$.

Let us now consider the estimator of the population variance to determine its characteristics relative to biasedness.

1 When sampling is with replacement, $s^2 = \Sigma (x_i - \bar{x})^2/(n-1)$ is an unbiased estimator of the population variance defined by $\sigma^2 = \Sigma (x_i - \mu)^2/N$, since with replacement $E(s^2) = \sigma^2$.

2 When sampling is without replacement, $s^2 = \Sigma (x_i - \bar{x})^2/(n-1)$ is an unbiased estimator of the population variance defined by $S^2 = \Sigma (x_i - \mu)^2/(N-1)$, since without replacement $E(s^2) = S^2$.

3 The estimator $s^{2*} = \Sigma (x_i - \bar{x})^2/n$ is a biased estimator of both σ^2 and S^2, since $E(s^{2*}) \neq \sigma^2$ and $E(s^{2*}) \neq S^2$. Note that

$$s = \sqrt{\frac{\Sigma (x_i - \bar{x})^2}{n-1}}$$

is a biased estimator of σ and S regardless of whether sampling is with or without replacement, although generally s is used to estimate these parameters.

5.3 CONFIDENCE INTERVAL FOR A POPULATION MEAN

We noted that when estimation is the inferential objective, two types of estimates are available: a *point estimate* and an *interval estimate*. There is an obvious problem associated with the use of point estimates. Although only one parameter is involved, the number of estimates available is generally quite large. Each of the possible samples that can be drawn from the population of interest yields an estimate. From our study of sampling distributions in Chapter 4, we know that some estimates will be closer to the parameter being estimated than others. We do not know, however, how close our single-point estimate is to the true parameter. In a given situation, we consider it highly unlikely that our point estimate is exactly equal to the parameter, but we are unable to say by how much we have missed it.

We attempt to overcome this problem associated with point estimates by constructing an interval estimate of the parameter of interest. We construct this interval in such a way that we can state the degree of *confidence* we have that the interval includes, within its end points, the parameter being estimated. Because of the associated statement of confidence, we call such an interval estimate a *confidence interval.*

In this section we present methods for constructing confidence intervals for a population mean in three different situations: (1) when the population is normal and the population variance is known, (2) when the population is normal and the population variance is unknown, and (3) when the population is not normal.

The concept of confidence intervals and their interpretation are important aspects of statistical inference. Consequently, in the discussion that

Figure 5.1 Sampling distribution of $\bar{x}$

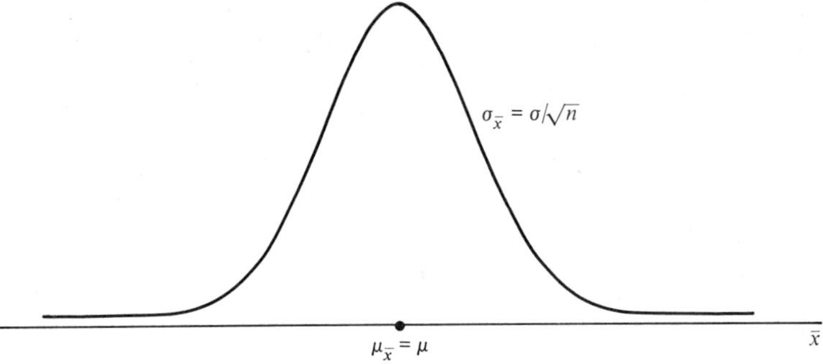

follows, we shall emphasize concepts and interpretation, as well as methodology.

Single mean of a normally distributed population, σ^2 known

Let us first consider what is involved in constructing a confidence interval for the mean of a normally distributed population. Although, in the typical situation, the population variance is unknown, the discussion will be facilitated if we assume a known population variance. We shall discuss the case in which σ^2 is unknown later.

From our discussion of sampling distributions in Chapter 4, we know that the sampling distribution of $\bar{x}$, computed from simple random samples of size n drawn from normally distributed populations, is normally distributed with mean μ and variance σ^2/n, where μ and σ^2 are the mean and variance of the population from which the samples are drawn, and the fpc is not applicable or may be ignored. The sampling distribution of $\bar{x}$, as we noted earlier, may be shown graphically as in Figure 5.1.

We may express the probability, say $1 - \alpha$, that a single simple random sample of size n will yield a mean $\bar{x}$ falling between two points, say $\bar{x}_a$ and $\bar{x}_b$ on the $\bar{x}$ axis, as follows.

$$P(\bar{x}_a \leq \bar{x} \leq \bar{x}_b) = 1 - \alpha \tag{5.1}$$

Let us choose $\bar{x}_a$ and $\bar{x}_b$ in such a way that they are equidistant from the mean $\mu_{\bar{x}} = \mu$, and let us express this distance in terms of standard errors. Suppose we locate $\bar{x}_a$ a distance of k standard errors to the left of the mean $\mu_{\bar{x}} = \mu$, and $\bar{x}_b$ a distance of k standard errors to the right of $\mu_{\bar{x}} = \mu$. We may now express this distance as $k(\sigma/\sqrt{n})$, and portray it graphically as in Figure 5.2.

Now that we know the distance of $\bar{x}_a$ and $\bar{x}_b$ from the mean $\mu_{\bar{x}} = \mu$, we may write the following equalities.

$$\bar{x}_a = \mu - k\frac{\sigma}{\sqrt{n}}, \qquad \bar{x}_b = \mu + k\frac{\sigma}{\sqrt{n}}$$

Figure 5.2 Sampling distribution of $\bar{x}$, showing $\bar{x}_a$ and $\bar{x}_b$

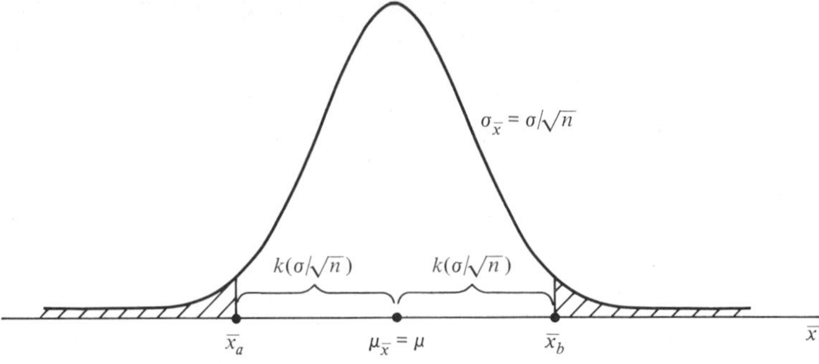

Consequently, we may use these equivalents of $\bar{x}_a$ and $\bar{x}_b$ to rewrite Equation 5.1. Thus we have

$$P\left(\mu - k\frac{\sigma}{\sqrt{n}} \leq \bar{x} \leq \mu + k\frac{\sigma}{\sqrt{n}}\right) = 1 - \alpha \tag{5.2}$$

By this equation, we are saying that for a simple random sample of size n drawn from a normally distributed population with mean μ and variance σ^2, the probability that $\bar{x}$ will be between $\mu - k(\sigma/\sqrt{n})$ and $\mu + k(\sigma/\sqrt{n})$ is $1 - \alpha$. Alternatively, we may say that the probability that $\bar{x}$ will be within k standard errors of the mean $\mu_{\bar{x}} = \mu$ is $1 - \alpha$.

Since, in the present situation, the statistic $\bar{x}$ is normally distributed, we know that when $1 - \alpha$ is specified, k can be replaced by a value of z, the standard normal variable. For example, if $1 - \alpha = 0.95$, $k = 1.96$, and Equation 5.2 becomes

$$P\left(\mu - 1.96\frac{\sigma}{\sqrt{n}} \leq \bar{x} \leq \mu + 1.96\frac{\sigma}{\sqrt{n}}\right) = 0.95$$

That is, we say that the probability is 0.95 that $\bar{x}$ is between a point equal to $\mu - 1.96$ standard errors and a point equal to $\mu + 1.96$ standard errors. Figure 5.3 shows these points.

Figure 5.3 Sampling distribution of $\bar{x}$, showing location of $1 - \alpha$

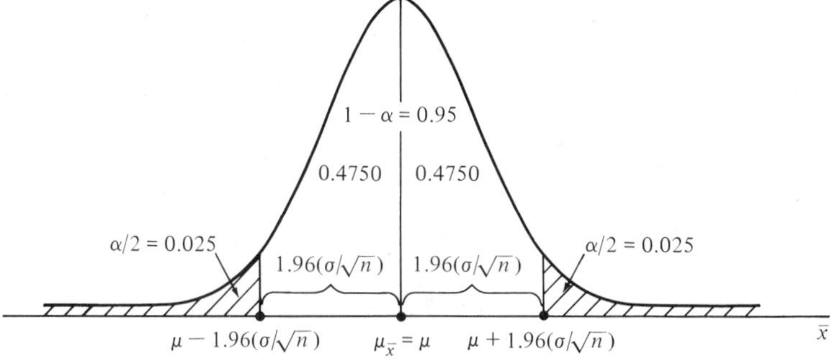

We will get closer to our objective of finding a confidence interval for the unknown parameter μ if we can obtain a probability statement similar to Equation 5.2 that has μ in the center rather than $\bar{x}$. In fact, by means of some algebraic manipulation, we can rewrite Equation 5.2 as

$$P\left(\bar{x} - k\frac{\sigma}{\sqrt{n}} \leq \mu \leq \bar{x} + k\frac{\sigma}{\sqrt{n}}\right) = 1 - \alpha \tag{5.3}$$

If, for example, $1 - \alpha = 0.95$, we may write Equation 5.3 as

$$P\left(\bar{x} - 1.96\frac{\sigma}{\sqrt{n}} \leq \mu \leq \bar{x} + 1.96\frac{\sigma}{\sqrt{n}}\right) = 0.95$$

and we say that the probability that the unknown parameter is between a point equal to $\bar{x} - 1.96$ standard errors and a point equal to $\bar{x} + 1.96$ standard errors is 0.95. Different values of $1 - \alpha$ will result in the substitution of different values of z for k. For example, if $1 - \alpha = 0.99$, we substitute $z \approx 2.58$ for k.

We interpret Equation 5.3 as follows. Suppose that a large number of simple random samples of size n are drawn from a normally distributed population with unknown mean μ and known variance σ^2. For each sample, we construct an interval of the form suggested by the expression in parentheses in Equation 5.3 by adding to and subtracting from the sample mean the quantity $k(\sigma/\sqrt{n})$. This procedure would lead to a series of intervals such as the following.

$$\bar{x}_1 \pm k\frac{\sigma}{\sqrt{n}}, \qquad \bar{x}_2 \pm k\frac{\sigma}{\sqrt{n}}, \qquad \bar{x}_3 \pm k\frac{\sigma}{\sqrt{n}}, \ldots$$

Equation 5.3 tells us that, in the long run, $100(1 - \alpha)\%$ of the intervals so constructed will contain the unknown population mean μ. All these intervals will be the same width, and this width will be equal to the width of the interval $\mu \pm k(\sigma/\sqrt{n})$ that could be placed about μ if μ were known. Figure 5.4 shows the relationship to μ of the intervals obtained through repeated sampling.

From our interpretation of Equation 5.2, we know that, in repeated sampling, $100(1 - \alpha)\%$ of the computed $\bar{x}$'s will be within the interval $\mu \pm k(\sigma/\sqrt{n})$. If an interval of the same width is centered on $\bar{x}$ [through addition and subtraction of $k(\sigma/\sqrt{n})$], we know that all $\bar{x}$'s within $k(\sigma/\sqrt{n})$ of μ will have intervals about them which in turn will include μ. This fact is illustrated in Figure 5.4, in which we see that only those $\bar{x}$'s between the vertical lines have intervals that include μ.

In practice, of course, when we wish to estimate a population mean, we do not draw a large number of simple random samples from the population; we draw only one sample. If we denote by $\bar{x}_0$ the mean of the single sample that might be drawn in a typical situation, we may construct the following interval estimate of μ.

Figure 5.4 The relationship to μ of a large number of interval estimates of μ

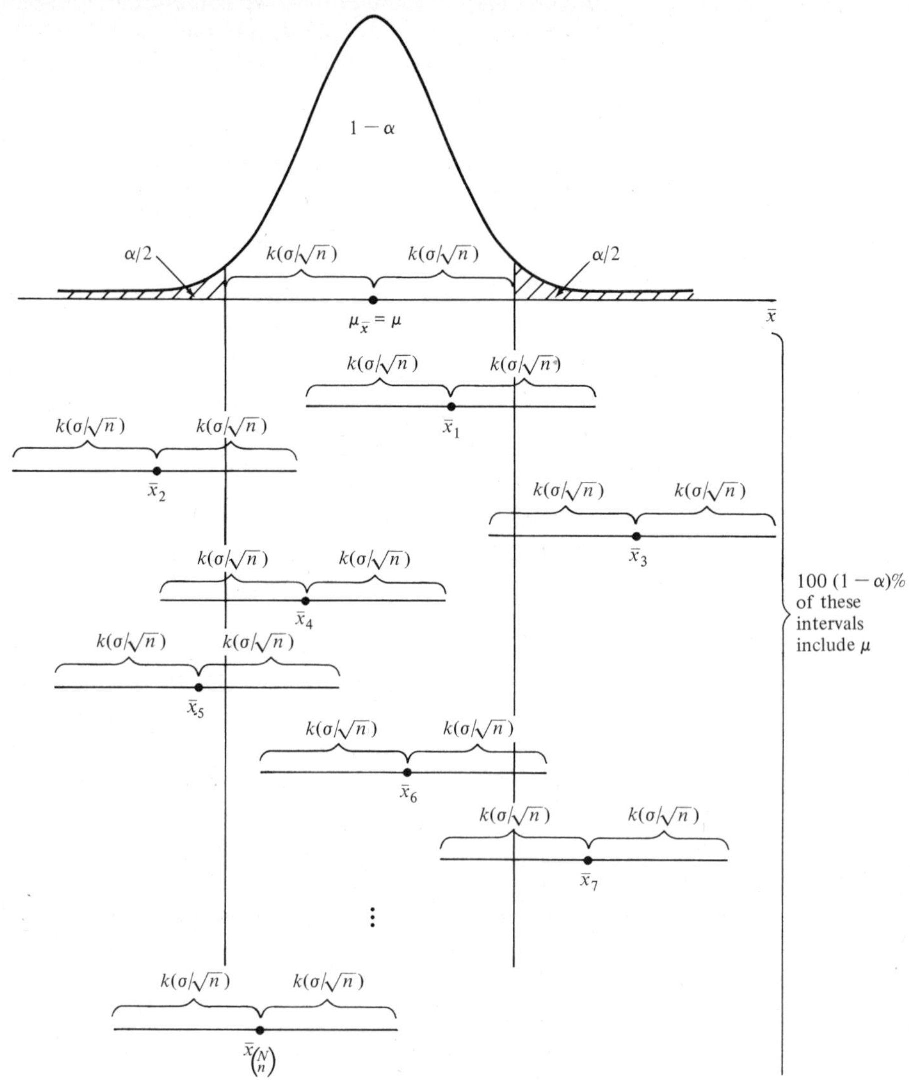

$$\bar{x}_0 \pm k\frac{\sigma}{\sqrt{n}}$$

This interval is called the $100(1-\alpha)\%$ confidence interval for μ because it is one of the large number of intervals of which $100(1-\alpha)\%$ contain μ. To put this single interval in the same form as Equation 5.3, we may write

$$C\left(\bar{x}_0 - k\frac{\sigma}{\sqrt{n}} \leq \mu \leq \bar{x}_0 + k\frac{\sigma}{\sqrt{n}}\right) = 1 - \alpha \qquad (5.4)$$

where C indicates that the interval is a confidence interval and that the statement is a confidence statement, rather than a probability statement. In Equation 5.4, $1 - \alpha$, which is called the *confidence coefficient*, indicates the degree or amount of confidence that we have that our single interval is one that contains μ. The confidence coefficient expressed as a percentage is called the *level of confidence* or *confidence level*.

Let us emphasize the difference between Equation 5.3 and Equation 5.4. In Equation 5.3, $\bar{x}$ is an unspecified value of a random variable, and the statement is therefore a legitimate probability statement. In Equation 5.4, $\bar{x}_0$ is a constant, and therefore the end points we obtain by adding and subtracting the known quantity $k(\sigma/\sqrt{n})$ are constants. The unknown μ either is in this known interval, or it isn't. For this reason, Equation 5.4 is not a probability statement. This equation, however, may be interpreted as a confidence interval. Since the probability is $1 - \alpha$ that a single random sample will yield an interval that includes μ, we are "confident" that that is what happened. The degree of our confidence depends on the size of $1 - \alpha$. The larger $1 - \alpha$, the greater our confidence.

Now let us summarize how we would construct a confidence interval in practice.

1 We draw a simple random sample of size n.

2 We calculate $\bar{x}_0$ and $\sigma_{\bar{x}} = \sigma/\sqrt{n}$.

3 We select a confidence coefficient, $1 - \alpha$. Based on $1 - \alpha$, we obtain the appropriate value of z from the standard normal table. We find this value of z from Appendix Table E by first locating $(1 - \alpha)/2$ in the body of the table. The value of z to be used is the one we obtain by adding the labels of the row and column which intersect at $(1 - \alpha)/2$. For example, if $(1 - \alpha) = 0.95$, we locate $0.95/2 = 0.4750$ in the body of Table E. We note that 0.4750 appears at the intersection of the row labeled 1.9 and the column headed 0.06. Hence the appropriate value of z is 1.96.

4 We construct the interval by adding and subtracting $z(\sigma/\sqrt{n})$ as follows.

$$\bar{x}_0 \pm z\frac{\sigma}{\sqrt{n}}$$

This interval is based on the assumption that the fpc either is not applicable or may be ignored.

5 We present the interval as a confidence statement as follows.

$$C\left[\bar{x}_0 - z_{\alpha/2}\frac{\sigma}{\sqrt{n}} \leq \mu \leq \bar{x}_0 + z_{\alpha/2}\frac{\sigma}{\sqrt{n}}\right] = 1 - \alpha \tag{5.5}$$

Note that once we have selected $1 - \alpha$, we replace k of Equation 5.4 by the value of z from the standard normal distribution that has $\alpha/2$ of the area under the curve to its right (hence the subscript on z). We see that, in general, a confidence interval of the form specified in Equation 5.5 is composed of three parts: $\bar{x}_0$, $z_{\alpha/2}$, and $\sigma/\sqrt{n}$. In this equation, $\bar{x}_0$ is the estimator, $z_{\alpha/2}$ is called the *reliability factor*, and $\sigma/\sqrt{n}$ is the standard error of the estimator. We may then express a confidence interval of this type, in general terms, as

Estimator $\pm$ (reliability factor)(standard error of the estimator)

Let us illustrate with an example.

Example 5.1 A research biologist wished to estimate, with a 95% confidence interval, the mean amount of water consumed per day by a certain species of animal under the conditions of an experiment. The investigator believed that the population of daily water-consumption values was normally distributed, and from past experience, felt that the population variance was 4 grams squared. A random sample of 25 animals yielded a mean of 16.5 grams.

From the data given, the biologist was able to construct the following 95% confidence interval.

$$C\left(16.5 - 1.96\frac{2}{\sqrt{25}} \leq \mu \leq 16.5 + 1.96\frac{2}{\sqrt{25}}\right) = 0.95$$

$$C(16.5 - 0.784 \leq \mu \leq 16.5 + 0.784) = 0.95$$

$$C(15.7 \leq \mu \leq 17.3) = 0.95$$

The biologist, then, is 95% confident that the true mean amount of water consumed per day by this species of animal, under the conditions of the experiment, is somewhere between 15.7 and 17.3 grams. The researcher is able to make this statement because it is known that, in repeated sampling, 95% of the intervals constructed in a similar manner would, in the long run, contain the unknown population mean μ.

Single mean of a normally distributed population, σ^2 unknown

Example 5.1 is somewhat unrealistic in that, in most practical situations, it is not likely that the variance of a population is known while the mean is unknown. In the typical situation, both are unknown. When the population variance σ^2 is unknown, one encounters a problem in attempting to substitute numerical values into Equation 5.5 when a confidence interval for μ is desired.

In the first place, the problem of what to do about the unknown σ arises. As you might suspect, the sample standard deviation, $s = \sqrt{\Sigma(x_i - \bar{x})^2/(n-1)}$ is used to estimate σ in Equation 5.5. Thus $s/\sqrt{n}$, the estimated standard error of $\bar{x}$, replaces $\sigma/\sqrt{n}$ in the equation.

The second problem in constructing a confidence interval for μ arises from the fact that it is $(\bar{x} - \mu)/(\sigma/\sqrt{n})$, not $(\bar{x} - \mu)/(s/\sqrt{n})$, that is distributed as the standard normal variable z. Therefore we cannot determine a z value accurately when we use s to estimate σ. We learned in Chapter 4, however, that, when certain assumptions are met, $(\bar{x} - \mu)/(s/\sqrt{n})$ follows a distribution known as Student's t distribution, with $n - 1$ degrees of freedom. Instead of z, then, we use a value from the t distribution in constructing the confidence statement of Equation 5.5. Thus, when sampling is from a normally distributed population with unknown variance, the $100(1 - \alpha)\%$ confidence interval for μ is given by

$$C\left(\bar{x}_0 - \frac{t_{1-\alpha/2,n-1}s_0}{\sqrt{n}} \leq \mu \leq \bar{x}_0 + \frac{t_{1-\alpha/2,n-1}s_0}{\sqrt{n}}\right) = 1 - \alpha \tag{5.6}$$

where s_0 is the standard deviation of the particular sample drawn. In a given situation, the appropriate value of t is located in Appendix Table F at the intersection of the row corresponding to $n - 1$ and the column labeled $t_{1-\alpha/2}$. For example, if the sample size is 11 and a 95% confidence interval is desired, the appropriate value of t is 2.2281, which is located at the intersection of the row labeled 10 and the column labeled $t_{0.975}$.

Now let us illustrate the construction of a confidence interval for μ when sampling is from a normally distributed population with unknown variance.

Example 5.2 A research psychologist who wants to estimate the mean response time of young adults to a certain sound selects a simple random sample of 25 healthy college sophomores to participate in the experiment. The mean response time for the sample is 160 msec (milliseconds), with a standard deviation of 5 msec. On the assumption that response times in all similar subjects are normally distributed, the researcher constructs the following 99% confidence interval by making proper numerical substitutions into Equation 5.6.

$$C\left(160 - 2.7969\frac{5}{\sqrt{25}} \leq \mu \leq 160 + 2.7969\frac{5}{\sqrt{25}}\right) = 0.99$$

$$C(160 - 2.7969 \leq \mu \leq 160 + 2.7969) = 0.99$$

$$C(157 \leq \mu \leq 163) = 0.99$$

The psychologist is 99% confident that the true mean response time for all subjects similar to those used in the experiment is between 157 and 163.

As noted in Chapter 4, the statistic

$$\frac{\bar{x} - \mu}{s/\sqrt{n}}$$

is approximately distributed as the standard normal (z) distribution when n is large, even though s^2 is used to estimate the population variance. In many applications of statistical methodology, therefore, researchers prefer to use z instead of t as the reliability factor when constructing confidence intervals for μ when they have a large sample. The fact that s^2 is computed from a large sample makes its use as a substitute for σ^2 defensible.

Single mean of a nonnormally distributed population

Frequently we wish to estimate parameters of populations that are so nonnormal that the procedures discussed so far in this section are not generally applicable. When sampling is from nonnormal populations, we may distinguish between two cases: the case in which the sample is small and the case in which the sample is large. When a small sample is drawn from a nonnormally distributed population, one cannot construct a meaningful confidence interval for μ by Equation 5.5 even though one knows the population variance σ^2. The reason is that the sampling distribution of $\bar{x}$ is not normal when based on small samples drawn from nonnormally distributed populations.

In fact, the sampling distribution of $\bar{x}$ is not truly normal for large samples drawn from nonnormal populations. We learned in Chapter 4, however, that the central-limit theorem assures us that when sample means are computed from large samples, the quantity $z = (\bar{x} - \mu)/(\sigma/\sqrt{n})$ is approximately normally distributed with mean 0 and variance 1, regardless of the form of the sampled population. Chapter 4 also pointed out that, in most cases, a sample of size $n = 30$ is large enough to warrant the application of the central-limit theorem. When large samples are drawn from nonnormally distributed infinite populations with known variances (or when sampling is with replacement), then one may use Equation 5.5 to construct confidence intervals for μ.

If sampling is from a nonnormally distributed finite population, and if, as is generally the case, sampling is without replacement, the fpc is appropriate, so a $100(1 - \alpha)\%$ confidence interval for μ, when n is large, is given by

$$C\left(\bar{x}_0 - z_{\alpha/2}\frac{\sigma}{\sqrt{n}}\sqrt{\frac{N-n}{N-1}} \leq \mu \leq \bar{x}_0 + z_{\alpha/2}\frac{\sigma}{\sqrt{n}}\sqrt{\frac{N-n}{N-1}}\right) = 1 - \alpha \qquad (5.7)$$

Known population variances are just as rare among nonnormal populations as among normal populations, so Equation 5.7 is rarely used. When the population variance σ^2 is unknown, it is estimated by $s^2 = \Sigma(x_i - \bar{x})^2/(n-1)$. Then the large-sample formula for the confidence interval for μ, when sampling is from finite populations, becomes

$$C\left(\bar{x}_0 - z_{\alpha/2}\frac{s_0}{\sqrt{n}}\sqrt{\frac{N-n}{N-1}} \leq \mu \leq \bar{x}_0 + z_{\alpha/2}\frac{s_0}{\sqrt{n}}\sqrt{\frac{N-n}{N-1}}\right) = 1 - \alpha \qquad (5.8)$$

When the functional form of a population, as well as its variance, is not known, and n is large, the investigator constructing a confidence interval for μ must decide whether to use Equation 5.6 or Equation 5.8. If the investigator is willing to assume that the population is at least approximately normal, Equation 5.6 may be used. If the investigator is unwilling to make that assumption, Equation 5.8 is used.

When the sample constitutes only a small proportion of the population, the fpc may be ignored. Usually, if n/N is less than or equal to 0.05, n is considered to constitute a small proportion of N.

Example 5.3 A counselor with a state department of corrections wishes to estimate, for the 5500 persons admitted to correctional institutions in the state during a certain year, the mean score made on an aptitude test. A simple random sample of 250 admissions yields a mean of 65 and a standard deviation of 15. Since the form of the population is not known, and since the population is finite, we use Equation 5.8 to construct a confidence interval for the unknown population mean μ. Let us assume that a 95% confidence interval is satisfactory. By Equation 5.8, we have

$$C\left(65 - 1.96\frac{15}{\sqrt{250}}\sqrt{\frac{5500-250}{5500-1}} \leq \mu \leq 65 + 1.96\frac{15}{\sqrt{250}}\sqrt{\frac{5500-250}{5500-1}}\right) = 0.95$$

$$C[65 - (1.96)(0.9487)(0.9547) \leq \mu \leq 65 + (1.96)(0.9487)(0.9547)] = 0.95$$

$$C(65 - 2 \leq \mu \leq 65 + 2) = 0.95$$

$$C(63 \leq \mu \leq 67) = 0.95$$

The interpretation of this interval is the same as that for those constructed earlier. The counselor is 95% confident that the single interval constructed includes the population mean, since, in repeated sampling, about 95% of the intervals constructed in the same manner will include the population mean.

Sample size for estimating population means

In any investigation that has statistical inference as one of its objectives, the question of how large a sample to take arises early in the planning stage. It is important in any investigation that a sample of the proper size be drawn. If too large a sample is taken, money and other resources are wasted. On the other hand, a sample that is too small yields useless results.

When estimation of the population mean is the inferential goal of an investigation, we know that, once the sample has been taken and the data have been made available for analysis, a confidence interval will be constructed by the general formula

Estimator $\pm$ (reliability factor)(standard error of the estimator)

Let us focus on the case in which the population to be sampled is normally distributed. The standard error will be equal to $\sigma/\sqrt{n}$, and if we look at the quantity to be added to and subtracted from our estimator, we see that we have $z(\sigma/\sqrt{n})$. Since this quantity contains n, the size of the sample, perhaps we can use it in some way to find the value of n needed in a given situation.

We note that $z(\sigma/\sqrt{n})$ is equal to one-half the width of the confidence interval. If, prior to the drawing of a sample, we can specify the width of the confidence interval we would ultimately like to construct, we will have specified the desired magnitude of $z(\sigma/\sqrt{n})$. In specifying the desired width of the confidence interval we seek, we are specifying how close to the true mean we would like our estimate to be. Suppose, for example, that we wish to estimate the mean IQ of some population. If we specify that we would like an interval that is ten units wide, we are saying that we want $z(\sigma/\sqrt{n})$ to be equal to five units. This is another way of saying that we would like our estimate to be within five units of the true mean.

Suppose that we are able to specify how close we want our estimate to be to the true mean. Suppose, in addition, that we know the population variance and can specify the desired level of confidence. Then we can set up the following equation and solve it for n to determine the necessary sample size.

$$d = z\frac{\sigma}{\sqrt{n}} \tag{5.9}$$

In Equation 5.9, d is one-half the width of the desired confidence interval (or how close we want our estimate to be to the true mean), z is a value from the standard normal table corresponding to the desired level of confidence, and σ is the standard deviation of the population to be sampled. Solving this equation for n will give the sample size needed to estimate, with $100(1-\alpha)\%$ confidence, the mean of a population with variance σ^2. The width of the confidence interval will be $2d$ units.

Solving Equation 5.9 for n gives

$$n = \frac{z^2\sigma^2}{d^2} \tag{5.10}$$

In most situations an investigator will be able to readily specify the level of confidence he or she wants to attach to a confidence interval and how close he or she would like an estimate to be to the true population mean. The specification of σ^2 will generally present more of a problem, since it is unlikely that the population variance will be known. In most cases σ^2 will have to be estimated. Possible sources of estimates include the following.

1 *A pilot sample* A researcher may draw a small pilot sample from the population of interest in order to compute an estimate of σ^2.

2 *Previous studies* Studies involving similar subjects and objectives may have been carried out in the past. If so, it may be possible to use sample variances from these studies to estimate σ^2.

3 *Similar studies* A review of the literature may reveal that studies similar to the one proposed have been carried out by other investigators. Such studies may provide satisfactory estimates of σ^2.

Note the effect on n of varying σ, z, and d, one at the time, while holding the others constant. The greater the population variance, the greater the sample size for fixed z and d. In other words, when sampling is from highly variable populations, larger samples are needed. An investigator who wants to have a great deal of confidence in an estimate must pay the price of an increased sample size. Finally, narrow confidence intervals (small values of d) require large samples.

Let us illustrate the use of Equation 5.10 to determine sample size.

Example 5.4 A researcher with a physical education department wishes to estimate the oxygen consumption (in liters per minute) of normal male students between the ages of 17 and 21 following a particular type of exercise. The researcher wants to be within 0.10 liter of the true mean with 95% confidence. Previous studies indicate that the variance of the oxygen consumption of this type of subject, under the specified conditions, is about 0.09 liter per minute squared. What size sample will this investigator need?

From the information given, we have $d = 0.10$, $z = 1.96$, and $\sigma = \sqrt{0.09} = 0.3$. Substituting these values into Equation 5.10 gives

$$n = \frac{(1.96)^2(0.3)^2}{(0.10)^2} = 34.57$$

Since the sample size must be an integer, the investigator will need to take a sample of size 35 to achieve the desired confidence and interval width.

If a sample is to be drawn from a finite population, it may be desirable to incorporate the fpc into the formula for n. When this is the case, the formula becomes

$$n = \frac{Nz^2\sigma^2}{z^2\sigma^2 + d^2(N-1)} \tag{5.11}$$

Example 5.5 A researcher at a community college with an enrollment of 2500 students wishes to estimate the mean travel time between school and the students' homes. The investigator wants a 99% confidence interval, and an estimate that will be within 1 minute of the true mean. A small pilot sample yields a variance of 25 minutes squared. What size sample should the researcher draw?

If we substitute the given information into Equation 5.11, we have

$$n = \frac{(2500)(2.58)^2(25)}{(2.58)^2(25) + (1)^2(2500-1)} = 156.08$$

A sample of size 157 should be adequate.

EXERCISES

1 A sample survey was conducted in a certain section of a large metropolitan area to determine the mean family income for the 3000 households in the area. A simple random sample of 200 households yielded a mean of \$12,500 and a standard deviation of \$3000. Construct a 95% confidence interval for μ.

2 Suppose that the investigator in Exercise 1 wants to estimate the mean to within \$300 with 99% confidence. A sample standard deviation of \$3000 obtained in a previous survey is to be used as an estimate of the population standard deviation. What size sample should be drawn?

3 What size sample would be required to estimate the mean of the population described in Exercise 1 if the investigator wished to be within \$300 of the true mean with 90% confidence? Use \$3000 as an estimate of σ.

4 As part of an experiment conducted in a physical education department, the mean muscular endurance score of a random sample of 16 subjects was found to be 145, with a standard deviation of 40. Assume that muscular endurance scores for all similar subjects are normally distributed, and construct the 90% confidence interval for the true mean score.

5 What size sample would be required to estimate the mean of the population described in Exercise 4 if the investigator wanted to be within 5 of the true mean score with 95% confidence? Use 40 as an estimate of σ.

6 Refer to Exercise 4. What size sample would be needed to estimate the population mean to within 10 with 95% confidence? Let $\sigma = 40$.

7 During a student survey, a random sample of 250 high school freshmen were asked to record the average amount of time per day spent studying. The sample yielded a mean of 45 minutes, with a standard deviation of 20 minutes. Construct the 95% confidence interval for the population mean.

8 A researcher selected a simple random sample of 300 from the 10,000 holders of a certain \$100 deductible medical insurance policy who hadn't filed a claim during the past year. Those individuals in the sample were asked to report the total amount spent for medical expenses during the past year. The mean and standard deviation were \$65 and \$12, respectively. Construct the 99% confidence interval for the population mean.

9 As part of an experiment, a large manufacturing firm found that the mean time required for 16 randomly selected assembly-line employees to complete a certain task was 26 minutes. The standard deviation was 5 minutes. Construct the 90% confidence interval for μ. What assumptions are necessary in order for a valid confidence interval to be constructed?

10 The mean weight of the adrenal glands of a sample of 25 rats was 46 mg. The standard deviation computed from the sample was 2.5. Construct the 95% confidence interval for μ. Assume that weights of rat adrenal glands are approximately normally distributed and that the present sample constitutes a simple random sample from that population.

11 A random sample of 16 flowers of a certain species had a mean

diameter of 20 mm (millimeters). Assume that this sample constitutes a simple random sample from all such flowers and that the diameter measurements of these flowers are normally distributed. Construct the 95% confidence interval for the population mean. ($s^2 = 16$.)

12 A random sample of 100 high school seniors made a mean score of 75 on a certain standardized test. The sample variance was 196. Construct the 99% confidence interval for the population mean.

13 Given a population of size 5000 with a variance of 256, construct the 95% confidence interval for μ from the following sample data.

$$n = 64, \qquad \bar{x} = 90$$

14 Given the following information, construct the 90% confidence interval for the population mean.

$$N = 25{,}000, \qquad n = 400, \qquad \bar{x} = 100, \qquad s^2 = 900$$

15 Given the following information, construct the 95% confidence interval for the population mean.

$$N = 500, \qquad n = 49, \qquad \bar{x} = 50, \qquad s^2 = 196$$

16 In an experiment designed to assess the effectiveness of a certain method of teaching arithmetic, 25 seventh-grade students selected at random from among the seventh-grade students of a certain school were taught arithmetic by the experimental method for one semester. At the end of the semester the 25 students were given an achievement test. The mean and standard deviation were 85 and 10, respectively. Assume that the 25 achievement scores constitute a simple random sample from a normally distributed population of achievement scores. Construct the 95% confidence interval for the population mean.

5.4 CONFIDENCE INTERVAL FOR THE DIFFERENCE BETWEEN TWO POPULATION MEANS

In many practical situations it is of interest to obtain a confidence interval for the difference between two population means. Two methods of teaching some subject may be tried, with each method being used with a different sample of students. As part of the evaluation procedure, it may be helpful to construct a confidence interval for the difference between population means. In studying the differences between two groups, we may find useful a confidence interval for the difference between the mean IQs of the two groups. One could list numerous situations and variables in which such confidence intervals would be of interest. In all cases in which confidence intervals for the difference between two population means are constructed, it is assumed that the samples on which the confidence intervals are based are independently and randomly drawn from the populations of interest. In Chapter 4 we discussed the appropriate sampling distributions for constructing confidence intervals for the difference between population means in a variety of situations.

We shall discuss these situations separately in this section, and give examples to illustrate methods of constructing confidence intervals for each.

Differences between means of two normally distributed populations, σ_1^2 and σ_2^2 known

In Chapter 4 we learned that when a simple random sample is drawn independently from each of two normally distributed populations with known variances, the sampling distribution of $\bar{x}_1 - \bar{x}_2$ is normal, with mean $\mu_1 - \mu_2$ and variance $(\sigma_1^2/n_1) + (\sigma_2^2/n_2)$. When this situation prevails, a $100(1 - \alpha)\%$ confidence interval for the difference between the two population means is given by

$$C\left[(\bar{x}_1 - \bar{x}_2) - z_{\alpha/2}\sqrt{\frac{\sigma_1^2}{n_1} + \frac{\sigma_2^2}{n_2}} \leq (\mu_1 - \mu_2) \leq (\bar{x}_1 - \bar{x}_2) + z_{\alpha/2}\sqrt{\frac{\sigma_1^2}{n_1} + \frac{\sigma_2^2}{n_2}}\right] = (1 - \alpha) \tag{5.12}$$

where $\bar{x}_1 - \bar{x}_2$ is an observed difference between two sample means. Understand that, in Equation 5.12, $\bar{x}_1$ and $\bar{x}_2$ are specific numerical values. The use of the subscript 0 to indicate this (as used previously) has been dropped here in an effort to simplify the notation. We shall follow this practice throughout our discussion of confidence intervals for the difference between two means.

Example 5.6 A study was made of the differences between first-grade students who had attended kindergarten and those who had not. Researchers examined a simple random sample of 50 first-grade students who had attended kindergarten and an independent simple random sample of 60 first-grade students who had not. At the end of the first semester of first grade, the students were given a reading achievement test. Those who had attended kindergarten had a mean score of 4.50. The mean score for the group that had not attended kindergarten was 3.75. The investigators wished to construct a 95% confidence interval for the difference between population means. Based on previous experience, they were willing to assume that the two populations were approximately normally distributed, with variances of 1.8 for the kindergarten population and 2.1 for the nonkindergarten population.

Substituting the given information into Equation 5.12, we obtain the following 95% confidence interval.

$$C\left[(4.50 - 3.75) - 1.96\sqrt{\frac{1.8}{50} + \frac{2.1}{60}} \leq (\mu_1 - \mu_2) \leq (4.50 - 3.75) + 1.96\sqrt{\frac{1.8}{50} + \frac{2.1}{60}}\right] = 0.95$$

$$C[0.75 - 0.52 \leq (\mu_1 - \mu_2) \leq 0.75 + 0.52] = 0.95$$

$$C[0.23 \leq (\mu_1 - \mu_2) \leq 1.27] = 0.95$$

We are 95% confident that this interval includes the true parameter $\mu_1 - \mu_2$, since, in repeated sampling, we would expect, in the long run, 95% of the intervals constructed in a similar manner to include $\mu_1 - \mu_2$.

Differences between means of two normally distributed populations, σ_1^2 and σ_2^2 unknown but equal

In the typical practical situation in which one wishes to estimate the difference between two population means, the variances, as well as the means, of the populations of interest will be unknown. When this is the case, one must estimate the population variances from sample data. Two situations may be distinguished: the case in which the two population variances are equal and the case in which they are not equal. We shall discuss only the procedure for constructing confidence intervals for the case in which the population variances are known to be (or can be assumed to be) equal. When one knows that the variances of two normally distributed populations are equal, or when one is willing to make that assumption, a $100(1-\alpha)\%$ confidence interval for $\mu_1 - \mu_2$, the difference between the population means, is given by

$$C\left[(\bar{x}_1 - \bar{x}_2) - t_{1-\alpha/2, n_1+n_2-2}\sqrt{\frac{s_p^2}{n_1} + \frac{s_p^2}{n_2}} \leq (\mu_1 - \mu_2) \leq (\bar{x}_1 - \bar{x}_2) + t_{1-\alpha/2, n_1+n_2-2}\sqrt{\frac{s_p^2}{n_1} + \frac{s_p^2}{n_2}}\right] = (1-\alpha) \tag{5.13}$$

In Equation 5.13, t is the value from Student's t distribution corresponding to the degrees of freedom and the desired value of $1-\alpha$; $\bar{x}_1$ and $\bar{x}_2$ are the means of independent random samples of size n_1 and n_2, respectively, drawn from the two populations; and s_p^2 is the pooled estimate of the common population variance described in Chapter 4. One obtains the appropriate degrees of freedom for determining what value of t to use in the confidence interval by adding the weights used to compute s_p^2; that is, the degrees of freedom are equal to $(n_1 - 1) + (n_2 - 1)$.

Example 5.7 A biologist wished to study the effects of certain drugs on the water consumption of a particular species of laboratory animals. Drug A, containing a thirst-inducing agent, was administered to a simple random sample of $n_A = 25$ animals. Drug B, which did not contain the thirst-inducing agent, was administered to an independent simple random sample of $n_B = 22$ similar animals. The biologist kept a record of the amount of water consumed by each animal during a specified period following administration of the drugs. The mean amounts of water consumed per animal by the two groups were $\bar{x}_A =$

50 ml and $\bar{x}_B = 25$ ml, and the standard deviations were $s_A = 5.3$ and $s_B = 5.6$. The biologist, who was willing to assume that the two samples of responses available were equivalent to independent simple random samples from normally distributed populations with equal variances, wished to construct a 95% confidence interval for $\mu_A - \mu_B$.

The first step in the construction of the desired interval is to compute s_p^2 as follows.

$$s_p^2 = \frac{(25-1)(5.3)^2 + (22-1)(5.6)^2}{(25-1)+(22-1)} = 29.616$$

By Equation 5.13, the desired confidence interval is

$$C\left[(50-25) - 2.0141\sqrt{\frac{29.616}{25} + \frac{29.616}{22}} \le (\mu_A - \mu_B) \le (50-25) + 2.0141\sqrt{\frac{29.616}{25} + \frac{29.616}{22}}\right] = 0.95$$

$$C[25 - (2.0141)(1.59) \le (\mu_A - \mu_B) \le 25 + (2.0141)(1.59)] = 0.95$$

$$C[22 \le (\mu_A - \mu_B) \le 28] = 0.95$$

We are 95% confident that the true difference in population means is between 22 and 28, since, in the long run in repeated sampling, we would expect 95% of the intervals constructed in this manner to include $\mu_A - \mu_B$.

Appropriate procedures to follow when the population variances are known to be unequal or when the assumption that they are equal is not justified are given in the books by Snedecor and Cochran (1) and Dixon and Massey (2).

Differences between means of two nonnormally distributed populations

Frequently, it is unrealistic for an investigator who wishes to make inferences about the difference between two population means, to assume, in the absence of certain knowledge, that the populations are even approximately normally distributed. In other situations the investigator knows for certain that the populations of interest are not normally distributed.

In such situations the investigator may draw from the two populations independent simple random samples of sufficient size for the central-limit theorem to apply. If the population variances are known, a $100(1-\alpha)\%$ confidence interval for the difference between population means is constructed by

$$C\left[(\bar{x}_1 - \bar{x}_2) - z_{\alpha/2}\sqrt{\frac{\sigma_1^2}{n_1} + \frac{\sigma_2^2}{n_2}} \le (\mu_1 - \mu_2) \le (\bar{x}_1 - \bar{x}_2) + z_{\alpha/2}\sqrt{\frac{\sigma_1^2}{n_1} + \frac{\sigma_2^2}{n_2}}\right] = 1 - \alpha \tag{5.14}$$

In the typical situation, however, the population variances are unknown. When this is the case, one uses the sample variances s_1^2 and s_2^2 as estimates of σ_1^2 and σ_2^2, respectively, and constructs the desired confidence interval by replacing the latter with the former in Equation 5.14. Samples large enough to justify use of the central-limit theorem are large enough to give sample variances that are good approximations of the population variances. When sample variances are used to estimate population variances, the $100(1-\alpha)\%$ confidence interval for $\mu_1 - \mu_2$ is given by

$$C\left[(\bar{x}_1 - \bar{x}_2) - z_{\alpha/2}\sqrt{\frac{s_1^2}{n_1} + \frac{s_2^2}{n_2}} \le (\mu_1 - \mu_2) \le (\bar{x}_1 - \bar{x}_2) + z_{\alpha/2}\sqrt{\frac{s_1^2}{n_1} + \frac{s_2^2}{n_2}}\right] = 1 - \alpha \tag{5.15}$$

Example 5.8 In a survey conducted to compare certain characteristics of two types of elementary schools, one item included in the survey was the amount of playground area per pupil. For the 100 type A schools, the mean and standard deviation were 240 and 30 square feet, respectively. The 110 type B schools in the survey reported a mean of 175 square feet, with a standard deviation of 25 square feet. The investigator wished to construct a 90% confidence interval for the difference between population means.

Since the population variances and the functional forms of the populations are unknown, and since the sample sizes are large, the desired confidence interval is given by Equation 5.15. Substituting sample data into the equation gives

$$C\left[(240 - 175) - 1.645\sqrt{\frac{(30)^2}{100} + \frac{(25)^2}{110}} \le (\mu_A - \mu_B) \le (240 - 175) + 1.645\sqrt{\frac{(30)^2}{100} + \frac{(25)^2}{110}}\right] = 0.90$$

$$C[65 - (1.645)(3.8317) \le (\mu_A - \mu_B) \le 65 + (1.645)(3.8317)] = 0.90$$

$$C[59 \le (\mu_A - \mu_B) \le 71] = 0.90$$

EXERCISES

17 In a sample survey of a certain population, 125 adult females with blood type A had a mean weight of 61 kg, with a standard deviation of 10 kg. The 130 adult females with blood type B had a mean weight of 56 kg, with a standard deviation of 7 kg. Construct the 95% confidence interval for $\mu_A - \mu_B$.

18 The following are the means and standard deviations of the nonverbal IQ scores obtained from independent simple random samples drawn from two groups of elementary school children.

Group	n	$\bar{x}$	s
I	20	110	10
II	25	95	15

Assume that the populations of nonverbal IQ scores are each approximately normally distributed with equal variances. Construct the 95% confidence interval for the difference between the two population means.

19 The verbal IQ scores for the students in the study referred to in Exercise 18 yielded the following means and standard deviations.

Group	n	$\bar{x}$	s
I	20	100	12
II	25	90	14

Assume that verbal IQ scores are also approximately normally distributed with equal variances. Construct the 95% confidence interval for the difference between the two population means.

20 A study was made of the differences in attention span in two groups of young children. While each child chosen in independent simple random samples from the two groups watched a 30-minute television program, a research team recorded the number of minutes of eye contact. The team obtained the following data.

Group	n	$\bar{x}$	s
A	15	23.5 min	5 min
B	20	18.1 min	3 min

Construct the 99% confidence interval for $\mu_A - \mu_B$. What assumptions are necessary?

21 Students registering for a course in educational research were randomly assigned to one of two classes. Class A used numerous techniques and activities designed to enrich the course. Class B was taught by the traditional lecture method. Achievement test scores given at the completion of the course yielded the following results.

Class	n	$\bar{x}$	s
A	10	80	8
B	12	72	10

Construct the 90% confidence interval for $\mu_A - \mu_B$. What assumptions must be made?

22 In a study of creativity, a research team drew two independent simple random samples from among college freshmen. Sample 1 was drawn from students who had attended one type of high school, and sample 2 was drawn from students who had attended another type of high school. Creativity tests administered to the students in the two samples gave the following results.

Sample	n	$\bar{x}$	s
1	75	200	25
2	90	160	30

Construct a 95% confidence interval for $\mu_1 - \mu_2$.

23 Students entering the seventh grade of a certain elementary school were randomly assigned to one of two classes. Class A used a computer-assisted instruction technique in teaching arithmetic. Arithmetic instruction in class B followed traditional patterns. Arithmetic achievement tests were administered at the end of the year, with the following results.

Class	n	$\bar{x}$	s
A	35	85	10
B	32	71	15

Construct a 95% confidence interval for $\mu_A - \mu_B$.

24 An agricultural research team carried out a study to assess the effect of a new fertilizer on a certain type of bean. A random sample of 100 mature beans grown in soil to which the new fertilizer had been added had a mean weight of 1.32 grams, with a standard deviation of 0.18 gram. An independent random sample of 100 beans grown in soil to which the standard fertilizer had not been added had a mean and standard deviation, respectively, of 1.10 grams and 0.20 gram. Construct a 95% confidence interval for the difference in population means.

5.5 PAIRED COMPARISONS

Frequently, the data available for analysis are obtained from two samples (one from each of two populations) that are not independent. Usually such data are the result of deliberate attempts to make the inferential procedures more meaningful.

A commonly used procedure that results in two non-independent samples is the so-called before-after test. Measurements are taken on a sample of subjects both prior and subsequent to the introduction of some phenomenon. The blood pressures of high school athletes may be taken before and after a period of exercise. Or elementary school students may take a reading comprehension test just before entering an experimental reading course, and may receive the same test at the conclusion of it. In such situations, interest centers on the magnitude of the change indicated by the two measurements.

Paired observations also may arise when an entity is somehow divided in such a way that half experiences one condition and the other half experiences another. In testing the effectiveness of a detergent, an investigator may divide soiled fabric specimens in half and randomly assign each half to one of two washing machines. One machine may contain the experimental detergent, and the other a competing detergent or no detergent at all. Two suntan lotions may be compared by applying one to the left shoulders and the other to the right shoulders of several individuals. The two sets of observations that result from procedures such as these are certainly related, since each entity has generated an observation for each of the two groups.

Subjects who share a common hereditary or environmental experience may be used to generate paired observations. One member of each of several pairs of individuals, where each pair shares a common experience, may be designated to receive some type of treatment, while the other member receives another type of treatment or acts as a control. One twin, for example, may take an experimental mathematics course, while the other twin attends the regular class. In the laboratory, pairs of experimental animals from several litters may be used in such a way that one member of each pair experiences an experimental condition, while the other experiences a different condition or acts as a control. In psychological or sociological studies, wives' responses to questionnaires may be paired with those of their husbands to obtain data for husband-wife comparisons.

The methods discussed in the preceding section are not applicable when the two samples generating the data are related, since those methods depend on the condition that the two samples be independent. In Section 5.4, interest was focused on estimating $\mu_1 - \mu_2$, the difference between two population means, and a point estimator of this parameter is provided by $\bar{x}_1 - \bar{x}_2$, where $\bar{x}_1$ and $\bar{x}_2$ are the means of two independent simple random samples. The construction of confidence intervals was also explained in Section 5.4. We accomplished this by adding to, and subtracting from, the point estimate some multiple of the standard error of the estimate. When the observations come in pairs, $\bar{x}_1 - \bar{x}_2$ still provides a point estimate of $\mu_1 - \mu_2$. However, when we come to the construction of confidence intervals, a complication arises because when the samples are related, the standard error of $\bar{x}_1 - \bar{x}_2$ is not the same as it is when the samples are independent. We avoid having to deal with this problem by taking the difference between the measurements taken on each pair and concentrating on the *mean of the differences* rather than on the *difference of the means*, which was our concern in the preceding section.

Suppose, for some random variable X, we have a sample of n pairs of measurements $(x_{11}, x_{12}), (x_{21}, x_{22}), (x_{31}, x_{32}), \ldots, (x_{n1}, x_{n2})$. We then obtain n differences by taking the difference d between each pair. For example, $d_1 = x_{11} - x_{12}$, $d_2 = x_{21} - x_{22}$, $d_3 = x_{31} - x_{32}, \ldots, d_i = x_{i1} - x_{i2}, \ldots, d_n = x_{n1} - x_{n2}$. The mean difference for the sample of n differences is given by

$$\bar{d} = \frac{\Sigma\, d_i}{n} \tag{5.16}$$

The sample variance is given by

$$s_{\bar{d}}^2 = \frac{\Sigma\,(d_i - \bar{d})^2}{n - 1} = \frac{n\,\Sigma\, d_i^2 - (\Sigma\, d_i)^2}{n(n-1)} \tag{5.17}$$

If our n differences are a simple random sample of differences from a population of differences with mean μ_d, then $\bar{d}$ provides us with a point estimate of μ_d. When the population of differences is normally distributed with mean μ_d and variance $\sigma_{\bar{d}}^2$, we may construct a $100(1 - \alpha)\%$ confidence interval for μ_d as follows.

$$C\left(\bar{d}_0 - z_{\alpha/2}\frac{\sigma_d}{\sqrt{n}} \leq \mu_d \leq \bar{d}_0 + z_{\alpha/2}\frac{\sigma_d}{\sqrt{n}}\right) = 1 - \alpha \qquad (5.18)$$

where $\bar{d}_0$ is a specific numerical value computed from a sample of differences. We see that Equation 5.18 is analogous to the equation for the confidence interval for a population mean given in Section 5.3. In fact, all the ideas relating to estimating a single mean presented in Section 5.3 are applicable in the present case. In other words, our single $\bar{d}$ is one of many $\bar{d}$'s that repeated sampling would generate, and the distribution of $\bar{d}$ has mean μ_d and variance σ_d^2/n. Furthermore, if the population of d's is normally distributed, the distribution of $\bar{d}$'s will also be normally distributed.

When, as is usually the case, σ_d^2 is unknown, it is estimated by the sample variance s_d^2. If the d's are normally distributed, the $100(1 - \alpha)\%$ confidence interval is given by

$$C\left(\bar{d}_0 - t_{1-\alpha/2,n-1}\frac{s_{d_0}}{\sqrt{n}} \leq \mu_d \leq \bar{d}_0 + t_{1-\alpha/2,n-1}\frac{s_{d_0}}{\sqrt{n}}\right) = 1 - \alpha \qquad (5.19)$$

When n, the number of paired observations, is large, one may use the central-limit theorem. And if the population variance σ_d^2 is known, one uses Equation 5.18 to construct the $100(1 - \alpha)\%$ confidence interval for μ_d, regardless of the functional form of the distribution of d's. If σ_d^2 is not known, it is estimated by s_d^2.

Example 5.9 Children enrolling in kindergarten at a certain school were paired on the basis of careful matching on such criteria as intelligence, chronological age, socioeconomic status of parents, and health status. One member of each pair (selected at random) was assigned to a kindergarten class in which there were three teacher aides. The other member of each pair was assigned to the regular kindergarten class, which did not have teacher aides. At the end of the year, each child was given a reading readiness test, with the results shown in Table 5.1.

The researcher, who was willing to assume that the population of differences was normally distributed, wished to construct a 95% confidence interval for μ_d, the mean difference in reading readiness scores.

From the sample data, we use Equation 5.16 to compute

$$\bar{d} = \frac{9 + 9 + \cdots + 3}{25} = \frac{89}{25} = 3.56$$

By Equation 5.17, we have

$$s_d^2 = \frac{25(941) - (89)^2}{25(24)} = 26.0067$$

and $s_d = \sqrt{26.0067} = 5.10$.

Table 5.1 Scores made on reading readiness test by 25 pairs of kindergarten children

Reading readiness scores of children											
Pair number	1	2	3	4	5	6	7	8	9	10	11
WITH TEACHER AIDES	31	33	30	25	36	27	39	38	36	24	29
WITHOUT TEACHER AIDES	22	24	28	32	29	21	32	27	33	25	22
$\mathbf{d}_i$	9	9	2	−7	7	6	7	11	3	−1	7
Pair number	12	13	14	15	16	17	18	19	20	21	22
WITH TEACHER AIDES	26	28	25	30	35	36	34	32	31	26	30
WITHOUT TEACHER AIDES	33	33	22	28	33	24	30	27	31	23	31
$\mathbf{d}_i$	−7	−5	3	2	2	12	4	5	0	3	−1
Pair number	23	24	25								
WITH TEACHER AIDES	29	39	33								
WITHOUT TEACHER AIDES	20	33	30								
$\mathbf{d}_i$	9	6	3								

By Equation 5.19, the 95% confidence interval for μ_d is

$$C\left[3.56 - 2.0639\left(\frac{5.10}{\sqrt{25}}\right) \leq \mu_d \leq 3.56 + 2.0639\left(\frac{5.10}{\sqrt{25}}\right)\right] = 0.95$$

$$C(3.56 - 2.11 \leq \mu_d \leq 3.56 + 2.11) = 0.95$$

$$C(1.45 \leq \mu_d \leq 5.67) = 0.95$$

We are 95% confident that this interval contains μ_d, since, in repeated sampling, about 95% of the intervals constructed in this manner would contain μ_d.

EXERCISES

25 Table 5.2 gives the weights, in kilograms, of 10 pairs of 15-year-old monozygotic male twins. Construct the 99% confidence interval for μ_d. What assumptions are necessary?

26 Hemoglobin determinations were made on 15 specimens of blood by each of two laboratory technicians, A and B. Table 5.3 shows the results,

Table 5.2 Data for Exercise 25

Pair number	1	2	3	4	5	6	7	8	9	10
WEIGHT OF HEAVIER TWIN	55	64	54	40	77	41	62	54	54	50
WEIGHT OF LIGHTER TWIN	53	63	50	39	75	40	59	51	53	48

Table 5.3 Data for Exercise 26

Specimen number	Technician A	Technician B	Specimen number	Technician A	Technician B
1	15.38	15.71	9	12.47	12.74
2	17.78	17.40	10	12.95	13.78
3	16.77	16.94	11	11.28	11.65
4	16.05	16.75	12	10.65	10.08
5	17.67	16.24	13	10.80	10.15
6	13.16	13.85	14	15.70	14.92
7	13.42	12.02	15	12.23	12.27
8	18.85	18.64			

Table 5.4 Data for Exercise 27

Pulse rate												
SUBJECT	1	2	3	4	5	6	7	8	9	10	11	12
BEFORE	75	61	62	68	58	70	59	79	68	80	64	75
AFTER	82	70	74	80	65	80	70	88	77	90	75	87

in grams per 100 cc (cubic centimeters) of blood. Construct the 95% confidence interval for μ_d.

27 Table 5.4 gives the pulse rates, in beats per minute, recorded for 12 subjects before and after they had smoked a uniform amount of marijuana. Construct the 95% confidence interval for the mean difference in pulse rate.

5.6 CONFIDENCE INTERVAL FOR A POPULATION PROPORTION

Frequently one wishes to estimate what proportion of subjects composing a population possess some characteristic of interest. What proportion of the children in a certain school system are underachieving? What proportion of adolescents in a certain community have exhibited some type of deliquent behavior? What proportion of households in some community are without indoor plumbing? These are only a few of countless questions regarding population proportions that one may ask. Generally, it is impractical to examine an entire population in order to determine p, the true proportion possessing the characteristic of interest. Instead, one draws a random sample from the population, and uses the sample proportion $\hat{p}$ to estimate p. Since, as we pointed out in Chapter 4, the mean of the distribution of $\hat{p}$ computed from all possible samples of size n is equal to p, the sample proportion $\hat{p}$ is an unbiased estimator of p. We also learned in Chapter 4 that the distribution of $\hat{p}$ is approximately normal when n is large and neither p nor $(1-p) = q$ is too close to 1 or 0. These conditions usually are considered to be met if np and $n(1-p)$ are both greater than 5. The variance of $\hat{p}$, which is equal to $p(1-p)/n$, is estimated by $\hat{p}(1-\hat{p})/n$ when p is unknown. It does not seem likely that in a practical situation p will ever be known, since if p is known, there is usually little incentive for obtaining its estimate.

When sampling is from an infinite population, a $100(1-\alpha)\%$ confidence interval for p is given by

$$C\left[\hat{p}_0 - z_{\alpha/2}\sqrt{\frac{\hat{p}(1-\hat{p})}{n}} \le p \le \hat{p}_0 + z_{\alpha/2}\sqrt{\frac{\hat{p}(1-\hat{p})}{n}}\right] = 1-\alpha \tag{5.20}$$

where $\hat{p}_0$ is a specific numerical value of $\hat{p}$ computed from a sample.

When sampling is without replacement (the usual case) from a finite population, use of the fpc is appropriate, and the $100(1-\alpha)\%$ confidence interval for p is given by

$$C\left[\hat{p}_0 - z_{\alpha/2}\sqrt{\frac{\hat{p}(1-\hat{p})}{n}}\sqrt{\frac{N-n}{N-1}} \le p \le \hat{p}_0 + z_{\alpha/2}\sqrt{\frac{\hat{p}(1-\hat{p})}{n}}\sqrt{\frac{N-n}{N-1}}\right]$$
$$= 1-\alpha \tag{5.21}$$

When $n/N \le 0.05$, the fpc is usually sufficiently close to 1 to justify its omission, so that Equation 5.20 may be used to obtain a confidence interval for p. Both equations assume that p is unknown.

Example 5.10 In a study of the reasons for high school dropouts, an investigator drew a sample of 200 students from a population of 1500 students who had dropped out of high school. Of the 200 dropouts interviewed, 140 said that they had dropped out because of family financial difficulties. The investigator therefore wished to construct an interval estimate of the true proportion who dropped out for this reason. A 95% confidence coefficient was selected.

The point estimate of p is $\hat{p} = 140/200 = 0.70$. Since $200/1500 = 0.13 > 0.05$, the fpc is appropriate, and we can construct the desired confidence interval by means of Equation 5.21, as follows.

$$C\left[0.70 - 1.96\sqrt{\frac{(0.70)(0.30)}{200}}\sqrt{\frac{1500-200}{1499}} \le p \le 0.70 + 1.96\sqrt{\frac{(0.70)(0.30)}{200}}\sqrt{\frac{1500-200}{1499}}\right] = 0.95$$

$$C(0.70 - 0.06 \le p \le 0.70 + 0.06) = 0.95$$

$$C(0.64 \le p \le 0.76) = 0.95$$

We are 95% confident that this interval contains p, since, in repeated sampling, about 95% of the intervals constructed in this manner from all possible samples of size 200 from the population of 1500 dropouts would include p.

Example 5.11 A survey to determine attitudes on certain social issues was conducted among the adult female residents of a particular community. From a population of about 5000 potential respondents, the researcher selected a random sample of 225 adult females for personal interviews.

One of the questions asked during the interview was, "Do you think mothers of preschool children should work outside the home?" Seventy-five of the 225 respondents answered no. The researcher wished to construct a 95% confidence interval for the true proportion in the population who feel that mothers of preschool children should not work outside the home.

The point estimate of p is $\hat{p} = 75/225 = 0.33$. Since $225/5000 = 0.045 < 0.05$, we shall ignore the fpc and use Equation 5.20 to obtain the following 95% confidence interval for p.

$$C\left[0.33 - 1.96\sqrt{\frac{(0.33)(0.67)}{225}} \leq p \leq 0.33 + 196\sqrt{\frac{(0.33)(0.67)}{225}}\right] = 0.95$$

$$C(0.33 - 0.06 \leq p \leq 0.33 + 0.06) = 0.95$$

$$C(0.27 \leq p \leq 0.39) = 0.95$$

Sample size for estimating population proportions

In determining the sample size needed to estimate a population proportion with a confidence interval, we follow essentially the procedure that was described for determining the sample size needed to estimate a population mean. We must specify how close we want our estimate to be to the true value being estimated, we must indicate what level of confidence we desire, and finally we must estimate the variability present in the population. Estimates of population variability may be obtained from other studies or from a pilot sample.

Two formulas for determining sample size are available, and the choice depends on whether or not the fpc is deemed necessary.

When sampling from an infinite population or a finite population with replacement, we shall not need the fpc, and the formula for sample size is

$$n = \frac{z^2pq}{d^2} \tag{5.22}$$

where z is the value from the standard normal distribution corresponding to the chosen level of confidence, p is the population proportion with the characteristic of interest (in practice, p is estimated), $q = 1 - p$, and d is equal to one-half the desired confidence interval (indicating how close an estimate is desired).

In practice, if the population to be sampled is quite large, so that n is sufficiently small that n/N is less than or equal to 0.05, we may use Equation 5.22 even though the population is finite and sampling is carried out without replacement.

When an estimate of p is not available from previous or similar studies, and when it is impossible or impractical to take a pilot sample, we may obtain the maximum value of n from Equation 5.22 by letting $p = 0.5$. Although the sample size obtained in this manner will be large enough, it may be too large, and hence the resulting sampling costs may be greater than necessary.

Example 5.12 A social worker with a state welfare department wished to conduct a study to determine the attitudes of citizens toward the state's welfare program. As part of the proposed interview schedule, respondents would be asked to indicate whether they agreed or disagreed with the statement, "Too many people who should be working are receiving welfare benefits." The social worker wished to know how large a sample to take in order to estimate the true proportion of citizens who agreed with the statement. The worker wished to be within 0.025 of the true value with 95% confidence. Researchers who conducted a similar study in an adjacent state found that 60% of the people interviewed agreed with the statement.

By Equation 5.22, we find the necessary sample size to be

$$n = \frac{(1.96)^2(0.6)(0.4)}{(0.025)^2} = \frac{0.921984}{0.000625} = 1475.17$$

Thus the social worker should take a sample of size 1476.

If no previous estimate of p had been available, we could have used $p = 0.5$ to yield a sample size of

$$n = \frac{(1.96)^2(0.5)(0.5)}{(0.025)^2} = \frac{0.9604}{0.000625} = 1536.64 \approx 1537$$

When sampling is without replacement from a finite population, it is always appropriate, though not necessary when $n/N \leq 0.05$, to use the fpc. When $n/N > 0.05$, the fpc can make a difference, and should be used in determining sample size in such cases. When the fpc is used, the formula for n becomes

$$n = \frac{Npqz^2}{(N-1)d^2 + z^2pq} \tag{5.23}$$

Example 5.13 A market research analyst wishes to know how large a sample of the households in a certain community should be drawn in order to determine in what proportion of the households at least one member regularly watches a certain television program. There are a total of 500 households in the community. The analyst wants to be within 0.04 of the true proportion with 90% confidence. In a pilot sample of 15 homes, 35% of the respondents indicated that someone in the home regularly watched the program.

From this information, we may use Equation 5.23 to compute

$$n = \frac{(500)(0.35)(0.65)(1.645)^2}{(499)(0.04)^2 + (1.645)^2(0.35)(0.65)} = 217.68$$

Thus a sample of size 218 is needed. Since 15 households have already been drawn in the pilot sample, only 203 additional households are needed to complete the sample.

EXERCISES

28 Of a simple random sample of 300 respondents drawn from a large population of adult males, 55% reported that their favorite spectator sport was football. Construct a 95% confidence interval for the true proportion who consider football their favorite spectator sport.

29 A random sample of 200 juveniles, from a population of 3000 juveniles committed to a training school, revealed that for 40 the offense for which they had been apprehended was car theft. Construct a 90% confidence interval for the true proportion who had been apprehended for car theft.

30 From a population of 1500 employees of a certain firm, a random sample of 150 were selected to participate in a survey. Of those in the sample, 120 stated that they were completely happy with overall working conditions. Construct a 95% confidence interval for the true proportion who share this feeling.

31 A certain community is composed of 1200 single-family dwelling units. Of a random sample of 200 dwelling units, 50 were found to be in severe need of repair. Construct a 90% confidence interval for the true proportion in severe need of repair.

32 Of a random sample of 400 selected from a large population of adults, 240 answered no in response to the question, "Do you think the police should arrest skid-row drunks for drunkenness alone?" Construct a 95% confidence interval for the true proportion who share this opinion.

33 In a survey to assess the attitudes of employees toward the monthly employee newsletter, a simple random sample of 500 employees of a large national organization were asked to indicate how often they read the newsletter. Of the 500, 375 reported that they read every issue. Construct the 95% confidence interval for the true proportion who read every issue.

34 Of a random sample of 500 individuals selected from a large population, 30 were found to have blood type AB. Construct the 95% confidence interval for the true proportion in the population with blood type AB.

35 A parasitologist wants to estimate, for a certain population of wild animals, the proportion infested with a certain intestinal parasite. What size sample should be drawn if the parasitologist wants to be within 0.05 of the true proportion with 95% confidence? No previous estimate of p is available and a pilot sample is not feasible.

36 Refer to Exercise 35. Assume that in a previous study 30% of the animals were found to be infested. Use this information to compute n.

37 Refer to Exercise 35. If the parasitologist wants to be within 0.02 of the true proportion, what is n?

38 A high school counselor wishes to estimate the proportion of the 2000 high school seniors in the school system who plan to go to college. How large a sample should the counselor take if the estimate is to be within 0.05 of the true value with 99% confidence? Last year 70% of the seniors polled said they were planning to go to college.

39 A research team wants to draw a sample of the records of juvenile arrests by a city police department in order to estimate the proportion arrested for committing offenses after midnight. How large a sample should be drawn if the research team wants to be within 0.05 of the true value with 95% confidence? A similar study several years ago yielded a proportion of 0.06.

5.7 CONFIDENCE INTERVAL FOR THE DIFFERENCE BETWEEN TWO POPULATION PROPORTIONS

The need to compare two population proportions frequently arises in practice. An educator may wish to know the magnitude of the difference between the proportion of students who go to college from two different school systems. A medical research team may be interested in knowing by how much the proportion of favorable responses to drug *A* differs from the proportion of favorable responses to drug *B* when each is given to subjects suffering from the same disease. A social worker may draw independent random samples of households in two communities in order to see whether they differ with respect to the proportion in which the head of the household is female.

A point estimate of the difference between two population proportions, $p_1 - p_2$, is provided by $\hat{p}_1 - \hat{p}_2$, the difference between proportions computed from independent random samples drawn from the two populations.

A $100(1 - \alpha)\%$ confidence interval for $p_1 - p_2$, based on the sampling distribution of $\hat{p}_1 - \hat{p}_2$ as discussed in Chapter 4, is given by

$$C\left[(\hat{p}_1 - \hat{p}_2) - z_{\alpha/2}\sqrt{\frac{p_1(1-p_1)}{n_1} + \frac{p_2(1-p_2)}{n_2}} \leq (p_1 - p_2) \leq (\hat{p}_1 - \hat{p}_2) + z_{\alpha/2}\sqrt{\frac{p_1(1-p_1)}{n_1} + \frac{p_2(1-p_2)}{n_2}}\right] = 1 - \alpha \tag{5.24}$$

where $\hat{p}_1$ and $\hat{p}_2$ are specific numerical values computed from samples. Again, in the interests of notational simplicity, we have omitted the subscript 0 to indicate this. In situations of practical interest, p_1 and p_2 will be unknown and will have to be estimated by $\hat{p}_1$ and $\hat{p}_2$. The interval estimate of $p_1 - p_2$ then becomes

$$C\left[(\hat{p}_1 - \hat{p}_2) - z_{\alpha/2}\sqrt{\frac{\hat{p}_1(1-\hat{p}_1)}{n_1} + \frac{\hat{p}_2(1-\hat{p}_2)}{n_2}} \leq (p_1 - p_2) \leq (\hat{p}_1 - \hat{p}_2) + z_{\alpha/2}\sqrt{\frac{\hat{p}_1(1-\hat{p}_1)}{n_1} + \frac{\hat{p}_2(1-\hat{p}_2)}{n_2}}\right] = 1 - \alpha \tag{5.25}$$

The use of Equation 5.24 or Equation 5.25 is valid only if n_1 and n_2 are large. Each of the quantities, $n_1 p_1$, $n_1(1-p_1)$, $n_2 p_2$, and $n_2(1-p_2)$ should be greater than 5. We estimate these quantities by using $\hat{p}_1$ and $\hat{p}_2$ in place of p_1 and p_2.

Example 5.14 In a study of the types of crimes committed by juveniles admitted to two state correctional institutions over a ten-year period, researchers observed the following facts.

State	n	**Percentage of admissions for assault or murder**
A	200	32
B	225	25

The investigators wished to construct a 95% confidence interval for $p_A - p_B$.

By Equation 5.25, we have

$$C\left[(0.32-0.25) - 1.96\sqrt{\frac{(0.32)(0.68)}{200} + \frac{(0.25)(0.75)}{225}} \le (p_A - p_B) \le (0.32-0.25) + 1.96\sqrt{\frac{(0.32)(0.68)}{200} + \frac{(0.25)(0.75)}{225}}\right] = 0.95$$

$$C[0.07 - (1.96)(0.0438) \le p_A - p_B \le 0.07 + (1.96)(0.0438)] = 0.95$$

$$C[0.07 - 0.09 \le (p_A - p_B) \le 0.07 + 0.09] = 0.95$$

$$C[-0.02 \le (p_A - p_B) \le 0.16] = 0.95$$

EXERCISES

40 In a certain penal system, criminologists identified two types of inmates. A random sample of 250 type *A* inmates revealed that 45% were alcoholics. Of a random sample of 230 type *B* inmates, 30% were alcoholics. Construct the 95% confidence interval for the difference between population proportions.

41 A team of botanists infected 200 plants with a certain disease. They then treated half the plants with chemical *A*, and half with chemical *B*. Of the plants treated with chemical *A*, 75 survived, and of those treated with chemical *B*, 64 survived. Construct a 90% confidence interval for $p_A - p_B$.

42 In a study of the characteristics of students in grades 10 through 12 in two public school systems, investigators collected the following data by means of simple random samples.

System	n	**Number who smoke**
A	250	150
B	300	150

Construct the 95% confidence interval for the difference in population proportions who smoke.

43 A random sample of 300 blue-collar workers revealed that 75% regularly watch a certain television program. Of a random sample of 200 white-collar workers, 66% stated that they watch the program regularly. Construct a 95% confidence interval for the difference between the two population proportions.

44 Of a random sample of 150 college students, 105 said they believe there is life elsewhere in the universe. Of a random sample of 200 persons in the same age group who did not go to college, 120 held the same belief. Construct a 95% confidence interval for the difference between the two population proportions.

5.8 CONFIDENCE INTERVAL FOR A POPULATION VARIANCE

In Chapter 4 we learned that if $s^2 = \Sigma (x_i - \bar{x})^2/(n-1)$ is the variance of a random sample of size n drawn from a normally distributed population with variance σ^2, the ratio $(n-1)s^2/\sigma^2$ follows a chi-square distribution with $n-1$ degrees of freedom. We may use this fact in constructing confidence intervals for population variances.

Note that the method of constructing a confidence interval for σ^2 described in this section is quite sensitive to departures from normality on the part of the sampled population. That is, if the sampled population is not normally distributed, the validity of the method described here for constructing confidence intervals for σ^2 is jeopardized.

We may begin the construction of our confidence interval for σ^2 by making the following probability statement.

$$P\left[\chi^2_{\alpha/2,n-1} \le \frac{(n-1)s^2}{\sigma^2} \le \chi^2_{1-\alpha/2,n-1}\right] = 1 - \alpha \tag{5.26}$$

In this expression, $\chi^2_{\alpha/2,n-1}$ is the value of χ^2 from Appendix Table G that corresponds to $n-1$ degrees of freedom and has $\alpha/2$ of the area under the curve of the chi-square distribution to its left. The symbol $\chi^2_{1-\alpha/2,n-1}$ is the value of χ^2 for $n-1$ degrees of freedom that has $1-\alpha/2$ of the area under the curve to its left.

Appropriate algebraic manipulation of Equation 5.26 leads to

$$P\left[\frac{(n-1)s^2}{\chi^2_{1-\alpha/2,n-1}} \le \sigma^2 \le \frac{(n-1)s^2}{\chi^2_{\alpha/2,n-1}}\right] = 1 - \alpha \tag{5.27}$$

When we substitute for s^2 in Equation 5.27 a specific numerical value of s^2, call it s_0^2, the statement becomes a confidence statement rather than a probability statement. Hence we may write the $100(1-\alpha)\%$ confidence interval for σ^2 as

$$C\left[\frac{(n-1)s_0^2}{\chi^2_{1-\alpha/2,n-1}} \le \sigma^2 \le \frac{(n-1)s_0^2}{\chi^2_{\alpha/2,n-1}}\right] = 1 - \alpha \tag{5.28}$$

Since $(n-1)s^2 = \Sigma (x_i - \bar{x})^2$, we may construct a confidence interval for σ^2 by using the sum of squared deviations of the sample values from their mean (the numerator of the sample variance); i.e., we may write the $100(1-\alpha)\%$ confidence interval for σ^2 as

$$C\left[\frac{\Sigma (x_i - \bar{x}_0)^2}{\chi^2_{1-\alpha/2, n-1}} \leq \sigma^2 \leq \frac{\Sigma (x_i - \bar{x}_0)^2}{\chi^2_{\alpha/2, n-1}}\right] = 1 - \alpha \tag{5.29}$$

where $\bar{x}_0$ is a specific numerical value of $\bar{x}$ computed from an observed sample.

We obtain a $100(1-\alpha)\%$ confidence interval for the population standard deviation σ by taking the square root of each term in Equation 5.29. We get

$$C\left[\sqrt{\frac{\Sigma (x_i - \bar{x}_0)^2}{\chi^2_{1-\alpha/2, n-1}}} \leq \sigma \leq \sqrt{\frac{\Sigma (x_i - \bar{x})^2}{\chi^2_{\alpha/2, n-1}}}\right] = 1 - \alpha \tag{5.30}$$

Example 5.15 A research biologist measured the carapace width of 15 crabs of a certain species and obtained $\Sigma (x_i - \bar{x})^2 = 508.06$. The biologist, wishing to construct a 95% confidence interval for the population variance, used Equation 5.29 to obtain

$$C\left(\frac{508.06}{26.119} \leq \sigma^2 \leq \frac{508.06}{5.629}\right) = 0.95$$

$$C(19.45 \leq \sigma^2 \leq 90.26) = 0.95$$

The point estimate of σ^2 is given by $s^2 = 508.06/14 = 36.29$. Remember that the assumptions underlying the construction of this confidence interval are: (1) The 15 observations constitute a simple random sample of size 15 from the population of interest, and (2) the population is normally distributed.

The 95% confidence interval for σ is, by Equation 5.30,

$$C(\sqrt{19.45} \leq \sigma \leq \sqrt{90.26}) = 0.95$$

$$C(4.41 \leq \sigma \leq 9.50) = 0.95$$

The point estimate of σ is $\sqrt{36.29} = 6.02$.

The method of constructing confidence intervals for σ^2 described above is called the *equal-tails method*, since the values of σ^2 used as divisors in Equations 5.28 and 5.29 are selected in such a way that the area left in each of the two tails of the appropriate chi-square distribution is equal to $\alpha/2$. It can be shown that the equal-tails method does not yield the shortest possible confidence intervals, and hence may not always give satisfactory results. Methods for obtaining "shortest" intervals have been proposed by Tate and Klett (3) and Lindley *et al.* (4).

We may obtain a "shortest" unbiased interval from

$$C\left[\frac{\Sigma (x_i - \bar{x}_0)^2}{\chi^2_2} \leq \sigma^2 \leq \frac{\Sigma (x_i - \bar{x}_0)^2}{\chi^2_1}\right] = 1 - \alpha \tag{5.31}$$

where χ_1^2 and χ_2^2 are given in Table I of the Appendix for $n-1$ degrees of freedom. The interval given by Equation 5.31 is shortest in the sense that the ratio of the upper limit to the lower limit is a minimum. Let us use Equation 5.31 and Appendix Table I to obtain a 95% confidence interval for σ^2 for Example 5.15. Since $n = 15$, we have 14 degrees of freedom, and χ_1^2 and χ_2^2 are 5.9477 and 27.263. Thus our 95% confidence interval for σ^2 is

$$C\left[\frac{508.06}{27.263} \leq \sigma^2 \leq \frac{508.06}{5.9477}\right] = 0.95$$

$$C(18.64 \leq \sigma^2 \leq 85.42) = 0.95$$

We see that, whereas the length of the equal-tails interval is equal to $90.26 - 19.45 = 70.81$ and yields a ratio of $90.26/19.45 = 4.64$, the width of the interval by Equation 5.31 is $85.42 - 18.64 = 66.78$, with a ratio of $85.42/18.64 = 4.58$.

EXERCISES

45 As part of an experiment, scientists removed the adrenal glands from a random sample of 20 mice. The sample yielded $\Sigma\,(x_i - \bar{x})^2 = 72.25$. Construct a 95% confidence interval for σ^2, using both methods described in the text. What assumptions must you make?

46 The following are the muzzle velocities of ten specimens of ammunition selected at random from a day's production at an ammunition factory. The data have been coded for ease of computation.

66, 37, 18, 31, 85, 63, 73, 83, 65, 80

Assume that muzzle velocities are normally distributed, and construct a 99% confidence interval for σ^2. Use both methods.

47 As part of a larger study, investigators recorded the heights (in centimeters) of a random sample of 25 18-year-old males. They found the sample variance to be 44. Construct the 95% confidence interval for the population variance by both methods. What assumptions must you make?

48 The following are the scores made on a reading comprehension test by 15 inmates of a juvenile correctional institution.

74, 98, 94, 55, 66, 55, 78, 71, 53, 90, 53, 85, 74, 74, 58

Assume that these scores constitute a random sample from a normally distributed population, and construct a 95% confidence interval for the population variance. Use both methods.

49 A random sample of 20 scores made by high school seniors on a college entrance examination yielded $\Sigma\,(x_i - \bar{x})^2 = 2230$. Assume the population of scores to be normally distributed, and construct the 99% confidence interval for σ^2. Use both methods.

50 The following are the pulse rates (in beats per minute) recorded for ten 15-year-old males.

63, 50, 75, 59, 51, 53, 48, 72, 82, 75

Assume that these data constitute a random sample from an approximately normally distributed population, and construct the 95% confidence interval for the population variance. Use both methods.

51 A random sample of 25 first-grade students were given a social-skills test. The sample variance computed from the scores was 0.08. Assume that the population from which the scores were drawn is normally distributed, and construct the 95% confidence interval for the population variance. Use both methods.

5.9 CONFIDENCE INTERVAL FOR THE RATIO OF TWO POPULATION VARIANCES

A point estimator of σ_1^2/σ_2^2, the ratio of two population variances, is provided by the ratio of corresponding sample variances, s_1^2/s_2^2. Such an estimator is useful when one is interested in assessing the comparative magnitudes of two variances. There are many situations in which one would want to know whether or not two population variances are equal. For example, two manufacturing procedures may yield two populations of items that are identical with respect to the mean of some critical measurement, but for one population the variability of the measurement may be much larger than for the other. This means that the procedure resulting in a larger variance yields a product that is less homogeneous with respect to the measurement of interest. All other things being equal, homogeneity is usually preferred to heterogeneity. If two population variances σ_1^2 and σ_2^2 are equal, their ratio σ_1^2/σ_2^2 is equal to 1. If the ratio of two sample variances s_1^2 and s_2^2 is vastly different from 1, one is inclined to question the equality of the corresponding population variances σ_1^2 and σ_2^2.

We may construct confidence intervals for σ_1^2/σ_2^2, the ratio of the variances of two normally distributed populations, by using the F distribution. Recall that in Chapter 4 we noted that if s_1^2 and s_2^2 are the variances computed from independent simple random samples of size n_1 and n_2 drawn from normally distributed populations with variances σ_1^2 and σ_2^2, then the ratio $(s_1^2/\sigma_1^2)/(s_2^2/\sigma_2^2)$ follows the F distribution, with $n_1 - 1$ and $n_2 - 1$ degrees of freedom.

To construct a confidence interval for σ_1^2/σ_2^2, we begin with the following probability statement.

$$P\left(F_{\alpha/2} \leq \frac{s_1^2/\sigma_1^2}{s_2^2/\sigma_2^2} \leq F_{1-\alpha/2}\right) = 1 - \alpha \tag{5.32}$$

where $F_{\alpha/2}$ and $F_{1-\alpha/2}$ are values from the appropriate F distribution that have to their left and right, respectively, $\alpha/2$ of the area under the curve of F. Since

$$\frac{s_1^2}{\sigma_1^2} \div \frac{s_2^2}{\sigma_2^2} = \frac{s_1^2}{\sigma_1^2} \cdot \frac{\sigma_2^2}{s_2^2} = \frac{s_1^2}{s_2^2} \cdot \frac{\sigma_2^2}{\sigma_1^2}$$

we may rewrite Equation 5.32 as

$$P\left(F_{\alpha/2} \leq \frac{s_1^2}{s_2^2} \cdot \frac{\sigma_2^2}{\sigma_1^2} \leq F_{1-\alpha/2}\right) = 1 - \alpha \tag{5.33}$$

Dividing the three terms inside the parentheses by s_1^2/s_2^2, we have

$$P\left(\frac{F_{\alpha/2}}{s_1^2/s_2^2} \leq \frac{\sigma_2^2}{\sigma_1^2} \leq \frac{F_{1-\alpha/2}}{s_1^2/s_2^2}\right) = 1 - \alpha \tag{5.34}$$

Taking the reciprocal of each term leads to

$$P\left(\frac{s_1^2/s_2^2}{F_{\alpha/2}} \geq \frac{\sigma_1^2}{\sigma_2^2} \geq \frac{s_1^2/s_2^2}{F_{1-\alpha/2}}\right) = 1 - \alpha \tag{5.35}$$

Finally, if we reverse the order of terms inside the parentheses to get the smallest quantity on the left and the largest on the right, we have

$$P\left(\frac{s_1^2/s_2^2}{F_{1-\alpha/2}} \leq \frac{\sigma_1^2}{\sigma_2^2} \leq \frac{s_1^2/s_2^2}{F_{\alpha/2}}\right) = 1 - \alpha \tag{5.36}$$

When we substitute specific numerical values for s_1^2 and s_2^2 into Equation 5.36, we have the $100(1 - \alpha)\%$ confidence interval for σ_1^2/σ_2^2.

Since Appendix Table H contains only upper-tail percentiles of the F distribution (that is, only values of $F_{1-\alpha/2}$), an explanation of how to obtain values of $F_{\alpha/2}$ is in order.

To find $F_{\alpha/2}$, we use the identity

$$F_{\alpha/2,\nu_1,\nu_2} = \frac{1}{F_{1-\alpha/2,\nu_2,\nu_1}}$$

where ν_1 and ν_2 are the numerator and denominator degrees of freedom, respectively. Thus, if $\nu_1 = 15$, $\nu_2 = 20$, and $\alpha = 0.05$,

$$F_{1-\alpha/2,\nu_2,\nu_1} = F_{1-0.025,20,15} = 2.76$$

$$F_{\alpha/2,\nu_1,\nu_2} = F_{0.025,15,20} = \frac{1}{2.76} = 0.36$$

Example 5.16 A team of physical education instructors gave two groups of college males muscular endurance tests following an exercise program. The scores of group 1, consisting of 16 subjects, yielded a sample variance of 4685.40. For group 2, consisting of 25 subjects, the sample variance was 1193.70. The instructors, who were willing to assume that the two groups of scores constituted independent simple random samples from normally distributed populations, wished to construct a 95% confidence interval for σ_1^2/σ_2^2.

From the F distribution with 15 and 24 degrees of freedom, we find that

$$F_{0.975,24,15} = 2.70 \qquad \text{and} \qquad F_{0.025,15,24} = \frac{1}{2.70} = 0.370$$

Through the use of Equation 5.36, we have

$$C\left[\frac{4685.40/1193.70}{2.44} \leq \frac{\sigma_1^2}{\sigma_2^2} \leq \frac{4685.40/1193.70}{0.370}\right] = 0.95$$

$$C[1.61 \leq (\sigma_1^2/\sigma_2^2) \leq 10.61] = 0.95$$

We are 95% confident, then, that the ratio of the two population variances is between 1.61 and 10.61.

Note that the interval does not include 1. Therefore 1 is not a candidate for the parameter being estimated. It seems reasonable to conclude, then, that the ratio of the two population variances is not equal to 1, and hence that σ_1^2 is not equal to σ_2^2.

EXERCISES

52 A botanist divided a quantity of seeds of a certain species of annual flower into two parts, and assigned each part for cultivation under a different set of experimental conditions. The diameters, in millimeters, of 25 mature flowers from plants grown under condition 1 yielded a variance of 8.50. Twenty-one mature flowers from plants grown under condition 2 gave a variance of 4.25. Assume that the data constitute independent simple random samples from two normally distributed populations. Construct a 95% confidence interval for σ_1^2/σ_2^2.

53 Of 29 applicants for a clerical position with a certain firm, 13 had just completed a six-month clerical course at a private secretarial school, and 16 had just graduated from high school, where they had taken the commercial curriculum. Each applicant was given the same clerical proficiency test. The variance of the scores for the first group was 525; for the second group, 350. Construct a 90% confidence interval for the ratio of the two population variances. What assumptions are necessary?

54 The scholastic aptitude test scores of a random sample of 21 seventh-grade students from school district 1 and of a random sample of 16 seventh-grade students from school district 2 yielded variances of $s_1^2 = 176$ and $s_2^2 = 110$. Assume that the samples are independent and that the scores in each population are normally distributed. Construct a 95% confidence interval for σ_1^2/σ_2^2.

55 The heights (in centimeters) of 25 12-year-old girls yielded a variance of 64 cm². For a sample of 21 12-year-old boys, the variance of their

heights was 36 cm^2. Construct a 95% confidence interval for σ_G^2/σ_B^2. What assumptions must you make?

56 An eighth-grade arithmetic class of 21 students was randomly divided into two groups. Each group was taught by a different method for a semester. At the end of the semester, the students were given the same achievement test, with the following results: $n_1 = 11$, $s_1^2 = 280$, $n_2 = 10$, $s_2^2 = 200$. Assume that the data constitute independent simple random samples from normally distributed populations, and construct a 90% confidence interval for the ratio of the two population variances.

CHAPTER SUMMARY

In this chapter we have used the material learned in the first four chapters to help you understand the basic concepts of statistical inference, with particular emphasis on the estimation of population parameters. We have discussed both point and interval estimators, along with some basic ideas regarding unbiasedness, one of the criteria used to evaluate the "goodness" of estimators. You have learned the basic concepts and techniques for constructing interval estimates of the following population parameters: the mean, the difference between two means, the population proportion, the difference between two proportions, the variance, and the ratio of two variances.

In Chapter 6 we shall cover the second type of statistical inference: hypothesis testing.

SOME IMPORTANT CONCEPTS AND TERMINOLOGY

Statistical inference
Estimation
Point estimation
Interval estimation
Unbiasedness
Confidence coefficient
Reliability factor
Confidence interval
Degrees of freedom
Paired comparison

REVIEW EXERCISES

57 The quality-control department of a plastics manufacturer periodically selects a sample of specimens in order to test for breaking strength. Past experience has shown that the breaking strengths of a certain type of plastic are normally distributed with a standard deviation of 196 pounds. A random sample of 16 specimens yielded a mean of 6000 pounds. The quality-control supervisor wants a 95% confidence interval for the mean breaking strength of the population.

58 An official with a manufacturing company wished to know the mean length of metal rods received in a large shipment. The official collected a random sample of 225, and measured the rods. The arithmetic

mean of the lengths was found to be 8.8 feet. The population has a standard deviation of 1.5 feet. What are the 95% confidence limits for μ for this sample result?

59 A sample of 100 apparently normal 25-year-old males had a mean systolic blood pressure of 120. The population standard deviation is 20.

(a) Find the 90% confidence interval for μ.

(b) Find the 95% confidence interval for μ.

60 A professional organization wished to know the average age of its members. A random sample of 100 members yielded a mean age of 35 years, with a standard deviation of 5 years. Construct the 95% confidence interval for μ.

61 The average number of heartbeats per minute for a sample of 36 subjects was found to be 100. Assume that the population is normally distributed with a standard deviation of 10.

(a) Find the 90% confidence interval for μ.

(b) Find the 95% confidence interval for μ.

62 In a dowel factory, two machines are used in the production of dowels. A random sample of 11 dowels from machine A and a random sample of 21 dowels from machine B gave the following results with respect to the lengths of dowels produced.

$$\bar{x}_A = 4.95 \text{ feet}, \qquad \bar{x}_B = 5.01 \text{ feet}$$
$$s_A^2 = 0.018, \qquad s_B^2 = 0.020$$

The populations are assumed to be approximately normally distributed. Assume that the population variances are equal, and construct the 95% confidence interval for $\mu_A - \mu_B$.

63 Sixteen experimental animals subjected to a type of stress had a mean endurance score of 98, with a standard deviation of 12. Construct the 95% confidence interval for the population mean.

64 In a laboratory experiment, ten animals were forced to breathe air contaminated with a harmful chemical, while ten control animals breathed clean air for the same length of time. At the end of the experiment, researchers conducted hemoglobin determinations on the animals, with the following results.

	$\bar{x}$	s^2
Control animals:	18	25
Experimental animals:	14	20

Construct the 95% confidence interval for the difference between the population means.

65 A large company wished to estimate the proportion of its employees who favored a new insurance plan. Of a random sample of 300 employees, 75 said they favored the plan. Construct the 95% confidence interval for the true proportion favoring the plan.

66 In a survey, a random sample of 350 working wives and an independent random sample of 325 nonworking wives were questioned about their reading habits. Of the 350 working wives, 105 indicated that they subscribed to a certain type of magazine; 130 of the nonworking wives said they subscribed to the same type of magazine. Construct the 99% confidence interval for the difference in the true proportions subscribing to this type of magazine.

REFERENCES

1 George W. Snedecor and William G. Cochran, *Statistical Methods*, sixth edition, Ames, Iowa: Iowa State University Press, 1967

2 Wilfred J. Dixon and Frank J. Massey, *Introduction to Statistical Analysis*, third edition, New York: McGraw-Hill, 1969

3 R. F. Tate and G. W. Klett, "Optimal Confidence Intervals for the Variance of a Normal Distribution," *Journal of the American Statistical Association*, 54 (1959), 674–682

4 D. V. Lindley, D. A. East, and P. A. Hamilton, "Tables for Making Inferences about the Variance of a Normal Distribution," *Biometrika*, 47 (1960), 433–437

Testing Hypotheses about Population Parameters

In Chapter 5 the discussion of the concepts and techniques of point and interval estimation served to introduce you to statistical inference. In this chapter we consider the other approach to statistical inference: hypothesis testing. Although we treat the topics of interval estimation and hypothesis testing in separate chapters, they are not as different as such treatment might suggest. Both ideas are based on the concepts of probability and the sampling distribution as presented in previous chapters. And both make it possible for one to make decisions about a population on the basis of the information contained in a sample from that population.

6.1 HYPOTHESES

The word *hypothesis* has been defined as: 1. An assertion subject to verification or proof, ... 2. An assumption used as the basis for action.*

**The American Heritage Dictionary of the English Language*, American Heritage Publishing Co. and Houghton Mifflin Company, Boston, 1970.

The key point in these definitions is that a hypothesis is an assertion or assumption, and not an established fact. Thus, in the absence of knowledge regarding its true effectiveness, a researcher may propose the hypothesis that, in the teaching of reading to first-grade students, method *A* is superior to method *B*. A drug manufacturer may hypothesize that a new drug will be more effective than the standard one in the treatment of a certain disease. A plastics manufacturer may hypothesize that sheets of a certain type of plastic have a mean tensile strength of 75 pounds. Hypotheses such as these may be based on experience and observation, experimentation, or an intuitive feeling. Hypotheses stated in this manner frequently provide the motivation for research. For this reason, we may refer to them as *research hypotheses.*

Generally, we have to restate research hypotheses before testing them statistically. When stated in a form suitable for testing by the statistical methods discussed in this chapter, hypotheses are called *statistical hypotheses.* Statistical hypotheses are statements about one or more populations, or, as is more often the case, statements about one or more parameters of one or more populations.

Statistical hypotheses are of two types. First, there is the *null hypothesis,* designated by H_0, which is the hypothesis to be tested. The null hypothesis is called the hypothesis of no difference (hence the term *null*). It is a statement to the effect that there is no difference between two populations, between two parameters of two populations, or between the true value of some parameter and the hypothesized value.

Let us look again at the three research hypotheses stated previously, and state for each the corresponding null hypothesis. For the research hypothesis regarding methods of teaching reading to first-grade students, let us assume that the criterion of effectiveness is the score made on a reading achievement test given at the end of the year. An appropriate null hypothesis (H_0) would be that there is no difference in the effectiveness of the two methods of teaching reading, or, more specifically, that the mean score on the test of students taught by method *A* is equal to (no different from) the mean score of students taught by method *B*. We may state the null hypothesis in more compact form as

$$H_0\colon \mu_A = \mu_B$$

Suppose that the effectiveness of the new drug and the standard drug mentioned earlier are measured in terms of the proportion of cases responding favorably to treatment by each. The appropriate null hypothesis would be that the proportion of cases responding favorably to the new drug is equal to the proportion of cases responding favorably to the standard drug, or

$$H_0\colon p_{\text{new}} = p_{\text{standard}}$$

Finally, for the research hypothesis that sheets of a certain type of plastic have a mean tensile strength of 75 pounds, an appropriate null hypothesis might be that the mean tensile strength is 75 pounds, or

$$H_0: \mu = 75$$

To test a null hypothesis, we examine sample data from the relevant population to determine whether or not they are compatible with the null hypothesis. If the sample data are not compatible with the null hypothesis, H_0 is *rejected*. If the sample data are compatible with the null hypothesis, H_0 is not rejected. We shall explain in Section 6.2 the criterion for determining whether or not the sample data are compatible with the null hypothesis.

If a null hypothesis is not rejected, we say that these particular sample data do not provide sufficient evidence for us to conclude that the null hypothesis is false. If the null hypothesis is rejected, we say that these particular sample data do provide sufficient evidence to enable us to conclude that the null hypothesis is false and that some other hypothesis is true. This other hypothesis, the one that we conclude is true if the null hypothesis is rejected, is called the *alternative hypothesis* (or *alternate hypothesis*), and is designated by the symbol H_1. Usually the alternative hypothesis and the research hypothesis are the same.

Let us refer again to the research hypotheses that we stated earlier, and indicate what the appropriate null and alternative hypotheses would be.

1 Research hypothesis: Method A is superior to method B for teaching reading to first-grade students.

H_0: $\mu_A = \mu_B$ (The two population means are equal.)

H_1: $\mu_A > \mu_B$ (Method A results in a higher mean score than method B.)

2 Research hypothesis: The new drug is more effective than the standard drug in the treatment of disease X.

H_0: $p_{\text{new}} = p_{\text{standard}}$ (The two population proportions are equal.)

H_1: $p_{\text{new}} > p_{\text{standard}}$ (The true proportion of cases responding favorably to the new drug is greater than the proportion responding favorably to the standard drug.)

3 Research hypothesis: The true mean tensile strength of sheets of type A plastic is 75 pounds.

H_0: $\mu = 75$ (The population mean is 75 pounds.)

H_1: $\mu \neq 75$ (The population mean is not 75 pounds.)

Note that in the first two sets of hypotheses, the research hypothesis and the alternative hypothesis are the same, while in the third set the research hypothesis is the same as the null hypothesis.

In stating hypotheses of the type shown in (1) and (2), one usually makes the null and alternative hypotheses complementary by including an inequality in the null hypothesis that is in a direction opposite to that of the inequality in the alternative hypothesis. For example, one would write hypotheses (1) and (2) above as

$$H_0\colon \mu_A \leq \mu_B, \qquad H_1\colon \mu_A > \mu_B$$

and

$$H_0\colon p_{\text{new}} \leq p_{\text{standard}}, \qquad H_1\colon p_{\text{new}} > p_{\text{standard}}$$

This method of stating the null and alternative hypotheses emphasizes the fact that when the alternative hypothesis specifies a deviation from equality in one direction, deviations from equality in the opposite direction are of no interest. For example, the director of quality control for a manufacturing firm may establish the following hypotheses as part of a procedure in accepting and rejecting shipments of raw material from various suppliers.

$H_0\colon p \leq p_0$ (The proportion of defective items in a shipment is less than or equal to some prescribed level, p_0.)

$H_1\colon p > p_0$ (The proportion of defective items in a shipment is greater than p_0.)

The quality-control director wants to detect those shipments in which the proportion of defective items is greater than p_0, the maximum acceptable level, so they can be rejected. If the proportion that is defective is less than the acceptable level, so much the better.

6.2 HYPOTHESIS-TESTING PROCEDURE

As an illustration of hypothesis-testing procedures, consider the following example. On the basis of years of experience, a team of psychologists believes that individuals who are nonconformists tend to have higher levels of self-esteem than individuals who are conformists. Although the psychologists can recall many cases to support their contention, they realize that, in order to lend greater weight to their conjecture, they must use a more scientific method of analyzing the evidence. A statistical hypothesis-testing procedure seems to them to be appropriate. Accordingly, they set up the following null and alternative hypotheses.

$$H_0\colon \mu_N \leq \mu_C, \qquad H_1\colon \mu_N > \mu_C$$

where μ_N is the population mean score made by nonconformists on a test designed to measure level of self-esteem and μ_C is the population mean score made by conformists on the same test.

The population about which the psychologists wish to make an inference is the population of all persons who can be characterized as either conformists or nonconformists. The psychologists obtain independent samples of conformists and nonconformists which they believe may be treated as random samples from the population of interest. They administer tests for measuring self-esteem to the individuals in the two samples, and compute the mean score for each sample. They find that $\bar{x}_N = 80$ and $\bar{x}_C = 75$. Although the direction of the difference in sample means is compatible with their research (and alternative) hypothesis, the psychologists know that there are at least two explanations for this difference: (1) The true mean self-esteem score for the population of nonconformists may not be higher than that for conformists. The observed sample results may be merely the result of chance. (2) The observed sample results may reflect the true state of affairs, making it appropriate for them to conclude that the true mean self-esteem score for nonconformists is higher than that for conformists. A knowledge and understanding of the fine points of hypothesis testing will enable the psychologists to choose between the two explanations. We shall devote the remainder of this section to the specific concepts and techniques of hypothesis testing.

We can formalize the procedure that is followed in testing a hypothesis by setting down, in sequence, the several steps composing the procedure. In this section we list and explain each of these steps in the order in which they typically occur. Nine major steps may be identified.

1 Statement of hypotheses

2 Selection of significance level

3 Description of population of interest and statement of necessary assumptions

4 Selection of relevant statistic

5 Specification of test statistic and consideration of its distribution

6 Specification of rejection and acceptance regions

7 Collection of data and calculation of necessary statistics

8 Statistical decision

9 Conclusion

Now let us describe each of these steps in general terms. Later we shall illustrate them by specific examples.

1 *Statement of hypotheses* We discussed the various hypotheses that may be identified and the manner in which they are stated in Section 6.1. Since the beginning student of statistics frequently finds it difficult to decide how to state the null and alternative hypotheses, we shall pursue the subject

further at this point. Generally, we would like to base a conclusion (step 9 above) on a rejected null hypothesis. That is, ordinarily we would prefer that our sample data support the alternative hypothesis. (We shall explain the reason for this in Section 6.4.) Consequently, in deciding what the alternative hypothesis should be, we ask ourselves, "What do I wish to conclude?" or "What do I believe is true?" The answer to these questions provides the statement for the alternative hypothesis. A statement that is the complement of the alternative hypothesis, then, serves as the null hypothesis.

For example, consider the researcher whose research hypothesis states that, in the teaching of reading to first-grade students, method A is superior to method B. In answer to the question, "What do I wish to conclude?" the researcher would answer that he or she wishes to conclude that method A is superior to method B. Hence the alternative hypothesis is $\mu_A > \mu_B$, and the null hypothesis, $\mu_A \leq \mu_B$, is the complement of this statement. This example illustrates the fact that, typically, the alternative hypothesis is formulated first.

2 *Selection of significance level* On the basis of the results of the analysis of the sample data, we either reject the null hypothesis or do not reject it. Rejection of a null hypothesis doesn't prove that it is false. Regardless of how incompatible the sample evidence is with the null hypothesis, there remains the possibility that the null hypothesis is indeed true. Similarly, failure to reject the null hypothesis doesn't prove that it is true and the alternative is false. Again, although the null hypothesis is not rejected, there is the possibility that it is false. Consideration of these facts leads to the realization that both rejection and nonrejection of a null hypothesis are accompanied by risks of being wrong, or in error. Although we usually never know whether or not a particular action (rejection or nonrejection of H_0) is in error, we may designate the two possible errors as follows.

(a) *Rejection of a true null hypothesis.* This error is called a *Type I error.*

(b) *Nonrejection of a false null hypothesis.* This error is called a *Type II error.*

We may illustrate the relationship between the true condition of the null hypothesis (that is, whether it is true or false) and the statistical decision (to reject or not to reject H_0) as shown in Table 6.1.

We shall follow statistical custom and represent the probability of com-

Table 6.1 Relationship between the true condition of the null hypothesis and the statistical decision

Statistical decision	**True condition of H_0**	
	TRUE	FALSE
Reject H_0	Type I error	Correct decision
Do not reject H_0	Correct decision	Type II error

mitting a Type I error by α and the probability of committing a Type II error by β. Thus

$$\alpha = P(\text{Type I error}), \qquad \beta = P(\text{Type II error})$$

For a particular hypothesis test, we would like both α and β to be small. Because of the nature of the relationship between the two probabilities, however, we find that for a given sample size, a decrease in α results in an increase in β, and vice versa.

This being true, it would seem prudent, in a given situation, to try to minimize the probability of committing the more serious error. Unfortunately, in many areas of research it is difficult or impossible to evaluate the two types of error relative to the seriousness of each. What one does in such a situation, then, is to preselect some small value of α, say 0.10, 0.05, or 0.01. The choice of α reflects the investigator's belief about the seriousness of Type I errors. The more serious the consequences of committing a Type I error are perceived to be, the smaller the preselected value of α.

Frequently α is referred to as the *level of significance*. When one chooses a level of significance equal to α, and rejects the null hypothesis, we say that the sample results are *significant*.

3 *Description of population of interest and statement of necessary assumptions* Hypothesis-testing procedures depend on the characteristics of the underlying sampling distribution. And the characteristics of the sampling distribution depend, in part, on the nature of the sampled population. For this reason, we must investigate the nature of the sampled population in order to justify our choice of testing procedure. We are usually interested in knowing the approximate size of the population and whether or not it can be considered approximately normally distributed. We shall also want to establish the fact that it is reasonable to assume that the sample drawn constitutes a simple random sample from the population of interest.

4 *Selection of relevant statistic* The particular statistic that enters into the hypothesis-testing procedure is determined by the parameter with which the hypothesis is concerned. Thus, if we are testing a hypothesis about a population mean, the relevant statistic is $\bar{x}$, the sample mean. We may wish to consider the sampling distribution of the relevant statistic. Generally, we shall want to know the mean, variance (or standard deviation), and functional form of the applicable sampling distribution. If, for example, we are testing a hypothesis about a population mean, and if sampling is from a population that is normally distributed, we know that the distribution of $\bar{x}$, the sample mean, will be normally distributed with mean μ and variance σ^2/n.

5 *Specification of test statistic and consideration of its distribution*

DEFINITION ***A test statistic** is a numerical quantity, computed from the data of a sample used in reaching a decision on whether or not to reject a null hypothesis.*

One determines the appropriate test statistic by the parameter about which the hypothesis is made and by the nature of the sampling distribution of the relevant statistic. When sampling is from a normally distributed population with known variance, for the purpose of testing a hypothesis about the population mean, the test statistic is

$$z = \frac{\bar{x} - \mu_0}{\sigma/\sqrt{n}}$$

where $\bar{x}$ is the mean of a sample of size n drawn from the population, μ_0 is the hypothesized value of the population mean μ, and σ is the population standard deviation. This test statistic is distributed as the standard normal distribution. When sampling is from a normally distributed population with unknown variance, the test statistic for testing a hypothesis about the population mean is

$$t = \frac{\bar{x} - \mu_0}{s/\sqrt{n}}$$

where $\bar{x}$, μ_0, and n are defined as above and s is the sample standard deviation. This test statistic follows Student's t distribution with $n - 1$ degrees of freedom. In later sections we shall discuss additional test statistics that are frequently encountered.

6 *Specification of rejection and acceptance regions*

DEFINITION *In a hypothesis test, the* ***rejection region*** *consists of all those values of the test statistic which are of such magnitude that, if the observed value of the test statistic is equal to one of them, the null hypothesis is rejected.*

The ***acceptance region*** *is the complement of the rejection region. If the observed value of the test statistic is equal to one of the values composing the acceptance region, the null hypothesis is not rejected.*

As we shall see, the sizes of the rejection and acceptance regions are determined by α.

To illustrate the manner in which the rejection and acceptance regions are determined, let us consider the case in which, for testing a hypothesis about a population mean, sampling is from a normally distributed population with known variance. As we have already pointed out, the appropriate test statistic is z.

Suppose we wish to test the null hypothesis that the population mean μ is equal to some particular value μ_0, against the alternative that μ is not equal to μ_0. We can state our null and alternative hypotheses symbolically as

$$H_0\colon \mu = \mu_0, \qquad H_1\colon \mu \neq \mu_0$$

Let us choose α, the probability of rejecting a true null hypothesis, to be 0.05.

Now let us consider the distribution of sample means computed from

Figure 6.1 Sampling distribution of the sample mean, based on samples of size n from a normally distributed population

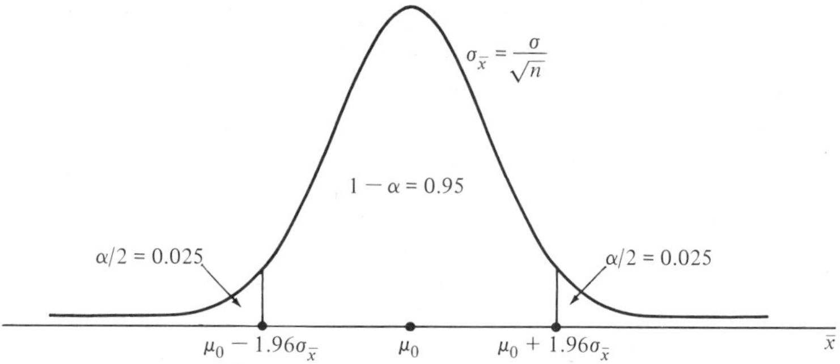

samples of size n drawn from our specified population. We know from previous discussion that the sampling distribution of $\bar{x}$ is normally distributed. If the null hypothesis is true, the mean of the sampling distribution is equal to μ_0. We also know that $1 - \alpha = 0.95$ of all $\bar{x}$'s will fall within 1.96 standard errors of the mean which, when H_0 is true, is equal to μ_0. We may express this by means of the following probability statement.

$$P(\mu_0 - 1.96\sigma_{\bar{x}} \leq \bar{x} \leq \mu_0 + 1.96\sigma_{\bar{x}}) = 0.95$$

Figure 6.1 depicts this equation and the sampling distribution graphically.

The probability that a single simple random sample of size n will yield a value of $\bar{x}$ equal to or larger than $\mu_0 + 1.96\sigma_{\bar{x}}$ is equal to $\alpha/2 = 0.025$. The probability that a single random sample will yield a value of $\bar{x}$ equal to or smaller than $\mu_0 - 1.96\sigma_{\bar{x}}$ is also equal to $\alpha/2 = 0.025$. With a specified numerical value for μ_0, we may compute actual numerical values of $\mu_0 \pm 1.96\sigma_{\bar{x}}$. For example, suppose that $\mu_0 = 100$ (that is, suppose we are hypothesizing that μ is equal to 100), $\sigma = 30$, and $n = 25$. The numerical values of $\mu_0 \pm 1.96(30/\sqrt{25})$ are 88.24 and 111.76. We may state that the probability of observing a value of $\bar{x}$ between 88.24 and 111.76, when H_0 is true, is equal to 0.95. When H_0 is true, the probability that a single simple random sample of size 25 drawn from the population will yield a mean equal to or greater than 111.76 is 0.025. And the probability that a single simple random sample will yield a mean equal to or less than 88.24 is also 0.025.

Suppose, in fact, that we do observe a value of $\bar{x}$ that is either equal to or greater than 111.76, or is equal to or less than 88.24. We must be willing either to conclude that a rare event (one with a 0.05 out of 100 chance of occurring) has taken place or to offer an alternative explanation. In a hypothesis-testing procedure, the only other alternative offered is that the null hypothesis is false; that is, that the sample was not drawn from a population with the hypothesized mean. In fact, it is this latter explanation that is accepted when the hypotheses are H_0: $\mu = \mu_0$ and H_1: $\mu \neq \mu_0$, the significance level is α, and there is a value of $\bar{x}$ that is equal to or greater than $\mu_0 + z_{\alpha/2}(\sigma/\sqrt{n})$ or one that is less than or equal to $\mu_0 - z_{\alpha/2}(\sigma/\sqrt{n})$. By accepting this explanation, we are rejecting the null hypothesis. In deciding in advance to reject H_0 under these circumstances, we run a risk, α, of making the wrong decision. Therefore, we choose α to be some small number (say 0.10, 0.05, or 0.01), so that the probability of being wrong (rejecting a true null hypothesis) will be small.

Since we shall reject H_0: $\mu = \mu_0$ in favor of H_1: $\mu \neq \mu_0$ if our single sample yields a mean $\bar{x}$ that either is equal to or greater than $\mu_0 + z_{\alpha/2}(\sigma/\sqrt{n})$, or is equal to or less than $\mu_0 - z_{\alpha/2}(\sigma/\sqrt{n})$, these values of $\bar{x}$ constitute the rejection region for our hypothesis test. Their complement,

$$\mu_0 - z_{\alpha/2}\frac{\sigma}{\sqrt{n}} < \bar{x} < \mu_0 + z_{\alpha/2}\frac{\sigma}{\sqrt{n}}$$

therefore makes up the acceptance region.

We may express the rejection and acceptance regions in terms of the test statistic z by noting that the numbers

$$\bar{x} = \mu_0 - z_{\alpha/2}\frac{\sigma}{\sqrt{n}} \qquad \text{and} \qquad \bar{x} = \mu_0 + z_{\alpha/2}\frac{\sigma}{\sqrt{n}}$$

transform to $-z_{\alpha/2}$ and $z_{\alpha/2}$, respectively, when we use the formula $z = (\bar{x} - \mu_0)/(\sigma/\sqrt{n})$.

Figure 6.2 shows the acceptance and rejection regions, in terms of both $\bar{x}$ and z, for testing, at the α significance level, H_0: $\mu = \mu_0$ against the alternative H_1: $\mu \neq \mu_0$.

If we compute, from the sample data, a value of

$$z = \frac{\bar{x} - \mu_0}{\sigma/\sqrt{n}}$$

that either is greater than or equal to $z_{\alpha/2}$, or is less than or equal to $-z_{\alpha/2}$, we reject H_0. Otherwise, we do not reject H_0. A computed value of z that leads to the rejection of a null hypothesis is said to be *significant*.

We call values of a test statistic, such as $z_{\alpha/2}$ and $-z_{\alpha/2}$ in Figure 6.2(b), that separate a rejection region from an acceptance region *critical values* of the test statistic. They tell us when to stop believing that the null hypothesis is true and start believing that it is false.

We call alternative hypotheses of the form H_0: $\mu \neq \mu_0$ *two-sided alternative hypotheses*, since they usually lead to a rejection region that is divided between the two sides, or tails, of the distribution of the test statistic. And we call the procedure appropriate for testing a null hypothesis with a two-sided alternative hypothesis as described above a *two-sided hypothesis test*.

Frequently, as we have noted, the null hypothesis may be of the form H_0: $\mu \leq \mu_0$, with the alternative H_1: $\mu > \mu_0$. We call a null hypothesis of this type a *one-sided hypothesis*, since only large computed values of the test statistic cause rejection of the null hypothesis, and hence the rejection region is located in only the upper tail of the distribution of the test statistic. That is, the entire α probability is located in one tail, rather than being split in half as

Figure 6.2 Acceptance and rejection regions in terms of $\bar{x}$ and z for the testing of H_0: $\mu = \mu_0$ against the alternative H_1: $\mu \neq \mu_0$

(a)

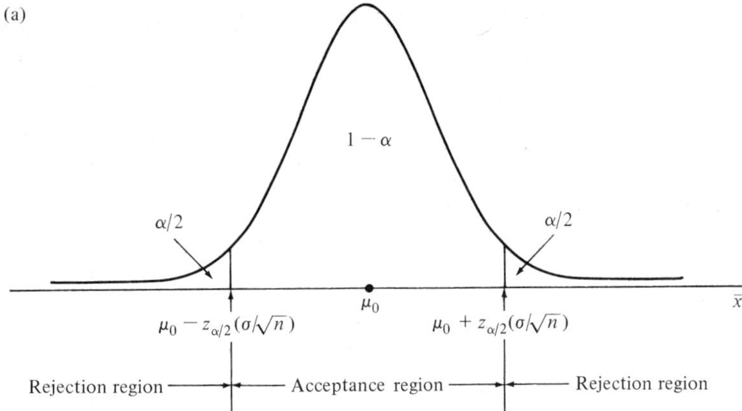

(b)

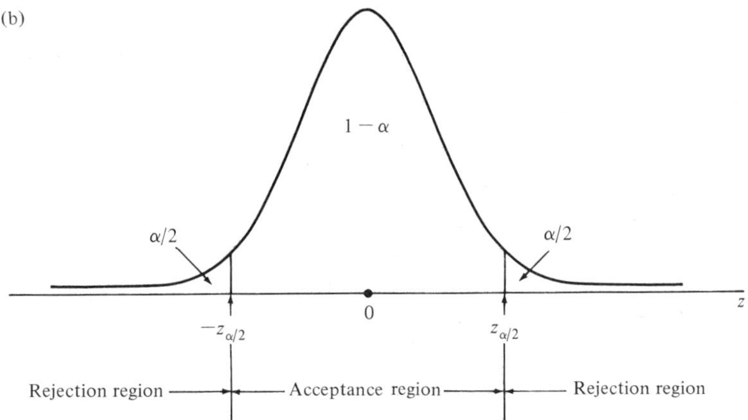

it is in the two-sided test. For example, the team of psychologists, described earlier, who are interested in self-esteem scores of conformists and nonconformists, use a one-sided test with the rejection region located only in the upper tail. If they select a significance level (probability of rejecting a true null hypothesis) of 0.05, the entire 0.05 will be the area in the upper tail of the sampling distribution. For the one-sided alternative hypothesis of the form H_1: $\mu < \mu_0$, only small computed values of the test statistic will cause rejection of the null hypothesis, and consequently the rejection region will all lie in the lower tail of the distribution.

Up to this point, our examples of hypothesis testing have been restricted to tests about the population mean. In later sections we shall discuss hypothesis testing for the case in which sampling is from populations that are not normally distributed, as well as for cases involving other population parameters.

7 *Collection of data and calculation of necessary statistics* Data that are relevant to the formulated hypotheses and that meet the necessary assumptions of the test must be collected in an appropriate manner. After the data have been collected, one computes the appropriate statistic and test statistic.

8 *Statistical decision* One compares the computed value of the test statistic with the critical value of the test statistic. If the computed value falls in the rejection region, one rejects H_0. Otherwise one does not reject it.

9 *Conclusion* Whereas the decision is stated in terms of the test statistic, the conclusion is stated in terms of the parameter and/or population with which the test is concerned. For example, when we reject H_0: $\mu = \mu_0$, we conclude that "the population mean is not equal to μ_0." When we do not reject the null hypothesis, our conclusion lacks the strength of conviction present in the conclusion reached when a null hypothesis is rejected. This is so because, although we know in advance that the probability of rejecting a true null hypothesis is small (we make it small by our choice of α), we usually do not know the value of β, the probability of accepting (failing to reject) a false null hypothesis. It very well may be, and frequently is, quite large. (We shall discuss this point in more detail in Section 6.4.)

Therefore, if we do not reject H_0: $\mu = \mu_0$, we conclude that "the population mean *may be* equal to μ_0."

In the following sections, we shall illustrate the general hypothesis-testing procedure described in this section with specific examples. We shall illustrate hypothesis testing when the parameters of interest are the population mean, the difference between two population means, the population proportion, the difference between two population proportions, the population variance, and the ratio of two population variances.

6.3 TESTING A HYPOTHESIS ABOUT A SINGLE POPULATION MEAN

In this section we shall illustrate the hypothesis-testing procedure that is appropriate when the parameter of interest is the population mean. We shall consider three cases: (1) the case in which sampling is from a normally distributed population with known variance, (2) the case in which sampling is from a normally distributed population with unknown variance, and (3) the case in which sampling is from a population that is not normally distributed.

Normally distributed population, σ^2 known

To illustrate the testing of hypotheses about population means, we consider first the case in which the population of interest is normally distributed and the population variance is known.

Example 6.1 The mean and standard deviation of the weights of the men who played football at a state university during the school's first ten seasons are $\mu = 162.5$ pounds and $\sigma = 18.0$ pounds. The athletic department wishes to know whether there is reason to believe that the mean weight of those who played varsity football during the most recent ten seasons is different from the mean weight of those who played during the first ten seasons.

Members of the department wish to base their conclusion on a sample of size $n = 25$. The hypothesis-testing procedure is as follows.

1 *Statement of hypotheses* The investigators, who want to know whether they can conclude that the mean weight of varsity football players during the most recent ten seasons differs from 162.5, reason that such a conclusion would be justified if they could reject the null hypothesis that the mean weight of the population of interest is equal to 162.5. The appropriate null and alternative hypotheses, then, are

$$H_0: \mu = 162.5, \qquad H_1: \mu \neq 162.5$$

2 *Significance level* The investigators set the probability of committing a Type I error at $\alpha = 0.05$.

3 *Description of population and assumptions* The population consists of the weights of all varsity football players who played during the most recent ten years. The investigators feel that the weights of this population are approximately normally distributed with a standard deviation equal to 18.0, the standard deviation of the weights of those who played during the first ten seasons.

4 *The relevant statistic* Since the hypotheses are concerned with a population mean, the appropriate statistic is $\bar{x}$, the sample mean. Since it is assumed that the population is approximately normally distributed, the sampling distribution of $\bar{x}$, for all practical purposes, may be considered to be approximately normally distributed. If the null hypothesis is true, $\mu_{\bar{x}}$, the mean of the sampling distribution, is equal to 162.5. If, as the investigators believe, the population standard deviation is 18.0 pounds, the standard deviation of the sampling distribution (the standard error of $\bar{x}$) is $\sigma_{\bar{x}} = \sigma/\sqrt{n} = 18.0/\sqrt{25} = 3.6$.

5 *The test statistic and its distribution* Since the relevant statistic is $\bar{x}$, σ is known, and $\bar{x}$ is assumed to be approximately normally distributed; the test statistic is z, which is normally distributed with mean 0 and standard deviation 1.

6 *Rejection and acceptance regions* Since $\alpha = 0.05$, and since we have a two-sided test, the rejection region consists of two parts. The first part, located in the right tail of the distribution of z, consists of all values of z such that, when H_0 is true, the probability of the random occurrence of a z that large or larger is equal to or less than 0.025. The second half of the rejection region, located in the left tail of the distribution of z, consists of all values

of z such that, when H_0 is true, the probability of the random occurrence of a z that small or smaller is equal to or less than 0.025. Reference to Appendix Table E shows that the critical values for this rejection region are $z = +1.96$ and $z = -1.96$. The acceptance region consists of all values of z that are less than +1.96, but greater than −1.96. If, from the sample data, we compute a value of z equal to or greater than +1.96 or less than or equal to −1.96, we will reject the null hypothesis. Otherwise, we will not reject the null hypothesis.

The rejection and acceptance regions may also be described in terms of $\bar{x}$. The rejection region consists of two sets of values of $\bar{x}$ (computed from samples of size 25 drawn from the population of interest): those so large that the probability of the occurrence of values that large or larger, when H_0 is true, is equal to or less than 0.025, and those so small that the probability of the occurrence of values that small or smaller is equal to or less than 0.025. Critical values for this rejection region are values of $\bar{x}$ that are located a distance of 1.96 standard errors on each side of the hypothesized mean. The critical values are

$$\mu_0 \pm 1.96\frac{\sigma}{\sqrt{n}} = 162.5 \pm 1.96\frac{18}{\sqrt{25}}$$

$$= 162.5 \pm 7.1 = 155.4 \text{ and } 169.6$$

If the sample yields a value of $\bar{x}$, a distance of 1.96 standard errors or more from the hypothesized mean (that is, if the computed $\bar{x}$ either is greater than or equal to 169.6 or is equal to or less than 155.4), we will reject H_0. Otherwise, we will not reject. Figure 6.3 shows the rejection and acceptance regions in terms of z and $\bar{x}$.

7 *Collection of data and calculations* A simple random sample of the records of 25 players during the past ten years is selected. The mean of the weights in the sample is 178.7.

8 *Statistical decision* From the sample data we compute

$$z = \frac{178.7 - 162.5}{3.6} = 4.50$$

Since 4.50 is greater than 1.96, it falls in the rejection region, and we reject H_0.

Also note that 178.7 is greater than 169.6, the upper critical value expressed in terms of $\bar{x}$. We may therefore reject the null hypothesis, on this basis, without the necessity of computing a z value.

9 *Conclusion* Since we reject H_0, we look to the alternative hypothesis for our conclusion. In this example we are able to conclude, on the basis of the present sample data, that the mean weight of football players at the university during the past decade is different from the mean weight of players during the school's first decade.

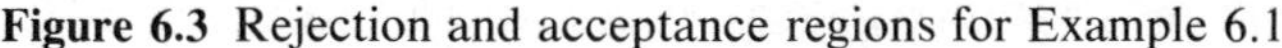

Figure 6.3 Rejection and acceptance regions for Example 6.1

(a)

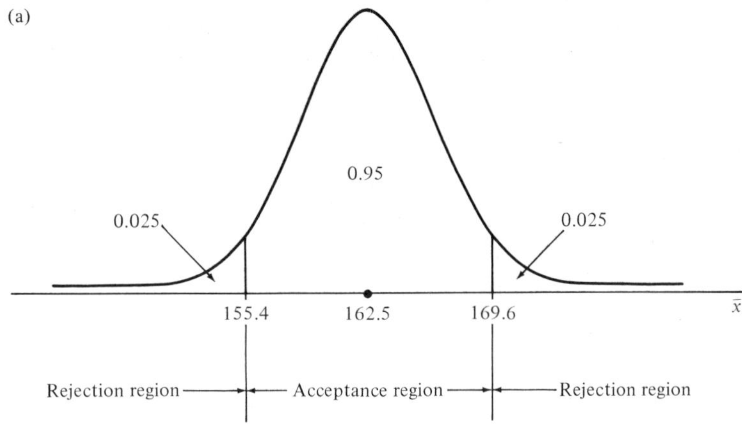

(b)

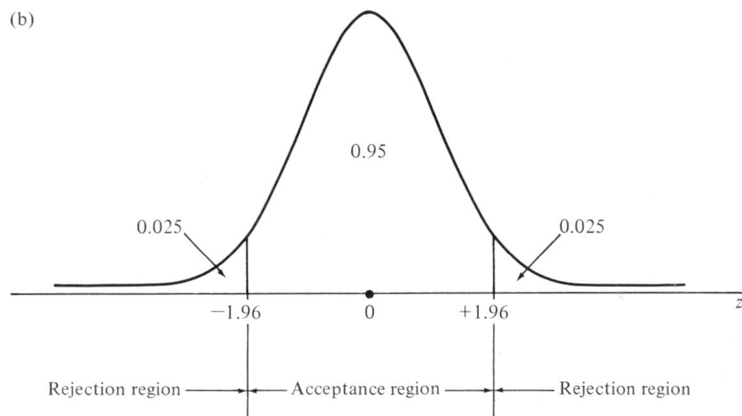

Reporting the results In journal articles containing statistical analyses of research projects, one encounters a variety of methods for presenting the results. Sometimes the value of the test statistic, or the sample statistic, is reported along with a statement as to whether or not it was significant at the chosen level of significance. Thus, by this method, we would report the results of the present example as "$z = 4.50$, significant at 0.05 level," or as "$\bar{x} = 178.7$, significant at 0.05 level."

When a result is significant at both the 0.05 level and the 0.01 level, many authors indicate this by the use of asterisks. A result that is significant at the 0.05 level, but not significant at the 0.01 level, is given one asterisk(*), and a statistic that is significant at the 0.01 level receives two asterisks (**). In the present example, since 4.50 is greater than 2.58 (the z value for a two-sided test with $\alpha = 0.01$), we would report the result as $z = 4.50^{**}$ or $\bar{x} = 178.7^{**}$. Perhaps the most common way of reporting results in the literature is by means of p values.

DEFINITION ***A p value** is the smallest value of α for which the null hypothesis may be rejected.*

It is the probability of obtaining, when H_0 is true, a value of the test statistic as extreme as or more extreme than that actually observed. If the statistical results are reported in a table, the p value is usually given as a footnote to the table. If the results are reported in the text of the article, the p value is usually similarly reported, sometimes in parentheses.

In determining a p value, we must take into account whether or not the test is one sided or two sided. If the test is two sided, the p values will be twice as large as they would be for a one-sided test, since we must allow for the probability of obtaining an extreme value of the test statistic in either direction.

To obtain the p value for the present example, which is two sided, we find the probability of observing a value of z as extreme as or more extreme than 4.50 in either direction when H_0 is true. Consulting Appendix Table E, we find that the largest tabulated value of z is 3.09, and the probability of obtaining a value this large or larger is $0.5 - 0.4990 = 0.001$. Since 4.50 is much further to the right of 0 than 3.09 is, the probability of observing a value of z as large as or larger than 4.50, when H_0 is true, is less than 0.001. Since $z = 4.50$ was computed as part of a two-sided test, we must allow for a value as extreme as 4.50 in the opposite direction. Consequently, the p value we seek is less than $2(0.001) = 0.002$. We would report this result as "$p < 0.002$." Figure 6.4 shows the p value for this example.

A one-sided hypothesis test Frequently the nature of the research hypothesis will be such that it will lead to a one-sided alternative hypothesis which will, in turn, lead to a one-sided test involving a one-sided rejection region. When either only extremely large values of the test statistic or only small values will cause rejection of the null hypothesis, a one-sided alternative is appropriate. We test the null hypothesis by means of a one-sided test, and a one-sided rejection region results. Suppose, for example, that sampling is from a

Figure 6.4 Calculation of p value for Example 6.1

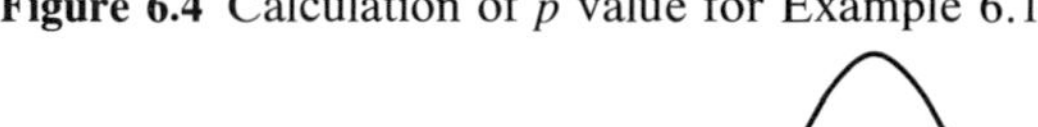

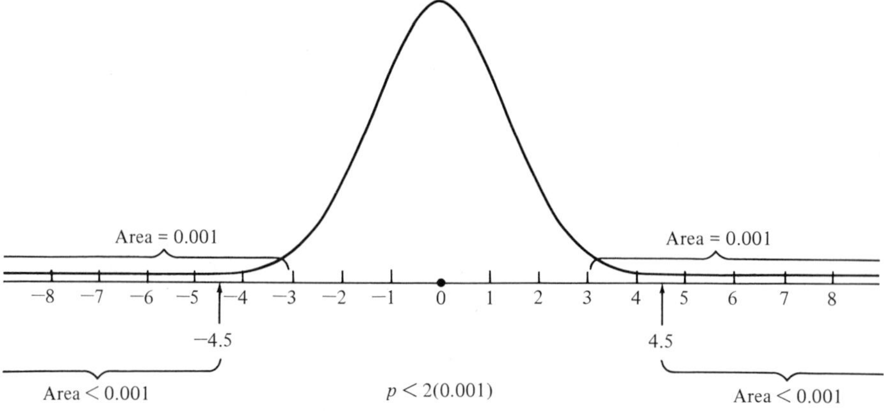

normally distributed population with a known population variance, and that the nature of the research hypothesis is such that the statistical hypotheses are

$$H_0: \mu \leq \mu_0, \qquad H_1: \mu > \mu_0 \qquad \text{with a significance level of } \alpha.$$

Since only large values of the test statistic would cause rejection of H_0 (small values would tend to support the null hypothesis), the rejection region must consist of large values of the test statistic, and hence must be located in the upper tail of the distribution of the test statistic. In fact, the rejection region must consist of those values of the test statistic that are so large that the probability of observing values that large or larger, when H_0 is true, is equal to or less than α. On the other hand, if the statistical hypotheses are

$$H_0: \mu \geq \mu_0, \qquad H_1: \mu < \mu_0$$

with a significance level of α, the rejection region must be located in the lower tail of the distribution of the test statistic, since only small values of the test statistic will cause rejection of the null hypothesis. Figure 6.5 shows rejection and acceptance regions for these two situations.

Figure 6.5 Rejection and acceptance regions for two sets of one-sided statistical hypotheses when sampling is from a normally distributed population with known population variance

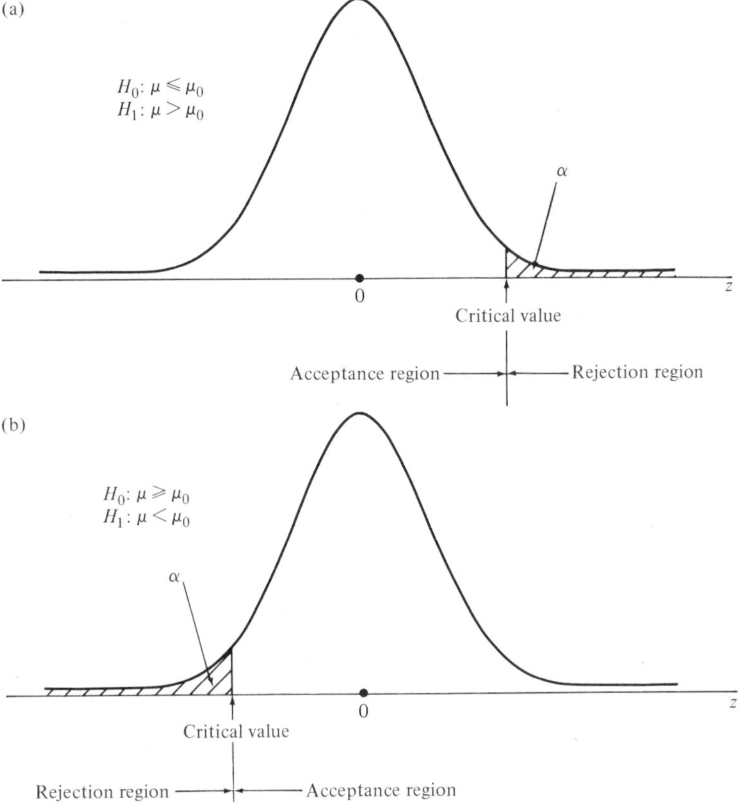

Note that for H_0: $\mu \leq \mu_0$ and H_0: $\mu \geq \mu_0$, there are a large number of hypothesized values for μ. The form of the hypotheses suggests that a hypothesis-testing procedure might be appropriate for each such hypothesized value. As a matter of practice, however, one tests the null hypothesis accompanying a one-sided alternative only at the point of equality. A little calculation should convince you that if H_0 is rejected when the test is made at the point of equality, H_0 will be rejected for any other hypothesized value of μ suggested by the null hypothesis.

Example 6.2 Experience has shown that for normal subjects within a certain age range, the mean reaction time to a certain stimulus is 65 msec (milliseconds) with a standard deviation of 15 msec. A psychological research team believes that subjects who receive a certain type of training will, on the average, show a shorter response time. To see whether it could substantiate this belief, the team carried out the following hypothesis-testing procedure.

1 *Statement of hypotheses* We may formally state the research hypothesis for this example as follows: "The mean reaction time of normal subjects who receive the experimental training is shorter than that of normal subjects who do not undergo such training." This research hypothesis leads to the following statistical hypotheses.

$$H_0: \mu \geq 65, \qquad H_1: \mu < 65$$

The alternative hypothesis is one sided since only "small" values of the test statistic will cause rejection of the null hypothesis. Note also that the research and alternative hypotheses are the same.

2 *Significance level.* We let $\alpha = 0.01$.

3 *Description of population and assumptions* The population, which consists of all response-time values for the given response which may ever be recorded for normal subjects, is hypothetical, since it does not actually exist at the moment. The researchers felt that it would be reasonable to assume that this population of hypothetical values, if available, would be normally distributed with a standard deviation of 15, the standard deviation of normal subjects not receiving the training. A sample of 20 subjects participated in the experiment.

4 *The relevant statistic* The relevant statistic is $\bar{x}$, and under the assumption of a normally distributed population, the researchers were able to assume that the sampling distribution of $\bar{x}$ would be normally distributed with a mean of 65, if the null hypothesis is true, and a standard deviation of $\sigma_{\bar{x}} = 15/\sqrt{20} = 3.35$.

5 *The test statistic and its distribution* Since the relevant statistic, $\bar{x}$, is normally distributed, and since σ is presumed to be known, the appropriate test statistic is z.

6 *Rejection and acceptance regions* Since only "small" values of the computed test statistic will cause rejection of the null hypothesis, the rejection region must be located in the left tail of the distribution of z. In other words, the rejection region will consist of all values of z so small that the probability of obtaining a value that small or smaller, when H_0 is true, is equal to or less than 0.01, the chosen level of significance. From Appendix Table E we find the critical value of z to be -2.33. We may obtain a critical value in terms of $\bar{x}$ by noting that it is located a distance of 2.33 standard errors to the left of the assumed mean of the sampling distribution of $\bar{x}$. Since $\sigma_{\bar{x}} = 3.35$, this distance is equal to $2.33 \times 3.35 = 7.81$. The critical value in terms of $\bar{x}$, then, is $65 - 7.81 = 57.19$. Figure 6.6 shows the rejection and acceptance regions in terms of both z and $\bar{x}$.

Figure 6.6 Rejection and acceptance regions for Example 6.2

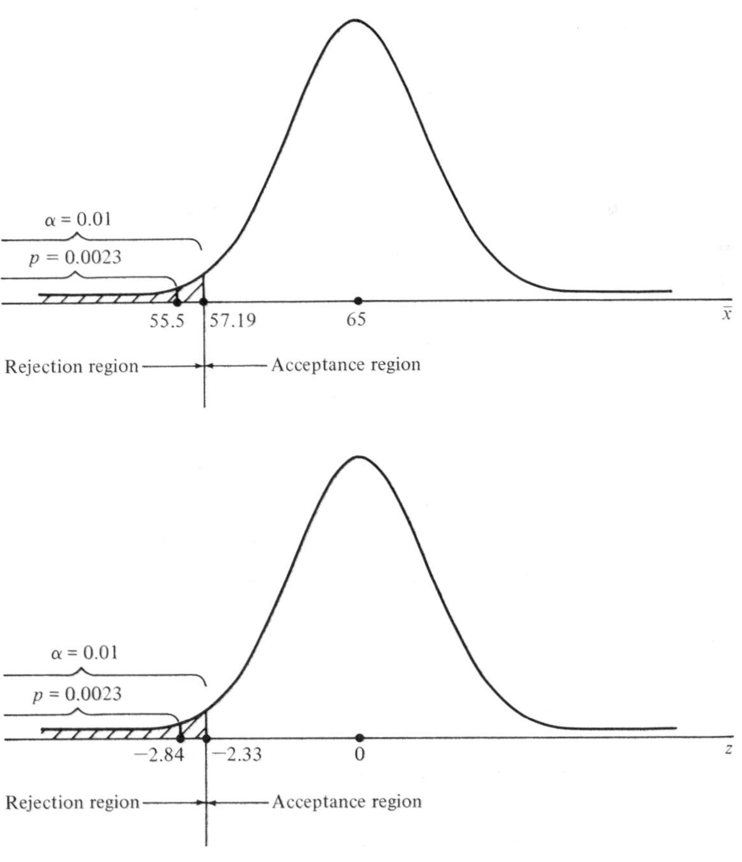

7 *Collection of data and calculations* The 20 normal subjects received the training, and were subsequently tested to determine their reaction times to the stimulus. Researchers recorded a mean reaction time of 55.5 msec. From these data we may compute $z = (55.5 - 65)/3.35 = -2.84$.

8 *Statistical decision* Since the computed z of -2.84 is less than -2.33 (that is, since -2.84 falls in the rejection region), we reject H_0. We note also that $\bar{x} = 55.5$ falls in the rejection region defined in terms of $\bar{x}$. Regardless of whether the relevant statistic (in this case $\bar{x}$) or the test statistic is used in deciding whether or not to reject H_0, in a given situation the decision will always be the same in each case. Figure 6.6 shows the locations of the computed values of $\bar{x}$ and z, relative to the critical values. From Appendix Table E we find that the probability of obtaining a z value as small as or smaller than -2.84, when H_0 is true, is 0.0023. The probability, then, of observing a value of $\bar{x}$ as small as or smaller than 55.5, when H_0 is true, is 0.0023. Thus the p value for this example is 0.0023, as indicated in Figure 6.6.

9 *Conclusion* Since we reject H_0, we conclude that H_1 is true. That is, in the present example, we conclude that the mean reaction time of subjects who receive the training is shorter than that of comparable subjects who do not receive the training.

Normally distributed population, σ^2 unknown

In the typical situation in which a test of a hypothesis about a population mean is appropriate, the population variance σ^2 is unknown, and consequently one cannot determine exactly $\sigma/\sqrt{n}$, the standard error of the relevant statistic $\bar{x}$. When the sample is large, one may compute a satisfactory estimate of σ^2 from sample data. When the population of interest is normally distributed, the sample means are normally distributed, and one can use the test statistic z. Even when the sampled population is not normally distributed, the sampling distribution of the sample mean is approximately normally distributed, as a result of the central-limit theorem, and therefore one may use z as the test statistic. However, when the sample size is small, the central-limit theorem does not apply, and one must seek a test statistic other than z. If it is known that the sampled population is at least approximately normally distributed, or if, in the absence of certain knowledge, this seems to be a reasonable assumption, the t statistic, discussed in Chapters 4 and 5, is the best choice for a test statistic. In Chapter 11 you will be introduced to hypothesis-testing procedures that are appropriate when the sample size is small and it cannot be assumed that the sampled population is normally distributed.

Example 6.3 A drug manufacturer claims that the mean time required for the contents of a certain capsule to dissolve is 50 minutes. A research team with a competing firm does not believe the claim. This research team tested a random sample of 20 capsules and, from the results, computed a sample mean of 54 minutes and a standard deviation of 15 minutes. The research team wished to know whether it could conclude that the mean time required for the contents of the capsule to dissolve is greater than 50 minutes. The team carried out the following hypothesis-testing procedure.

1 *Statement of hypotheses* The research hypothesis is as follows: "The mean time required for the contents of the capsule to dissolve is greater than 50 minutes." The statistical hypotheses are

$$H_0: \mu \leq 50. \qquad H_1: \mu > 50$$

2 *Significance level* The probability of committing a Type I error is set at $\alpha = 0.05$.

3 *Description of population and assumptions* The research team assumes that the population of dissolving times is approximately normally distributed.

4 *The relevant statistic* The relevant statistic is $\bar{x}$, the sample mean.

5 *The test statistic and its distribution* Since n is small (less than 30), σ is unknown, and it is assumed that the sampled population is normally distributed, the appropriate test statistic is

$$t = \frac{\bar{x} - \mu_0}{s/\sqrt{n}}$$

which follows the Student's t distribution with $n - 1$ degrees of freedom.

6 *Rejection and acceptance regions* From Appendix Table E we find the critical value of t for a one-sided test with $\alpha = 0.05$ and $20 - 1 = 19$ degrees of freedom to be 1.7291. The critical value in terms of $\bar{x}$ is given by $50 + (1.7291)(15/\sqrt{20}) = 55.8$. Figure 6.7 shows the acceptance and rejection regions in terms of t.

Figure 6.7 Acceptance and rejection regions for Example 6.3

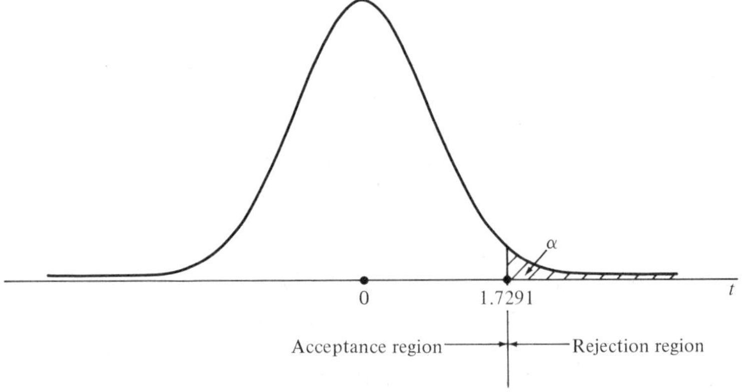

7 *Collection of data and calculations* As we have noted, a random sample of 20 observations yielded a mean of 54 and a standard deviation of 15. From these data, we may compute

$$t = \frac{54 - 50}{15/\sqrt{20}} = 1.19$$

8 *Statistical decision* Since the computed t of 1.19 is less than 1.7291 (that is, it falls in the acceptance region), we cannot reject H_0. We reach the same decision by noting that $\bar{x} = 54$ is less than the critical $\bar{x}$ of 55.8. By referring to Appendix Table F, we may obtain some notion of the magnitude of the p value for this test. We note that, for 19 degrees of freedom, the probability of obtaining a t value as large as or larger than 1.328, when H_0 is true, is 0.10. Since the computed t of 1.19 is less than 1.328, we see that for this test $p > 0.10$. To obtain a more exact value for p, we would need to consult a more complete table of the t distribution.

9 *Conclusion* Since we do not reject H_0, we conclude that H_0 may be true; that is, the mean time required for the contents of the capsule to dissolve may be 50 minutes or less.

We noted in Chapter 5 that, when the sample size is large, many users of statistics prefer to use the z distribution rather than the t distribution when constructing confidence intervals for μ, even though σ is unknown. It is also true that many practitioners prefer z to t in testing hypotheses when they have large samples, despite the fact that σ is unknown. This practice is justified by the fact that, when H_0 is true ($\mu = \mu_0$) and n is large, the statistic

$$\frac{\bar{x} - \mu_0}{s/\sqrt{n}}$$

is distributed approximately as the standard normal distribution. When one follows this practice, one compares the computed value of the test statistic for significance with an appropriate value of the z distribution.

Sampling from a nonnormally distributed population

Frequently the population of interest is not normally distributed. In other cases an investigator may not be willing to assume that a population is normally distributed when she or he doesn't know its functional form. In situations such as these, the t statistic is not appropriate as a test statistic, and the z statistic is appropriate only if the sample size is large. In the following example we shall illustrate the hypothesis-testing procedure that is appropriate when sampling is from a nonnormally distributed population with an unknown variance (the typical case), and when the sample size is large enough for us to use the central-limit theorem.

Example 6.4 A group of research professors at a school of education at a state university hypothesized that enrichment of the high school curriculum would increase the verbal scores made by students on college board examinations. To see whether they could obtain evidence to support their hypothesis, the professors introduced a curriculum-enrichment program into the freshman class at a local high school. The program continued for this class through the senior year. At the end of their senior year, 125 members of this class took the college board examinations. The mean verbal score was 590, with a standard deviation of 35. The mean verbal score of students taking the examination during the five previous years was 580. The professors wished to know whether they could conclude that curriculum enrichment had increased the mean verbal score. The following hypothesis-testing procedure may be carried out.

1 *Statement of hypotheses* Research hypothesis: "Enrichment of the high school curriculum improves verbal scores made on college board examinations."

H_0: $\mu \leq 580$, H_1: $\mu > 580$

2 *Significance level* We let $\alpha = 0.05$.

3 *Description of population and assumptions* Since the sample size is large, the central-limit theorem applies, and the functional form of the population is of no concern. It is assumed that the 125 scores constitute a random sample of scores from a large population of scores.

4 *The relevant statistic* Since the hypotheses are concerned with a population mean, the relevant statistic is $\bar{x}$, the sample mean. The distribution of $\bar{x}$ is approximately normally distributed, since n is large. If H_0 is true, the mean of the sampling distribution of $\bar{x}$ is 580 or less. Since the test will be performed at the point of equality, the relevant distribution has a mean of 580 if H_0 is true. The estimated standard error of $\bar{x}$ is given by $s/\sqrt{n} = 35/\sqrt{125} = 3.13$.

5 *The test statistic and its distribution* The appropriate test statistic is z, which is normally distributed with a mean of 0 and a standard deviation of 1.

6 *Rejection and acceptance regions* The critical value of z is 1.645, so the rejection region consists of all values of z equal to or greater than 1.645, and the acceptance region consists of all values of z less than 1.645. The critical value of $\bar{x}$ is $580 + (1.645)(3.13) = 585.15$. In terms of $\bar{x}$, the rejection region consists of all values of $\bar{x}$ greater than or equal to 585.15, and the acceptance region consists of all values of $\bar{x}$ less than 585.15.

7 *Collection of data and calculations* As we have already noted, $n = 125$, $\bar{x} = 590$, and $s = 35$. From these data we may compute

$$z = \frac{590 - 580}{35/\sqrt{125}} = 3.19$$

8 *Statistical decision* Since computed $z = 3.19$ is greater than critical $z = 1.645$, we reject H_0. Alternatively, we may reject H_0 because observed $\bar{x} = 590$ is greater than critical $\bar{x} = 585.15$. The p value for this test is less than 0.001.

9 *Conclusion* Since H_0 is rejected, the professors may conclude that enrichment of the high school curriculum does improve verbal scores on college board examinations.

EXERCISES

1 From a normally distributed population with a standard deviation of 32, a simple random sample of size 16 yields a mean and standard deviation of 520 and 40, respectively. From these data, can one conclude, at the 0.05 significance level, that μ is greater than 516? Draw a picture to illustrate the location of the rejection and acceptance regions in terms of both the relevant statistic and the test statistic. What is the p value for this test?

2 A simple random sample of size nine drawn from a normally distributed population yielded a mean and standard deviation of 150 and 30, respectively. Do these data provide sufficient evidence to warrant the conclusion that the population mean is less than 160? What is the p value for this test?

3 A simple random sample of 100 junior high school students, selected from the junior high schools in a certain city, reported a mean weekly allowance of \$3.25, with a standard deviation of \$1.00. Do these data provide sufficient evidence to indicate that the population mean is different from \$3.00? What is the p value for this test?

4 A reading specialist feels that students in ungraded classes score higher on reading comprehension tests than students in graded classes. The mean reading comprehension test score of students in graded classes entering the fourth grade during the past five years is 4.25. A group of 81 students who attended ungraded classes during their first three years of school made a mean reading comprehension test score of 5.30, with a standard deviation of 1.8. Do these data provide sufficient evidence to support the reading specialist's hypothesis? Let $\alpha = 0.01$. What is the p value for this test?

5 An agricultural researcher believed that the mean number of acres farmers in a given state devoted to a certain crop was less than six acres. The researcher mailed a questionnaire to a simple random sample of 25 farmers in the state, asking them to report the number of acres planted in that crop. The sample mean and standard deviation were 5 and 1.5 acres, respectively. At the 0.05 level of significance, do these data support the researcher's belief? What is the p value for this test?

6 A high school counselor has found that, during the past five years, seniors with no vocational counseling who took a vocational maturity test had a mean score of 190. The counselor believes that students who receive individual vocational counseling will score higher than this on the average.

The mean score of 64 seniors who received individual vocational counseling during their last year of high school was 205, with a standard deviation of 24. Do these data support the counselor's belief? Let $\alpha = 0.05$. What is the p value for this test?

7 A social worker believes that the mean number of years of school completed by adults who are on welfare is less than five. A random sample of 169 of these adults yielded a mean of 4.6 years, with a standard deviation of 3.9 years. Do these data provide sufficient evidence for the social worker to conclude that $\mu < 5$ years? Let $\alpha = 0.05$. What is the p value for this test?

8 A random sample of 100 households selected from a certain area yielded a mean annual total family income of \$9700 and a standard deviation of \$1000. Do these results provide sufficient evidence to indicate that the true mean is less than \$10,000? Let $\alpha = 0.05$.

9 A random sample of 16 adult females of a small species of mammal was selected from a certain geographic region. The mean tail length for the sample was 94 mm, with a standard deviation of 12 mm. The mean tail length for adult females of this species in another geographic region was 81 mm. Do these data provide sufficient evidence to indicate that the sample came from a population with a mean greater than 81? Let $\alpha = 0.05$. What is the p value for this test? What assumptions are necessary?

10 A random sample of 25 persons engaged in a certain occupation had a mean spatial aptitude score of 89, with a standard deviation of 20. Do these data provide sufficient evidence to indicate that the true mean score for the population represented is less than 100? Let $\alpha = 0.05$. What is the p value for this test?

11 A social worker believes that the mean weight of ten-year-old boys residing in a rural area is less than 34 kg. A random sample of 25 boys selected from the population had a mean weight of 30 kg and a standard deviation of 10 kg. Do these data support the social worker's belief at the 0.05 level of significance? State any assumptions necessary for the application of your testing procedure, and compute the p value for the test.

12 A nutritionist believes that the mean daily protein consumption in a certain population is less than 75 grams. A random sample of 16 subjects yielded a mean of 73.8 grams, with a standard deviation of 2.4 grams. Do these data support the nutritionist's belief? Let $\alpha = 0.05$. Calculate the p value for this test, and state any assumptions that are necessary.

13 A survey of 64 professional employees at a correctional institution revealed the mean length of employment in the correctional field to be five years, with a standard deviation of four years. Do these data support the hypothesis that the mean length of employment of all employees of this type is less than six years? Let $\alpha = 0.05$.

14 A survey of 100 male students enrolled at an urban university revealed that, during a certain spring quarter, the mean amount of money spent on clothes was \$55, with a standard deviation of \$20. Test the null hypothesis that $\mu = \$60$. Let $\alpha = 0.05$. Calculate the p value for this test.

6.4 THE TYPE II ERROR AND THE POWER OF A TEST

In the hypothesis tests that we have discussed, α, the probability of making a Type I error (rejecting a true null hypothesis), has been under the control of the investigator and has been set at some small value, either 0.05 or 0.01. In this section we shall discuss in more detail β, the probability of committing a Type II error (accepting a false null hypothesis).

The Type II error

Consider the hypotheses H_0: $\mu = \mu_0$ and H_1: $\mu \neq \mu_0$, with $\alpha = 0.05$. We assume that the relevant population is normally distributed with known variance σ^2. With $\alpha = 0.05$, the rejection region is defined and consists of values of $\bar{x}$ greater than or equal to $\mu_0 + 1.96\sigma_{\bar{x}}$ and less than or equal to $\mu_0 - 1.96\sigma_{\bar{x}}$, where μ_0 is the mean, under H_0, of the sampling distribution of $\bar{x}$. Figure 6.8 shows this sampling distribution, which forms the basis for the hypothesis test.

If H_0 is false, it will be false because μ is equal to some value other than μ_0. If H_0 is false, the true sampling distribution of $\bar{x}$ will not be centered over μ_0, as shown in Figure 6.8, but will be centered over the true population mean. If μ is equal to μ_1, for example, the sampling distribution of $\bar{x}$ will be centered over μ_1. The rejection and acceptance regions, however, remain fixed, since they are determined by α and H_0. If $\bar{x}$, computed from the single sample drawn from the population for the purpose of testing H_0, falls in the acceptance region when μ really equals μ_1, H_0 will be "accepted," and a Type II error will be committed. The probability, β, of this happening is equal to that portion of the area under the curve of $\bar{x}$, centered over μ_1, that overlaps the area under the curve of $\bar{x}$, centered over μ_0, that lies between the critical values of $\bar{x}$. See Figure 6.9.

Figure 6.8 Assumed sampling distribution of $\bar{x}$ for H_0: $\mu = \mu_0$, H_1: $\mu \neq \mu_0$, when sampling is from a normally distributed population with variance σ^2 ($a = 0.05$)

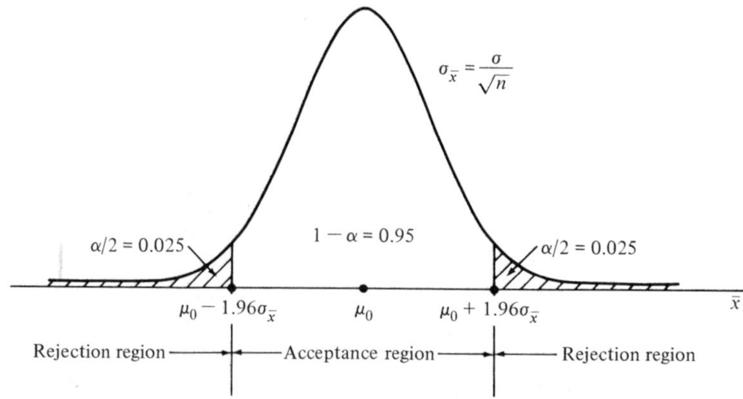

Figure 6.9 Sampling distribution of Figure 6.8, showing some possible alternatives and corresponding β's when H_0 is false

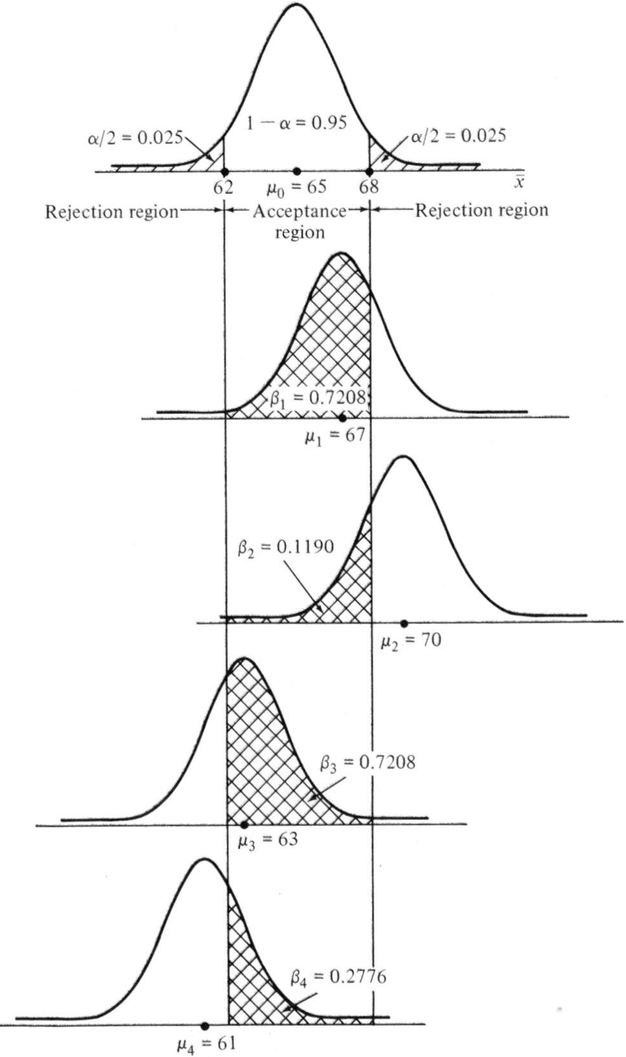

Under the hypothesis H_1: $\mu \neq \mu_0$, μ may assume an infinite number of values, so there are an infinite number of possible values of β. The one that applies in a given situation, when H_0 is false, depends on the true value of μ. In practice we do not know the true value of μ when H_0 is false, and hence we never know the true value of β. Figure 6.9 shows several possible alternatives for μ, when H_0 is false, and the corresponding β's. In this figure the sampling distributions for alternative values of μ are displayed vertically for clarity. You should realize that in reality the alternative values of μ are all located on the same $\bar{x}$ axis, and consequently the corresponding sampling distribution curves have the same $\bar{x}$ line as their horizontal axes. You should also realize that although only six alternatives to H_0: $\mu = \mu_0$ are shown in Figure 6.9, there are an infinite number of alternatives. By observing Figure 6.9, one can see that alternatives to μ_0 that are located near μ_0 yield larger values of β than alternatives that are far from μ_0. For example, the distance between μ_1 and μ_0 is shorter than that between μ_3 and μ_0, and consequently β_1 is greater than β_3.

Let us now illustrate, by means of an example, the calculation of a Type II error.

Example 6.5 A clinical psychologist wished to test, at the 0.05 level of significance, the hypothesis that the mean performance IQ of a certain group of mental retardates was 65. A random sample of 50 subjects yielded a standard deviation of 12. The psychologist also wanted to calculate the probability of committing a Type II error. Alternative values of μ for which β was computed were $\mu_1 = 67$, $\mu_2 = 70$, $\mu_3 = 63$, and $\mu_4 = 61$.

Critical $\bar{x}$ values for the hypothesis test are

$$65 \pm 1.96\frac{12}{\sqrt{50}} = 65 \pm 3 = 62 \text{ and } 68$$

Figure 6.10 shows these critical values.

In determining β for the various alternative values of μ, we assume that $s = 12$, the sample estimate of σ, is an appropriate estimate in each case. We first compute β for the alternative $\mu_1 = 67$. If $H_0 = 65$ is false because μ is really equal to 67, the appropriate sampling distribution of $\bar{x}$ would be centered over 67. Since the critical values of $\bar{x}$ under H_0 are 62 and 68, we will "accept" H_0 for any observed value of $\bar{x}$ falling between 62 and 68. If μ is really equal to 67, however, we will be "accepting" a false null hypothesis. The probability of "accepting" a false null hypothesis (committing a Type II error) when $\mu = 67$ is equal to the area between 62 and 68 that is under the

Figure 6.10 Critical values for the hypothesis test of Example 6.5

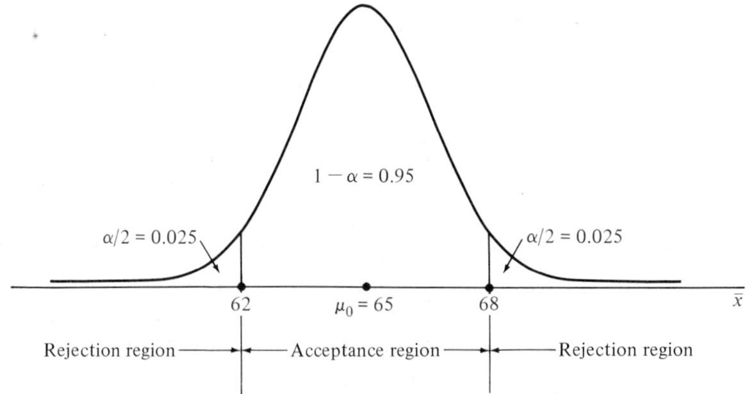

Figure 6.11 Sampling distribution of $\bar{x}$ under H_0, and sampling distribution of the alternative $\mu_1 = 67$ and β_1 for Example 6.5

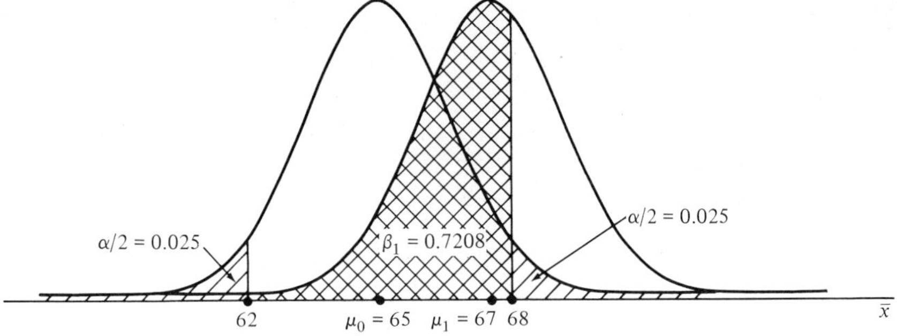

curve centered over 67. We may express this area in probabilistic terms as follows.

$$P(62 < \bar{x} < 68 \mid \mu = 67) = P(67 < \bar{x} < 68 \mid \mu = 67) + P(62 < \bar{x} < 67 \mid \mu = 67)$$

To evaluate this probability numerically, we convert $\bar{x}$ to the standard normal scale and obtain

$$P\left(\frac{62 - 67}{12/\sqrt{50}} < z < \frac{68 - 67}{12/\sqrt{50}}\right) = P\left(\frac{67 - 67}{12/\sqrt{50}} \leq z < \frac{68 - 67}{12/\sqrt{50}}\right) + P\left(\frac{62 - 67}{12/\sqrt{50}} < z \leq \frac{67 - 67}{12/\sqrt{50}}\right)$$

$$= P(-2.95 < z < 0.59) = P(0 \leq z < 0.59) + P(-2.95 < z \leq 0)$$

$$= 0.2224 + 0.4984 = 0.7208$$

Figure 6.11 shows this value of β, β_1 graphically.

Similar calculations for $\mu_2 = 70$, $\mu_3 = 63$, and $\mu_4 = 61$ yield, respectively, the values $\beta_2 = 0.1190$, $\beta_3 = 0.7208$, and $\beta_4 = 0.2776$. Figure 6.12 shows these values of β, along with β_1.

Note that β decreases as the distance between μ_0 and the alternative value of μ for which β is computed increases. Also note that in Example 6.5, all computed values of β, as shown in Figure 6.12, are larger than the preselected value of $\alpha = 0.05$. In fact, one has to select an alternative value of μ of about 70.8 or 59.2 before the associated value of β is less than 0.05. It seems safe to say, then, that β, the probability of "accepting" a false null hypothesis, is always larger than α except when the null hypothesis is false because the true value of μ is "a long way" from μ_0.

In many practical situations, we are not motivated to test hypotheses about a population mean when, if H_0 is false, the true value of μ is a long way

Figure 6.12 Various values of β for Example 6.5

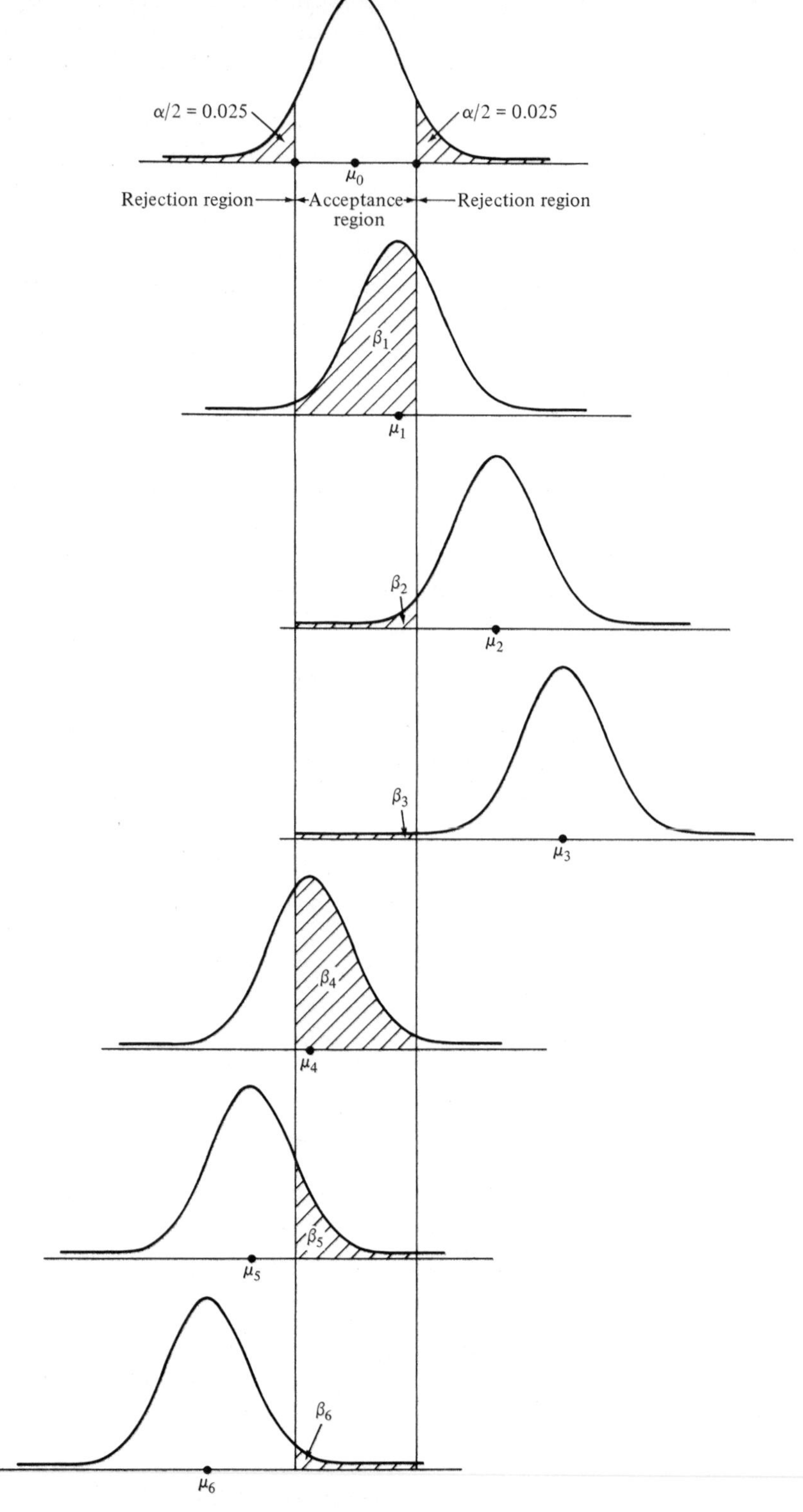

from μ_0. For example, we cannot imagine anyone being interested in testing statistically the null hypothesis that the mean height of six-year-old boys is equal to the mean height of adult males. On the other hand, we would not question someone's interest in testing the null hypothesis that the mean height of some particular group of adult females is equal to the mean height of another group of adult females. In other words, in many practical situations, if H_0 is false, it is false because the true value of μ is close to μ_0. Furthermore, the closer the true μ is to μ_0, the larger the value of β, the probability of "accepting" a false null hypothesis. It is for this reason that we say that a conclusion based on a rejected null hypothesis is more decisive than one based on an "accepted" null hypothesis.

It is for this reason, also, that when we reject a null hypothesis, we say that H_1 is true, yet when we "accept," or fail to reject, a null hypothesis, we say that H_0 *may be* true.

The power of a test

A concept that is useful in evaluating hypothesis tests is the power of a test. The *power of a test* is the probability of rejecting a false null hypothesis. It is generally expressed as $1 - \beta$. For a given α, we say that one test is *more powerful* than another if the value of $1 - \beta$ for the one is greater than it is for the other for all values of μ.

It is frequently useful to have available, for a particular test, what is known as a power function.

DEFINITION *A* ***power function*** *is a function that shows the relationship between the probability of rejecting a null hypothesis and the various values which the parameter may assume under the null and alternative hypotheses for a given level of significance.*

Table 6.2 gives some values of the power function for Example 6.5.

One may obtain a *power curve* by graphing the power function. Possible values of the parameter on the horizontal axis are plotted against the values of $1 - \beta$ on the vertical axis. Figure 6.13 shows the graph of the power function of Table 6.2.

Table 6.2 Values of the power function for Example 6.5

Alternative value of μ	β	Value of power function $(1 - \beta)$
58	0.0091	0.9909
59	0.0384	0.9616
60	0.1190	0.8810
61	0.2776	0.7224
63	0.7208	0.2792
$\mu_0 = 65$	$1 - \alpha = 0.9500$	$\alpha = 0.0500$
66	0.8716	0.1284
67	0.7208	0.2792
70	0.1190	0.8810
71	0.0384	0.9616

Figure 6.13 Power function for Example 6.5

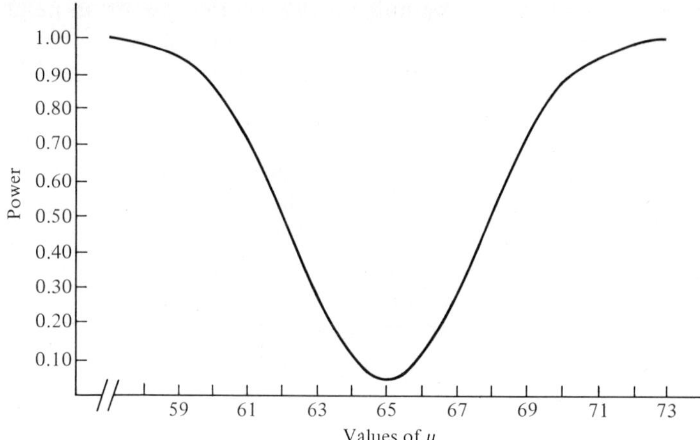

The power function is used to determine the magnitude of $1 - \beta$ when specified values of the alternative hypothesis are true.

Figure 6.13 illustrates the general V-shaped appearance of a power curve for a two-sided test. In general, a two-sided test that discriminates well between the value of the parameter in H_0 and values in H_1 (except those lying close to the value specified in H_0) will result in a narrow V-shaped power curve. A widespread V-shaped curve indicates that the test discriminates poorly over a relatively wide interval of alternative values of the parameter.

The power curve for a one-sided test with the rejection region in the upper tail appears as an elongated S. A one-sided test with the rejection region in the lower tail of the distribution results in a power curve that appears as a reverse elongated S. Figure 6.14 shows the power curve for Ex-

Figure 6.14 Power curve for one-sided test, rejection region in lower tail (Example 6.2)

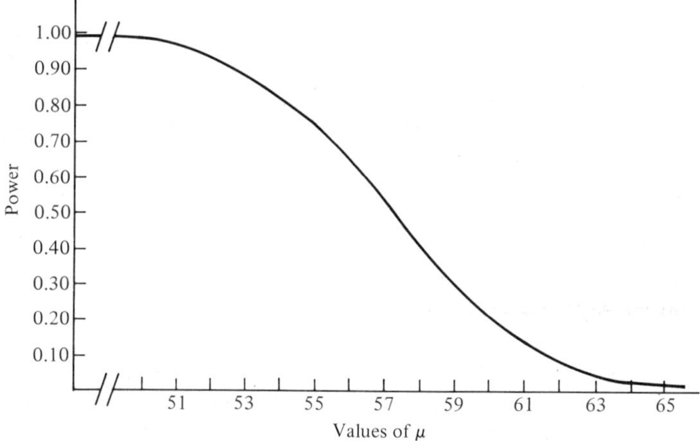

ample 6.2, which uses a one-sided test with the rejection region in the lower tail of the sampling distribution.

EXERCISES

15 Refer to Exercise 1. Construct and graph the power function.
16 Construct and graph the power function for Exercise 3.
17 Construct and graph the power function for Exercise 4.
18 Construct and graph the power function for Exercise 6.
19 Construct and graph the power function for Exercise 7.
20 Construct and graph the power function for Exercise 8.
21 Construct and graph the power function for Exercise 13.
22 Construct and graph the power function for Exercise 14.

6.5 TESTING A HYPOTHESIS ABOUT THE DIFFERENCE BETWEEN TWO POPULATION MEANS

In Chapter 5 we discussed the construction of confidence intervals for the difference between two population means. In this chapter we consider the problem of testing hypotheses about the difference between two population means. The example, discussed earlier, of the team of psychologists interested in the self-esteem scores of conformists and nonconformists is an illustration of this kind of hypothesis test. In this example the psychologists want to know whether they can conclude that the mean self-esteem scores of nonconformists is higher than that of conformists. We can cite other examples. A biologist may want to know whether it is possible to conclude that the mean life span of some animal is lower in one type of environment than in another. A sociologist may want to know whether the mean number of years of education is different in two populations. An economist may wish to know whether the mean family income is different in two groups. We shall discuss both two-sided tests and one-sided tests in each of three situations: (1) when sampling is from two at least approximately normally distributed populations with known population variances, (2) when sampling is from two at least approximately normally distributed populations with unknown but equal population variances, and (3) when sampling is from two nonnormally distributed population.

We discussed the appropriate sampling distributions for these situations in Chapter 4.

Normally distributed populations, σ_1^2 and σ_2^2 known

Example 6.6 In a suburban school system a random sample of 25 fifth-grade students (group A) was selected from the population of students whose parents both work. A random sample of 15 students (group B) in the same grade and school system was selected from among those whose fathers, only, work. Analysis of the school achievement scores of the two groups gave the following results.

	Mean score ($\bar{x}$)
Group A	78
Group B	85

Experience shows that the populations of scores for both groups are approximately normally distributed with variances of $\sigma_A^2 = 81$ and $\sigma_B^2 = 25$. In order to determine whether they can conclude, on the basis of these data, that the mean of the population from which group A was selected is lower than the mean of the population from which group B was selected, researchers may carry out the following hypothesis test.

1 *Statement of hypotheses*

$$H_0\colon \mu_A \geq \mu_B \quad \text{or, equivalently,} \quad \mu_A - \mu_B \geq 0$$

$$H_1\colon \mu_A < \mu_B \quad \text{or, equivalently,} \quad \mu_A - \mu_B < 0$$

2 *Significance level* $\alpha = 0.05$.

3 *Description of populations and assumptions* As noted above, researchers felt that it was reasonable to assume approximately normally distributed populations. The samples are independent.

4 *The relevant statistic* Since a hypothesis about the difference between two population means is to be tested, the relevant statistic is the difference between sample means computed from samples drawn from the populations. The statistic may be designated by $\bar{x}_A - \bar{x}_B$. From Chapter 4 we know that in this situation, we can consider the sampling distribution of $\bar{x}_A - \bar{x}_B$ to be normally distributed with a variance of

$$\sigma^2_{\bar{x}A - \bar{x}B} = \frac{\sigma_A^2}{n_A} + \frac{\sigma_B^2}{n_B}$$

and, if H_0 is true, a mean of 0.

5 *The test statistic and its distribution* Since we assume the populations to be normally distributed, and since we know the population variances, the appropriate test statistic is z, which follows the standard normal distribution.

6 *Rejection and acceptance regions* The critical value of z is -1.645. The critical value of $\bar{x}_A - \bar{x}_B$ is

$$0 - 1.645\sqrt{\frac{81}{25} + \frac{25}{15}} = -3.64$$

7 *Collection of data and calculations* From the results given earlier,

we find $\bar{x}_A - \bar{x}_B = 78 - 85 = -7$. The z value that may be computed from the data is

$$z = \frac{-7 - 0}{\sqrt{(81/25) + (25/15)}} = -3.16$$

8 *Statistical decision* Since $-7 < -3.64$ and $-3.16 < -1.645$, we can reject H_0.

9 *Conclusion* Researchers concluded that, in this school system on the average, the overall achievement scores of fifth-grade students whose parents both work are lower than those for students whose fathers, only, work.

We may test, in a manner similar to that just described, two-sided hypothesis tests in which the hypotheses are of the form

$$H_0\colon \mu_1 = \mu_2 \qquad \text{or, equivalently,} \qquad \mu_1 - \mu_2 = 0$$

$$H_1\colon \mu_1 \neq \mu_2 \qquad \text{or, equivalently,} \qquad \mu_1 - \mu_2 \neq 0$$

and one-sided tests in which the hypotheses are of the form

$$H_0\colon \mu_1 \leq \mu_2 \qquad \text{or, equivalently,} \qquad \mu_1 - \mu_2 \leq 0$$

$$H_1\colon \mu_1 > \mu_2 \qquad \text{or, equivalently,} \qquad \mu_1 - \mu_2 > 0$$

Normally distributed populations, σ_1^2 and σ_2^2 unknown but equal

Example 6.7 Two research professors in a university school of education wish to compare the overall achievement scores of eighth-grade students who have been mobile (population 1) during their elementary school years with the scores of students who have not (population 2). Specifically, they want to see whether they can conclude, on the basis of sample data ($n_1 = 15$, $n_2 = 22$), that the mean achievement scores in the two groups are different. The professors defined mobile students as those who have attended two or more elementary schools. They classified as nonmobile students who have attended the same school throughout their elementary school years. The researchers carry out the following hypothesis-testing procedure.

1 *Statement of hypotheses* Since the researchers have no basis for specifying the direction of any difference that might exist in the two population means, they make their alternative hypothesis two-sided. Thus

$$H_0\colon \mu_1 = \mu_2 \qquad \text{or, equivalently,} \qquad \mu_1 - \mu_2 = 0$$

$$H_1\colon \mu_1 \neq \mu_2 \qquad \text{or, equivalently,} \qquad \mu_1 - \mu_2 \neq 0$$

2 *Level of significance* Let $\alpha = 0.05$.

3 *Description of populations and assumptions* The professors

assume that both populations are approximately normally distributed. The population variances are unknown, but they are assumed to be equal. The samples are independent.

4 *The relevant statistic* The relevant statistic is $\bar{x}_1 - \bar{x}_2$, which, because the two populations are assumed to be approximately normally distributed, we may consider normally distributed. If H_0 is true, the mean of the sampling distribution is $\mu_1 - \mu_2 = 0$, and its variance is $(\sigma_1^2/n_1) + (\sigma_2^2/n_2)$. Since σ_1^2 and σ_2^2 are unknown, however, we cannot compute the true variance of $\bar{x}_1 - \bar{x}_2$, and consequently we rule out z as the test statistic.

5 *The test statistic* As noted in step 4, z is not the appropriate test statistic. Since the two populations are assumed to be approximately normally distributed, with unknown but equal variances, Student's t statistic with $n_1 + n_2 - 2$ degrees of freedom, as discussed in Chapter 4, is the appropriate test statistic.

6 *Rejection and acceptance regions* Since the degrees of freedom are $15 + 22 - 2 = 35$ and $\alpha = 0.05$, we find from Appendix Table F that the critical values of t are ± 2.0301. We cannot compute the critical values of $\bar{x}_1 - \bar{x}_2$ until we have computed the sample variances.

7 *Collection of data and calculations* We compute the following sample means and variances.

$$\bar{x}_1 = 85, \qquad \bar{x}_2 = 87, \qquad s_1^2 = 30, \qquad s_2^2 = 25$$

The pooled estimate of the common population variance is

$$s_p^2 = \frac{14(30) + 21(25)}{35} = 27$$

We may now compute

$$s_{\bar{x}_1 - \bar{x}_2} = \sqrt{\frac{27}{15} + \frac{27}{21}} = 1.76$$

the standard error of $\bar{x}_1 - \bar{x}_2$.

The value of t that can be computed from the sample data is

$$t = \frac{(85 - 87) - 0}{1.76} = -1.14$$

8 *Statistical decision* Since $-2.0301 < -1.14 < 2.0301$, that is, since -1.14 falls in the acceptance region, we cannot reject H_0.

Alternatively, we could have based our decision to reject or not to reject H_0 on the magnitude of the observed difference, $\bar{x}_1 - \bar{x}_2 = 85 - 87 = -2$. The critical values of $\bar{x}_1 - \bar{x}_2$ are given by

$$0 \pm (2.0301)(1.76) = \pm 3.57$$

Since $-3.57 < -2 < 3.57$, we cannot reject H_0.

9 *Conclusion* On the basis of these data, researchers may conclude that there may be no difference between the two population means.

Sampling from nonnormally distributed populations

Example 6.8 A team of juvenile-rehabilitation counselors with a state department of corrections has the impression that repeaters and nonrepeaters differ with respect to the average age at which they first become involved with authorities. Based on their experience, the counselors feel that nonrepeaters tend to become involved with authorities at a later age than repeaters. To see whether they can secure evidence to corroborate this belief, the team draws a random sample of $n_R = 50$ records of repeaters and a random sample of $n_N = 60$ records of nonrepeaters. They carry out the following hypothesis-testing procedure.

1 *Statement of hypotheses*

$$H_0\colon \mu_N \leq \mu_R \qquad \text{or, equivalently,} \qquad \mu_N - \mu_R \leq 0$$

$$H_1\colon \mu_N > \mu_R \qquad \text{or, equivalently,} \qquad \mu_N - \mu_R > 0$$

where μ_N is the mean age at time of first encounter with authorities for nonrepeaters, and μ_R is the mean age at time of first encounter with authorities for repeaters.

2 *Significance level* Let $\alpha = 0.05$

3 *Description of populations and assumptions* The functional forms of the populations are not known, but this causes no problem in the determination of a test statistic, since the samples are large. The central-limit theorem is applicable. We can assume that the sample sizes are sufficiently large to provide acceptable estimates of σ_R^2 and σ_N^2. The samples are independent.

4 *The relevant statistic* The relevant statistic is $\bar{x}_N - \bar{x}_R$, which, because of the central-limit theorem, is approximately normally distributed, with a standard error of

$$\sigma_{\bar{x}_N - \bar{x}_R} = \sqrt{\frac{\sigma_N^2}{n_N} + \frac{\sigma_R^2}{n_R}}$$

and a mean, if H_0 is true, of 0. Since σ_N^2 and σ_R^2 are unknown, we can estimate them by s_N^2 and s_R^2 to give

$$s_{\bar{x}_N - \bar{x}_R} = \sqrt{\frac{s_N^2}{n_N} + \frac{s_R^2}{n_R}}$$

an estimate of $\sigma_{\bar{x}_N - \bar{x}_R}$.

5 *The test statistic* In light of the considerations stated in step 4, the appropriate test statistic is z.

6 *The rejection and acceptance regions* The critical value of z is 1.645.

7 *Collection of data and calculations* The following sample means and variances are computed.

$$\bar{x}_N = 14.9, \qquad \bar{x}_R = 12.3, \qquad s_N^2 = 4.0, \qquad s_R^2 = 6.25$$

From these data we compute

$$z = \frac{(14.9 - 12.3) - 0}{\sqrt{(4.0/60) + (6.25/50)}} = 5.94$$

8 *Statistical decision* Since $5.94 > 1.645$, we reject H_0.

Alternatively, we could have based our decision on the magnitude of the sample difference, $\bar{x}_N - \bar{x}_R = 14.9 - 12.3 = 2.6$, as compared to the critical value of $\bar{x}_N - \bar{x}_R$, which is given by

$$0 + 1.645\sqrt{\frac{4.0}{60} + \frac{6.25}{50}} = 0.72$$

Since $2.6 > 0.72$, we can reject H_0.

9 *Conclusion* The mean age at which nonrepeaters have their initial contact with authorities is later than that for repeaters.

EXERCISES

23 An occupational therapist conducted a study to evaluate the relative merits of two prosthetic devices designed to facilitate manual dexterity. The therapist assigned 21 patients with identical handicaps to wear one or the other of the two devices while performing a certain task. Eleven patients wore device A, and ten wore device B. The researcher recorded the time each patient required to perform a certain task, with the following results.

$$\bar{x}_A = 65 \text{ seconds}, \qquad s_A^2 = 81$$

$$\bar{x}_B = 75 \text{ seconds}, \qquad s_B^2 = 64$$

Do these data provide sufficient evidence to indicate that device A is more effective than device B? Let $\alpha = 0.05$.

24 As part of a larger study concerned with the behavior of a species of wild animal, zoologists conducted an experiment to determine whether this type of animal, on the average, had different response times under two different conditions, condition I and condition II. Researchers subjected a random sample of 15 animals to condition I. For each animal they recorded

the time elapsing between introduction of a stimulus and the response. They made similar recordings on the random sample of 17 animals that were subjected to condition II. Their results were as follows.

$$\bar{x}_{\text{I}} = 10 \text{ minutes}, \qquad s_{\text{I}}^2 = 9$$

$$\bar{x}_{\text{II}} = 16 \text{ minutes}, \qquad s_{\text{II}}^2 = 16$$

Do these data provide sufficient evidence to indicate that the mean response times are different under the two conditions? Let $\alpha = 0.01$.

25 As part of a research project, a psychologist selected a random sample of 12 adolescent girls and a random sample of 9 adolescent boys. Each subject was asked to draw a male figure. The mean time spent on the drawing by girls was 8 minutes, with a variance of 18. For the boys the mean time spent on the drawing was 13 minutes, with a variance of 22.5. Do these data suggest that adolescent boys, on the average, spend more time when drawing the male figure than do adolescent girls? Let $\alpha = 0.05$.

26 A survey was conducted among elderly persons of a certain community to compare the levels of self-esteem among those elderly persons living in nursing homes and those not living in nursing homes (living alone or with relatives). Each person was given a test to measure his or her self-esteem. The following results were obtained.

	n	**Mean score**	**Sample variance**
Persons living in nursing homes	50	65	100
Persons not living in nursing homes	30	88	90

Do these data provide sufficient evidence to indicate that elderly persons not living in nursing homes have a higher mean self-esteem score than those who do live in nursing homes? Let $\alpha = 0.01$.

27 A study was conducted to evaluate the effects of crowding on learning among elementary school children. A random sample of 50 children were taught a certain skill under crowded conditions, and a random sample of 45 children were taught the same skill by the same teachers under uncrowded conditions. At the end of the experiment, each child was given a test to determine her or his level of mastery of the skill. The following results were obtained.

Conditions	$\bar{x}$	**Sample variance**
Crowded	70	100
Uncrowded	80	90

Do these data provide sufficient evidence to indicate that teaching is less effective under crowded conditions? Let $\alpha = 0.05$.

28 At the beginning of the school year, the members of a high school senior class were randomly assigned to one of two groups, each containing 50 students. Group A received individual vocational counseling. Group B received none. At the end of the year, each senior was given a test to measure his/her level of knowledge about vocations. The results were as follows.

Group	Mean	Variance
Individual counseling (A)	42	150
No counseling (B)	35	250

Do these data provide sufficient evidence to indicate that individual counseling is effective in increasing knowledge of vocations? Let $\alpha = 0.05$.

29 In a study designed to evaluate the effects of noise on ability to learn, 24 students were randomly assigned to two groups. Group 1 was taught a certain skill under noisy conditions. Group 2 was taught the skill by the same teacher under quiet conditions. At the end of the experiment, each student was given a test to measure her/his level of mastery of the skill. The results were as follows.

Group	n	Mean	Variance
Noisy conditions (1)	12	80	65
Quiet conditions (2)	12	89	55

Do these data provide sufficient evidence to indicate that noise is a deterrent to learning? Let $\alpha = 0.05$.

30 In a psychology laboratory, researchers administered a toxic substance to experimental animals by one of two pathways to the central nervous system. The variable of interest was the length of time, in hours, between administration of the toxin and onset of symptoms. The following results were obtained.

Pathway	n	$\bar{x}$	Variance
A	11	40	10.0
B	7	31	20.9

Do these data provide sufficient evidence to indicate that, on the average, the onset of symptoms appears sooner when the toxin is administered via pathway B? Let $\alpha = 0.05$.

6.6 PAIRED COMPARISONS

In Chapter 5 we discussed the construction of confidence intervals for a population mean difference based on data from nonindependent random samples. We also discussed the rationale and advantages of using this type of data, which are called *paired data* or *paired observations*. On the basis of the same theory that underlies the construction of confidence intervals for a population mean difference, we may also test hypotheses about a population mean difference. A two-sided test is appropriate when the null hypothesis states that the true mean of the differences between two sets of paired observations is equal to 0, with no specification that the alternative should be in one direction rather than the other. If the alternative hypothesis states that one set of observations in a population of matched pairs is larger (or smaller) than the other set, a one-sided test is appropriate. Let us illustrate with an example.

Example 6.9 A medical team measured the level of a certain chemical in the blood of 15 adult subjects before and after they were placed in an anxiety-producing situation. Table 6.3 shows the results.

The investigators wished to know whether they could conclude, on the basis of these data, that anxiety-producing situations tend to raise the level of this chemical in the blood. We may carry out the following hypothesis-testing procedure.

1 *Statement of hypotheses*

$$H_0: \mu_d \leq 0, \qquad H_1: \mu_d > 0$$

2 *Level of significance* Let $\alpha = 0.05$.

3 *Description of population and assumptions* The population consists of differences between before and after $(A - B)$ measurements of the level of the chemical in the blood of subjects placed in an anxiety-producing situation. The population is hypothetical. One can make it real by carrying out the procedure with a large number of subjects. We assume that the sample available for analysis, consisting of 15 differences, constitutes a simple random sample from that population. We also assume that the population of differences is normally distributed.

4 *The relevant statistic* The relevant statistic is the sample mean difference, $\bar{d} = \Sigma\, d_i/n$. Since the population of differences is assumed to be normally distributed, we consider the sampling distribution of $\bar{d}$ to be normal with a variance of σ_d^2/n and a mean, if H_0 is true, of 0.

5 *The test statistic* Since we assume that the population is normally distributed, μ_d is unknown, and n is small, the appropriate sampling distribution is

$$t = \frac{\bar{d} - \mu_d}{s_d/\sqrt{n}}$$

Table 6.3 Levels of a certain chemical in the blood of 15 subjects before and after they were placed in an anxiety-producing situation

	Level of chemical									
SUBJECT	1	2	3	4	5	6	7	8	9	10
BEFORE (B)	9	17	14	8	15	20	18	12	10	22
AFTER (A)	10	22	24	28	10	15	14	12	21	25
DIFFERENCE $(A - B)$	+1	+5	+10	+20	−5	−5	−4	0	+11	+3
SUBJECT	11	12	13	14	15					
BEFORE (B)	18	7	14	7	20					
AFTER (A)	22	11	16	10	27					
DIFFERENCE $(A - B)$	+4	+4	+2	+3	+7					

where s_d is the standard deviation of the n sample differences and t is distributed as Student's t with $n - 1$ degrees of freedom.

6 *Rejection and acceptance regions* We obtain the critical value of t by entering Appendix Table H with $\alpha = 0.05$ (one-sided test) and df $= 15 - 1 = 14$. We find $t = 1.7613$. If the t computed from the sample exceeds 1.7613, we will reject H_0.

7 *Collection of data and calculations* From the data given in Table 6.3, we compute

$$\bar{d} = \frac{\Sigma d_i}{n} = \frac{1 + 5 + \cdots + 7}{15} = 3.73$$

$$s_d = \sqrt{\frac{15(1 + 25 + \cdots + 49) - (1 + 5 + \cdots + 7)^2}{15(14)}} = 6.58$$

$$t = \frac{3.73 - 0}{6.58/\sqrt{15}} = 2.20$$

8 *Statistical decision* Since the computed value of $t = 2.20$ is greater than the tabulated value of 1.7613, we reject H_0.

9 *Conclusion* We conclude that anxiety-producing situations tend to raise the level of the chemical in the blood.

EXERCISES

31 Table 6.4 shows the full-scale IQ scores of 12 children diagnosed as having learning disabilities before, and nine months after, initiation of a remedial program. Do these data provide sufficient evidence to indicate that a remedial program is effective in increasing IQ scores in this type of child? Let $\alpha = 0.05$.

32 Table 6.5 shows the concentration of a certain chemical in the

Table 6.4 Data for Exercise 31

	Full-scale IQ											
CHILD	1	2	3	4	5	6	7	8	9	10	11	12
BEFORE	97	103	99	100	105	107	101	108	90	96	90	105
AFTER	103	106	112	116	118	125	123	116	100	106	101	108

Table 6.5 Data for Exercise 32

	Route of administration									
SUBJECT	1	2	3	4	5	6	7	8	9	10
INTRAVENOUS	5.5	8.4	6.3	3.3	5.7	5.2	3.7	7.0	5.6	3.1
INTRAMUSCULAR	7.4	7.6	11.8	5.6	5.9	6.0	6.2	12.2	12.1	4.2

Table 6.6 Data for Exercise 33

	Score of:														
WIFE	33	57	32	54	52	34	60	40	59	39	40	59	44	32	55
HUSBAND	44	60	55	68	40	48	57	49	47	52	58	51	66	60	68

urine of ten adults following administration, by two routes, of a drug containing the chemical. Do these data provide sufficient evidence to indicate that intramuscular administration of the drug results in a higher concentration of the chemical in the urine? Let $\alpha = 0.05$.

33 A psychologist randomly selected 15 wives and their husbands from among the residents of an urban area, and asked them to complete a questionnaire designed to measure the level of their satisfaction with the community in which they lived. Table 6.6 shows the results. Do these data indicate that the husbands in the area are better satisfied with the community than the wives? Let $\alpha = 0.05$.

6.7 TESTING A HYPOTHESIS ABOUT A SINGLE POPULATION PROPORTION

As we have seen, we often want to be able to make inferences about population proportions. In Chapter 5 we discussed the construction of confidence interval estimates of population proportions. In this section we shall illustrate the testing of hypotheses about population proportions. In Chapter 4 we discussed the appropriate sampling distribution underlying the test.

Example 6.10 A social worker believes that fewer than 25% of the couples in a certain area have ever used any form of birth control. In order to see whether this is a reasonable assumption, the social worker selects a random sample of 120 couples from the area, and carries out the following hypothesis-testing procedure.

1 *Statement of hypotheses*

H_0: $p \geq 0.25$, $\quad H_1$: $p < 0.25$

2 *Significance level* Let $\alpha = 0.05$.

3 *Description of population and assumptions* The population is binomial, consisting of a series of yes and no answers, by couples living in the area, to the question, "Have you ever used any form of birth control?" The population is sufficiently large relative to the sample size for us to ignore the fpc. The sample is also large enough for us to apply the normal approximation to the binomial procedure in testing the hypothesis.

4 *The relevant statistic* The relevant statistic is $\hat{p}$, the proportion of

couples in the sample who have used some form of birth control. Under H_0, the sampling distribution of $\hat{p}$ is approximately normally distributed, with a mean of $p = p_0 = 0.25$ (testing at the point of equality) and a standard error of

$$\sigma_{\hat{p}} = \sqrt{\frac{(p_0)(1-p_0)}{n}} = \sqrt{\frac{(0.25)(0.75)}{120}}$$

Note that we use the hypothesized value of p, p_0, in the formula for $\sigma_{\hat{p}}$. This is a logical practice, since one assumes a null hypothesis to be true until one has enough evidence to cause one to reject it.

5 *The test statistic* Since the distribution of $\hat{p}$ is considered to be approximately normal, the appropriate test statistic is z, which is distributed as the standard normal.

6 *Rejection and acceptance regions* The critical value of z is -1.645, so the rejection region consists of values of z equal to or less than -1.645. The acceptance region consists of values of z greater than -1.645. The critical value of $\hat{p}$ is given by

$$0.25 - 1.645\sqrt{\frac{(0.25)(0.75)}{120}} = 0.18$$

7 *Collection of data and calculations* Of the 120 couples in the sample, 20 said they had used some method of birth control. From this information we calculate

$$\hat{p} = \tfrac{20}{120} = 0.17$$

and

$$z = \frac{\hat{p} - p_0}{\sigma_{\hat{p}}} = \frac{0.17 - 0.25}{\sqrt{(0.25)(0.75)/120}} = -\frac{0.08}{0.0395} = -2.03$$

8 *Statistical decision* We reject the null hypothesis, since $-2.03 < -1.645$ (or, alternatively, $0.17 < 0.18$).

9 *Conclusion* We conclude that fewer than 25% of the couples in the area have ever used some form of birth control.

When a population proportion is the parameter of interest, one may also carry out two-sided tests and one-sided tests with an upper-tail rejection region when appropriate.

EXERCISES

34 State the appropriate statistical hypotheses for a researcher who wants to test the null hypothesis that a population proportion is equal to 0.40. If a sample of size 240 yields a sample proportion of 0.48, would H_0 be rejected at the 0.05 level of significance? Support your answer with an appropriate hypothesis-testing procedure.

35 An official with a state department of corrections believes that 20% of the juveniles admitted to the state's training schools are admitted as a

result of convictions for car theft. In a random sample of 100 admission records, 16 of the juveniles had been admitted because of convictions for car theft. Do these data contradict the official's belief? Let $\alpha = 0.01$.

36 An official with a state department of agriculture believes that more than 20% of the farmers in the state have part-time jobs in addition to farming. A survey of 200 randomly selected farmers revealed that 60 had part-time jobs. Do these data support the official's belief? Let $\alpha = 0.05$.

37 A candidate for a state office believes that fewer than 25% of the persons eligible to vote for the office favor the passage of a certain bill on which the candidate must take a stand. Of a random sample of 200 voters, 30 stated that they favored passage of the bill. Do these data support the candidate's belief at the 0.05 level of significance?

38 A sociologist believes that more than 70% of the adults in a certain low-income area would be in favor of the establishment of a community recreation center. Of a random sample of 200 adults from the area, 144 said they favored the idea. Do these data support the sociologist's belief? Let $\alpha = 0.05$.

39 A specialist with a state college of agriculture believed that three-month weight-gain goals would be achieved by more than 80% of the hogs fed a certain diet. On an experimental farm, 400 hogs were randomly assigned to be fed the diet. At the end of three months, 340 of the hogs had achieved the three-month goals. Do these data support the specialist's belief? Let $\alpha = 0.05$.

40 A random sample of 225 apartment dwellers in a certain metropolitan area revealed that 18 owned dogs. Do these data provide sufficient evidence to indicate that fewer than 10% of the apartment dwellers in the area own dogs? Let $\alpha = 0.05$.

41 The mayor of a certain city believes that more than 60% of the adult residents of an adjacent suburb favor annexation by the city. In a random sample of 120 adults, 76 said they favored annexation. Do these data provide sufficient evidence to support the mayor's belief? Let $\alpha = 0.05$.

42 It is estimated that fewer than 10% of the students of a certain urban university use public transportation in commuting to and from classes. In a random sample of 225 students, 20 said they used public transportation. In light of the sample evidence, does the estimate seem realistic? Let $\alpha = 0.05$.

43 Of a random sample of 255 young adults living in a certain area, 25 said they believed that most mental illness is inherited. Do these data support the hypothesis that fewer than 15% of the young adults in the area hold such a belief? Let $\alpha = 0.05$.

6.8 TESTING A HYPOTHESIS ABOUT THE DIFFERENCE BETWEEN TWO POPULATION PROPORTIONS

In practice, situations often arise in which people want to test the null hypothesis that two population proportions, p_1 and p_2, are equal, or that they differ by some specified amount. For example, we may want to test the hypothesis that two groups of individuals do not differ with respect to the proportion who favor the passage of some city ordinance. Or we may want to know whether it is reasonable to conclude that the proportion of women who regularly watch a particular television program exceeds, by some specified fraction, the proportion of men who watch the program. We discussed the appropriate sampling distribution for testing these hypotheses in Chapter 4.

Let us now illustrate the method of testing each of these two types of hypotheses.

Case 1. $p_1 - p_2 = 0$

In Chapter 4 we learned that the sampling distribution of the difference between two sample proportions is approximately normally distributed, if n_1 and n_2 are sufficiently large, with mean $\mu_{\hat{p}_1-\hat{p}_2}$ and standard deviation

$$\sigma_{\hat{p}_1-\hat{p}_2} = \sqrt{\frac{p_1(1-p_1)}{n_1} + \frac{p_2(1-p_2)}{n_2}}$$

Since in practical situations p_1 and p_2 are unknown, we estimate $\sigma_{\hat{p}_1-\hat{p}_2}$ by

$$s_{\hat{p}_1-\hat{p}_2} = \sqrt{\frac{\hat{p}_1(1-\hat{p}_1)}{n_1} + \frac{\hat{p}_2(1-\hat{p}_2)}{n_2}}$$

where $\hat{p}_1$ and $\hat{p}_2$ are the observed proportions in independent random samples from population 1 and population 2, respectively.

When it is hypothesized that p_1 and p_2 are equal, a reasonable procedure would be to pool the data from the two samples to obtain estimates of $p = p_1 = p_2$, the common population proportion.

If we let x_1 be the number having the characteristic of interest in the sample from population 1, and x_2 the number having that characteristic in the sample from population 2, we may find a pooled estimate of $p = p_1 = p_2$ by

$$\bar{p} = \frac{x_1 + x_2}{n_1 + n_2}$$

We may then rewrite the standard-error formula as

$$s_{\hat{p}_1-\hat{p}_2} = \sqrt{\frac{\bar{p}(1-\bar{p})}{n_1} + \frac{\bar{p}(1-\bar{p})}{n_2}}$$

The following example illustrates the situation in which the null hypothesis specifies that two population proportions are equal.

Example 6.11 An anthropologist believes that the proportion of individuals in two populations with double occipital hair whorls is the same. To see whether there is any reason to doubt this hypothesis, the anthropologist takes independent random samples from each of the two populations and determines the number in each sample with this characteristic. The results are as follows.

Population	n	Number with characteristic
1	100	23
2	120	32

The investigator may carry out the following hypothesis-testing procedure.

1 *Statement of hypotheses*

H_0: $p_1 = p_2$ or, equivalently, $p_1 - p_2 = 0$

H_1: $p_1 \neq p_2$ or, equivalently, $p_1 - p_2 \neq 0$

2 *Significance level* We let $\alpha = 0.05$.

3 *Description of populations and assumptions* The anthropologist may classify each person in the two populations as either possessing or not possessing the double occipital hair whorl. The two samples are independent.

4 *The relevant statistic* Since the hypothesis is concerned with $p_1 - p_2$, the difference between two population proportions, the relevant statistic is $\hat{p}_1 - \hat{p}_2$, which is approximately normally distributed (since n_1 and n_2 are large), with a standard error obtained by pooling and a mean, if the null hypothesis is true, of 0.

5 *The test statistic and its distribution* Since the distribution of the relevant statistic is approximately normally distributed, the test statistic is

$$z = \frac{(\hat{p}_1 - \hat{p}_2) - 0}{\sqrt{\dfrac{\bar{p}(1-\bar{p})}{n_1} + \dfrac{\bar{p}(1-\bar{p})}{n_2}}}$$

which is distributed approximately as the standard normal when H_o is true.

6 *Rejection and acceptance regions* The critical values of z are ± 1.96. In terms of $\hat{p}_1 - \hat{p}_2$, the critical values are given by

$$0 \pm 1.96\sqrt{\frac{\bar{p}(1-\bar{p})}{n_1} + \frac{\bar{p}(1-\bar{p})}{n_2}}$$

7 *Collection of data and calculations* From the sample data given earlier, we compute $\hat{p}_1 = 23/100 = 0.23$, $\hat{p}_2 = 32/120 = 0.27$, and pooling gives

$$\bar{p} = \frac{23 + 32}{100 + 120} = \frac{55}{220} = 0.25$$

The pooled standard error is

$$s_{\hat{p}_1-\hat{p}_2} = \sqrt{\frac{(0.25)(0.75)}{100} + \frac{(0.25)(0.75)}{120}} = 0.06$$

The computed value of z, then, is

$$z = \frac{(0.23 - 0.27) - 0}{0.06} = -0.67$$

We may also compute the following critical values in terms of $\hat{p}_1 - \hat{p}_2$.

$$0 \pm (1.96)(0.06) = \pm 0.12$$

8 *Statistical decision* Since the computed z of -0.67 falls between -1.96 and $+1.96$, we cannot reject H_0. Alternatively, we may base our decision on the observed magnitude of $\hat{p}_1 - \hat{p}_2$. Since $\hat{p}_1 - \hat{p}_2 = 0.23 - 0.27 = -0.04$ falls between -0.12 and $+0.12$, we cannot reject H_0.

9 *Conclusion* Since we do not reject H_0, we conclude that the two population proportions may be equal. That is, the proportion of individuals with double occipital hair whorls may be the same in population 1 and population 2.

Case 2. $p_1 - p_2 \neq 0$

In case 2 the null hypothesis specifies that $p_1 - p_2$ is something other than 0. Consequently, there is no justification for pooling the data from the two samples in order to estimate $\sigma_{\hat{p}_1-\hat{p}_2}$.

In case 2, just as in case 1, $\hat{p}_1 - \hat{p}_2$ is approximately normally distributed, if n_1 and n_2 are large independent random samples. In case 2, $\hat{p}_1 - \hat{p}_2$ has a mean of $p_1 - p_2$ and an estimated standard error of

$$\sqrt{\frac{\hat{p}_1(1 - \hat{p}_1)}{n_1} + \frac{\hat{p}_2(1 - \hat{p}_2)}{n_2}}$$

Example 6.12 A political scientist at a university believes that the proportion of voters in area A who will vote in the upcoming primary exceeds by more than 0.05 the proportion of voters in area B who will vote in the same primary.

In order to see whether the facts support this hypothesis, the professor conducts a survey among area A and area B voters, with the following results.

Area	**Sample size**	**Number who say they plan to vote in primary**
A	$n_A = 150$	113
B	$n_B = 160$	104

The investigator may carry out the following hypothesis-testing procedure to determine whether the observed data provide sufficient evidence to support the hypothesis. (It is assumed that voters will do what they say they plan to do.)

1 *Statement of hypotheses*

H_0: $p_A - p_B \leq 0.05$, $\quad H_1$: $p_A - p_B > 0.05$

2 *Level of significance* We let $\alpha = 0.05$.

3 *Description of populations and assumptions* The two populations consist of voters in area A and voters in area B, respectively. We assume that the two samples are independently and randomly drawn from the respective populations.

4 *The relevant statistic* The relevant statistic is $\hat{p}_A - \hat{p}_B$, which is considered to be approximately normally distributed (since n_1 and n_2 are large). If H_0 is true, the mean of the distribution is 0.05 or less (we test at 0.05).

5 *The test statistic* The test statistic is

$$z = \frac{(\hat{p}_A - \hat{p}_B) - 0.05}{\sqrt{\dfrac{\hat{p}_A(1 - \hat{p}_A)}{n_A} + \dfrac{\hat{p}_B(1 - \hat{p}_B)}{n_B}}}$$

which, when H_0 is true, is distributed approximately as the standard normal.

6 *Rejection and acceptance regions* The critical value of z is $+1.645$.

7 *Collection of data and calculations* From the sample data we compute $\hat{p}_A = 113/150 = 0.75$ and $\hat{p}_B = 104/160 = 0.65$.

The standard error is

$$s_{\hat{p}_A - \hat{p}_B} = \sqrt{\frac{(0.75)(0.25)}{150} + \frac{(0.65)(0.35)}{160}} = 0.05$$

which allows us to compute

$$z = \frac{(0.75 - 0.65) - 0.05}{0.05} = 1.00$$

8 *Statistical decision* Since the computed z of 1.00 is less than 1.645, we do not reject H_0.

9 *Conclusion* We may not conclude, on the basis of these data, that the political science professor's hypothesis is true.

EXERCISES

44 An anthropologist believes that the proportion of individuals with type A blood is the same in two populations, I and II. A survey of the two populations yields the following information, based on independent random samples.

Population	Sample size	Number with type A blood
I	$n_A = 150$	87
II	$n_B = 200$	100

Do these data provide sufficient evidence to indicate that the proportions in the two populations are not equal? Let $\alpha = 0.05$.

45 A sociologist wishes to test the null hypothesis that the proportion of married couples participating in informal group activities is the same in two communities. Independent random samples of couples from the two communities give the following information.

Community	Sample size	Number participating in informal group activities
A	175	88
B	225	101

Do these data provide sufficient evidence to indicate that the two proportions are not equal? Let $\alpha = 0.05$.

46 A researcher with a state department of corrections believes that, among youths incarcerated for acts of violence, the percentage reared in overcrowded homes exceeds by more than 10% the percentage reared in overcrowded homes among those incarcerated for all other crimes. To gain evidence relative to this theory, the researcher obtained independent random samples of the records for the past five years of the two types of offenders with the following results.

Type of offense	Sample size	Number reared in overcrowded homes
Acts of violence	200	132
All others	300	147

Do these data provide sufficient evidence, at the 0.05 level of significance, to support the researcher's belief?

47 A sociologist believes that the proportion of adult males in one socioeconomic group (group *A*) who regularly watch wrestling on television exceeds by more than 0.10 the proportion in a second socioeconomic group (group *B*) who regularly watch wrestling. Independent random samples of adult males from the two groups gave the following information.

Group	Sample size	Number who regularly watch wrestling on television
A	$n_A = 150$	98
B	$n_B = 200$	80

Do these data provide sufficient evidence to support the sociologist's belief? Let $\alpha = 0.05$.

48 In a study of the relationship between adolescent behavior and various religious factors and influences, researchers studied a random sample of high school students identified by their teachers as troublemakers, and an independent random sample of nontroublemakers. Each student in the two samples was rated by his or her peers as to how religious the peers perceived him or her to be. The results were as follows.

Group	Sample size	Number receiving high religious ratings by peers
Troublemakers	200	68
Nontroublemakers	250	140

Do these data provide sufficient evidence to indicate that the proportion of students receiving high religious ratings is higher among the nontroublemakers than among the troublemakers? Let $\alpha = 0.05$.

49 A social worker who wishes to compare two communities on several variables selects a random sample of 120 households from community A and an independent random sample of 100 households from community B. Of the 120 households from community A, 36 are found to be receiving public assistance. Of the 100 households from community B, 35 are receiving public assistance. Do these data provide sufficient evidence to indicate that the proportions of households receiving public assistance in the two communities are different? Let $\alpha = 0.01$.

50 A rehabilitation counselor with a state department of corrections believes that a certain rehabilitation program will reduce the recidivism rate among released prisoners by more than 15%. During a certain year, by random assignment, 100 released prisoners were designated to participate in the rehabilitation program. Another 100 released prisoners, by random assignment, were designated as a control group. Both groups were followed for a period of five years. At the end of the follow-up period, 22 persons in the experimental group and 45 in the control group had been reconvicted. Is the counselor's belief about the rehabilitation program justified? Let $\alpha = 0.05$.

51 A biologist investigating the effects of two methods of making a certain plant resistant to a disease obtained the following results.

Method	Number of plants treated	Number of resistant plants
A	200	50
B	250	88

On the basis of these data, can the biologist conclude that the proportions of plants surviving are different? Let $\alpha = 0.05$.

6.9 TESTING A HYPOTHESIS ABOUT THE VARIANCE OF A NORMALLY DISTRIBUTED POPULATION

In order to test hypotheses about the variance of a normally distributed population, we make use of the distribution of

$$\chi^2 = \frac{(n-1)s^2}{\sigma^2}$$

which, as we discussed in Chapter 4, has a chi-square distribution with

$n - 1$ degrees of freedom. If we replace σ^2 in the denominator by σ_0^2, the hypothesized value of σ^2, we have

$$\chi^2 = \frac{(n-1)s^2}{\sigma_0^2}$$

which, when $\sigma_0^2 = \sigma^2$, also has a chi-square distribution with $n - 1$ degrees of freedom.

From a random sample of size n drawn from the population of interest, we compute the sample variance s^2, which is entered in the numerator of χ^2. We place the hypothesized value of σ^2, σ_0^2, in the denominator to give the following computed value of χ^2.

$$\chi^2 = \frac{(n-1)s^2}{\sigma_0^2}$$

If the null hypothesis is

$$H_0: \sigma^2 = \sigma_0^2$$

with the two-sided alternative

$$H_1: \sigma^2 \neq \sigma_0^2$$

and significance level α, we obtain the two critical values of χ^2 by entering Appendix Table G with α and the appropriate degrees of freedom. The lower critical value of χ^2 is the one with $\alpha/2$ of the area under the curve of χ^2 to its left, and the upper critical value is the χ^2 with $1 - \alpha/2$ of the area under the curve of χ^2 to its left. We may designate these by $\chi^2_{\alpha/2,n-1}$ and $\chi^2_{1-\alpha/2,n-1}$, respectively.

The rejection region, then, consists of values of χ^2 greater than or equal to $\chi^2_{1-\alpha/2,n-1}$ and values less than or equal to $\chi^2_{\alpha/2,n-1}$. Figure 6.15 shows the rejection and acceptance regions for a two-sided alternative.

Figure 6.15 Rejection and acceptance regions for H_0: $\sigma^2 = \sigma_0^2$, H_1: $\sigma^2 \neq \sigma_0^2$

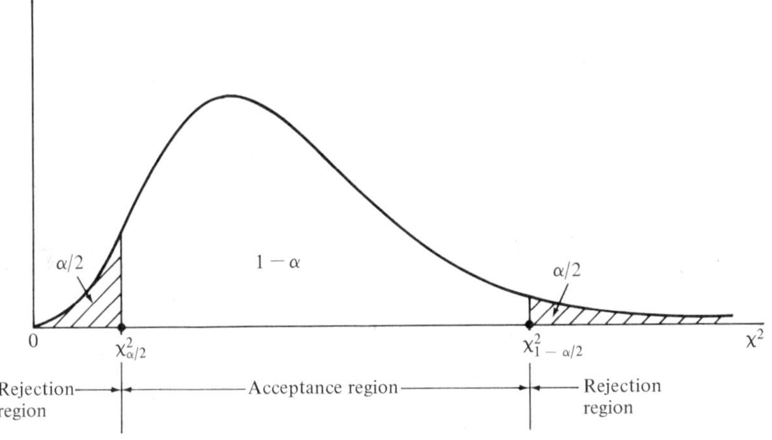

For a one-sided alternative of the form

$$H_1\colon \sigma^2 < \sigma_0^2$$

with a level of significance equal to α, only small computed values of χ^2 will cause rejection of the H_0. Consequently, the rejection region consists of values of χ^2 less than or equal to $\chi^2_{\alpha,n-1}$. When the alternative is

$$H_1\colon \sigma^2 > \sigma_0^2$$

only large computed values of χ^2 will cause rejection of H_0. Hence the rejection region consists of all values of χ^2 equal to or greater than $\chi^2_{1-\alpha,n-1}$. Figure 6.16 shows rejection and acceptance regions for these two one-sided alternatives.

Figure 6.16 Rejection and acceptance regions for one-sided alternatives in tests involving the variance of a normally distributed population

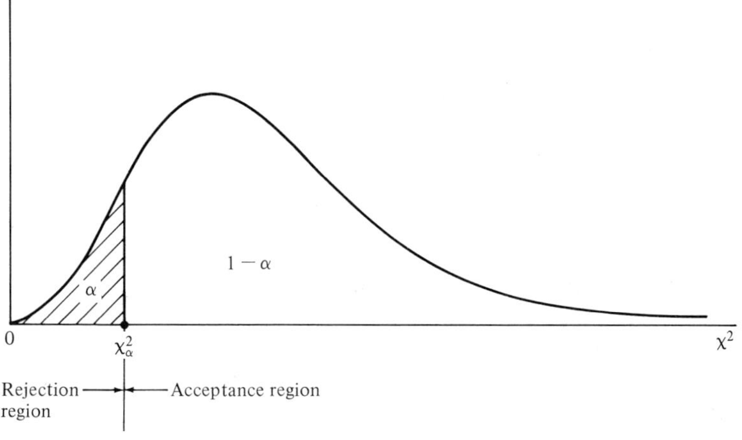

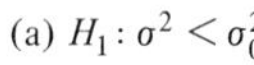
(a) $H_1 : \sigma^2 < \sigma_0^2$

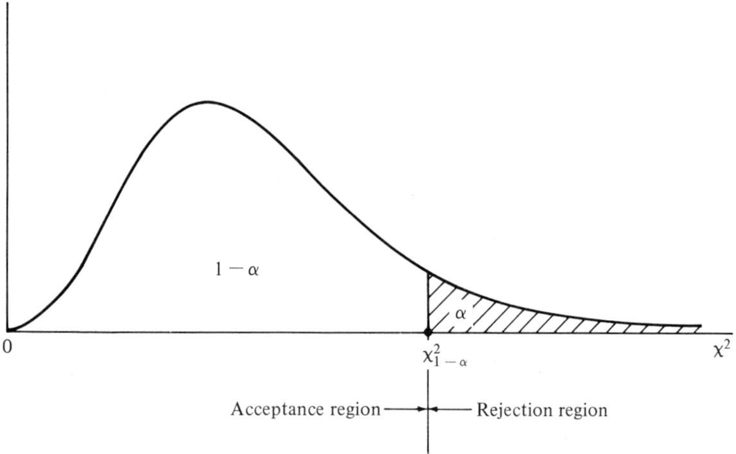

(b) $H_1 : \sigma^2 > \sigma_0^2$

The method described in this section for testing hypotheses about population variances, like the parallel method of constructing confidence intervals for σ^2 described in Chapter 5, is highly sensitive to departures from normality on the part of the sampled population. For this reason, the method presented in this section is limited in its applicability.

Example 6.13 The variance computed from reading scores of third-grade students of school system A over a ten-year period is 1.44. A random sample of 21 third-grade students from another school system (B) who were given the same reading test yielded a variance of $s^2 = 1.05$. Do these data provide sufficient evidence to indicate, at the 0.05 level of significance, that the scores of third-grade students from school system B are less variable than those of students from school system A?

The hypothesis-testing procedure is as follows.

1 *Statement of hypotheses*

$$H_0\colon \sigma^2 \geq 1.44, \qquad H_1\colon \sigma^2 < 1.44$$

2 *Significance level* Let $\alpha = 0.05$.

3 *Description of population and assumptions* It is assumed that the scores of third-grade students of school system B are normally distributed.

4 *The relevant statistic* The relevant statistic is the sample variance s^2.

5 *The test statistic and its distribution* The test statistic is $\chi^2 = (n-1)s^2/\sigma_0^2$, which, when H_0 is true, is distributed as the chi-square distribution with $n-1$ degrees of freedom.

6 *Rejection and acceptance regions* The critical value of χ^2 from Appendix Table G is 10.851. If the computed χ^2 is less than or equal to 10.851, we will reject H_0.

7 *Collection of data and calculations* From the information given, we have $n = 21$ and $s^2 = 1.05$. From these data we compute

$$\chi^2 = \frac{20(1.05)}{1.44} = 14.58$$

8 *Statistical decision* Since 14.58 is greater than the critical value of 10.851, we cannot reject H_0.

9 *Conclusion* Since we do not reject H_0, we conclude that the reading test scores of third-grade students from school system B may be no less variable than those of third-grade students from school system A.

52 It is known that the variance of nonverbal IQ scores of a certain population of children is 144. A random sample of ten children of the same age from another population yielded a sample variance of 289. On the basis of these data, can one conclude that the population from which this sample was drawn is more variable with respect to nonverbal IQ scores than the other population? Let $\alpha = 0.05$. What assumption is necessary?

53 A professor of biology believes that the variance of the life of a certain organism exposed to a killing agent is 625 minutes squared. A random sample of 11 organisms yielded a variance of 1225. Do these data provide sufficient evidence to indicate that the professor's assessment of the variability is incorrect? Let $\alpha = 0.05$.

54 A psychologist believes that the variance of the times assembly-line employees require to perform a certain task is 9 minutes squared. A random sample of six employees who perform the task yielded a sample variance of 25. Do these data provide sufficient evidence to indicate that the variance is greater than the psychologist believes? Let $\alpha = 0.05$.

55 An anthropologist drew a sample of 15 females from a certain population. The variance of the weights of these subjects was 81 kg^2. On the basis of these data, can the anthropologist conclude that the true population variance is greater than 25? Let $\alpha = 0.05$.

56 A physical education instructor gives a random sample of 12 male college students a muscular endurance test at the end of a program of intensive exercise. The variance of the test scores is 1225. Do these data provide sufficient evidence to cast doubt on the hypothesis that the true variance is not less than 2500? Let $\alpha = 0.05$.

6.10 TESTING HYPOTHESES ABOUT THE RATIO OF THE VARIANCES OF TWO NORMALLY DISTRIBUTED POPULATIONS

There are many situations in which people want to be able to reach a decision based on sample data, relative to the equality of two population variances. We encountered one such instance earlier in this chapter when, in testing hypotheses about the difference between two population means, we found it convenient to assume that the population variances were equal in order to pool them. The procedure we discuss in this section provides a useful test for determining whether or not such an assumption is justified.

As an example, consider a reading specialist who must evaluate the effectiveness of two remedial reading programs, each of which is tried with a group of slow readers. The specialist may find that the mean improvement scores for the two groups are not significantly different, but that the variances are. The researcher, then, would consider the program yielding the smaller variance among scores the better of the two.

If independent random samples of size n_1 and n_2, drawn from two normally distributed populations (population 1 and population 2, respectively), yield variances of s_1^2 and s_2^2, respectively, the appropriate test statistic for testing

$$H_0\colon \sigma_1^2 = \sigma_2^2$$

against the alternative

$$H_1\colon \sigma_1^2 \neq \sigma_2^2 \quad \text{is} \quad \text{VR} = \frac{s_1^2}{s_2^2}$$

which is distributed as F with $\nu_1 = n_1 - 1$ numerator degrees of freedom and $\nu_2 = n_2 - 1$ denominator degrees of freedom. The test is called the *variance ratio* (VR) *test*.

For a significance level of α, the upper critical value is $F_{1-\alpha/2,\nu_1,\nu_2}$, which appears in Appendix Table H at the intersection of the column corresponding to ν_1 and the row corresponding to ν_2 in the table for the $(1 - \alpha/2)^{\text{th}}$ percentile of F. Since Appendix Table H does not give lower-tail percentiles of F, we obtain the lower critical value for the test, $F_{\alpha/2,\nu_1,\nu_2}$, from the relationship

$$F_{\alpha/2,\nu_1,\nu_2} = \frac{1}{F_{1-\alpha/2,\nu_2,\nu_1}}$$

Figure 6.17 shows rejection and acceptance regions for this test. When the hypotheses are

$$H_0\colon \sigma_1^2 \leq \sigma_2^2, \qquad H_1\colon \sigma_1^2 > \sigma_2^2$$

Figure 6.17 Rejection and acceptance regions for $H_0\colon \sigma_1^2 = \sigma_2^2$, $H_1\colon \sigma_1^2 \neq \sigma_2^2$

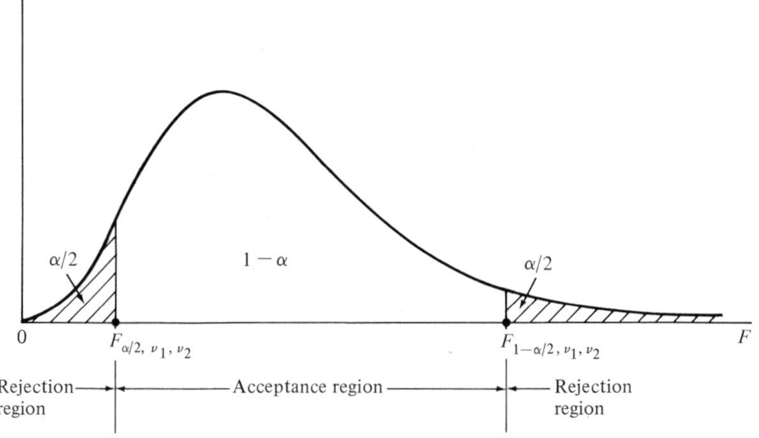

only large values of VR will cause rejection of H_0, and all of α is in the upper tail of the distribution. We obtain the single critical value $F_{1-\alpha,\nu_1,\nu_2}$ directly from Appendix Table H.

When the hypotheses are

$$H_0\colon \sigma_1^2 \geq \sigma_2^2, \qquad H_1\colon \sigma_1^2 < \sigma_2^2$$

only small values of VR will cause rejection of H_0, and all of α is in the left tail of the distribution. We obtain the single critical value F_{α,ν_1,ν_2} from the formula given earlier. Figure 6.18 shows rejection and acceptance regions for the one-sided alternatives.

Figure 6.18 Rejection and acceptance regions for one-sided alternatives in the variance ratio test

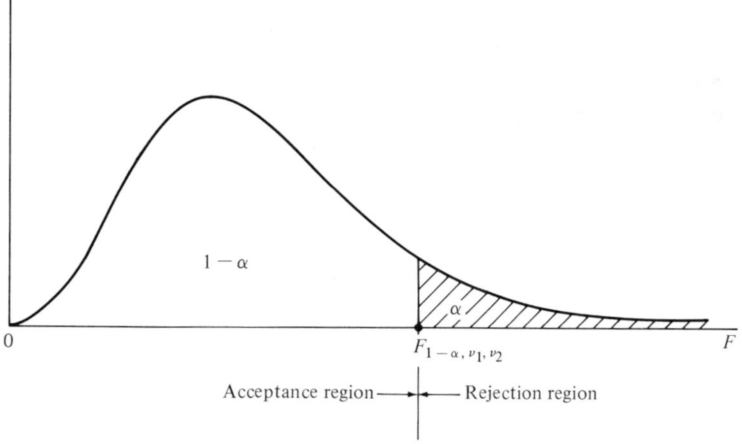

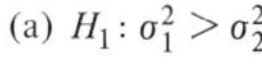
(a) $H_1: \sigma_1^2 > \sigma_2^2$

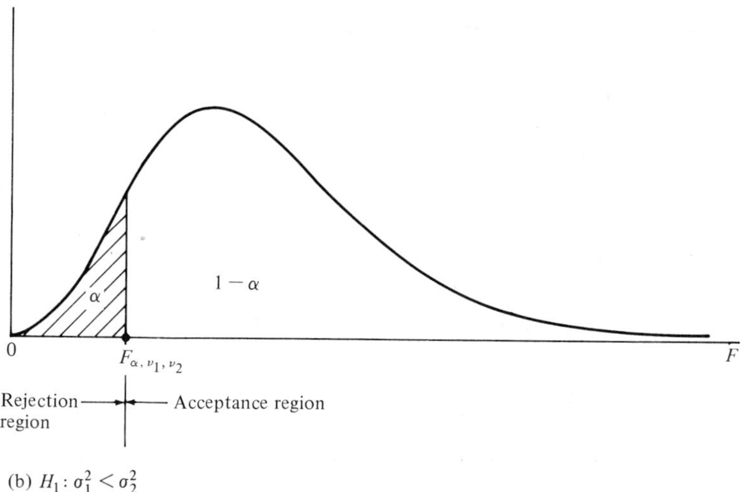

(b) $H_1: \sigma_1^2 < \sigma_2^2$

Example 6.14 A team of researchers gave a social skills test to a random sample of $n_A = 25$ sixth-grade students of school system A and to an independent random sample of $n_B = 21$ sixth-grade students of system B. The team conducting the research project wished to know, among other things, whether they could conclude that the two groups differ with respect to the variability of the scores. They carried out the following hypothesis test.

1 *Statement of hypotheses*

$$H_0: \sigma_A^2 = \sigma_B^2, \qquad H_1: \sigma_A^2 \neq \sigma_B^2$$

2 *Significance level* Let $\alpha = 0.05$.

3 *Description of populations and assumptions* It is assumed that the two populations of scores from which the observed data were drawn are normally distributed. The samples are independent.

4 *The relevant statistic* The relevant statistics are the two sample variances s_A^2 and s_B^2.

5 *The test statistic and its distribution* $\text{VR} = s_A^2/s_B^2$, which is distributed as F with $\nu_A = n_A - 1$ and $\nu_B = n_B - 1$ degrees of freedom.

6 *Rejection and acceptance regions* We obtain the upper critical value by consulting Appendix Table H for $F_{1-\alpha/2} = F_{0.975}$. With $\nu_A = n_A - 1 = 24$ and $\nu_B = n_B - 1 = 20$, we find this critical value to be 2.41. For the lower critical value, we obtain

$$F_{0.025,24,20} = \frac{1}{F_{0.975,20,24}} = \frac{1}{2.33} = 0.43$$

The rejection region, then, consists of values of VR equal to or greater than 2.41 and values of VR equal to or less than 0.43.

7 *Collection of data and calculations* Researchers found the two sample variances to be $s_A^2 = 64$ and $s_B^2 = 190$. From these data we compute

$$\text{VR} = \frac{64}{190} = 0.34$$

8 *Statistical decision* Since the computed VR of 0.34 is less than the lower critical value of 0.43 (that is, since 0.34 falls in the rejection region), we reject H_0.

9 *Conclusion* Since we reject H_0, we conclude that the social-skills scores of sixth-grade students in school system A and those of sixth-grade students in school system B do not have equal variances.

57 In a study designed to compare two methods of treating alcoholics, a medical team collected data on the levels of anxiety of the patients in the two treatment groups at the end of the first two weeks of treatment. The variances computed from the samples were $s_{\mathrm{I}}^2 = 2916$ and $s_{\mathrm{II}}^2 = 4624$. There were eight patients in each group. At the 0.05 level of significance, do these data provide sufficient evidence to suggest that the variability in level of anxiety is different in the two populations represented by the samples? State all necessary assumptions.

58 Prior to the initiation of a remedial reading program, two groups of slow-reading ninth-grade students were asked to read a passage of prose. Group A consisted of 10 students who had attended elementary school A, and group B was composed of 13 students who had attended elementary school B. The sample variances computed from the times each student required to read the passage in order to attain a specified level of comprehension were $s_A^2 = 100$ and $s_B^2 = 676$. On the basis of these data, can one conclude that the population represented by group A is less variable than the population represented by group B? Let $\alpha = 0.05$, and state all necessary assumptions.

59 Two groups of college freshmen were given a test to measure their levels of extraversion. Group I consisted of 25 students who had been reared in rural environments and group II was made up of 31 students who had been reared in urban environments. The variances computed from the sample data were $s_{\mathrm{I}}^2 = 81$ and $s_{\mathrm{II}}^2 = 36$. Do these data suggest, at the 0.05 level of significance, that the population of scores represented by group I is more variable than that represented by group II? Let $\alpha = 0.05$, and state all necessary assumptions.

60 A random sample of 16 college freshmen who stated that they planned to major in the arts and a random sample of 13 who said that they planned to major in business administration were given a musical aptitude test. The variance of the scores of the aspiring arts majors was 7.29. The variance of the scores of those planning to major in business administration was 39.69. Do these data provide sufficient evidence to indicate, at the 0.01 level of significance, that the two population variances are different? State all necessary assumptions.

61 In a study of human reaction times in response to a certain stimulus, psychologists used two independent samples of subjects. One sample was composed of 11 males between the ages of 25 and 45, and the second sample consisted of 13 females in the same age group. The variances of the reaction times were 12 msec^2 for the males and 4 msec^2 for the females. Should the psychologist conclude, on the basis of these data, that the population represented by the reaction times of the males is more variable than that represented by the reaction times of the females? Let $\alpha = 0.05$, and state all necessary assumptions.

6.11 THE RELATIONSHIP BETWEEN INTERVAL ESTIMATION AND HYPOTHESIS TESTING

Although our treatment of interval estimation and hypothesis testing tends to give the impression that they are two separate and unrelated topics, this is really not the case. In this section we shall point out the relationship between these two procedures and show that the two are alternative procedures for testing hypotheses of the type discussed in this chapter.

To illustrate, let us consider the case in which the parameter of interest is the population mean. Recall from Chapter 5 that we obtain upper and lower confidence limits for a population mean by adding to and subtracting from the point estimate of the parameter the product of a reliability factor and the standard error of the estimate. If, for example, a random sample of size n, drawn from a normally distributed population with known variance σ^2, yields a mean of $\bar{x}$, we obtain an interval estimate of the population mean by the expression

$$\bar{x} \pm z\frac{\sigma}{\sqrt{n}}$$

where z, the reliability factor, depends on the desired level of confidence.

Given the same population and the same sample, we have seen in this chapter that we test H_0: $\mu = \mu_0$ against the alternative H_1: $\mu \neq \mu_0$ by obtaining upper and lower critical values by the expression

$$\mu_0 \pm z\frac{\sigma}{\sqrt{n}}$$

where the numerical value of z depends on α, the chosen level of significance.

If the computed value of $\bar{x}$ falls outside the acceptance region defined by these critical values, we reject H_0 at the α significance level. Otherwise we are unable to reject H_0.

We may also use the confidence limits to test H_0: $\mu = \mu_0$ against the alternative H_1: $\mu \neq \mu_0$. If the hypothesized value of μ, μ_0, falls outside the $100(1-\alpha)\%$ confidence limits, we reject H_0 at the α level of significance.

Let us illustrate by means of an example.

Example 6.15 In Example 6.1 we wished to test, on the basis of the data from a simple random sample of size $n = 25$, the hypothesis H_0: $\mu = 162.5$ against the alternative H_1: $\mu \neq 162.5$, where the population was assumed to be normally distributed with a standard deviation of $\sigma = 18.0$, and α was set at 0.05.

The computed critical values were

$$162.5 \pm 1.96\frac{18}{\sqrt{25}} = 162.5 \pm 7.1 = 155.4 \text{ and } 169.6$$

Since the sample mean of 178.7 falls outside these limits, we rejected H_0.

Now let us show that we may reach the same decision by means of a $100(1-\alpha)\% = 100(1-0.05)\% = 95\%$ confidence interval. The upper and lower confidence limits are

$$178.7 \pm 1.96\frac{18}{\sqrt{25}} = 178.7 \pm 7.1 = 171.6 \text{ and } 185.8$$

Since the hypothesized value of the mean, 162.5, does not fall within these limits, we conclude that it is not a candidate for what is being estimated, and we reject H_0: $\mu = 162.5$.

An advantage of confidence intervals in testing hypotheses is the fact that, in addition to enabling us to reject or not reject the null hypothesis, they also give us some notion of the probable magnitude of the unknown parameter. In the present example we are able to say not only that we reject H_0: $\mu = 162.5$, but also that we are 95% confident that μ is between 171.6 and 185.8. On the other hand, the hypothesis-testing procedure described in this chapter merely allows us to reject or not reject H_0, and unless we carry out further calculations, we can make no distinction between results that are barely significant and those that are highly significant.

By the construction of one-sided confidence intervals, we may extend the confidence-interval approach to hypothesis testing to the case in which the alternative is one sided. For example, if we have H_0: $\mu \geq \mu_0$ versus H_1: $\mu < \mu_0$, we compute $\mu_U = \bar{x} + z(\sigma/\sqrt{n})$. If the hypothesized mean exceeds this value, we reject H_0. Similarly, for H_0: $\mu \leq \mu_0$ versus H_1: $\mu > \mu_0$, we compute $\mu_L = \bar{x} - z(\sigma/\sqrt{n})$. If μ_0 is less than μ_L, we reject H_0.

Example 6.16 In Example 6.2 we have H_0: $\mu \geq 65$ versus $\mu < 65$, $\alpha = 0.01$, a population assumed to be normally distributed with $\sigma = 15$, a sample of size 20, and a computed sample mean of 55.5.

The critical value of the statistic is

$$65 - 2.33\frac{15}{\sqrt{20}} = 65 - 7.82 = 57.18$$

Since $\bar{x} = 55.5$ is less than 57.18, we reject H_0.

The appropriate one-sided confidence interval for testing H_0 is given by

$$\mu_U = 55.5 + 2.33\frac{15}{\sqrt{20}} = 55.5 + 7.82 = 63.32$$

Since 65 is greater than 63.32, we reject H_0 as before.

EXERCISES

62 Refer to Exercise 1. Test H_0 by means of a confidence interval.
63 Refer to Exercise 2. Test H_0 by means of a confidence interval.
64 Refer to Exercise 4. Test H_0 by means of a confidence interval.

65 Refer to Exercise 6. Test H_0 by means of a confidence interval.
66 Refer to Exercise 7. Test H_0 by means of a confidence interval.
67 Refer to Exercise 9. Test H_0 by means of a confidence interval.
68 Refer to Exercise 10. Test H_0 by means of a confidence interval.
69 Refer to Exercise 23. Test H_0 by means of a confidence interval.
70 Refer to Exercise 24. Test H_0 by means of a confidence interval.
71 Refer to Exercise 34. Test H_0 by means of a confidence interval.
72 Refer to Exercise 36. Test H_0 by means of a confidence interval.

CHAPTER SUMMARY

In this chapter we have covered the topic of hypothesis testing in considerable detail. We have presented procedures for testing hypotheses about population means, the difference between means, proportions, the difference between proportions, variances, and the ratio of two variances. We have also discussed and illustrated the difference between one-sided and two-sided alternatives.

In addition, we have treated the concept of the power of a test, the calculation of p values, and the relationship between interval estimation and hypothesis testing.

SOME IMPORTANT CONCEPTS AND TERMINOLOGY

Research hypothesis
Statistical hypothesis
Null hypothesis
Alternative hypothesis
Type I error
Type II error
Significance level
Test statistic
Rejection region
Acceptance region
Critical value
One-sided test
Two-sided test
p value
Power of a test
Relationship between interval estimation and hypothesis testing

REVIEW EXERCISES

73 A firm specializing in direct-mail advertising has developed a new questionnaire which it believes will obtain a higher response rate than the standard one. The new questionnaire is sent to a sample of 200 potential respondents. The number of responses to the new questionnaire is 120. When the standard form was used with 250 potential respondents, 115 responded. Do these data provide sufficient evidence to indicate that the new questionnaire is better than the standard one? Let $\alpha = 0.05$.

74 A sample of 100 patients with disease A, admitted to a hospital for chronic diseases, remained in the hospital, on the average, 32 days. Another sample of 100 patients with disease B stayed, on the average, 35 days. The

sample variances were 50 and 60, respectively. Do these data provide sufficient evidence to indicate, at the 0.05 level of significance, that the two populations differ with respect to mean length of stay?

75 A frozen-food-processing firm specifies that the mean net weight per package of a certain food must be 30 ounces. Experience has shown that the weights are approximately normally distributed, with a standard deviation of 1.5 ounces. A random sample of nine packages yielded a mean weight of 29.4 ounces. Is this sufficient evidence to indicate that the true mean weight of the packages has decreased? Let $\alpha = 0.05$.

76 A test designed to measure a person's level of security was administered to a random sample of 25 alcoholics and an independent random sample of 20 nonalcoholics. The results were as follows.

	$\bar{x}$	s^2
Alcoholics	60	100
Nonalcoholics	72	100

Do these data provide sufficient evidence to indicate, at the 0.01 level of significance, that alcoholics, on the average, score lower on the test than nonalcoholics?

77 Nine pairs of truck-driver trainees were matched on age, years of automobile-driving experience, and other relevant variables. One member of each pair was randomly assigned to a training course that was taught by method A. The other member of each pair was assigned to the same type of training course taught by method B. At the end of the course, each trainee was given an examination designed to test retention of the material presented. Table 6.7 shows the results. Can one conclude from these data that method A is superior to method B? Let $\alpha = 0.05$.

78 A real estate agent claims that the average value of homes in a certain neighborhood is greater than \$65,000. A random sample of 36 homes had a mean value of \$67,000 and a standard deviation of \$12,000. Do these data support the agent's claim at the 0.05 level of significance?

79 The following data are based on random samples taken from two populations of children. Children in population 1 are educably mentally retarded. Those in population 2 are normal. The variable of interest is the length of time (in minutes) required to perform a certain task.

	Population 1	**Population 2**
n	10	8
$\bar{x}$	26.1	17.6
s^2	140	115

Table 6.7 Scores made by nine pairs of truck-driver trainees, each trained by a different method

Pair	1	2	3	4	5	6	7	8	9
Method *A*	90	95	87	85	90	94	85	88	92
Method *B*	85	88	87	86	82	82	70	72	80

Do these data provide sufficient evidence to indicate that the average time required by population 2 is less than that required by population 1? Let $\alpha = 0.05$. Specify all assumptions you have to make in order to validate your procedure.

80 A board of public health officials believes that fewer than 50% of the adults in their jurisdiction have dental checkups twice a year. Of 300 adults interviewed in a survey, 125 said that they regularly had dental checkups twice a year. Do these data support the health officials' belief at the 0.05 level of significance?

81 In a survey conducted in two sections of a city, the following results were obtained with respect to abnormal blood-sugar levels.

Section	Number of persons screened	Number abnormal on screening
1	200	20
2	250	40

Can one conclude from these data that the two sections differ with respect to the proportions of residents with abnormal blood-sugar levels? Let $\alpha = 0.05$.

82 In a survey of disabled adult males, a research team found that 35% of those who were veterans were disabled before the age of 55. Among nonveterans the percentage disabled before age 55 was 45. Do these data provide sufficient evidence, at the 0.05 level of significance, to indicate that veterans and nonveterans differ with respect to the proportion disabled before age 55? ($n_V = n_N = 100$.)

83 Determinations of the level of a certain enzyme were made on blood samples from 16 apparently normal subjects. The sample yielded a mean of 96 units/100 ml and a standard deviation of 36 units/100 ml. Do these data provide sufficient evidence to indicate that the mean for the sampled population is less than 100 units/100 ml? Let $\alpha = 0.05$.

7

Experimental Design

The three major sources of data available to the researcher are: (1) the routine operation of an organization, (2) sample surveys, and (3) designed experiments. The primary focus of this chapter is the third source, the designed experiment. But before we discuss this in detail, let us touch briefly on the general nature of the first two sources.

Most businesses, institutions, and other organizations generate, as by-products of their routine operation, data that the researcher may use. Businesses, for example, keep records on their employees, sales records, production records, and many other types of records one may use in conducting studies of importance to management, marketing, and other functions of the firm. As another example, consider the government agency whose function is to provide service to certain citizens. The agency keeps records of allocations and expenditures of funds, characteristics of clients, services performed, and results achieved.

All such records provide the raw material for important studies. When the data generated by an organization's routine operation are not sufficient or appropriate for answering the question under study, one alternative is to conduct a sample survey. For instance, the marketing department of a manufacturing firm may wish to determine the public's opinion of some new product. Of a government service agency may wish to compare the characteristics of its clients with those of citizens who are not on its rolls.

A characteristic of research that is based on data generated by the ongoing operation of an organization and data obtained from sample surveys is that the researcher is unable to manipulate the condition related to the research objectives. The data available from a business enterprise, for example, may be part of the records that are kept because they are required by law, or part of those that are motivated by the profit-making objectives of the firm. That the data from such records may also be used in research is usually incidental.

Typically, the investigator who obtains data by means of a sample survey must be content with the situation as it exists naturally, since it is usually impossible for him or her to impose any type of control on the subjects of the inquiry in order to observe their reaction to such manipulation. For example, suppose you want to compare the opinions of young adults who are married with those of young adults who are not married. You can't select a sample of comparable, single young adults and instruct half to get married and half to say single so that you can study the opinions of the two groups. Instead, you must select a sample from each of the two groups as they exist, that is, as they have been formed as a result of forces unrelated to the investigation.

In contrast to the data generated by an organization's ongoing operation and that obtained through sample surveys, data obtained from a designed experiment arise from a situation that has been under the control of the investigator. Our purpose in this chapter is to learn more about the nature of such experiments. In Section 7.1 we introduce some basic concepts and vocabulary that will help you in the discussion that follows. In the remaining sections we discuss some important considerations in the design of an experiment, and explore in some detail the nature of some standard experimental designs.

7.1 SOME BASIC CONCEPTS IN EXPERIMENTAL DESIGN

Here we shall look at some basic concepts with which an investigator who is to engage in meaningful experimentation must be familiar. We do not present these concepts in any particular order, so you should not infer any particular importance from the sequence of the discussion that follows.

Experiment The definition of *experiment* used in this chapter is more limiting than the definition in Chapter 2, in which we defined an experiment as any process or activity that yields an outcome or observation. A better expression to use in this chapter might be the term *controlled experiment*. We shall, however, comply with general practice, and use the word *experiment* instead.

One can find many definitions of the word experiment in the literature. For our purposes, we define an *experiment* as a sequence of activities planned for the purpose of acquiring information. To distinguish an experiment from other types of investigation in which information is obtained, we require that, in order to qualify as an experiment, a sequence of activities must be designed in such a way that the experimenter has the ability to manipulate at least one relevant variable.

Independent variable The *independent variable* in an experiment is the variable that the experimenter has the ability to manipulate. In other words, the independent variable is under the control of the experimenter. Consider, for example, an experiment designed to acquire information about the effectiveness of different methods of teaching arithmetic to fourth-grade pupils. The experimenter has the ability to manipulate the methods of teaching employed in the experiment, in that the selection of the methods for inclusion is under his or her control. In this example, then, "method of teaching arithmetic" is called the independent variable. As we shall mention later, the independent variable is also referred to as the *treatment*.

Dependent variable The *dependent variable* in an experiment reflects any effects that may accompany manipulation of the independent variable. In the experiment designed to acquire information about the effectiveness of different methods of teaching arithmetic, arithmetic achievement, during the experimental period, of the pupils participating in the experiment is the dependent variable. The dependent variable is sometimes called the *response variable*.

Nuisance variable In addition to dependent and independent variables, in every experiment the investigator must contend with one or more *nuisance variables*. Nuisance variables are so called because of their potential effect on the dependent variable. They are generally of no interest to the experimenter, in the sense in which the dependent variable is of interest. The experimenter may control nuisance variables, since they are sources of unwanted variation in the experiment. In the example concerning methods of teaching arithmetic, possible nuisance variables are teachers (it may be necessary to have more than one teacher participate in the experiment), abilities of pupils, and time of day (it may be necessary to conduct parts of the experiment at different times during the school day). The complexity of an experiment usually increases as the number of nuisance variables the experimenter tries to control increases.

Let us illustrate more specifically how, in the example of teaching of arithmetic, the investigator might control some nuisance variables. To control for time of day alone, he or she could arrange for a class to be taught by each method in the morning, in the middle of the day, and in the afternoon. A representation of the design might look like Figure 7.1 for methods A, B, and C. Suppose the experimenter wants to control for teacher effects, and three teachers are available. It seems logical that each teacher should

Figure 7.1 Representation of a design to control time-of-day effects in the teaching of arithmetic by methods *A*, *B*, and *C*

Time of day			
MORNING	*A*	*B*	*C*
MIDDAY	*A*	*B*	*C*
AFTERNOON	*A*	*B*	*C*

teach each method. Figure 7.2 shows an arrangement whereby the experimenter can simultaneously control both variables, time of day and teacher effects, for methods *A, B,* and *C*. Note in Figure 7.2 that each teacher uses each method once in each time period. When one designs experiments as illustrated in Figures 7.1 and 7.2, one can control the effects of the nuisance variables in the statistical analysis, as we shall explain later. Nuisance variables are also referred to as *extraneous variables*.

Treatment The term *treatment* is used synonymously with independent variable. The word has a rather broad meaning in experimental design, and refers to any variable whose effect one desires to measure. A treatment in a given experiment may be a treatment in the usual sense, in that one uses some procedure (drug or other therapeutic procedure) for the purpose of treating some existing condition. In another experiment the term treatment may have an entirely different connotation. In the example on the teaching of arithmetic, "teaching methods" is the treatment.

Figure 7.2 Representation of a design to control time-of-day and teacher effects in the teaching of arithmetic by methods *A*, *B*, and *C*

		Teacher	
Time of day	**1**	**2**	**3**
MORNING	*A*	*B*	*C*
MIDDAY	*C*	*A*	*B*
AFTERNOON	*B*	*C*	*A*

In an experiment, the treatment occurs at different levels, which may be either qualitative or quantitative, depending on whether the treatment variable is quantitative or qualitative. The levels of a treatment that is quantitative indicate the *amounts* of the treatment that are present. The levels of a treatment that is qualitative differ in *type*. The treatment in the example concerning the teaching of arithmetic is qualitative, so the levels differ according to the type of method employed. In this example, the treatment levels are the different ways of teaching that are employed in the experiment. These might be (1) class demonstration only, (2) class demonstration with homework assignments, (3) class demonstration with group work in class, and (4) class demonstration with individual work in class. An example of an experiment in which the treatment is quantitative is a drug experiment in which one wishes to measure the effects of different dose levels, which might be 1 mg, 2 mg, 3 mg, and so on. We may also refer to a level itself as a treatment.

Experimental unit The *experimental unit* is the smallest entity to which a treatment is applied. In the example about the teaching of arithmetic, the pupils are the experimental units. Depending on the objectives of the experiment, the classes of students could also be the experimental units. When the experimental units are people or animals, they are frequently referred to as *subjects*. In a biology experiment the experimental unit might be a specimen of tissue, a mouse, or a colony of bacteria. The experimental unit in an agricultural experiment might be a plot of land, a plant, or an insect.

Measurement A *measurement* is a value of the dependent variable, and represents the attempt to determine the effect of the treatment on an experimental unit. In the example on the teaching of arithmetic, the measurements might consist of scores made by the pupils (experimental units) on an arithmetic achievement test administered at the conclusion of the experiment. Measurements are frequently called *observations*.

Experimental error Experimental units exposed to the same level of the treatment generally respond differently. That is, experimental units exposed to the same treatment level generally yield different measurements. The measure of this variability in response is called *experimental error*. There are two primary sources of experimental error: (1) the inherent differences in the experimental units, and (2) any lack of uniformity that may exist in the experimental procedure. In the example about the teaching of arithmetic, pupils differ both genetically and with respect to past experiences; these differences contribute to the experimental error. In the course of the experiment, pupils sit in different locations in the classroom, so that the temperature, humidity, lighting, and social conditions may not be the same for all pupils exposed to the same level of the treatment. These sources of experimental error constitute a subset of the nuisance variables. An important objective of the experimenter is the reduction of the experimental error whenever possible.

Replication When, in an experiment, a given level of a treatment is applied to more than one experimental unit, the level is said to be *replicated*. Replication may be partial, when only some levels are replicated, or complete, when all levels are replicated. Replication is desirable for the following reasons.

1 Replication makes possible the computation of an estimate of experimental error.

2 As we shall see in Chapter 8, the availability of an estimate of experimental error allows for the reduction of the standard error of the treatment mean, thereby improving the precision of the experiment.

3 Replication increases the capability of the experimental results to be generalized. In the example on the teaching of arithmetic, if only one student is taught by each method under consideration, there is no replication. On the other hand, if two or more pupils are exposed to each teaching method, there is complete replication of the experiment.

Randomization We discussed randomization in general in Chapter 1. In the present context, *randomization* refers to the random assignment of experimental units to the treatment levels. When proper randomization is employed, every experimental unit has an equal opportunity of being assigned to any treatment level. The effect of proper randomization is to average out, among treatment levels, any systematic effects that may be present. The result is that, when we make comparisons between treatment levels, we assess only the pure level effects. Randomization may also refer to the initial selection of experimental units or to the selection of treatments.

Power We discussed the general concept of the *power* of a hypothesis test in detail in Chapter 6. In designing an experiment, the investigator is interested in the power of any statistical test that is conducted. We shall discuss this concept in more detail in Chapter 8.

External validity *External validity* refers to the capability of the results of an experiment to be generalized. The external validity of the results depends on the employment of proper randomization and replication. In other words, the random selection of experimental units from a population of such units allows one to generalize to that population.

Internal validity An experiment that yields results that are free from bias is said to be *internally valid*. Effective control of the sources of experimental error and other nuisance variables contributes to the internal validity of an experiment.

EXERCISES

1 Name and discuss the three major sources of data available to the researcher.

2 Describe a research project in your particular area of interest in which you could use data generated by the ongoing activities of some organization.

3 Describe a research project in your area of interest that would require acquisition of data by means of a sample survey.

4 Describe a research project in your area of interest that would require a designed experiment.

5 Define the following terms.

Experiment	Experimental error
Independent variable	Replication
Dependent variable	Randomization
Nuisance variable	Power
Treatment	External validity
Experimental unit	Internal validity
Measurement	

6 Describe three experiments in your particular field of interest that a researcher in that field might wish to conduct. For each experiment, name the dependent variables, the independent variables, the nuisance variables, and the experimental units.

7 In Exercise 6, explain how you would measure the effects of the independent variables.

7.2 SOME IMPORTANT CONSIDERATIONS IN THE DESIGN OF EXPERIMENTS

In this section we discuss some of the aspects of experimental design that should be given particular consideration by the experimenter or potential experimenter. In Section 7.1 we mentioned some of the matters on which the experimenter should focus attention. In addition to these, the experimenter should give careful attention to the following aspects of the experiment.

1 *Selection of the independent variable* (*or treatment*) The choice of the independent variable is dictated by the experimental objectives, which presumably are carefully formulated by the experimenter. It may seem superflous to point out what you may feel is self-evident, namely, that the independent variable should be one that will shed light on the phenomenon under investigation. In the example on the teaching of arithmetic, introduced in Section 7.1, the phenomenon under investigation is the learning of arithmetic skills and concepts. It is presumed that the independent variable "method of teaching" will provide some insight into the acquisition of arithmetic skills and concepts. Other potential independent variables

that one might consider in this example include methods of changing pupils' attitudes toward arithmetic, methods of improving the learning environment, and learning aids in which the treatment levels consist of different aids, such as audio and/or visual materials of various kinds.

2 *Selection of the dependent variable* When possible, the experimenter selects the dependent variable that, so far as one can tell, is the best indicator of the treatment effects. The dependent variable must be one that can be measured. Other characteristics of the dependent variable that the experimenter must consider are sensitivity and reliability. A dependent variable that will not detect any differences between treatment effects, when they are present, is said to be *insensitive*. *Reliability* refers to the consistency of results when repeated measurements are taken on the same experimental unit. In the example concerning methods of teaching arithmetic, we assume that the dependent variable, arithmetic achievement, is sensitive. That is, we assume that it is capable of reflecting any differences among teaching methods that may exist. We also assume that it is reliable. That is, we assume that if the achievement test were administered to the same pupil several times, the pupil's scores would be approximately the same each time.

3 *Methods of controlling nuisance or extraneous variables* Methods of controlling nuisance or extraneous variables include the following.

(a) Holding the nuisance variable constant for all experimental units. In the example on teaching arithmetic, we could control the nuisance variable "intelligence" by using in the experiment only pupils with approximately the same intelligence scores.

(b) Randomly assigning experimental units to the treatment levels. As we mentioned in the preceding section, randomization tends to "average out," among treatment levels, any systematic effects that may be present.

(c) Matching experimental units on the basis of the nuisance variable. So long as only one or two nuisance variables have to be controlled, this method is usually feasible. Like other methods of control, however, the difficulties increase as the number of nuisance variables increases. We discussed this technique in Chapters 5 and 6 under the heading of paired comparisons. We shall discuss an extension of the technique later in this chapter, and again in Chapter 8.

(d) Using statistical control. We may control the effects of nuisance variables by means of a statistical technique known as *analysis of covariance*. This method of control is covered in many of the books on experimental design mentioned at the end of this chapter.

4 *Selection of experimental units* In most experiments, an objective of the experimenter is the ability to infer from the results obtained on a sample of experimental units to the population from which the sample was drawn. One should select experimental units with this objective in

mind. One usually meets this objective by selecting subjects randomly from the population of interest.

5 *Selection of a design* The nature of a given experiment may be such that the experimenter may use one of many standard experimental designs. The theory underlying these designs and appropriate procedures and statistical analyses are well known, and may be found in standard textbooks on experimental design. In this chapter we discuss the basic ideas of three of these designs, and in Chapter 8 we cover the statistical analysis for each. You may find other standard designs in the books listed at the end of this chapter.

6 *Power and relative efficiency* We discussed the concept of power in Chapter 7, and we shall discuss it further in Chapter 8. In planning an experiment, the experimenter wants to use a design that leads to statistical tests with high power.

7 *The cost of the experiment* One of the most important aspects of an experiment is cost. Unless you have the funds necessary for carrying out an experiment, all other considerations will come to naught. Before progressing very far with the other aspects of planning, you should consider the costs of the experiment in terms of equipment, experimental units, and personnel.

8 *Analysis and reporting of results* An investigator should not postpone consideration of the analysis and reporting of results until after the experiment has been completed. Bear in mind, from the beginning, the statistical analyses you plan to use on the results and the nature of the report you plan to write.

EXERCISE

8 List and discuss the aspects of an experiment that should be considered by anyone designing an experiment.

7.3 THE COMPLETELY RANDOMIZED EXPERIMENTAL DESIGN

In this section and the two that follow, we discuss three of the standard experimental designs that are available. We shall discuss the appropriate statistical analyses of these designs in Chapter 8. The first experimental design we discuss is the completely randomized design. We discuss two other designs, the randomized complete block design and the Latin square design, in the sections that follow.

The *completely randomized design* gets its name from the fact that, in this design, the experimental units on which the measurements are taken are randomly assigned to the treatments, or levels of the independent variable. As we noted earlier, the random assignment specification indicates that each experimental unit has an equal probability of receiving any one of the treatments.

As an example of a completely randomized design, consider an experiment designed to compare the effects of three different drugs (A, B, C) on the reaction times of normal adult humans (the subjects). Suppose 21 subjects are available for the experiment. To meet the requirements of a completely randomized design, we would randomly assign each subject to receive one of the drugs in such a way that all subjects have an equal probability of being assigned to each drug.

In general, we can assign experimental units to treatments by using a table of random numbers, such as Appendix Table A. For our example above, we have a total of $n = 21$ subjects, with $S = 7$ subjects to be assigned to each of $d = 3$ drugs. The subjects are numbered 1, 2, . . . , 21. We then select random numbers between 1 and 21 from the table of random numbers. The numbers comprising the first 7 numbers selected from the table correspond to the numbers of the subjects to be assigned to drug A. The numbers comprising the second set of 7 random numbers correspond to the numbers of subjects to be assigned to drug B. And so on.

Figure 7.3 Representation of the completely randomized design for comparing three drugs used on 21 subjects

	Drug	
A	B	C
S_2	S_9	S_{19}
S_6	S_{14}	S_{11}
S_{21}	S_{10}	S_5
S_3	S_{17}	S_{20}
S_7	S_{18}	S_{13}
S_{15}	S_1	S_{12}
S_{16}	S_4	S_8

Figure 7.3 shows the completely randomized design for three drugs and 21 subjects. In this figure we see that an equal number of subjects is assigned to each treatment. Although it is not necessary to have this type of balance in a design, one usually uses it if it is convenient.

In the example on assigning subjects to drugs, we have three levels represented by the three drugs under investigation. This example is an illustration of an independent variable whose levels vary according to type. If the effects of different dosages of a drug were of interest, the independent variable would be quantitative, and the levels would vary according to amount. We think of each treatment level as designating a population of measurements of which the observed measurements constitute a sample.

We shall discuss the appropriate statistical analysis for the completely randomized design in Chapter 8.

EXERCISES

9 Describe an experiment in your field of interest in which you could use the completely randomized design. Name the appropriate dependent variable, and specify the levels of the independent variable that would be appropriate.

10 In Exercise 9, what would be the appropriate experimental units?

11 Do you think the experiment you described in Exercise 9 would be very expensive to carry out?

12 For the experiment you described in Exercise 9, specify as many nuisance variables as you can think of.

7.4 THE RANDOMIZED COMPLETE BLOCK DESIGN

When the objective of an experiment is to evaluate the effects of several treatments, the presence of some extraneous source of variation (nuisance variable) may mask any true effects of the treatment, if they exist. In comparing different methods of teaching arithmetic, for example, true differences in effectiveness may be obscured by variations in pupils' IQs, ages, physical health, or a number of other factors.

Through proper experimental design, we can deal with this problem in such a way that the effects of the extraneous source of variation are eliminated in the statistical analysis of the results. Before we apply the treatments, we can allocate subjects to subgroups that are homogeneous with respect to the variable whose effects we wish to eliminate. If age is a source of extraneous variation, we may divide subjects into subgroups in such a way that those within each subgroup are approximately the same age. The objective is to make the variation among subjects within subgroups smaller than the variation among subjects when there is no subgrouping.

Figure 7.4 Representation of the randomized complete block design with four treatments, three blocks, and 12 subjects

	Treatments			
Blocks	*A*	*B*	*C*	*D*
1	S_{10}	S_2	S_{11}	S_5
2	S_9	S_1	S_6	S_4
3	S_7	S_8	S_{12}	S_3

Once we have established the homogenous subgroups, we randomly assign treatments to subjects within each subgroup, with each treatment occurring once in each subgroup.

We call an experimental design that has these characteristics a *randomized complete block design*. A *block* is one of the homogeneous subgroups. We call the variable that defines the block or subgroup the *blocking variable*. The design is complete in the sense that each treatment appears in each block.

The paired-comparisons test discussed in Chapter 6 is a special case of the randomized complete block design. Here there are two treatments, and the blocks are the matched or paired subjects.

Figure 7.4 shows the randomized complete block design. You will also recognize Figure 7.1 as a representation of the randomized complete block design.

We shall discuss the appropriate statistical analysis of the randomized complete block design in Chapter 8.

EXERCISES

13 Describe an experiment in your field of interest in which you could use the randomized complete block design. Name the dependent, independent, and blocking variables.

14 In Exercise 13, what would be the appropriate experimental units?

15 Do you think the experiment you described in Exercise 13 would be very expensive to carry out?

16 For the experiment you described in Exercise 13, specify as many additional nuisance or extraneous variables as you can think of.

7.5 THE LATIN SQUARE DESIGN

In Section 7.4 we learned that we can eliminate variability resulting from a nuisance variable by using the randomized complete block design. There may be instances in which we wish to remove the effects of two extraneous sources of variation. In such cases, we accomplish our goal by using the *Latin square design*.

A characteristic of the Latin square design is that the number of levels of each extraneous variable must equal the number of treatments. If, for example, there are five treatments, the Latin square design necessitates the use of five levels of each extraneous variable.

As an example of a Latin square design, let us refer to the example concerning the assignment of subjects to drugs. The experimenter wishes to compare the effects of three drugs on the reaction times of normal adult humans. Suppose the experimenter wants to control the effects of the two extraneous or nuisance variables "age of subject" and "time of day." The investigator could use three age groups (say young, middle-aged, and elderly) and three times of day (say morning, midday, and afternoon). Figure 7.5 is a representation of the Latin square design for this example. As the figure indicates, with this design, subjects would be assigned at random to the nine age–time–drug combinations, thus necessitating the use of nine subjects. You will also recognize Figure 7.2 as a representation of the Latin square design.

Advantages and disadvantages of the Latin square design

The advantages that may be realized when the Latin square design is used include the following.

Figure 7.5 Representation of the Latin square design for three treatments, A, B, and C, with blocking on age and time of day

	Age group of subject		
Time of day	YOUNG	MIDDLE-AGED	ELDERLY
MORNING	A (S_1)	B (S_3)	C (S_9)
MIDDAY	C (S_5)	A (S_6)	B (S_2)
AFTERNOON	B (S_4)	C (S_8)	A (S_7)

1 One may investigate several variables with only a relatively small number of subjects.

2 One may eliminate the effects of two extraneous sources of variation from the error term.

3 The analysis is relatively simple compared to what potentially may be accomplished in the way of reducing the error variance.

There are also some possible disadvantages of using the Latin square design. These include the following.

1 There must be the same number of levels of each of the nuisance variables as there are treatments. In some situations this may be impractical or impossible.

2 Latin squares with fewer than five treatments may be of limited value, or no value at all, because of the small number of degrees of freedom available for the error term.

3 The randomization procedure is somewhat tedious.

4 As we shall point out later, an assumption underlying the use of the Latin square is that the nuisance variables and the treatments do not interact. This may be an unrealistic assumption in some cases.

We shall discuss the appropriate statistical analysis of the Latin square design in Chapter 8.

EXERCISES

17 Describe an experiment in your field of interest in which you could use the Latin square design. Name the dependent variable and the two extraneous variables that you can control.

18 In Exercise 17, what experimental units would be appropriate?

19 Do you think the experiment you described in Exercise 17 would be very expensive to carry out?

20 For the experiment you described in Exercise 17, specify as many additional nuisance or extraneous variables as you can think of.

CHAPTER SUMMARY

In this chapter we talked about the basic concepts of experimental design. The first section introduced and defined some basic terms. Then came a discussion of certain aspects of experimental design that the experimenter should consider. In the last three sections we discussed three standard experimental designs: the completely randomized design, the randomized complete block design, and the Latin square design.

A complete discussion of the philosophy and techniques of experimental design is beyond the scope of this book. For a more extensive coverage of experimental design, see the books on the subject by Cochran and Cox (1), Dayton (2), Edwards (3), Hicks (4), and Kirk (5). Experimental design is discussed by Neter and Wasserman (6).

SOME IMPORTANT CONCEPTS AND TERMINOLOGY

Experiment	Randomization
Independent variable	Power
Dependent variable	External validity
Nuisance variable	Internal validity
Treatment	Completely randomized experimental design
Experimental unit	Randomized complete block design
Measurement	Latin square design
Experimental error	
Replication	

REFERENCES

1 William G. Cochran and Gertrude M. Cox, *Experimental Designs,* New York: Wiley, 1957

2 C. Mitchell Dayton, *The Design of Educational Experiments,* New York: McGraw-Hill, 1970

3 Allen L. Edwards, *Experimental Design in Psychological Research,* third edition, New York: Holt, Rinehart, and Winston, 1968

4 Charles R. Hicks, *Fundamental Concepts in the Design of Experiments,* New York: Holt, Rinehart, and Winston, 1964

5 Roger E. Kirk, *Experimental Design: Procedures for the Behavioral Sciences,* Belmont, Calif.: Brooks/Cole, 1968

6 John Neter and William Wasserman, *Applied Linear Statistical Models,* Homewood, Ill.: Richard D. Irwin, 1974

8 Analysis of Variance

An industrial psychologist wished to study the effects on production of five different incentives in the manufacturing plants of a large business. An experiment was carried out in a single department in each of five plants. Employees on the 11:00 P.M. to 7:00 A.M. shift took part in the experiment, which lasted for a week. Ten employees were involved at each plant; they were promised certain rewards for increased production during the experimental week. The rewards were (1) money, (2) transfer to shift or department of choice, (3) time off with pay, (4) an additional coffee break, and (5) special recognition. The psychologist's specific objective was to determine whether or not the different rewards would have differing effects on volume of production. The effects of the rewards were measured by the difference between the number of items produced during the experimental week and the mean number of items produced per week during the month preceding the experiment.

The psychologist designed this experiment as a completely randomized experiment with five treatment levels and with ten employees randomly assigned to each of the treatments. The analysis of this experiment poses a problem in statistical inference involving the comparison of five sample means, where a sample mean is computed from the data for each group of ten employees.

Although Chapters 5 and 6 introduced the concepts and techniques of statistical inference, including both interval estimation and hypothesis testing, in these chapters we were concerned with the construction of confidence intervals for, and the testing of hypotheses about, the difference between *two* population means. When we are interested in inferences about three or more populations, the techniques presented in Chapters 5 and 6 do not in themselves provide the tools necessary for handling the required statistical analysis. The techniques of Chapter 6 alone, for example, are not adequate for making inferences about the true relative effects of five methods of rewarding increased employee production. To investigate this question statistically, we must use concepts and techniques that go beyond those we have learned so far.

In this chapter we introduce *analysis of variance,* a technique that enables us to simultaneously make inferences about the parameters of three or more populations.

We may define analysis of variance, which was introduced by R. A. Fisher in the 1920s, as follows.

DEFINITION ***Analysis of variance*** *is an arithmetic procedure whereby the total variation in a set of data is partitioned into two or more components, each of which can be ascribed to an identifiable source.*

Although, in its application, analysis of variance is not limited to the analysis of data generated by a formally designed experiment, this is the area in which it is most frequently encountered.

8.1 ONE-WAY ANALYSIS OF VARIANCE

In this section we discuss *one-way analysis of variance,* which is suitable for the analysis of data resulting from a completely randomized experiment, such as the one described at the beginning of this chapter. The technique is called one-way analysis of variance because each response or observation is categorized according to one criterion of classification—the treatment to which it belongs.

Data display

In the experiment involving rewards to employees for increased production, the psychologist recorded the difference between the number of items produced during the week of the experiment and the mean number of items produced per week during the month preceding the experiment for each of the ten employees in each treatment group. Table 8.1 displays the 50 resulting observations (or measurements).

Table 8.1 Difference between number of items produced during experimental week and mean number of items produced per week during month preceding the experiment for each employee for each treatment

	Reward (Treatment)					
	MONEY	TRANSFER TO SHIFT OR DEPARTMENT OF CHOICE	TIME OFF WITH PAY	ADDITIONAL COFFEE BREAK	SPECIAL RECOGNITION	
	76	52	37	19	11	
	70	52	26	21	15	
	59	43	28	16	23	
	77	48	38	23	15	
	59	43	25	23	25	
	69	56	30	23	18	
	80	52	28	14	20	
	78	53	25	15	18	
	61	58	26	13	20	
	66	50	25	16	16	
Total	695	507	288	183	181	**1854**
Mean	69.5	50.7	28.8	18.3	18.1	**37.08**
Variance	65.17	24.23	23.73	15.79	16.99	**438.65**

In general, one may display the measurements from a completely randomized design consisting of k treatment levels in a table such as Table 8.2. The symbols used in Table 8.2 are defined as follows.

y_{ij} = ith measurement from treatment j, where $i = 1, 2, \ldots, n_j$ and $j = 1, 2, \ldots, k$

$$T_{.j} = \sum_{i=1}^{n_j} y_{ij} = \text{total of the measurements in the } j\text{th column}$$

$$\bar{y}_{.j} = \frac{T_{.j}}{n_j} = \frac{\sum_{i=1}^{n_j} y_{ij}}{n_j} = \text{mean of the measurements in the } j\text{th column}$$

Table 8.2 Measurements obtained from a completely randomized experiment

	Treatment					
	1	2	3	...	k	
	y_{11}	y_{12}	y_{13}	...	y_{1k}	
	y_{21}	y_{22}	y_{23}	...	y_{2k}	
	$\vdots$				$\vdots$	
	y_{i1}	y_{i2}	y_{i3}	...	y_{ik}	
	$\vdots$				$\vdots$	
	$y_{n_1 1}$	$y_{n_2 2}$	$y_{n_3 3}$	...	$y_{n_k k}$	
Total	$T_{.1}$	$T_{.2}$	$T_{.3}$	...	$T_{.k}$	$T_{..}$
Mean	$\bar{y}_{.1}$	$\bar{y}_{.2}$	$\bar{y}_{.3}$	...	$\bar{y}_{.k}$	$\bar{y}_{..}$
Variance	s_1^2	s_2^2	s_3^2	...	s_k^2	s^2

$$s_j^2 = \frac{\sum_{i=1}^{n_j} (y_{ij} - \bar{y}_{.j})^2}{n_j - 1} = \text{variance of the measurements in the } j\text{th column}$$

$$T_{..} = \sum_{j=1}^{k} T_{.j} = \sum_{j=1}^{k} \sum_{i=1}^{n_j} y_{ij} = \text{total of all measurements}$$

$$\bar{y}_{..} = \frac{T_{..}}{n} = \text{mean of all measurements, where } n = \sum_{j=1}^{k} n_j$$

$$s^2 = \frac{\sum_{j=1}^{k} \sum_{i=1}^{n_j} (y_{ij} - \bar{y}_{..})^2}{n - 1} = \text{variance of all measurements}$$

Note the relationship between the symbols in Table 8.2 and the numerical observations in Table 8.1. In Table 8.1 there are $k = 5$ treatments with $n_1 = n_2 = \cdot \cdot \cdot = n_5 = 10$ observations in each treatment group. The first observation in the first treatment (money), y_{11}, is 76; the total of all observations in treatment 1, $T_{.1}$, is 695. The sample mean for the first treatment, $\bar{y}_{.1}$, is $695/10 = 69.5$; the sample variance for the first treatment, s_1^2, is 65.17. A similar correspondence between symbols and numbers exists for the other treatments. Finally, the grand total of all observations, $T_{..}$, is 1854; the grand mean, $\bar{y}_{..}$, is $1854/50 = 37.08$; and the variance for all 50 observations, s^2, is 438.65. Note also that, although in Table 8.1 each treatment has the same number of observations, there can be unequal numbers of observations per treatment.

Partitioning the total sum of squares

Earlier we defined analysis of variance as an arithmetic procedure for partitioning the total variation in a set of data into two or more components. You will recall that variation in a set of sample data is generally measured by the sample variance. If we have available for analysis a sample of data that can be displayed as in Table 8.2, we may measure the total variation by

$$s^2 = \frac{\sum_{j=1}^{k} \sum_{i=1}^{n_j} (y_{ij} - \bar{y}_{..})^2}{n - 1}$$

We call the numerator of this equation the *total sum of squares*, since it is the sum of the squared deviations of the individual measurements from their mean, where the summation is over the total number of measurements. We can also think of the total sum of squares as a measure of variation, and it is this representation of variation that is partitioned in analysis of variance.

When we analyze the data from a completely randomized experiment, we can identify two sources of variability: variability among treatments and variability within treatments. *Variability among treatments* measures the extent to which the sample means differ from one another. *Variability within treatments* measures the extent to which observations within treatments vary about their individual treatment means. The total sum of squares, then, is partitioned into two sum-of-squares components, one associated with variability among treatments and one with variability within treatments. The sum of squares (abbreviated SS) ascribable to variability within treatments is generally referred to as the *error sum of squares*.

Using the symbols of Table 8.2, we may write the results of the partitioning as follows.

$$\sum_{j=1}^{k} \sum_{i=1}^{n_j} (y_{ij} - \bar{y}_{..})^2 = \sum_{j=1}^{k} n_j(\bar{y}_{.j} - \bar{y}_{..})^2 + \sum_{j=1}^{k} \sum_{i=1}^{n_j} (y_{ij} - \bar{y}_{.j})^2 \tag{8.1}$$

Total sum of squares = treatment sum of squares + error sum of squares

Later we shall present the algebraic details of the partitioning of the total sum of squares.

For convenience, we designate the total sum of squares by SST, the treatment sum of squares by SSTR, and the error sum of squares by SSE. We express the partitioned total sum of squares as

$$\text{SST} = \text{SSTR} + \text{SSE}$$

The formulas for the sums of squares given in Equation 8.1 are rather tedious to evaluate. The following computational formulas are more practical for calculation purposes.

$$\text{SST} = \sum_{j=1}^{k} \sum_{i=1}^{n_j} y_{ij}^2 - C$$

$$= (\text{sum of all squared observations}) - C$$

where

$$C = \frac{T_{..}^2}{n} = \frac{\left(\sum_{j=1}^{k} \sum_{i=1}^{n_j} y_{ij}\right)^2}{n} = \frac{(\text{grand total})^2}{\text{total number of observations}}$$

and $n = \sum_{j=1}^{k} n_j$.

$$\text{SSTR} = \sum_{j=1}^{k} \frac{T_{.j}^2}{n_j} - C = (\text{sum of all squared treatment totals divided by the corresponding group size}) - C$$

Finally, we obtain the error sum of squares by subtraction:

$$\text{SSE} = \text{SST} - \text{SSTR}$$

Although it is possible to compute SSE directly, the calculations are quite tedious, and consequently it is more practical to obtain this quantity by subtraction.

Let us illustrate the calculation of the sums of squares in one-way analysis of variance by continuing with the employee-reward example. Table 8.1

gives the necessary data for computing the sums of squares for this example. From these data, we compute

$$C = \frac{T_{..}^2}{n} = \frac{(1854)^2}{50} = 68746.32$$

$$\text{SST} = \sum_{j=1}^{5} \sum_{i=1}^{n_j} y_{ij}^2 - C$$

$$= 76^2 + 70^2 + \cdot \cdot \cdot + 16^2 - 68746.32 = 21493.68$$

$$\text{SSTR} = \sum_{j=1}^{5} \frac{T_{.j}^2}{n_j} - C = \frac{695^2}{10} + \frac{507^2}{10} + \cdot \cdot \cdot + \frac{181^2}{10} - 68746.32$$

$$= \frac{695^2 + 507^2 + \cdot \cdot \cdot + 181^2}{10} - 68746.32 = 20180.48$$

and $\quad \text{SSE} = \text{SST} - \text{SSTR} = 21493.68 - 20180.48 = 1313.20$

Partitioning the degrees of freedom

We saw in previous chapters that a sum of squares divided by the appropriate degrees of freedom (df) yields a variance. Thus, if we divide each of the sums of squares described above by the appropriate degrees of freedom, we obtain a variance, which we shall call a *mean square* (abbreviated MS). The degrees of freedom associated with the total sum of squares are $n - 1$, the total number of measurements in the set of data minus 1. We call $n - 1$ the *total degrees of freedom*. Dividing the total sum of squares by the total degrees of freedom gives the *total mean square*, which we denote by MST. Thus $\text{SST}/(n-1) = \text{MST}$.

The value for the degrees of freedom associated with the treatment sum of squares is equal to the number of treatments minus 1. If there are k treatments, the value for the degrees of freedom for treatments is $k - 1$. The *treatment mean square* (MSTR) is given by $\text{SSTR}/(k-1) = \text{MSTR}$.

We obtain the error degrees of freedom by subtraction. Thus the error degrees of freedom are equal to $(n-1) - (k-1) = n - k$. The *error mean square* (MSE) is given by $\text{SSE}/(n-k) = \text{MSE}$.

The F test

As we have already noted, an objective of analysis of variance is the testing of hypotheses about population parameters. Specifically, analysis of variance enables us to test the null hypothesis that several population (treatment) means are equal against the alternative hypothesis that they are not all equal. We may formally state the hypotheses as follows.

$$H_0\colon \mu_1 = \mu_2 = \cdot \cdot \cdot = \mu_k, \qquad H_1\colon \text{Not all } \mu_j \text{ are equal}$$

We obtain the appropriate test statistic for testing this hypothesis by taking the ratio of the treatment mean square to the error mean square. Designating this ratio as the *variance ratio* (VR), we have

$$\text{VR} = \frac{\text{MSTR}}{\text{MSE}}$$

VR is the test statistic we use to determine whether or not to reject H_0. The valid use of this test statistic rests on a set of well-defined assumptions, which we shall examine before we continue with our inferential procedure.

The assumptions underlying one-way analysis of variance

The following assumptions must be met in order for inference procedures in one-way analysis of variance to be valid.

1 Each observed set of n_j measurements is an independent simple random sample of measurements from a population of measurements, and each population is identified according to the treatment designation.

2 Each population of measurements (that is, the y_{ij} for each j) is normally distributed with mean μ_j.

3 Each population of measurements has the same variance.

In practice, all these assumptions will not be realized in a given situation. Fortunately, one can obtain useful results, provided the violations of the assumptions are not severe. One can overcome the ill effects of unequal variances to some extent by keeping the number of observations in each treatment group the same.

In performing the analysis of variance on the employee-reward data shown in Table 8.1, then, we are assuming that we have five independent random samples consisting of ten measurements each. For example, we assume that the first sample is drawn from a population of measurements that would be generated if a large number (a population) of employees were offered a reward of money for increased production. We assume that the measurements in each population are normally distributed with respective means of $\mu_1, \mu_2, \ldots, \mu_5$. Also we assume that each population of measurements has the same variance; that is, $\sigma_1^2 = \sigma_2^2 = \cdots = \sigma_5^2 = \sigma^2$, the common variance. Figure 8.1 is a representation of the populations for unequal means.

Figure 8.1 Representation of five sampled populations in a one-way analysis of variance when all means are unequal

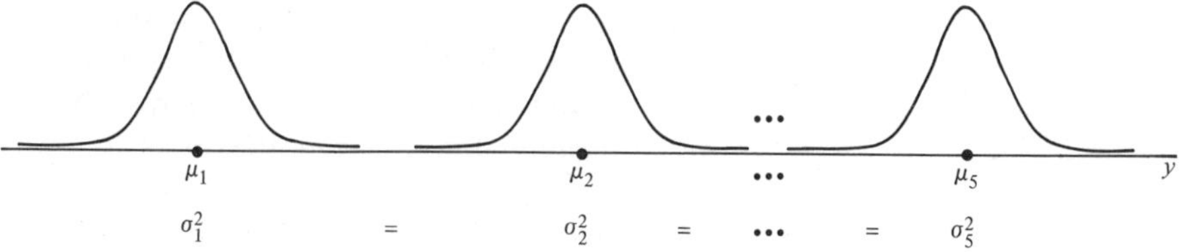

Figure 8.2 Representation of five sampled populations in a one-way analysis of variance when all means are equal

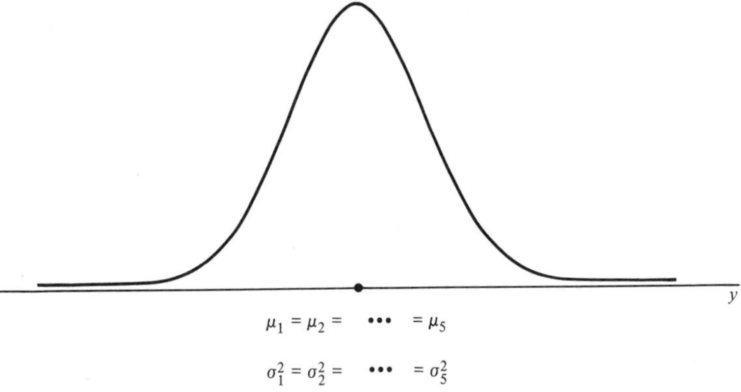

When the means are all equal, the populations all have the same location. Since the variances are assumed to be equal, we can superimpose the locations on each other, as illustrated in Figure 8.2.

The rationale for computing the variance ratio, as we shall discuss in more detail later, is based on the fact that if the assumptions underlying analysis of variance are met, MSTR and MSE are both independent estimates of σ^2, the common variance of the sampled populations. This being true, we would expect these two estimates to be close in magnitude when H_0 is true. We have seen in Chapter 6 that the VR is the appropriate statistic for comparing two sample variances.

If H_0 is true (that is, if all treatment means are equal), and if the assumptions of normally distributed populations, homogeneous population variances, and independent error terms are met, VR is distributed as F with ν_1 and ν_2 degrees of freedom. We introduced the F distribution in Chapter 4 and used it in Chapters 5 and 6 to make inferences about the ratio of two population variances. In the present case, the value for the numerator degrees of freedom is $\nu_1 = k - 1$, the degrees of freedom associated with SSTR. And the value for the denominator degrees of freedom is $\nu_2 = n - k$, the degrees of freedom associated with SSE.

We can choose a critical value of F for any level of significance, α, from Appendix Table H. If the computed VR is equal to or greater than the critical F, we reject H_0.

Suppose that, for the employee-reward example, we wish to test

$$H_0\text{: } \mu_1 = \mu_2 = \mu_3 = \mu_4 = \mu_5$$

against the alternative hypothesis

$$H_1\text{: Not all treatment means are equal}$$

at the $\alpha = 0.01$ level of significance. In other words, we want to test the null hypothesis that the true mean increase in number of items produced

Figure 8.3 Acceptance and rejection regions for variance-ratio test, $\nu_1 = 4$, $\nu_2 = 45$, $\alpha = 0.01$

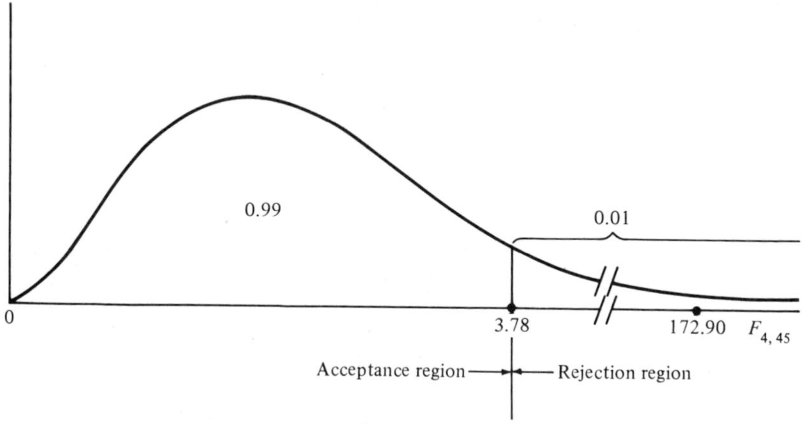

is the same regardless of the type of reward used. Since we have $k = 5$ treatments, the numerator degrees of freedom are $k - 1 = 4$. And since we have a total of $n = 50$ measurements, the denominator degrees of freedom are $n - k = 50 - 5 = 45$. From Appendix Table H, we find the critical F for $\alpha = 0.01$, $\nu_1 = 4$, and $\nu_2 = 45$ to be about 3.78 if we interpolate between $\nu_2 = 40$ and 60. From the results obtained earlier, we may compute

$$\text{MSTR} = \frac{20180.48}{4} = 5045.12, \qquad \text{MSE} = \frac{1313.20}{45} = 29.18,$$

and

$$\text{VR} = \frac{5045.12}{29.18} = 172.90$$

Since 172.90 is greater than 3.78, we reject H_0, and conclude that not all treatment means are equal. On the basis of the observed data, we conclude that different rewards have different effects on production.

By consulting Appendix Table H, we find that the p value associated with this hypothesis test is less than 0.005. That is, $p < 0.005$. The probability of observing a value of VR as large as or larger than 172.90, when H_0 is true, is less than 0.005.

Figure 8.3 shows the acceptance and rejection regions for this test.

The analysis of variance table

It is convenient to display the sources of variation, the associated sums of squares, the degrees of freedom, the mean squares, and the variance ratio in a summary table called the *analysis of variance table,* or ANOVA table for short. Table 8.3 is a general ANOVA table for one-way analysis of variance.

Table 8.3 ANOVA table for one-way analysis of variance

Source of variation	SS	df	MS	VR
Treatments	SSTR	$k-1$	$\text{MSTR} = \dfrac{\text{SSTR}}{k-1}$	$\dfrac{\text{MSTR}}{\text{MSE}}$
Error	SSE	$n-k$	$\text{MSE} = \dfrac{\text{SSE}}{n-k}$	
	SST	$n-1$		

Table 8.4 ANOVA table for employee-reward example

Source of variation	SS	df	MS	VR
Treatments	20180.48	4	5045.12	172.90
Error	1313.20	45	29.18	
	21493.68	49		

Table 8.4 is the ANOVA table for the employee-reward example.

The rationale underlying the F test

Although the use of the variance ratio VR for testing the null hypothesis of equal population means can be justified mathematically, we shall not present the proofs in this book. Instead we present a more intuitive argument in defense of our use of VR as the test statistic for testing H_0: $\mu_1 = \mu_2 = \cdots = \mu_k$.

Let us consider first MSE, the denominator of the variance ratio. As we have seen,

$$\text{MSE} = \frac{\text{SSE}}{n-k} = \frac{\sum_{j=1}^{k} \sum_{i=1}^{n_j} (y_{ij} - \bar{y}_{.j})^2}{n-k}$$

We obtain the numerator of MSE directly by first finding, for a given treatment group, the sum of the squared deviations of the measurements from the mean of that group. We then add the resulting sum of squares for the k groups. To obtain MSE, we divide this sum of the treatment sum of squares by $n - k$, which is the sum of the degrees of freedom of the individual treatments.

In symbols, we write the error mean squares as

$$\text{MSE} = \frac{\sum_{i=1}^{n_1} (y_{i1} - \bar{y}_{.1})^2 + \sum_{i=1}^{n_2} (y_{i2} - \bar{y}_{.2})^2 + \cdots + \sum_{i=1}^{n_k} (y_{ik} - \bar{y}_{.k})^2}{n-k}$$

$$= \frac{\sum_{j=1}^{k} \sum_{i=1}^{n_j} (y_{ij} - \bar{y}_{..})^2}{n-k}$$

Thus we see that the error mean square MSE is a pooled estimate of the common population variance that the treatment groups are assumed to share. You will recognize this pooling procedure as an extension of the procedure, discussed in Chapter 6, for computing s_p^2, the pooled estimate of σ^2, in the case of testing hypotheses about the difference between two population means.

We can demonstrate that MSE is a pooled estimate of the assumed common population variance of the treatment groups by means of the employee-reward example. To do this, we weight each of the sample variances shown in Table 8.1 by its respective degrees of freedom (thereby obtaining the individual SSTRs), sum these weighted variances, and divide the total by the sum of the treatment degrees of freedom. That is,

$$\text{MSE} = \frac{9(65.17) + 9(24.23) + 9(23.73) + 9(15.79) + 9(16.99)}{9 + 9 + 9 + 9 + 9} = 29.18$$

We see that this is the same result that we obtained by subtraction, as shown in Table 8.4

Now let us consider the numerator of VR, the treatment mean square. We want to show that MSTR is also an estimate of the common population variance of the treatments when the assumptions underlying the F test are met. We make use of the relationship, learned in Chapter 4,

$$\sigma_{\bar{y}}^2 = \frac{\sigma^2}{n}$$

where $\sigma_{\bar{y}}^2$ is the variance among sample means, n is the size of the samples, and σ^2 is the variance of the population from which the samples were drawn. For convenience, we present this argument for the case in which all sample sizes are equal, and we use n to designate the common sample size. Note that in this formula n is not the sum of all sample sizes, as it was in the previous discussion. From this equation we may compute

$$\sigma^2 = n\sigma_{\bar{y}}^2$$

We may write the variance among sample means, $\sigma_{\bar{y}}^2$, as the sum of squared deviations of the sample means from their mean, the true population mean μ, divided by the number of sample means involved. If there are k sample means, we may write

$$\sigma_{\bar{y}}^2 = \frac{\Sigma\,(\bar{y}_i - \mu)^2}{k}$$

and therefore

$$\sigma^2 = \frac{n\,\Sigma\,(\bar{y}_i - \mu)^2}{k}$$

or, equivalently,

$$\sigma^2 = \frac{\Sigma\, n(\bar{y}_i - \mu)^2}{k}$$

Using the notation introduced in Table 8.2,

$$\frac{\sum_{j=1}^{k} n_j(\bar{y}_{.j} - \bar{y}_{..})^2}{k-1}$$

gives an unbiased estimate of σ^2. The numerator of this, as we saw in Equation 8.1, is the SSTR (the treatment sum of squares). Division by the degrees of freedom, $k - 1$, yields MSTR, the treatment mean square.

Thus we see that, if the true variances of the treatment populations are equal, we have two estimates of this common variance, MSTR and MSE. We would expect, then, that in a large number of repetitions of an experiment, the average value of MSTR would be equal to the average value of MSE *if H_0 is true*, that is, *if the μ_j are equal.* If H_0 is not true—that is, if the μ_j are not equal—we would expect the average value of MSTR to be greater than the average value of MSE, reflecting the presence of treatment effects.

In a specific instance, we would not expect the observed value of MSTR to equal the observed value of MSE, even if the treatment means are equal. Sampling variability alone will operate to cause the two estimates to differ. If the treatment means are equal, we would expect, however, in a given situation, to find MSTR and MSE to be close in value. If the treatment means are not equal, we would expect, in a given situation, to observe a greater difference between MSTR and MSE. To compare the relative magnitudes of MSTR and MSE, we form their ratio, VR = MSTR/MSE, as already noted. If VR is near 1, we feel that the sample data are compatible with the null hypothesis of equal treatment means. A value of VR substantially greater than 1, however, casts doubt on the truth of the null hypothesis of equal treatment means. As we have seen, we decide to reject H_0 in favor of H_1 when the deviation of VR from 1 is so great that VR exceeds the tabulated value of F for some preselected level of significance α.

Unequal sample sizes

In the employee-reward example, the treatments were represented by samples of equal size. It is not necessary for the samples in each group to be the same size. The analysis of variance is basically the same for both equal and unequal sample sizes, and the calculations and interpretation of results are the same. Remember that having equal sample sizes provides some protection against the ill effects of possible departures from the assumption of homogeneity of population variances.

To illustrate the hypothesis-testing procedure for unequal samples, we present the following example.

Example 8.1 Table 8.5 shows the determinations (in microgram percentages) of plasma testosterone for three groups of mentally ill subjects. We wish to know whether one can conclude from these data that the mean values are not equal in all three groups. The hypotheses are

$$H_0\colon \mu_1 = \mu_2 = \mu_3, \qquad H_1\colon \text{Not all the means are equal}$$

Suppose we choose a significance level of $\alpha = 0.05$.

The next step is to compute the SST, SSTR, and SSE.

First we find

$$C = \frac{(18.29)^2}{26} = 12.8663$$

Then

$$\text{SST} = (0.66^2 + 0.63^2 + \cdots + 0.99^2) - 12.8663 = 0.6786$$

and

$$\text{SSTR} = \frac{5.78^2}{10} + \frac{4.85^2}{7} + \frac{7.66^2}{9} - 12.8663 = 0.3544$$

By subtraction, we obtain

$$\text{SSE} = 0.6786 - 0.3544 = 0.3242$$

Finally,

$$\text{MSTR} = \frac{0.3544}{2} = 0.1772, \qquad \text{MSE} = \frac{0.3242}{23} = 0.0141$$

Table 8.6 gives the sums of squares, along with the degrees of freedom, mean squares, and variance ratios.

Table 8.5 Plasma testosterone determinations on three groups of mentally ill subjects

	Group			
	I	II	III	
	0.66	0.65	0.93	
	0.63	0.60	0.99	
	0.65	0.69	0.96	
	0.69	0.73	0.74	
	0.44	0.52	0.81	
	0.63	0.85	0.93	
	0.61	0.81	0.63	
	0.42		0.68	
	0.59		0.99	
	0.46			
Total	5.78	4.85	7.66	**18.29**
Mean	0.58	0.69	0.85	**0.70**

Table 8.6 ANOVA Table for Example 8.1

Source	SS	df	MS	VR
Treatments	0.3544	2	0.1772	12.57
Error	0.3242	23	0.0141	
	0.6786	25		

Since the computed VR of 0.1772/0.0141 = 12.57 is greater than the critical F of 3.42, we reject H_0. We conclude that the true mean plasma testosterone levels are not the same in the three types of mentally ill patients.

SUMMARY OF ONE-WAY ANALYSIS OF VARIANCE

We may break down the hypothesis-testing procedure of one-way analysis of variance into the same series of steps that we suggested for hypothesis testing in Chapter 7. Thus, for one-way analysis of variance, we would have the following steps.

1 *Statement of hypotheses*

$$H_0: \mu_1 = \mu_2 = \cdots = \mu_k, \qquad H_1: \text{Not all } \mu_j \text{ are equal}$$

2 *Significance level* One chooses an appropriate value of α.

3 *Description of population and assumptions*

(a) The k observed samples are randomly and independently drawn.

(b) The k treatment populations are each normally distributed.

(c) The variances of the k treatment populations are all equal.

4 *The relevant statistic* The relevant statistics are the treatment mean square (MSTR) and the error mean square (MSE).

5 *The test statistic and its distribution* The test statistic is VR = MSTR/MSE, which, when H_0 is true and the previously stated assumptions are met, is distributed as F with $k - 1$ and $n - k$ degrees of freedom.

6 *Rejection and acceptance regions* The rejection region consists of values of F greater than or equal to $F_{1-\alpha,k-1,n-k}$, the F, for $k - 1$ and $n - k$ degrees of freedom, that has α of the area under the curve of F to its right. The acceptance region consists of values of F that are less than $F_{1-\alpha,k-1,n-k}$.

7 *Collection of data and calculations* After randomly assigning subjects to treatments, one takes the measurement of interest on each subject. The calculations consist of those necessary for obtaining MSTR and MSE. The results of the calculations may be displayed in an ANOVA table.

8 *Statistical decision* If the computed value of VR is equal to or greater than the critical value of F, H_0 is rejected; otherwise, H_0 is not rejected.

9 *Conclusion* If H_0 is rejected, one concludes that the treatment population means are not all equal. If H_0 is not rejected, one concludes that the treatment population means may all be equal.

Proof of partitioning of total sum of squares*

To show how the total sum of squares SST is partitioned, we begin by noting that, for a given measurement y_{ij},

$$y_{ij} - \bar{y}_{..}$$

denotes the amount by which this individual measurement deviates from the overall sample mean $\bar{y}_{..}$. For a given measurement, we may denote this deviation as the *total deviation*. In a similar manner, we may indicate the magnitude of the deviation of the individual measurement y_{ij} from its group mean $\bar{y}_{.j}$ as follows.

$$y_{ij} - \bar{y}_{.j}$$

The difference between the deviation $y_{ij} - \bar{y}_{..}$ and the deviation $y_{ij} - \bar{y}_{.j}$ gives the amount by which the jth treatment mean, $\bar{y}_{.j}$, deviates from the overall mean $\bar{y}_{..}$. That is,

$$(y_{ij} - \bar{y}_{..}) - (y_{ij} - \bar{y}_{.j}) = (\bar{y}_{.j} - \bar{y}_{..}) \tag{8.2}$$

If we move $(y_{ij} - \bar{y}_{.j})$ to the right-hand side of Equation 8.2, we have

$$(y_{ij} - \bar{y}_{..}) = (\bar{y}_{.j} - \bar{y}_{..}) + (y_{ij} - \bar{y}_{.j}) \tag{8.3}$$

which shows that the total deviation, $y_{ij} - \bar{y}_{..}$, is the sum of two components:

1 The deviation of the treatment mean from the overall mean

2 The deviation of the individual observation from its group (treatment) mean

Squaring both sides of Equation 8.3 yields

$$(y_{ij} - \bar{y}_{..})^2 = (\bar{y}_{.j} - \bar{y}_{..})^2 + 2(\bar{y}_{.j} - \bar{y}_{..})(y_{ij} - \bar{y}_{.j}) + (y_{ij} - \bar{y}_{.j})^2$$

Summing over all measurements in the data set, we get

$$\sum_{j=1}^{k}\sum_{i=1}^{n_j}(y_{ij} - \bar{y}_{..})^2 = \sum_{j=1}^{k} n_j(\bar{y}_{.j} - \bar{y}_{..})^2 + 2\sum_{j=1}^{k}\sum_{i=1}^{n_j}(\bar{y}_{.j} - \bar{y}_{..})(y_{ij} - \bar{y}_{.j}) + \sum_{j=1}^{k}\sum_{i=1}^{n_j}(y_{ij} - \bar{y}_{.j})^2 \tag{8.4}$$

We can rewrite the middle term of the right-hand side of Equation 8.4 as

$$2\sum_{j=1}^{k}\left[(\bar{y}_{.j} - \bar{y}_{..})\sum_{i=1}^{n_j}(y_{ij} - \bar{y}_{.j})\right]$$

*The instructor may omit this section without loss of continuity.

Since the sum of the deviations of a set of measurements from their mean is equal to 0, we have

$$2\sum_{j=1}^{k}(\bar{y}_{\cdot j}-\bar{y}_{\cdot\cdot})(0)$$

and we see that the middle term of the right-hand side of Equation 8.4 vanishes, leaving

$$\sum_{j=1}^{k}\sum_{i=1}^{n_j}(y_{ij}-\bar{y}_{\cdot\cdot})^2=\sum_{j=1}^{k}n_j(\bar{y}_{\cdot j}-\bar{y}_{\cdot\cdot})^2+\sum_{j=1}^{k}\sum_{i=1}^{n_j}(y_{ij}-\bar{y}_{\cdot j})^2$$

Thus the partition of the sum of squares is complete.

EXERCISES

1 In a study of the effects on children of films portraying aggression, a group of psychologists randomly assigned 30 second-grade students to view one of three films with varying degrees of aggressive content. After the children had viewed the film investigators observed each group separately for a period of one hour, and kept a record of the number of aggressive acts engaged in by each child. Table 8.7 shows the results.

Do these data provide sufficient evidence to indicate that films with different degrees of aggressive content have different effects on children? Let $\alpha = 0.05$.

2 Twenty-four employees of a particular firm were randomly assigned to one of three training groups to learn how to perform a certain assembly-line task. The groups differed with respect to the amount of instruction received. At the end of the training period, the 24 employees each performed the task. A record was kept of the time, in minutes, required for its successful completion, with results shown in Table 8.8. Do these data provide sufficient

Table 8.7 Data for Exercise 1

Content of film viewed										
HIGHLY AGGRESSIVE	24	16	16	16	10	15	19	17	25	13
MODERATELY AGGRESSIVE	13	9	8	8	9	15	15	14	14	13
NOT AGGRESSIVE	10	3	0	3	9	10	7	10	2	9

Table 8.8 Data for Exercise 2

Amount of instruction								
MINIMAL	39	35	35	50	40	37	42	42
MODERATE	25	23	22	23	22	27	33	27
EXTENSIVE	25	11	17	13	18	13	15	19

Table 8.9 Data for Exercise 3

Diagnosis			
A	*B*	*C*	*D*
3.21	9.18	10.59	7.60
6.92	8.35	13.62	5.68
1.95	3.26	5.56	10.12
4.51	3.82	15.95	7.00
3.31	1.75	5.33	11.58
8.09	7.24	4.22	8.89
7.97	10.85	13.66	13.38
1.86	6.57		9.67
7.40			17.11
5.41			

evidence to indicate that amount of instruction has an effect on speed of task completion? Let $\alpha = 0.05$.

3 Table 8.9 shows the urinary androsterone determinations (in milligrams per 24 hours) made on 34 patients in a state mental hospital. The patients are classified by diagnosis. On the basis of these data, can one conclude that the true mean urinary androsterone level differs by diagnosis? Let $\alpha = 0.05$.

4 In a random sample of 30 business executives, each was identified as being either a risk taker, a risk avoider, or neutral in terms of risks. Each executive was then given a test to measure his or her level of anxiety. Table 8.10 shows the results (which have been coded for ease of computation). What are your conclusions regarding this study? Let $\alpha = 0.05$.

5 In an experiment designed to study the effects of a certain drug on REM (rapid-eye-movement) sleep in normal humans, psychologists randomly assigned 28 apparently normal adults to receive one of four dose levels of the drug, and recorded the length, in minutes, of the first REM period of each subject. Table 8.11 shows the results. Do these data provide sufficient evidence to indicate a treatment effect? Let $\alpha = 0.05$.

6 A botanist conducted an experiment to compare the effects of three different treatments on the amount of water lost from the leaves of a certain plant. The researcher randomly assigned a total of 43 leaves to each of the three treatments. Table 8.12 shows the amounts (in milligrams per square centimeter) of water lost in three days. Test H_0: $\mu_A = \mu_B = \mu_C$ at the 0.05 level of significance.

Table 8.10 Data for Exercise 4

Risk takers	3	4	3	2	2	5	3	2	5	4	4	4
Risk avoiders	4	8	7	10	10	4	8	8				
Risk neutrals	4	5	3	4	3	3	4	4	6	5		

Table 8.11 Data for Exercise 5

Dose Level			
1	2	3	4
39	13	10	8
40	10	9	5
34	16	12	8
11	14	5	8
17	15	14	9
30	8	13	5
27	17	11	10

Table 8.12 Data for Exercise 6

Treatment															
A	66	85	63	73	83	88	65	80	74	69	91	80	44	94	95
B	76	80	86	87	77	66	68	85	95	67	97	73	75	82	95
C	68	90	91	74	70	76	87	83	83	80	88	98	86		

Table 8.13 Data for Exercise 7

Communication system											
A	15	6	13	8	8	8	11	5	5	13	10
B	8	11	8	10	13	7	10	9	13	11	11
C	14	15	19	13	16	16	13	13	12	20	18
D	16	17	15	18	21	20	24	24	16	21	20

7 In a study of the effectiveness of different personal communication systems, 44 adult subjects were randomly assigned to one of four groups. In each group a different system of communicating a complex message from member to member was used. At the end of a specified period, each subject was given a test to measure her or his comprehension of the message. Table 8.13 shows the results (coded to facilitate calculations). Do these data support the hypothesis that the communication systems differ in effectiveness? Let $\alpha = 0.01$.

8 Psychiatrists drew a random sample of patients from each of four groups of patients in a state psychiatric hospital, and defined the groups on the basis of diagnosis. Researchers gave each patient a test to measure awareness of current national events. Table 8.14 shows the results. Is there reason to believe that awareness of current national events is different among the four diagnostic groups? Let $\alpha = 0.01$.

9 Agricultural engineers conducted an experiment to assess the effects of three different fertilizers on the yields of strawberry plants. They planted 24 plots of equal size and treated them alike except for the type of fertilizer applied. Fertilizer *A* was applied to nine plots, fertilizer *B* to eight plots, and

Table 8.14 Data for Exercise 8

Diagnosis										
A	10	8	12	11	9	12	15	7	5	6
B	8	10	5	14	15	8	5	5		
C	17	19	14	17	17	14	11	16		
D	25	13	15	15	14	14	21	24		

Table 8.15 Data for Exercise 9

Fertilizer									
A	10	14	12	12	10	9	12	10	10
B	7	7	8	7	8	9	10	10	
C	7	6	9	7	8	9	5		

Table 8.16 Data for Exercise 10

		System		
1	2	3	4	5
0	3	1	−3	9
18	14	20	8	−8
−2	6	−6	−2	−4
−3	17	5	−10	3
8	17	8	13	15
11	14	7	9	6
19	11	17	−2	−8
9	16	2	8	7
2	−5	0	2	1
5	−3	5	8	15

fertilizer *C* to seven plots. Table 8.15 shows the yields, in pounds per plot. Do these data provide sufficient evidence to indicate a treatment effect? Let $\alpha = 0.05$.

10 A team of educators conducted a survey of five school systems to determine whether it could be concluded that the mean IQ scores of members of the senior classes were different. The team selected ten seniors from each system. Table 8.16 shows the IQ scores (coded) for the 50 seniors. Can the educators conclude from these data that the mean IQ scores of seniors differ among school systems? Let $\alpha = 0.05$.

11 In a study of the levels of adjustment of first-year foreign college students to their new environments, investigators obtained the data shown in Table 8.17. The investigators wished to know whether they could conclude that the three populations, represented by the samples, differ with respect to mean adjustment score.

Table 8.17 Data for Exercise 11

	Country of origin	
A	*B*	*C*
29	90	45
64	66	59
33	73	51
40	55	44
		55

Table 8.18 Data for Exercise 12

		Smokers	
Nonsmokers	LIGHT	MODERATE	HEAVY
24	36	30	56
17	49	33	55
13	41	27	72
18	37	29	73
13	26	30	74
15	45	38	71
24	29		

Table 8.19 Data for Exercise 13

Type of hearing aid		
A	*B*	*C*
15	17	10
17	18	12
16	20	14
15	22	11
13	21	

Table 8.20 Data for Exercise 14

Father's educational level			
College	High school	Elementary school	None
128	117	104	102
109	101	95	90
125	102	108	106
128	123	122	103
124	117	114	107
105	123	98	120
130	114	101	

Table 8.21 Data for Exercise 15

Normal subjects			Suicidal subjects			Nonsuicidal mentally ill subjects		
34	25	27	43	45	59	32	34	35
25	31	28	51	57	64	50	46	30
36			55			44		

12 Table 8.18 shows the anxiety scores obtained from independent random samples in a study of the characteristics of smokers and nonsmokers. Do these data provide sufficient evidence to indicate that the populations represented differ with respect to mean anxiety score?

13 Fourteen children with hearing disabilities were matched on the basis of their unaided speech-reception thresholds, and then randomly assigned to wear one of three types of hearing aid. Table 8.19 shows their aided speech-reception thresholds. Do these data provide sufficient evidence to indicate a difference among types of hearing aids? Let $\alpha = 0.05$.

14 A random sample of seventh-grade school children was selected from each of four groups of children grouped according to their fathers' educational levels. Table 8.20 shows the children's IQs. On the basis of these data, can we conclude that the groups differ with respect to IQ? Let $\alpha = 0.05$.

15 Random samples were selected from each of three populations: normal subjects, persons who had made one or more suicide attempts, and nonsuicidal mentally ill subjects. Table 8.21 shows the scores made by these subjects on a test designed to measure level of anxiety. Do these data suggest a difference in mean anxiety level among the three groups?

8.2 COMPARISONS AMONG TREATMENT MEANS

After an experimenter has carried out an analysis of variance procedure resulting in a significant VR, he or she is usually interested in obtaining additional information from the data. Generally, if one is able to conclude, on the basis of the F test, that not all means are equal, it is desirable to be able to determine where the differences occur. In other words, interest is usually focused on *pairwise comparisons*, where a pairwise comparison is the difference between two means without regard to the algebraic sign. Investigation of comparisons between pairs of means may take the form of a hypothesis test regarding the difference between two means, or the objective may be to construct a confidence interval for the difference. We may describe two situations involving comparisons.

Before conducting the experiment, the experimenter may decide that it will be worthwhile to compare only certain pairs of sample treatment means to see whether they are significantly different. Such comparisons are called *planned* or *a priori* comparisons. *A priori* comparisons may be made regardless of whether or not the computed VR in the analysis of variance is significant.

On the other hand, there will be times when the experimenter has no basis for planning comparisons among means before conducting the experiment. If the VR, computed in the analysis of variance, is not significant, indicating no evidence of a treatment effect, the experimenter most likely will not be interested in comparing individual pairs of means. If, however, the VR is significant, he or she will probably be motivated to determine which pairs of sample treatment means are significantly different. Comparisons that are made after the initial analysis of variance are called *a posteriori* or *post hoc* comparisons.

A number of procedures have been developed that allow the experimenter to make all possible pairwise comparisons among means, regardless of whether they were planned in advance. With these procedures available, the researcher either can make a large number of comparisons routinely, or, following the outcome of the experiment, can select those comparisons that appear most interesting.

Tukey's HSD test

Tukey (1) has proposed a procedure for making all pairwise comparisons among means; this procedure is now widely used. It is generally called the HSD (*honestly significant difference*) *test* or the *w procedure*. In this procedure, one computes a single value against which all differences are compared. This value is called the HSD, and is given by

$$\text{HSD} = q_{\alpha,k,n-k} \frac{1}{\sqrt{2}} \sqrt{\frac{2\text{MSE}}{n_j}}$$

where q is obtained from Appendix Table J, for significance level α, k means in the experiment, and $n - k$ error degrees of freedom. Any difference between pairs of means that exceeds HSD is declared significant at the α level. Let us illustrate the application of this procedure by referring again to the employee-reward example of Section 8.1.

Example 8.2 As a first step in applying Tukey's HSD procedure to the data of Table 8.1, for convenience we display the differences between means as in Table 8.22. In this table the sample treatment means arranged in descending order of magnitude provide the row and column labels. The corresponding differences are given in the body of the table. If we choose a significance level of $\alpha = 0.05$, we find $q_{0.05,5,45}$ in Appendix Table J to be 4.02, when we interpolate between 40 and 60 error degrees of freedom. From Table 8.4 we find MSE = 29.18, so we may compute

$$\text{HSD} = 4.02\frac{1}{\sqrt{2}}\sqrt{\frac{2(29.18)}{10}} = 6.87$$

When we compare the differences between means shown in Table 8.22 with 6.87, we find that only the difference between means for the rewards of an additional coffee break and special recognition are not significant. In all cases except these, then, we are able to conclude that there is a difference in population means.

We may use Tukey's procedure to construct confidence intervals for the difference between two treatment means. A $100(1 - \alpha)\%$ confidence interval for the difference $\mu_{.j} - \mu_{.j'}$ is given by

$$(\bar{y}_{.j} - \bar{y}_{.j'}) \pm q_{\alpha,k,n-k}\frac{1}{\sqrt{2}}\sqrt{\frac{2\text{MSE}}{n_j}}$$

where j and j' refer to any treatment group. Since Appendix Table J gives values of q that are exceeded $100\alpha\%$ of the time in the distribution of q (that is, the table is one-tailed), we shall be able to use the table to construct only 90 and 98% confidence intervals.

Example 8.3 Let us refer again to the employee-reward example and construct the 90% confidence interval for the difference between the true means of increased production in return for the rewards of time off with pay ($\bar{y}_{.3}$) and an additional coffee break ($\bar{y}_{.4}$). From the sample means given in Table 8.22, we have

$$28.8 - 18.3 \pm 4.02\frac{1}{\sqrt{2}}\sqrt{\frac{2(29.18)}{10}} = 10.5 \pm 6.87$$

$$= 3.63 \quad \text{and} \quad 17.37$$

Table 8.22 Differences between means for Example 8.2 (Means are taken from Table 8.1.)

	$\bar{y}_{.1}$	$\bar{y}_{.2}$	$\bar{y}_{.3}$	$\bar{y}_{.4}$	$\bar{y}_{.5}$
$\bar{y}_{.1} = 69.5$	—	18.8	40.7	51.2	51.4
$\bar{y}_{.2} = 50.7$		—	21.9	32.4	32.6
$\bar{y}_{.3} = 28.8$			—	10.5	10.7
$\bar{y}_{.4} = 18.3$				—	0.2
$\bar{y}_{.5} = 18.1$					—

Thus we are 90% confident that the true difference in the two treatment means is between 3.63 and 17.37 units.

EXERCISES

16 Refer to Exercise 1.

(a) Use Tukey's procedure to test all pairwise comparisons. Let $\alpha = 0.05$.

(b) Use Tukey's procedure to construct a 90% confidence interval for the difference between each possible pair of means.

17 Refer to Exercise 2. Use Tukey's procedure to test all possible pairwise comparisons at the 0.05 level of significance.

18 Refer to Exercise 5. Use Tukey's procedure to test all pairwise comparisons. Let $\alpha = 0.05$.

19 Refer to Exercise 7. Use Tukey's procedure to test all pairwise comparisons at the 0.05 level of significance.

8.3 CALCULATION OF POWER AND SAMPLE SIZE

In Chapter 6 we introduced the concept of the power of a test, and demonstrated the calculation of power for a simple case. We defined the power of a test as the probability of rejecting a false null hypothesis. This is exactly the objective when a hypothesis is tested.

We wish now to consider the concept of power as it relates to the more complex hypothesis-testing procedures encountered in analysis of variance. Whenever possible, it is important for the experimenter to know the power associated with the proposed design of an experiment before conducting the experiment. Experiments conducted in such a way that false hypotheses are rejected with low probability may be of little value in shedding light on the problems at hand, or in enabling the experimenter to make sound decisions. Analyses of the power associated with the test of hypotheses in analysis of variance are also useful, in that such analyses will point out those cases in which the proposed number of subjects for an experiment is greater than necessary to achieve a desired level of power. Thus analysis of power may result in the saving of valuable resources.

The calculation of power in analysis of variance is more complicated than in the type of hypothesis-testing procedures we covered in Chapter 6. Fortunately, the labor of the task has been greatly reduced by the introduction of special charts prepared by Pearson and Hartley (3). These charts appear in this book as Appendix Table K.

To use the charts, one must first calculate the quantity

$$\phi = \frac{\sqrt{\sum_{j=1}^{k} \alpha_j^2 / k}}{\sigma_e / \sqrt{n_j}} \tag{8.5}$$

where $\Sigma_{j=1}^{k} \alpha_j^2$ is the sum of the squared treatment effects, σ_e is the common standard deviation of the treatment populations, k is the number of treatment populations represented in the analysis, and n_j is the size of the sample from the jth treatment population. We assume here that n_j is the same for all j; that is, we assume that all sample sizes are equal. To enter the charts, we also need to know the level of significance, α, and the treatment and error degrees of freedom, v_1 and v_2, respectively.

In Equation 8.5, the true value of $\Sigma_{j=1}^{k} \alpha_j^2$ will, of course, be unknown. We estimate this quantity by

$$\sum_{j=1}^{k} \hat{\alpha}_j^2 = \frac{k-1}{n_j}(\text{MSTR} - \text{MSE}) \tag{8.6}$$

For proof that $[(k-1)/n_j](\text{MSTR} - \text{MSE})$ is an unbiased estimator of the sum of the squared treatment effects, see Kirk (2). An estimate of σ_e is provided by $\sqrt{\text{MSE}}$.

We shall now illustrate the calculation of the power of an F test in a one-way analysis of variance by means of an example.

Example 8.4 In an experiment designed to investigate the effects of different types of muscle tension on anxiety scores, psychologists randomly assigned 24 subjects to one of four treatment groups. Table 8.23 shows the analysis of variance of the results. Before the analysis, the psychologists selected a significance level of $\alpha = 0.05$.

Substituting the data from Table 8.23 into Equation 8.6, we compute the following estimate of the sum of the squared treatment effects.

$$\sum_{j=1}^{k} \hat{\alpha}_j^2 = \tfrac{3}{6}(76.8195 - 13.0750) = 31.87225$$

By Equation 8.5, we compute

$$\phi = \frac{\sqrt{31.87225/4}}{\sqrt{13.0750}/\sqrt{6}} = 1.91$$

To find the power associated with the F test of this example, we find, in Appendix Table K, the chart for $\nu_1 = 3$, the numerator degrees of freedom of the variance ratio VR. We next locate, at the bottom of the chart for $\nu_1 = 3$, the value of ϕ for $\alpha = 0.05$, which we note is in the top row of numbers.

Table 8.23 ANOVA Table, Example 8.4

Source of variation	SS	df	MS	VR
Treatment	230.4584	3	76.8195	5.88
Error	261.4999	20	13.0750	
	491.9583	23		

From $\phi = 1.91$ on the bottom scale, we move up until we reach the curve for $\nu_2 = 20$, the denominator degrees of freedom. Reading across to the vertical axis, we find that $1 - \beta$, the power of the test, is a little greater than 0.80.

One can increase the power in two ways: (1) by increasing the sample sizes, and (2) by selecting an alternative design that reduces the error variance. We shall discuss these alternatives in the following sections.

Determining sample sizes

We may use the power charts of Appendix Table K as an aid in determining the number of subjects that should be included in an experiment. In order to use the charts for this purpose, we must know or estimate the following information.

1 *The level of significance,* α This is the desired probability of rejecting a true null hypothesis. We select it with the seriousness of Type I errors in mind.

2 *The desired power* This is the probability of rejecting a false null hypothesis.

3 *The error variance,* σ_e^2 This quantity is generally unknown. We may use estimates from previous experiments or pilot experiments.

4 *The number of treatment groups* This information enables us to determine the numerator degrees of freedom in the variance ratio, so that we can consult the correct chart of Appendix Table K in determining the size of sample.

5 *The minimum value of* $\Sigma(\mu_{.j} - \mu)^2$ We obtain this quantity from our specification of the minimum differences among means that we consider it important to detect.

The following example illustrates the use of the charts in Appendix Table K for determining the size of sample needed for an experiment.

Example 8.5 Suppose the researchers who conducted the experiment described in Example 8.4 wish to repeat the experiment, but this time they want to achieve a power of 0.95. How large a sample will be required? We assume that the desired level of significance will again be 0.05. Let us also assume that the experimenters consider any difference $\mu_{.j} - \mu \geq 3$ worth detecting. We then have

$$\alpha = 0.05$$

$$\text{Power} = 0.95$$

$$\text{MSE} = 13.0750, \quad \text{an estimate of } \sigma_e^2 \text{ from the previous study}$$

$$k = 4$$

$$\Sigma(\mu_{.j} - \mu)^2 = 3^2 + 3^2 + 3^2 + 3^2 = 36$$

From these data and Equation 8.5, we compute

$$\phi' = \frac{\sqrt{36/4}}{\sqrt{13.0750/\sqrt{n_j}}} = 0.83\sqrt{n_j}$$

We seek a value of n_j such that when $\phi = 0.83\sqrt{n_j}$, $\alpha = 0.05$, $\nu_1 = 3$, and $\nu_2 = n - k = 4(n_j - 1)$, the power will be 0.95. We now turn to the chart in Appendix Table K for $\nu_1 = 3$, and we try different values of n_j in the above equation until we find one that will result in the desired power. Suppose we try $n_j = 7$. This will result in $\nu_2 = 4(6) = 24$ and $\phi' = 2.20$. We locate $\phi' = 2.20$ on the horizontal scale for $\alpha = 0.05$, and move up until we reach the point, just beyond the curve for $\nu_2 = 20$, where the curve for $\nu_2 = 24$ would be drawn. Reading across to the power axis, we see that the power is about 0.94, which is just short of the desired power of 0.95. If we try $n_j = 8$, we obtain $\phi' = 2.35$ and $\nu_2 = 4(7) = 28$. Entering the chart with these values, we find the power to be about 0.97, or slightly above the desired power of 0.95. In order to achieve the desired power, then, the experiment must use 8 subjects in each group, for a total of 32 subjects.

EXERCISES

20 To study the learning of a task by a certain experimental animal, researchers conducted an experiment designed to evaluate the effects of different types of punishment for slow performance. The data yielded the following ANOVA table.

Source	SS	df	MS	VR
Treatments	136.1713	4	34.0428	2.26
Error	452.0001	30	15.0667	
	588.1714	34		

Calculate the power for this test for $\alpha = 0.05$.

21 Suppose that the researchers in Exercise 20 want to repeat the experiment. What size sample would be required to achieve a power of 0.90 if the researchers want to detect any difference $(\mu_{.j} - \mu) \geq 2$?

22 Refer to Exercise 5. Calculate the power for $\alpha = 0.05$.

23 Refer to Exercise 10. How large a sample should one take if one wants to detect any difference $(\mu_{.j} - \mu) \geq 2$ at an $\alpha = 0.05$ level of significance? The desired power is 0.90.

24 A coach wants to design an experiment to evaluate three methods of teaching swimming. Following the period of instruction, investigators will apply a scoring system that incorporates measures of form, style, and endurance to each subject. From past experience, a good estimate of σ_e^2 is 200. A significance level of $\alpha = 0.05$ will be used. The investigators want to reject H_0 if successive means are five or more units apart.

(a) How many subjects should be in each group, given that the desired power is 0.80?
(b) How many subjects should be in each group, given that a power of 0.90 is desired?

25 Psychologists designed an experiment to evaluate the effects of four methods of communicating information on the time required for subjects to solve a problem that served as the content of the communicated information. They constructed the following ANOVA table from the data generated by the experiment.

Source	SS	df	MS
Treatments	1089.667	3	363.22
Error	1876.333	20	93.82
	2966.000	23	

(a) Calculate the power for $\alpha = 0.05$.
(b) Calculate the power for $\alpha = 0.01$.

8.4 ANALYSIS OF THE RANDOMIZED COMPLETE BLOCK DESIGN

In certain experimental situations, a person may want to select an alternative to the completely randomized design. By so doing, the experimenter may be able to reduce the error variance and thereby increase the power of the statistical tests used in the analysis. A widely used alternative to the completely randomized design is the randomized complete block design.

The following example illustrates the concepts and techniques involved in the randomized complete block design.

Example 8.6 Researchers conducted an experiment to evaluate the effectiveness of four different drugs in relieving the symptoms of depression in patients. The researchers, who wished to eliminate the effects of duration of illness, randomly selected four patients from each of three duration-of-illness categories—short, medium, and long—then randomly assigned the patients in each duration-of-illness group (block) to receive one of the four drugs (treatments). The measured response was the difference between the patients' before and after scores made on a test designed to measure the level of depression. Table 8.24 shows the results.

Table 8.25 shows the general table for displaying the data from a randomized complete block design.

Assumptions The assumptions underlying the randomized complete block design are as follows.

1 Each observation constitutes an independent random sample of size one from a population defined by a given treatment-block combination.

2 The kn populations represented in the experiment are normally distributed.

Table 8.24 Difference between before and after levels of depression in 12 patients receiving four different drugs

Duration of illness	Drugs *A*	*B*	*C*	*D*	Total	Mean
Short	12	11	16	15	54	13.50
Medium	10	13	15	17	55	13.75
Long	8	8	10	10	36	9.00
Total	30	32	41	42	145	
Mean	10.00	10.67	13.67	14.00		12.08

Table 8.25 Data layout for a randomized complete block design

Block	Treatment 1	2	3	...	j	...	k	Total	Mean
1	y_{11}	y_{12}	y_{13}	...	y_{1j}	...	y_{1k}	$T_{1.}$	$\bar{y}_{1.}$
2	y_{21}	y_{22}	y_{23}	...	y_{2j}	...	y_{2k}	$T_{2.}$	$\bar{y}_{2.}$
3	y_{31}	y_{32}	y_{33}	...	y_{3j}	...	y_{3k}	$T_{3.}$	$\bar{y}_{3.}$
⋮	⋮	⋮	⋮		⋮		⋮	⋮	⋮
i	y_{i1}	y_{i2}	y_{i3}	...	y_{ij}	...	y_{ik}	$T_{i.}$	$\bar{y}_{i.}$
⋮	⋮	⋮	⋮		⋮		⋮	⋮	⋮
n	y_{n1}	y_{n2}	y_{n3}	...	y_{nj}	...	y_{nk}	$T_{n.}$	$\bar{y}_{n.}$
Total	$T_{.1}$	$T_{.2}$	$T_{.3}$	...	$T_{.j}$	...	$T_{.k}$	$T_{..}$	
Mean	$\bar{y}_{.1}$	$\bar{y}_{.2}$	$\bar{y}_{.3}$	...	$\bar{y}_{.j}$	...	$\bar{y}_{.k}$		$\bar{y}_{..}$

3 Each of the kn populations has the same variance σ^2.

4 Treatments and blocks do not interact. This means that combinations of block effects and treatment effects do not act together to produce special effects. For the experiment in Example 8.6, the assumption of no interaction means that there is no particular combination of a drug and a duration-of-illness category that produces an effect different from other combinations of drug and duration-of-illness categories.

Computations The hypothesis to be tested is

$$H_0: \mu_{.1} = \mu_{.2} = \cdots = \mu_{.k}$$

Expressed in words, the null hypothesis states that there is no difference between the true treatment means. Or, equivalently, that all treatment effects are equal. The alternative hypothesis is

H_1: Not all $\mu_{.j}$ are equal

For our drug-depression example, the null and alternative hypotheses are

$H_0\colon \mu_A = \mu_B = \mu_C = \mu_D \qquad H_1\colon$ Not all population means are equal

Since we use blocks to eliminate an extraneous source of variation, we are usually not interested in testing a hypothesis regarding block effects.

We test the null hypothesis of no treatment effects by forming the ratio of the treatment mean square MSTR to the error mean square MSE. When the null hypothesis is true, and when the assumptions stated earlier are met, we can show that the ratio follows the F distribution, with $\nu_1 = k - 1$ and $\nu_2 = (k-1)(n-1)$ degrees of freedom.

We obtain the mean squares necessary for performing the test by first partitioning the total variability (as expressed by the sum of squares) present in the sample data into components attributable to treatments, blocks, and error. The resulting sums of squares, divided by their appropriate degrees of freedom, yield the necessary mean squares.

Partitioning the total sum of squares We may partition the total sum of squares, SST, as follows.

$$\text{SST} = \text{SSBL} + \text{SSTR} + \text{SSE}$$

where SSBL is the sum of squares for blocks, SSTR is the sum of squares for treatments, and SSE is the error sum of squares. The computational formulas for these components are as follows.

$$C = \frac{T_{..}^2}{kn} = \frac{(\text{grand total})^2}{\text{total number of observations}}$$

$$\text{SST} = \sum_{j=1}^{k}\sum_{i=1}^{n} y_{ij}^2 - C = (\text{sum of all squared observations}) - C$$

$$\text{SSBL} = \frac{\sum_{i=1}^{n} T_{i.}^2}{k} - C = \frac{\text{sum of the squared block totals}}{\text{number of treatments}} - C$$

$$\text{SSTR} = \frac{\sum_{j=1}^{k} T_{.j}^2}{n} - C = \frac{\text{sum of squared treatment totals}}{\text{number of blocks}} - C$$

$$\text{SSE} = \text{SST} - \text{SSBL} - \text{SSTR}$$

For our example about using drugs to relieve depression, we use the data in Table 8.24 to compute the following sums of squares.

$$\text{SST} = \sum_{j=1}^{4}\sum_{i=1}^{3} y_{ij}^2 - C$$

$$= 12^2 + 11^2 + \cdots + 10^2 - \frac{(12 + 11 + \cdots + 10)^2}{12}$$

$$= 1857.00 - 1752.0833 = 104.9167$$

Table 8.26 ANOVA table for the randomized complete block design

Source	SS	df	MS	VR
Treatments	SSTR	$k-1$	MSTR = SSTR/$(k-1)$	MSTR/MSE
Blocks	SSBL	$n-1$	MSBL = SSBL/$(n-1)$	
Error	SSE	$(k-1)(n-1)$	MSE = SSE/$[(n-1)(k-1)]$	
	SST	$kn-1$		

Table 8.27 ANOVA Table for Example 8.6

Source	SS	df	MS	VR
Treatments	37.5834	3	12.5278	7.394
Blocks	57.1667	2	28.5834	
Error	10.1666	6	1.6944	
	104.9167	11		

$$\text{SSBL} = \frac{\sum_{i=1}^{3} T_{i.}^2}{4} - C = \frac{54^2 + 55^2 + 36^2}{4} - 1752.0833 = 57.1667$$

$$\text{SSTR} = \frac{\sum_{j=1}^{4} T_{.j}^2}{3} - C = \frac{30^2 + 32^2 + 41^2 + 42^2}{3} - 1752.0833$$

$$= 37.5834$$

$$\text{SSE} = \text{SST} - \text{SSBL} - \text{SSTR} = 104.9167 - 57.1667 - 37.5834$$

$$= 10.1666$$

Table 8.26 shows the ANOVA table for the randomized complete block design. Table 8.27 is the ANOVA table for our drug-depression example.

The critical value of F for $\alpha = 0.05$ and 3 and 6 degrees of freedom is 4.76. Since 7.394 is greater than 4.76, we reject the null hypothesis at the 0.05 level. We conclude that, after eliminating variability resulting from duration of illness, there is a difference in treatment effects.

EXERCISES

26 A manufacturer of steel beams conducted an experiment to compare the effects of four different processing methods on the strength of the finished product. The investigator used the randomized complete block design, with the four sources of raw material serving as blocks. Table 8.28 shows the strengths (coded) of 16 specimens prepared during the experiment. After eliminating block effects, can the manufacturer conclude that the different processing methods do have different effects on product strength? Let $\alpha = 0.01$.

27 In an experiment in which age served as a blocking variable, an investigator evaluated three procedures for the treatment of persons diagnosed as having inferiority complexes. The response variable was the post-treatment score each individual made on a test designed to measure ego

Table 8.28 Data for Exercise 26

Block	Processing method A	B	C	D
1	11	10	13	14
2	12	10	16	14
3	16	17	18	18
4	17	15	18	18

Table 8.29 Data for Exercise 27

Age group	Therapeutic procedure A	B	C
1	10	9	10
2	9	10	12
3	10	11	12
4	11	10	14

strength. Table 8.29 shows the results. Once we have eliminated the effects of age, do these data provide sufficient evidence to indicate that the therapeutic methods have different effects? Let $\alpha = 0.05$.

28 A government agricultural experiment station conducted an experiment to compare the effects of four different fertilizers on the yield of a cane crop. Researchers divided four blocks of soil into four plots of equal size and shape and assigned the fertilizers to the plots at random, in such a way that each fertilizer was applied once in each block. Table 8.30 shows the yield of cane, in hundredweights per plot. After block effects have been removed, do these data provide sufficient evidence to indicate a difference among treatment effects? Let $\alpha = 0.01$.

29 A researcher conducted an experiment to evaluate the effects of four different drugs on human reaction times. Four subjects in each of five age groups were randomly assigned to receive one of the drugs. Table 8.31 shows the reduction in each subject's time of reaction to a certain stimulus after administration of the drugs. (The data have been coded for ease of computation.) After eliminating the effect of age, can the researcher conclude that the drugs have different effects? Let $\alpha = 0.05$.

30 Psychologists conducted an experiment to evaluate three methods of motivating patients in a state mental hospital. The researchers felt that they should use a design that would allow for blocking by diagnosis. Table 8.32 shows the scores (coded) made on a test designed to measure effectiveness of the treatment. After the effect of diagnosis has been eliminated, do these data indicate that there is a difference in treatment effects? Let $\alpha = 0.05$.

31 Refer to Exercise 32 in Chapter 6. Treat this exercise as a randomized complete block design, and solve it by means of two-way analysis of variance.

Table 8.30 Data for Exercise 28

Block	Fertilizer A	B	C	D
1	50	45	45	40
2	55	45	50	45
3	45	40	35	35
4	45	35	45	35

Table 8.31 Data for Exercise 29

Age group	Drug A	B	C	D
1	6	7	4	7
2	6	8	9	9
3	9	12	8	6
4	8	9	5	9
5	8	10	7	6

Table 8.32 Data for Exercise 30

Diagnostic group	Treatment group A	B	C
1	8	9	7
2	5	8	7
3	3	9	5
4	4	8	5

8.5 ANALYSIS OF THE LATIN SQUARE DESIGN

If it is possible and desirable to eliminate two sources of variation from the error variance, one may use the Latin square design. The following example illustrates a situation in which one can use this design.

Example 8.7 An educator wishes to conduct an experiment to evaluate the effectiveness of five methods of teaching a certain mathematical concept to eighth-grade students. The experimenter wishes to eliminate the effects of different teachers, as well as the time of day during which teaching takes place. Table 8.33 shows the results of the experiment.

In the data layout table for a Latin square design, the levels of one of the two blocking variables make up the rows of the table, and the levels of the other blocking variable make up the columns. Treatment are randomly assigned to the cells of the table in such a way that *each treatment appears once and only once in each row and each column.* Suppose we have five treatments, which we designate by the letters A, B, C, D, and E. The data layout for our experiment might look like Table 8.34.

All subscripts used in the Latin square analysis go from 1 to r; that is, $i = 1, 2, \ldots, r =$ the number of rows; $j = 1, 2, \ldots, r =$ the number of columns; and $k = 1, 2, \ldots, r =$ the number of treatments. We enclose the

Table 8.33 Scores made by subjects participating in the experiment described in Example 8.7

Teacher	Time of day 9 A.M.	9 A.M.	Noon	2 P.M.	4 P.M.	Total
1	*E*, 80	*D*, 90	*C*, 66	*A*, 64	*B*, 58	358
2	*B*, 75	*E*, 91	*A*, 71	*D*, 90	*C*, 67	394
3	*C*, 75	*A*, 79	*B*, 74	*E*, 78	*D*, 80	386
4	*D*, 99	*C*, 79	*E*, 91	*B*, 81	*A*, 73	423
5	*A*, 62	*B*, 85	*D*, 96	*C*, 61	*E*, 85	389
Total	391	424	398	374	363	1950

	Treatment totals and means *A*	*B*	*C*	*D*	*E*
Total	349	373	348	455	425
Mean	69.8	74.6	69.6	91.0	85.0

Table 8.34 Possible data layout for Latin square design with five treatments

Row	Column 1	2	3	4	5
1	*A*	*B*	*C*	*D*	*E*
2	*B*	*A*	*E*	*C*	*D*
3	*C*	*D*	*A*	*E*	*B*
4	*D*	*E*	*B*	*A*	*C*
5	E	C	D	B	A

subscript k, which is used to denote the treatment, in parentheses to show that it depends on the (i, j) subscript combination. For a given Latin square pattern, k is determined as soon as we have specified any ijth cell. In Table 8.34, for example, if the cell under consideration is cell $i=4, j=5$, we know that $k=3$ and that the treatment is treatment C.

The assumptions underlying the Latin square design are as follows.

1 The observation y_{ijk} constitutes an independent random sample of size one drawn from a population that is defined by the row-block-treatment combination ijk. In a given experiment, r^2 populations will be represented.

2 Each population represented in the experiment is normally distributed.

3 The variances of all represented populations are equal.

4 Rows, columns, and treatments do not interact.

Partitioning the total sum of squares

We may partition the total sum of squares (SST) in the Latin square design into four components, the sources of which are rows (SSR), columns (SSC), treatments (SSTR), and error (SSE). To summarize, we may write

$$\text{SST} = \text{SSR} + \text{SSC} + \text{SSTR} + \text{SSE}$$

If we let

$T_{i..}$ = total of the ith row

$T_{.j.}$ = total of the jth column

$T_{..k}$ = total of the kth treatment

$T_{...}$ = total of all sample observations

we may write the following computational formulas for the sums of squares for the various components.

$$C = \frac{T_{\ldots}^2}{r^2} = \frac{(\text{grand total})^2}{\text{total number of observations}}$$

$$\text{SST} = \sum_{i=1}^{r} \sum_{j=1}^{r} y_{ij(k)}^2 - C = (\text{sum of all squared observations}) - C$$

$$\text{SSR} = \frac{\sum_{i=1}^{r} T_{i..}^2}{r} - C = \frac{\text{sum of squared row totals}}{\text{number of columns}} - C$$

$$\text{SSC} = \frac{\sum_{j=1}^{r} T_{.j.}^2}{r} - C = \frac{\text{sum of squared column totals}}{\text{number of rows}} - C$$

$$\text{SSTR} = \frac{\sum_{k=1}^{r} T_{..k}^2}{r} - C = \frac{\text{sum of squared treatment totals}}{\text{number of rows}} - C$$

$$\text{SSE} = \text{SST} - \text{SSR} - \text{SSC} - \text{SSTR}$$

From the data in Table 8.33, we compute the following sums of squares for Example 8.7.

$$\text{SST} = \sum_{i=1}^{5} \sum_{j=1}^{5} y_{ij(k)} - C$$

$$= (80^2 + 90^2 + \cdots + 85^2) - \frac{1950^2}{5^2} = 155{,}122 - 152{,}100$$

$$= 3022$$

$$\text{SSR} = \frac{\sum_{i=1}^{5} T_{i..}^2}{5} - C = \frac{358^2 + 394^2 + \cdots + 389^2}{5} - 152{,}100 = 429.2$$

$$\text{SSC} = \frac{\sum_{j=1}^{5} T_{.j.}^2}{5} - C = \frac{391^2 + 424^2 + \cdots + 363^2}{5} - 152{,}100 = 441.2$$

$$\text{SSTR} = \frac{\sum_{k=1}^{5} T_{..k}^2}{5} - C = \frac{349^2 + 373^2 + \cdots + 425^2}{5} - 152{,}100$$

$$= 1836.8$$

$$\text{SSE} = \text{SST} - \text{SSR} - \text{SSC} - \text{SSTR} = 3022 - 429.2 - 441.2 - 1836.8$$

$$= 314.8$$

The analysis of variance table

We may display the results of the analysis of variance in an analysis of variance (ANOVA) table such as Table 8.35. Table 8.36 shows the ANOVA table for Example 8.7.

As we have already noted, the objective in using the Latin square design is to eliminate two sources of extraneous variation. Consequently, attention

Table 8.35 ANOVA table for Latin square design

Source	SS	df	MS	VR
Rows	SSR	$r-1$	MSR = SSR/$(r-1)$	
Columns	SSC	$r-1$	MSC = SSC/$(r-1)$	
Treatments	SSTR	$r-1$	MSTR = SSTR/$(r-1)$	MSTR/MSE
Error	SSE	$(r-1)(r-2)$	MSE = SSE/$[(r-1)(r-2)]$	
Total	SST	r^2-1		

Table 8.36 ANOVA table for Example 8.7

Source	SS	df	MS	VR
Rows	429.2	4	107.3	
Columns	441.2	4	110.3	
Treatments	1836.8	4	459.2	17.51
Error	314.8	12	26.23	
	3022.0	24		

is usually focused entirely on the mean square and the variance ratio for treatments. Specifically, we wish to test

$$H_0: \mu_1 = \mu_2 = \cdots = \mu_k$$

against the alternative hypothesis

$$H_1: \text{Not all } \mu_k \text{ are equal}$$

When H_0 is true and when the assumptions are met, we can show that

$$\text{VR} = \frac{\text{MSTR}}{\text{MSE}}$$

is distributed as F with $r-1$ and $(r-1)(r-2)$ degrees of freedom. If the computed value of VR is equal to or greater than the tabulated value of F for $(r-1)$ and $(r-1)(r-2)$ degrees of freedom and the chosen level of significance, α, we reject H_0 in favor of the alternative hypothesis.

For Example 8.7, we have VR = 17.51. Reference to Appendix Table H, with 4 and 12 degrees of freedom, indicates that the probability of observing a VR value as large as or larger than 17.51, when H_0 is true, is less than 0.005. Therefore we can reject H_0 at the 0.005 level of significance, and we conclude, after eliminating effects resulting from teacher and time of day, that not all methods of teaching arithmetic are equally effective.

EXERCISES

32 Psychiatrists in a mental hospital carried out an experiment to compare the effectiveness of five anxiety-reducing drugs. Researchers felt that a major source of variation that should be accounted for was the ward

Table 8.37 Data for Exercise 32

Row (Severity level)	Column (Ward) 1	2	3	4	5
1	D, 24	C, 21	A, 13	E, 12	B, 31
2	A, 20	B, 13	E, 7	C, 19	D, 27
3	B, 30	A, 16	C, 8	D, 23	E, 24
4	C, 22	E, 12	D, 18	B, 27	A, 24
5	E, 19	D, 22	B, 21	A, 25	C, 26

on which the patient being tested lived. Varying degrees of rapport between patients and staff existed among the wards, and researchers believed that this would have some effect on a patient's response to the experimental procedures. Another source of variation that experimenters felt would affect patient response was severity of illness. Since five wards were available, and since five levels of severity of illness could be identified, researchers decided that a Latin square design would be the appropriate experimental design. Accordingly, one patient representing each level of severity of illness was selected from each of the five wards, making a total of 25 patients who participated in the experiment. The wards were associated with the columns of the square, and severity of illness with the rows. The drugs were randomly labeled A, B, C, D, and E and were assigned to the patients according to the Latin square shown in Table 8.37. The patient with severity level 1 from ward 1 received drug D; the patient with severity level 2 from ward 2 received drug B; and so on. At the end of the experimental period, researchers gave the patients tests to measure their levels of anxiety. Table 8.37 shows the scores from these tests. After the effects of ward and severity of illness have been eliminated, do these data provide sufficient evidence to indicate a difference in effectiveness among the drugs? Let $\alpha = 0.05$.

33 Carpeting impregnated with a bactericidal agent was installed in five different areas of a general hospital. Five different bactericidal agents were used. Bacteria counts were made in each area on each of 5 days, with the results shown in Table 8.38. Do these data provide sufficient evidence to indicate a difference among the bactericidal agents? Let $\alpha = 0.05$.

Table 8.38 Data for Exercise 33

Area	Day MON.	TUES.	WED.	THURS.	FRI.
I	A, 58	B, 62	C, 77	D, 94	E, 66
II	B, 48	A, 76	D, 103	E, 58	C, 99
III	C, 74	D, 113	E, 70	A, 105	B, 93
IV	E, 66	C, 95	A, 111	B, 108	D, 150
V	D, 110	E, 63	B, 113	C, 126	A, 160

Table 8.39 Data for Exercise 34

Week	Day Mon.	Tues.	Wed.	Thurs.	Fri.
1	*A*, 13	*E*, 11	*C*, 19	*B*, 15	*D*, 10
2	*D*, 12	*C*, 20	*E*, 12	*A*, 14	*B*, 13
3	*E*, 9	*A*, 12	*B*, 14	*D*, 12	*C*, 17
4	*C*, 15	*B*, 10	*D*, 11	*E*, 10	*A*, 13
5	*B*, 11	*D*, 13	*A*, 13	*C*, 21	*E*, 17

34 Psychiatrists in a mental hospital conducted a study to evaluate the effectiveness of five treatments in reducing the destructive and aggressive behavior of patients housed on a certain ward. The treatments consisted of various manipulations of the environment, including lighting, ward decoration, and piped music. A different treatment was tested on each of the five weekdays. A Latin square design was used to rotate the treatments from week to week, so that over a five-week period, each treatment was tested on each of the five days and once during each week. The variable of interest was the number of destructive and hostile episodes occurring per day. Table 8.39 shows the resulting data. After we eliminate effects of day and week, do these data provide sufficient evidence to indicate a difference among treatments? Let $\alpha = 0.01$.

35 A team of biologists used 25 experimental animals in an experiment to compare five different foods. The variable of interest was weight gain. Experimenters placed each animal in a separate cage, arranged the cages in a 5×5 square, and gave the animals the foods according to a Latin square, with the letters *A* through *E* indicating the different foods. Table 8.40 shows the arrangement and the animals' weight gains during the experimental period. After row and column effects have been eliminated, do these data suggest a difference among the foods? Let $\alpha = 0.05$.

36 In a study to evaluate the effectiveness of five programs designed to help patients regain their strength following trauma, a physical therapist used a Latin square design in which the columns represented the ages of the patients, and the rows represented the types of trauma. At the end of the experimental period, the therapist obtained the strength scores shown in Table 8.41. Do these data provide sufficient evidence to indicate a difference among programs? Let $\alpha = 0.05$.

Table 8.40 Data for Exercise 35

Row	Column 1	2	3	4	5
I	*E*, 21	*A*, 21	*D*, 20	*C*, 21	*B*, 21
II	*B*, 21	*C*, 25	*A*, 22	*E*, 18	*D*, 19
III	*A*, 26	*D*, 22	*C*, 23	*B*, 24	*E*, 19
IV	*C*, 25	*B*, 23	*E*, 20	*D*, 22	*A*, 21
V	*D*, 24	*E*, 20	*B*, 23	*A*, 20	*C*, 21

Table 8.41 Data for Exercise 36

Type of trauma	Age 1	2	3	4	5
I	*A*, 61	*E*, 53	*D*, 61	*B*, 57	*C*, 58
II	*D*, 56	*B*, 70	*C*, 66	*A*, 61	*E*, 57
III	*E*, 57	*C*, 62	*B*, 66	*D*, 66	*A*, 68
IV	*B*, 69	*A*, 69	*E*, 65	*C*, 59	*D*, 62
V	*C*, 56	*D*, 62	*A*, 72	*E*, 57	*B*, 68

CHAPTER SUMMARY

In this chapter we introduced the basic concepts and techniques of analysis of variance. We discussed in detail the completely randomized design, the randomized complete block design, and the Latin square design. Additional topics we treated are comparisons among treatment means, power, and determination of sample size.

An introductory book such as this cannot begin to cover all aspects of such a specialized subject as analysis of variance. For elaboration of the various points raised in this chapter, as well as additional topics not mentioned, consult the References. In addition to those cited in Chapter 7, the books by Meddis (4) and Guenther (5) may be helpful.

SOME IMPORTANT CONCEPTS AND TERMINOLOGY

Analysis of variance
One-way analysis of variance
Sum of squares
Mean square
Variance ratio
Degrees of freedom
Analysis of variance (ANOVA) table
Rationale underlying the F test
Tukey's HSD test
Assumptions underlying a completely randomized design
Assumptions underlying a randomized complete block design
Assumptions underlying a Latin square design

REVIEW EXERCISES

37 A study was conducted to compare the abilities of three drugs to retard reaction times of experimental animals to a stimulus. Three strains of animals were used in the experiment. Table 8.42 shows the animals' response times, in seconds, following administration of the drug. After strain effects have been eliminated, do these data provide sufficient evidence to indicate a difference among the drugs? Let $\alpha = 0.05$.

38 In a treatment facility for the mentally retarded, a group of psychiatrists conducted an experiment to compare three methods of teaching basic self-care skills. To eliminate the effects of different levels of retardation, the investigators used that factor as a blocking variable. At the end of the experimental period, the subjects were tested to measure the degree to

Table 8.42 Data for Exercise 37

		Drug	
Strain	*A*	*B*	*C*
I	5	8	12
II	6	10	14
III	8	12	16

Table 8.43 Data for Exercise 38

Retardation level	Method A	B	C
I	52	60	94
II	68	70	76
III	24	49	70

Table 8.44 Data for Exercise 39

Formula										
A	10	12	10	16	12	14	12	10	12	14
B	16	18	16	14	18	18	20	16	16	18
C	20	20	18	16	16	18	20	18	16	18

which they had learned the skills. Table 8.43 shows the results. After eliminating block effects, did the investigators have sufficient evidence to conclude that there were differences among the three methods?

39 A paint manufacturer was interested in determining the effect on durability of paint of three formula ingredients. The manufacturer randomly assigned each of the ingredients to ten batches of experimental material. Table 8.44 shows the results of durability tests (in coded form) performed on each batch of the final product. The manufacturer wished to know whether the data indicate that the ingredients have different effects on durability. Perform the appropriate analysis of variance.

40 In a study designed to evaluate four brands of air freshener (*A*, *B*, *C*, and *D*), investigators used a Latin square design in which the columns represented the four seasons of the year and the rows represented four areas of the country. The variable of interest was the quality of the air following use of the freshener in rooms that were uniformly unpleasant before the experiment. Table 8.45 shows the data. Test, at the 0.05 level of significance, the null hypothesis of no difference among treatment means.

41 A study was conducted to compare the tensile strengths of plastics produced in four different factories at four different temperature settings. The effects of the factories were eliminated by blocking. Table 8.46 shows the results (coded). Do these data provide sufficient evidence, at the 0.05 level of significance, to indicate a temperature effect?

42 Researchers used four different methods to teach four groups of mice to perform a task. At the end of a specified time period, they gave each mouse a test to determine speed of performance. Table 8.47 indicates the results. Perform the analysis of variance, letting $\alpha = 0.05$.

Table 8.45 Data for Exercise 40

Area	Season: FALL	WINTER	SPRING	SUMMER
1	*A*, 12	*B*, 10	*C*, 10	*D*, 12
2	*B*, 10	*A*, 12	*D*, 12	*C*, 10
3	*C*, 10	*D*, 11	*B*, 10	*A*, 12
4	*D*, 11	*C*, 10	*A*, 13	*B*, 10

Table 8.46 Data for Exercise 41

Factory	Temperature: *A*	*B*	*C*	*D*
W	12	26	24	23
X	15	29	23	25
Y	15	27	25	24
Z	18	38	33	31

Table 8.47 Data for Exercise 42

Teaching method			
1	2	3	4
64	76	58	95
88	70	74	90
72	90	66	80
80	80	60	87
79	75	82	88
71	82	75	85

REFERENCES

1 John W. Tukey, "The Problem of Multiple Comparisons," Ditto, Princeton University, 1953; cited in Kirk (2)

2 Roger E. Kirk, *Experimental Design: Procedures for the Behavioral Sciences*, Belmont, Calif.: Brooks/Cole, 1968

3 E. S. Pearson and H. O. Hartley, "Charts of the Power Function for Analysis of Variance Tests, Derived from the Noncentral *F*-Distribution," *Biometrika*, 38 (1951), 112–130

4 Ray Meddis, *Elementary Analysis of Variance for the Behaviorial Sciences*, New York: Wiley, 1973.

5 William C. Guenther, *Analysis of Variance*, Englewood Cliffs, N.J.: Prentice-Hall, 1964

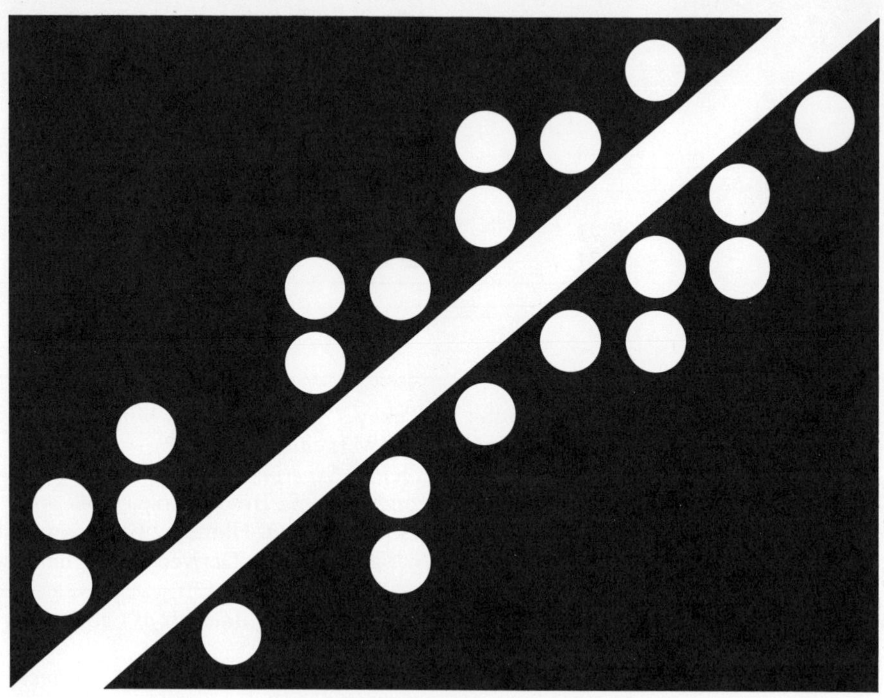

9 Simple Linear Regression and Correlation Analysis

Many situations arise in which a study of the relationship between two variables is of interest. A teacher may wish to know how well he or she can predict a student's performance in arithmetic on the basis of the pupil's score on an arithmetic aptitude test. A psychologist may wish to know whether there is a relationship between a student's self-concept and her or his grade point average. A sociologist may be interested in knowing the nature of the relationship that may exist between the rate of juvenile delinquency in a community and the extent of overcrowding in the homes of the community.

One may investigate relationships such as these by means of regression analysis and/or correlation analysis. *Regression analysis* is concerned with the *nature* of relationships among variables. *Correlation analysis* is concerned with the *strength* of relationships. The concepts of regression and correlation were introduced by the English scientist Sir Francis Galton (1822–1911) in connection with his research in heredity and other areas of biology.

When the investigation of relationships is limited to only two variables, we call the analytical procedures *simple regression analysis* and *simple correlation analysis*. Simple regression and correlation analysis are the topics of this chapter. When we are considering more than two variables, we refer to the analytical techniques as *multiple regression analysis and multiple correlation analysis*. We shall not cover these topics in this book.

9.1 THE SIMPLE LINEAR REGRESSION MODEL

The following example illustrates the type of problem that is appropriate for the application of regression analysis.

Example 9.1 A research team at a psychiatric hospital conducted an experiment to study the relationship, in schizophrenic patients, between reaction time to a particular stimulus and dose level of a certain drug. The researchers wished, specifically, to conduct the experiment with dose levels of 0.5, 1.0, 1.5, 2.0, 2.5, and 3.0 mg. They selected a random sample of 18 patients from the hospital population of schizophrenics, and assigned each patient at random to one of the dose levels. Thus each dose level was administered to a total of three patients.

In simple regression analysis, we label the two variables involved X and Y. We call the variable X the *independent variable*, since in many situations it can be manipulated by the investigator. For example, the researcher may select only certain values of X to use in the analysis. It is generally of interest to know what effect changes in the values of X have on the values that Y assumes. We call the variable Y the *dependent variable*. As we shall see, one of the uses of regression analysis is to construct a device, *called a prediction equation*, which enables one to predict what value Y is likely to assume when X assumes some particular value. For this reason, we sometimes call X the *predictor variable* and Y the *response variable*.

Now let us determine which is the independent variable and which the dependent variable in our example.

Since dose level was under the control of the investigators, we designate this variable the independent variable X, and treat reaction time as the dependent variable Y.

Table 9.1 shows the results of the experiment.

The study of the relationship between two variables should begin with the construction of a graph, called a *scatter diagram*, which portrays the nature of the relationship. One assigns values of the independent variable to the horizontal axis, and values of the dependent variable to the vertical axis. The body of the graph consists of dots placed at the intersections of imaginary lines extending vertically from each value of X and horizontally from corresponding values of Y. Figure 9.1 shows the scatter diagram for Example 9.1.

Figure 9.1 Scatter diagram for Example 9.1

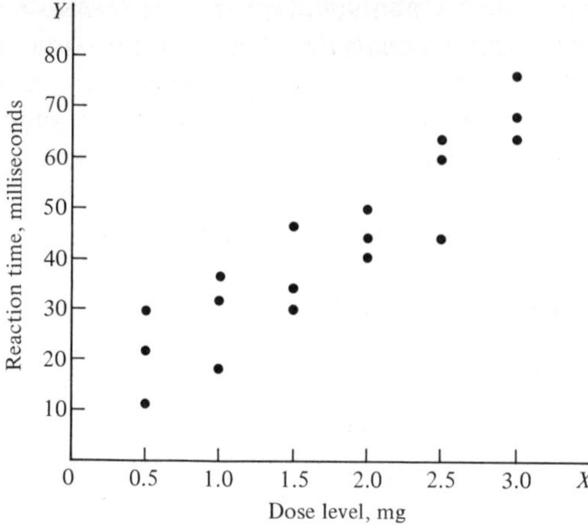

Scatter diagrams are extremely helpful in the study of the relationship between two variables. Examination of Figure 9.1, for example, suggests that, in general, reaction time tends to increase with an increase in dose level. When two variables move together in this manner, we say that there is a *direct* relationship between them.

The scatter of points in Figure 9.1 also suggests that a straight line would provide a good representation of the data, since the values of Y appear to increase over the observed interval of X. More specifically, our intuition is leading us to select a *mathematical model* that expresses the relationship between the independent variable X and the dependent variable Y. The particular model that seems appropriate in the present situation is called the *simple linear regression model.* We may express this model symbolically as follows.

$$y_i = \beta_0 + \beta_1 x_i + \xi_i \tag{9.1}$$

Table 9.1 Dose levels and reaction times for Example 9.1

Patient	**Dose level X** (mg)	**Reaction time Y** (msec)	**Patient**	**Dose level X** (mg)	**Reaction time Y** (msec)
1	0.5	12	10	2.0	40
2	0.5	22	11	2.0	44
3	0.5	30	12	2.0	50
4	1.0	18	13	2.5	44
5	1.0	32	14	2.5	60
6	1.0	36	15	2.5	64
7	1.5	30	16	3.0	64
8	1.5	34	17	3.0	68
9	1.5	46	18	3.0	76

where y_i is a typical value of the response variable Y, β_0 and β_1 are population parameters, x_i is a known constant or the ith value of the independent variable X, and ξ_i is a random error term. In Equation 9.1, the subscript i runs from 1 to n, the number of values of X under consideration. We call the parameter β_0 the *regression constant*, and we call β_1 the *regression coefficient*. We call this model a linear regression model, since the independent variable is raised to the first power only. We may distinguish one mathematical model from another on the basis of the assumptions that are made about the nature of the data being modeled.

Assumptions of the simple linear regression model

The basic assumptions of the simple linear regression model are as follows.

1 The variable X may be either a nonrandom variable (sometimes called a *fixed* or *mathematical* variable) or a random variable. The method of sample selection determines whether or not X is regarded as fixed or random. If the relationship between X and Y is of interest only for certain predetermined values of X, the usual procedure is to bring into the sample only subjects with these X measurements, and hence X is fixed. For example, if we are studying the relationship between height (X) and weight (Y) in adult males, we may decide, before drawing our sample, to study the relationship only in men who are 68, 70, 72, 74, and 76 inches tall. The sample, then, will consist only of men with these heights. After we have selected a man to be in the sample, we shall determine his weight.

If the investigator does not decide in advance what values of X will be admissible in the sample, but first draws a sample of subjects and then obtains whatever X measurements they exhibit, we consider the variable X to be random. In the height-weight study, for example, we consider height (X) to be a random variable if, first, a random sample of adult males is selected, and then their heights (as well as weights) are observed.

In Example 9.1, dose level is a fixed variable, since the investigators selected specific values of dose level in advance.

We call the entity (subject) on which a pair of X, Y measurements is taken the *unit of association*. In the height-weight example, the unit of association is an adult male. A sample of data, then, consists of a pair of X, Y measurements taken on each of n units of association. In Example 9.1, the unit of association is a schizophrenic patient, and the sample consists of $n = 18$ units of association (schizophrenic patients).

2 The variable Y is a random variable, and for each value of X there is a subpopulation of Y values.

3 The means of these subpopulations all lie on the same straight line. We may write the equation for this line as

$$\mu_{Y|x} = \beta_0 + \beta_1 x$$

Figure 9.2 Representation of the simple linear regression model

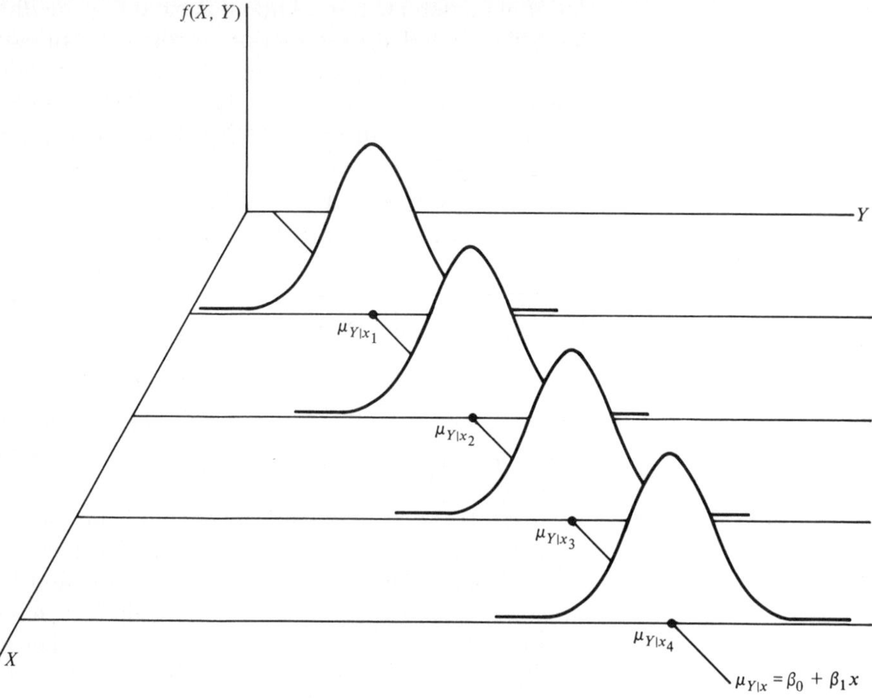

We may express a subpopulation mean for x_i, a given value of X, as

$$\mu_{Y|x_i}$$

4 The values of Y for a given value of X are chosen independently of those values of Y for any other value of X.

5 The variable X is measured without error.

These five assumptions are the minimum assumptions for the model. When they are met, certain descriptive analyses of the data are valid. In this chapter (as in most of this book), we shall be interested primarily in making inferences about the populations from which samples have been drawn. In simple linear regression analysis, valid inferential procedures are possible only when the following additional assumptions are tenable.

6 The subpopulations of Y values are all normally distributed.

7 The subpopulations of Y values all have the same variance.

Figure 9.2 shows a graphic representation of the simple linear regression model, including the last two assumptions.

The assumptions underlying the simple linear regression model may appear to be somewhat unrealistic. Fortunately, however, there are many practical situations in which the model of Equation 9.1 and the underlying assumptions fit the data sufficiently well to yield useful results. How well the model fits the real-world situation is a decision that the investigator must make primarily on the basis of his/her familiarity with the data and the processes by which they are generated. If the model under consideration does not appear to be a reasonable representation of reality, the investigator may consider other models, such as multiple regression models or curvilinear models.

9.2 OBTAINING THE SAMPLE REGRESSION EQUATION

In most situations of practical importance, the true population regression line illustrated in Figure 9.2 is unknown. One may obtain an estimate of the line, however, from sample data, and one may reach decisions regarding the usefulness of the line and its equation in studies of the relationship between X and Y, provided the assumptions stated in Section 9.1 are met.

In this section we shall present a method for obtaining a simple linear regression equation from sample data. In the next section we shall discuss methods for evaluating the sample equation as an analytical tool. In Section 9.4 we shall treat uses of the sample regression equation. We shall use Example 9.1 to illustrate the relevant concepts and procedures.

As we have noted, our scatter diagram (Figure 9.1) seems to support the choice of the simple linear regression model, specified in Equation 9.1, as a representation of the population from which the sample was drawn.

We could use any line drawn through the scatter of points in Figure 9.1 to estimate the true regression line of the population from which the sample was drawn. Intuition tells us that such a line should "represent" the sample data as much as possible. We usually express this idea by saying that the line should provide as good a *fit* to the data as possible. We can define *goodness of fit* in a number of ways, and can devise various methods for obtaining the line. One method is to draw a line, by eye, through the points (perhaps by means of a transparent ruler), being careful to achieve as good a fit as possible. The deficiency of such a procedure is obvious. For a given set of data, it would be rare for any two people to draw exactly the same line. What is needed is some objective method that yields a line that provides the best fit to the data according to some nontrivial criterion.

The method that is usually employed in fitting a straight line to data is the *method of least squares*. The line, called the *least-squares line,* obtained by this method is the line of best fit to the sample data in the following sense.

The sum of the squared vertical distances of the points on the scatter diagram from the least-squares line is smaller than any similar sum computed with reference to any other line.

Furthermore, as we shall see, the method of least squares yields estimates of population parameters that have very desirable properties.

We may express the sample simple linear regression equation as

$$y_c = b_0 + b_1 x \tag{9.2}$$

where y_c is a calculated value of Y obtained by substituting some value of X into the equation, b_0 is the Y intercept, and b_1 is the slope of the line. The Y intercept b_0 indicates the point at which the line crosses the Y axis, and the slope b_1 gives the amount by which y_c changes for a unit change in x. We measure the slope of a line by forming the ratio of the change in the Y coordinates to the change in the X coordinates for any two points on the line. That is,

$$\text{Slope} = \frac{\text{change in } Y}{\text{change in } X}$$

Figure 9.3 is a graphic illustration of a typical sample simple linear regression equation. We shall use b_0 and b_1, respectively, to estimate the parameters β_0 and β_1 in the simple linear regression model of Equation 9.1.

We may obtain the numerical values of b_0 and b_1 (the least-squares estimates of β_0 and β_1, respectively) from

$$b_1 = \frac{\Sigma (x_i - \bar{x})(y_i - \bar{y})}{\Sigma (x_i - \bar{x})^2} = \frac{\Sigma x_i y_i - \dfrac{(\Sigma x_i)(\Sigma y_i)}{n}}{\Sigma x_i^2 - \dfrac{(\Sigma x_i)^2}{n}} \tag{9.3}$$

$$b_0 = \frac{1}{n}(\Sigma y_i - b_1 \Sigma x_i) = \bar{y} - b_1 \bar{x} \tag{9.4}$$

Figure 9.3 A linear regression equation illustrating the geometric interpretations of β_0 and β_1

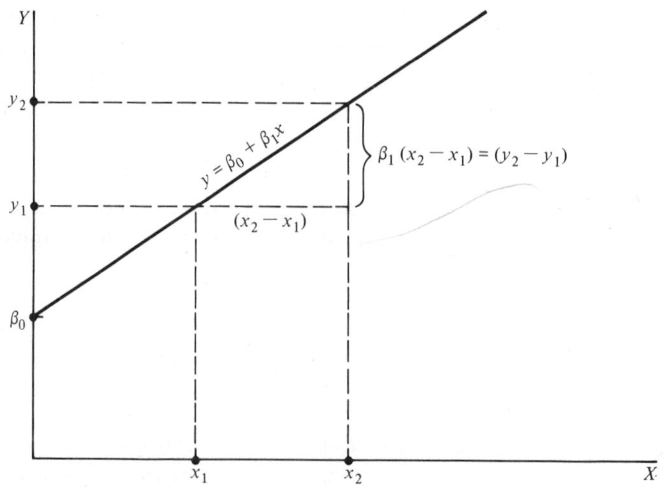

Table 9.2 Intermediate calculations for obtaining the least-squares regression equation for Example 9.1

	x_i	y_i	x_i^2	y_i^2	x_iy_i
	0.5	12	0.25	144	6.0
	0.5	22	0.25	484	11.0
	0.5	30	0.25	900	15.0
	1.0	18	1.00	324	18.0
	1.0	32	1.00	1024	32.0
	1.0	36	1.00	1296	36.0
	1.5	30	2.25	900	45.0
	1.5	34	2.25	1156	51.0
	1.5	46	2.25	2116	69.0
	2.0	40	4.00	1600	80.0
	2.0	44	4.00	1936	88.0
	2.0	50	4.00	2500	100.0
	2.5	44	6.25	1936	110.0
	2.5	60	6.25	3600	150.0
	2.5	64	6.25	4096	160.0
	3.0	64	9.00	4096	192.0
	3.0	68	9.00	4624	204.0
	3.0	76	9.00	5776	228.0
Total	31.5	770	68.25	38508	1595.0
Mean	1.75	42.7778			

Now let us obtain the sample linear regression equation for Example 9.1 by using the data of Table 9.1 and Equations 9.3 and 9.4. Table 9.2 gives the necessary intermediate calculations. Substituting numbers from Table 9.2 into Equations 9.3 and 9.4 gives

$$b_1 = \frac{1595 - \dfrac{(31.5)(770)}{18}}{68.25 - \dfrac{(31.5)^2}{18}} = 18.8571$$

$$b_0 = 42.7778 - (18.8571)(1.75) = 9.7779$$

The least-squares equation for our sample, then, is

$$y_c = 9.7779 + 18.8571x$$

Figure 9.4 shows the graph of this equation, along with the original observations.

As we have noted, the sample data of Example 9.1 suggest that there is a direct linear relationship between X and Y. In other situations Y may tend to decrease as X increases, and the scatter diagram of the data will look something like Figure 9.5(a).

A scatter diagram of this type suggests an *inverse* linear relationship between X and Y. When the linear model of Equation 9.1 does not apply, the relationship between X and Y may be *curvilinear*. Figures 9.5(b) and

Figure 9.4 Least-squares regression line and scatter diagram for Example 9.1

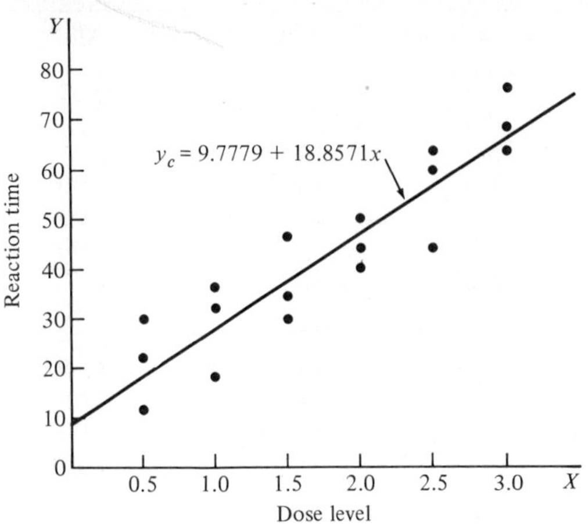

Figure 9.5 Scatter diagrams illustrating different types of relationships between X and Y

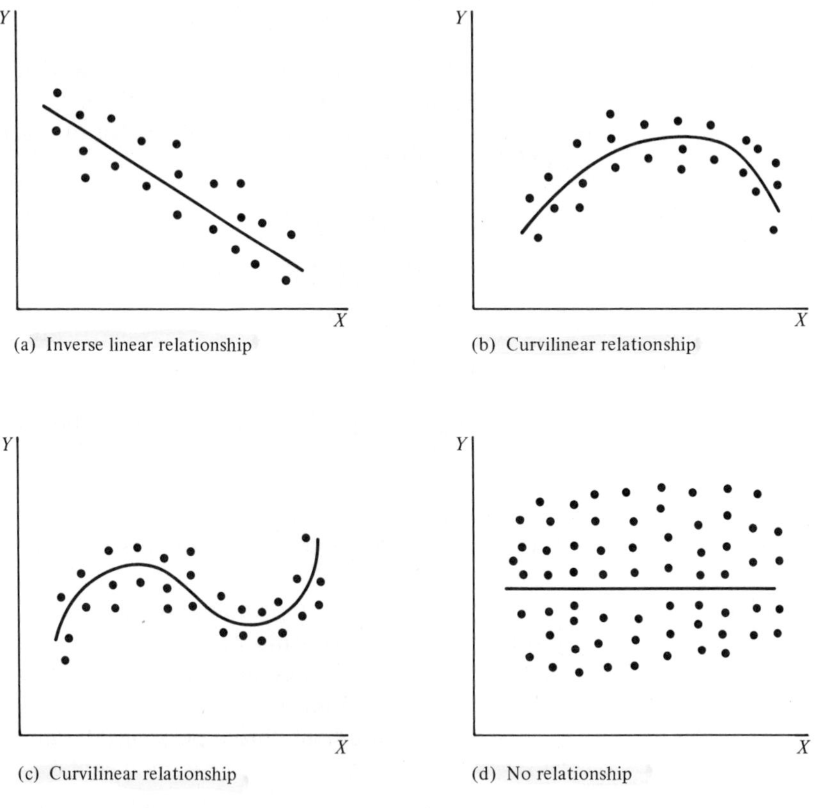

9.5(c) show scatter diagrams of curvilinear relationships. A straight line would provide a poor fit to these data, and one would therefore use a model other than that of Equation 9.1 as a basis for the analysis of the relationship between X and Y.

When large values of X are just as likely to be paired with large values of Y as with small values of Y, we say that there is no relationship (or at best a weak one) between X and Y. Figure 9.5(d) shows an example of a set of data for which X and Y are unrelated.

The calculations necessary to obtain a sample regression equation may become burdensome if they have to be made by hand or with a desk or pocket calculator. For those who have access to a computer, however, these calculations pose no particular problem, since "canned" programs are available which can perform all the calculations needed for complete regression analysis. The investigator with access to a computer can virtually forget about the calculations, and devote her or his attention to the collection of data and the interpretation of the results.

EXERCISES

For each of the following exercises,

(a) draw a scatter diagram of the data,

(b) determine the simple linear regression equation.

1 In a study of the relationship between job satisfaction and aptitude for the job, investigators collected data on ten professionals. Table 9.3 shows their job-satisfaction scores and the scores they made, when college freshmen, on an aptitude test for their profession.

Table 9.3 Data for Exercise 1

Job satisfaction score, *Y*	58	54	67	64	66	73	70	85	74	85
Aptitude score, *X*	50	55	60	65	70	75	80	85	90	95

2 Table 9.4 shows the weights of 11 female sheep and the weights of their mothers at the same age.

Table 9.4 Data for Exercise 2

Offspring's weight, *Y*	68	63	70	66	81	74	82	76	81	92	85
Mother's weight, *X*	60	64	68	72	76	80	84	88	92	96	100

3 Table 9.5 gives the IQs of 15 ghetto children and the scores they made on a personal-adjustment test. The data are coded for ease of calculation.

Table 9.5 Data for Exercise 3

Adjustment score, Y	4	5	4	6	5	7	8	9	13	11	15	14	13	16	17
IQ, X	−10	−8	−7	−5	−4	0	3	5	8	10	12	14	15	16	20

4 Table 9.6 shows the number of hours per week ten college sophomores spent studying and their cumulative grade point averages.

Table 9.6 Data for Exercise 4

Grade point average, Y	2.1	2.7	2.6	2.5	3.5	3.0	3.5	3.7	2.9	4.0
Hours of study, X	5	6	7	8	9	10	11	12	13	14

5 A random sample of 12 fifth-grade students were given two tests. One was designed to measure level of hostility, and the other was a reading-comprehension test. Table 9.7 shows the scores (coded).

Table 9.7 Data for Exercise 5

Reading comprehension score, Y	98	90	95	80	84	79	67	70	65	57	55	50
Hostility score, X	0	1	2	3	4	5	6	7	8	9	10	11

6 The data in Table 9.8 were collected on ten communities in a developing country.

Table 9.8 Data for Exercise 6

Incidence of goiter in population, Y (%)	60	75	50	55	45	33	50	52	35	30
Iodine content in water supply, X (μ g/liter)	2	2	3	5	7	8	8	8	10	12

7 Table 9.9 shows, for a random sample of 11 subjects, the length of time, in minutes, required for completion of a certain task and the number of minutes spent learning how to perform the task.

Table 9.9 Data for Exercise 7

Time required to complete task, Y	40	35	20	38	17	26	28	22	12	12	5
Time spent learning task, X	30	30	40	40	50	50	60	60	60	70	70

8 Researchers conducted an experiment in which each member of a random sample of 16 subjects was presented with a certain visual stimulus in the presence of one of various levels of noise. Table 9.10 shows the sub-

jects' reaction times to the visual stimulus at the various noise levels. (The data have been coded.)

Table 9.10 Data for Exercise 8

Reaction time, *Y*	5	8	5	3	7	5	8	6	7	12	8	7	10	7	8	10
Noise level, *X*	1	1	2	3	3	4	5	6	6	6	7	8	9	9	10	10

9 As part of a study conducted at a facility for the mentally retarded, researchers collected the data in Table 9.11 for a random sample of 15 young men between the ages of 20 and 25 who had been placed in jobs five years prior to the study.

Table 9.11 Data for Exercise 9

Number of months on job successfully completed, *Y*	7	14	25	20	33	31	37	46	45	49	54	58	60	60	60
IQ, *X*	50	50	55	60	60	65	65	65	70	70	70	70	75	75	75

10 Table 9.12 shows the data reported by a random sample of ten schools for exceptional children.

Table 9.12 Data for Exercise 10

Students withdrawing prior to completion of prescribed course, *Y* (%)	3	5	6	5	8	8	10	12	12	15
Pupil/teacher ratio, *X*	10	10	12	15	15	20	20	22	25	25

11 A first-grade teacher wished to know whether it was possible to use the scores children made on a reading-readiness test at the beginning of first grade (the independent variable) to predict the children's reading performance scores at the end of first grade (the dependent variable). A child was selected at random from all children with reading-readiness scores of 50. Another was selected from all those with scores of 55. And so on, until a sample of 11 children had been selected in such a way that every score between 50 and 100 that was a multiple of 5 was represented. The end-of-the-year reading performance score for each subject was also recorded. Table 9.13 shows the results.

Table 9.13 Data for Exercise 11

Child	1	2	3	4	5	6	7	8	9	10	11
Reading readiness score, *X*	50	55	60	65	70	75	80	85	90	95	100
Reading performance score, *Y*	65	72	68	83	77	79	92	88	98	94	100

9.3 EVALUATING THE REGRESSION EQUATION

As we shall explain in greater detail in Section 9.4, one of the primary objectives of regression analysis is to obtain an equation that will enable us to predict the value that Y is likely to assume for a particular value of X. In Example 9.1, for instance, we shall want to be able to predict the reaction time of a subject who is given a particular dose of the drug under consideration. The least-squares equation obtained in Section 9.2 is the one used for this purpose. Before using the regression equation for prediction, as well as for other purposes, we shall want to know how useful we can expect it to be in achieving these objectives. In other words, we want to know how accurately the equation will predict the value of Y for a particular value of X. One of our objectives in this section is to learn how to evaluate the sample regression equation in order to answer that question.

When one wishes to predict the value that some variable Y is likely to assume, in the absence of any consideration of some possibly related variable X, the most likely candidate is the mean of the available observations on Y. In Example 9.1, for instance, suppose that only the reaction times were available, and that we wished to predict the reaction time of some new individual taking the drug. In the absence of knowledge of the dosage level, the best we could do to predict Y would be to use the sample mean $\bar{y} = 42.7778$. If the values of Y are closely concentrated about $\bar{y}$, this may not be a bad choice. However, if the scatter of values about $\bar{y}$ is great, it can be a poor choice.

Suppose there is a direct linear relationship between dose level and reaction time, as suggested by the scatter diagram of Figure 9.1. And suppose we know the dose received by the new individual. If we take these facts into consideration, it seems reasonable to expect an improvement in our prediction. Since we have the equation $y_c = b_0 + b_1x$ and we know the dose received by the new subject, we would substitute this value of X into the equation and use y_c as our prediction. The extent to which this predicted value is an improvement over $\bar{y}$ depends on how much more closely the values of Y are concentrated about y_c than about $\bar{y}$. In the situation in which all the points in a scatter diagram fall on the regression line, any observed value of Y and the corresponding value of y_c coincide, and there is no scatter about y_c. Figure 9.6 shows examples of this situation.

The scatter diagrams in this figure depict situations in which the scatter of Y values about the regression line is zero, and hence is less than the scatter of the points about $\bar{y}$. In these cases y_c would be a more desirable predictor of reaction time than $\bar{y}$.

At the other extreme, consider the case, illustrated by the scatter diagram in Figure 9.5(d), in which there is no relationship between X and Y. Here the regression line and the line drawn through $\bar{y}$ are the same (that is, $y_c = \bar{y}$ for all x), and the scatter of Y values about y_c is equal to their scatter

Figure 9.6 Examples of perfect linear relationships between X and Y

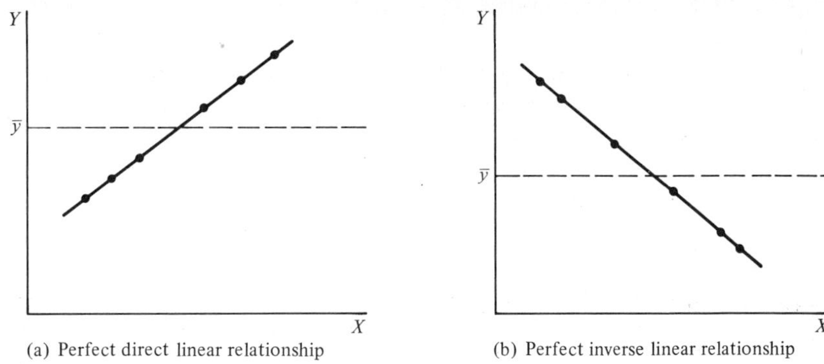

(a) Perfect direct linear relationship

(b) Perfect inverse linear relationship

about $\bar{y}$. In this case y_c has no advantage over $\bar{y}$ as a predictor, since they are equal.

In the typical situation the scatter of Y values about y_c is somewhere between these two extremes. Evaluation of the sample regression equation enables us to determine whether the measure of the scatter of Y values about the least-squares line is sufficiently close to zero to warrant its use in predicting the value Y is likely to assume for a given value of X. Taking measurements on X and determining the sample regression equation requires some expenditure of time and other resources. If the results do not enable us to obtain better estimates of Y, and to more satisfactorily achieve other objectives, we shall want to divert our resources to other activities. If, on the other hand, evaluation of the sample regression line indicates that it is useful in reaching the goals of regression analysis, it can be a powerful and valuable analytical tool.

The dispersion of Y values about $\bar{y}$ In Chapter 1 we learned that the scatter, or dispersion, of a set of values about their mean is usually measured by the variance and standard deviation. For a sample of n observations on the variable Y, the variance is given by

$$s^2 = \frac{\Sigma\,(y_i - \bar{y})^2}{n - 1} \tag{9.5}$$

where the denominator is the value for degrees of freedom and the numerator is the sum of squared deviations of the observations from their mean, or *sum of squares* for short. In the regression context, we call the numerator of Equation 9.5 the *total sum of squares*, and we call each individual deviation $y_i - \bar{y}$ a *total deviation*.

The dispersion of Y values about y_c In a similar manner, we can measure the dispersion of the Y values about the sample regression line by

$$s^2_{Y|x} = \frac{\Sigma (y_i - y_c)^2}{n-2} \tag{9.6}$$

where again the denominator is the value for degrees of freedom, and the numerator is called the *unexplained sum of squares*. We call an individual deviation $y_i - y_c$ an *unexplained deviation.*

This variance is an estimate of the variance of the subpopulation of Y values assumed to exist at each value of X. We call the numerator the unexplained sum of squares, since it is a measure of the variability present in a set of data that is not "explained" by the regression of Y on X. It is the unexplained sum of squares that is minimized when the least-squares method of fitting a line to sample data is used.

The explained sum of squares Instead of using variances (sums of squares divided by appropriate degrees of freedom) to measure variability, we may use the sums of squares (numerators of variances) for this purpose. Thus we can think of the total sum of squares as a measure of the total variability present in a set of data, and the unexplained sum of squares as a measure of the variability about the regression line.

Unlike variances, the sums of squares may be added and subtracted (that is, they may be partitioned into subcomponents, as explained in Chapter 8). For example, we may subtract the unexplained sum of squares in our regression analysis from the total sum of squares to give a sum-of-squares component that we call the *explained sum of squares.* We say that the difference between the total sum of squares and the unexplained sum of squares represents that part of the total variability that is "explained" by the regression of Y on X. We may write the explained sum of squares as $\Sigma (y_c - \bar{y})^2$. We see that the individual deviation $y_c - \bar{y}$ is the deviation of the computed value of Y (using the sample regression equation) from the mean of the sample Y values.

We may summarize the relationship between the various sum-of-squares components as follows.

Total sum of squares = explained sum of squares + unexplained sum of squares.

Using symbols, we may write this equation as

$$\sum (y_i - \bar{y})^2 = \sum (y_c - \bar{y})^2 + \sum (y_i - y_c)^2 \tag{9.7}$$

In the discussion that follows, it will be convenient for us to use a notation system similar to that used in Chapter 8 for the analysis of variance. Thus we shall refer to the total sum of squares as SST. Since the explained sum of squares is that component that is "explained" by the regression, we may refer to it as the *sum of squares due to regression* and designate it by the symbol SSR. Since the unexplained sum of squares is the residual or

Figure 9.7 Scatter diagram showing total, explained, and unexplained deviation in regression analysis

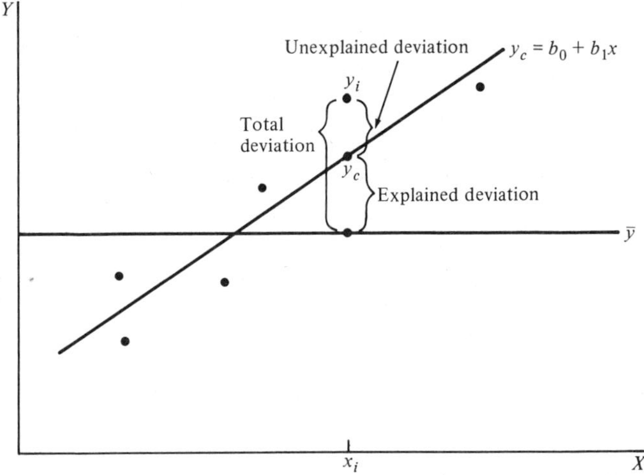

error sum of squares, we may use the symbol SSE to designate this component. Using this notation, we may write Equation 9.7 as

$$\text{SST} = \text{SSR} + \text{SSE}$$

We may express the relationship between the individual deviations that are squared and summed to obtain the sums of squares as

Total deviation = explained deviation + unexplained deviation

or

$$(y_i - \bar{y}) = (y_c - \bar{y}) + (y_i - y_c) \tag{9.8}$$

Figure 9.7 illustrates this relationship for a single observation of Y in a typical regression analysis.

The coefficient of determination It seems intuitive to measure the usefulness of an observed relationship between X and Y, as expressed by the regression equation, in terms of how much of the total variability in Y is "explained" by the regression of Y on X. It would seem that a regression equation based on a relationship in which the explained sum of squares constitutes a substantial proportion of the total sum of squares would be more useful than one based on a relationship in which the explained sum of squares is only a small proportion of the total sum of squares. To determine what proportion of the total sum of squares is explained by the regression of Y on X, we compute a measure known as the *coefficient of determination*, which is the ratio of the explained sum of squares to the total sum of squares. When we compute the coefficient of determination from sample data, we designate it by r^2 and obtain it by the following formula.

$$r^2 = \frac{\text{SSR}}{\text{SST}} = \frac{\Sigma\,(y_c - \bar{y})^2}{\Sigma\,(y_i - \bar{y})^2} \tag{9.9}$$

We may compute the numerator and denominator of Equation 9.9 from the following computational formulas.

$$\text{SSR} = \sum (y_c - \bar{y})^2 = b_1\left(\sum x_i y_i - \frac{\Sigma x_i\, \Sigma y_i}{n}\right)$$

$$\text{SST} = \sum (y_i - \bar{y})^2 = \sum y_i^2 - \frac{(\Sigma\, y_i)^2}{n}$$

The coefficient of determination can assume values between 0 and 1 inclusive. When there is no relationship between X and Y, their scatter diagram resembles that shown in Figure 9.5(d), and $r^2 = 0$. When X and Y are perfectly related, their scatter diagram resembles one of those shown in Figure 9.6, and $r^2 = 1$.

Let us illustrate the calculation of r^2 by referring again to Example 9.1, which deals with the analysis of the relationship between patients' reaction times and dose levels of a certain drug. First we use the data of Table 9.2 and the previously computed value of b_1 to obtain

$$\text{SSR} = 18.8571\left[1595 - \frac{(31.5)(770)}{18}\right] = 4667.13$$

$$\text{SST} = 38{,}508 - \frac{(770)^2}{18} = 5569.11$$

Using these results, we have

$$r^2 = \frac{4667.13}{5569.11} = 0.84$$

Thus we see that 84% of the total variability present in our sample data is explained by the regression.

Analysis of variance We may interpret r^2 as a measure of the strength of the linear relationship between the observed sample values of X and Y. Our primary interest, however, is in the true relationship between X and Y as it exists in the population. We therefore wonder whether the sample data provide sufficient evidence to indicate the existence of a linear relationship between X and Y in the population. In other words, we would like to know whether a sample value of r^2 as large as 0.84 could reasonably have come about as a result of chance alone, when X and Y are, in fact, not linearly related. Or is this too large a value of r^2 to be attributable to chance, so that we must give some other explanation, namely, that X and Y are linearly related?

When X and Y are not related, and when the assumptions given earlier are met, one can show that

$$\text{VR} = \frac{\text{SSR}/1}{\text{SSE}/(n-2)} = \frac{\text{MSR}}{\text{MSE}} \tag{9.10}$$

is distributed as F with 1 and $n-2$ degrees of freedom. In Equation 9.10, we call MSR = SSR/1 the *regression mean square,* and SSE/$(n-2)$ the *error mean square*. We see that VR is the ratio of two variances. The denominators 1 and $n-2$ are, respectively, the regression and error degrees of freedom. In general, in simple linear regression, the value of total degrees of freedom is equal to $n-1$, the value of regression degrees of freedom is equal to 1, and the value of error degrees of freedom is equal to $(n-1)-1=n-2$.

Equation 9.10 provides a test statistic for testing

H_0: X and Y are not linearly related

against the alternative

H_1: X and Y are linearly related

If the computed VR exceeds the tabulated F for 1 and $n-2$ degrees of freedom and the chosen level of significance α, we reject H_0, and conclude that X and Y are related.

We may summarize the calculations in an ANOVA table such as Table 9.14. Table 9.15 shows the ANOVA table for Example 9.1.

Since our computed VR of 82.79 is greater than the tabulated value of $F_{0.995,1,16} = 10.58$, we can reject the null hypothesis of no linear relationship between X and Y at the 0.001 level of significance. We conclude, then, that reaction time and dose level are linearly related.

We would use, with assurance, our sample regression equation to predict the value that Y is likely to assume for a given value of X, as well as for other purposes described in the sections that follow.

Table 9.14 ANOVA table for simple linear regression

Source of variation	SS	df	MS	VR
Regression	SSR	1	MSR = SSR/1	MSR/MSE
Error	SSE	$n-2$	MSE = SSE/$(n-2)$	
	SST	$n-1$		

Table 9.15 ANOVA table for Example 9.1

Source of variation	SS	df	MS	VR
Regression	4667.13	1	4667.13	82.79
Error	901.98	16	56.37	
	5569.11	17		

A hypothesis test for β_1

If, in some population of interest, X and Y are not linearly related, the slope of the population regression line, β_1, will be 0, and the means of the subpopulations of Y will all be equal. The scatter diagram of such a finite population might resemble Figure 9.5(d). On the other hand, if β_1 is not 0, there will be some relationship between X and Y, and the scatter diagram will tend to look more like Figure 9.1 or Figure 9.5(a).

Consequently, in evaluating a sample regression equation, we may, as an alternative to the analysis-of-variance procedure just described, employ a procedure based directly on the slope of the line. Specifically, if we can reject the null hypothesis that $\beta_1 = 0$, we can conclude that X and Y are linearly related.

If the underlying assumptions given in Section 9.1 are met, we can show that the sampling distribution of the sample slope b_1 is normally distributed with a mean of β_1 and a variance of

$$\sigma_{b_1}^2 = \frac{\sigma_e^2}{\Sigma (x_i - \bar{x})^2} \tag{9.11}$$

where σ_e^2 is the population error variance, that is, the common variance to which the Y subpopulation variances are assumed to be equal. The appropriate test statistic for testing

$$H_0\colon \beta_1 = \beta_{1_0}$$

against the alternative

$$H_1\colon \beta_1 \neq \beta_{1_0}$$

is

$$z = \frac{b_1 - \beta_{1_0}}{\sqrt{\sigma_e^2 / \Sigma (x_i - \bar{x})^2}}$$

which follows the standard normal distribution. In most practical situations σ_e^2 will be unknown, but we can estimate it by the sample error mean square, MSE. The test statistic then becomes

$$t = \frac{b_1 - \beta_{1_0}}{\sqrt{\text{MSE} / \Sigma (x_i - \bar{x})^2}} \tag{9.12}$$

which is distributed as Student's t distribution with $n - 2$ degrees of freedom.

To test $H_0\colon \beta_1 = 0$, we substitute 0 for β_{1_0} in Equation 9.12. If the computed t falls in the rejection region determined by the tabulated value of t (two-sided test) for $n - 2$ degrees of freedom and the chosen level of significance, we reject H_0, and conclude that X and Y are linearly related.

Let us illustrate the procedure by using the data of Example 9.1 to test

$$H_0\colon \beta_1 = 0$$

against the alternative

$$H_1\colon \beta_1 \neq 0$$

We can compute the quantity $\Sigma\,(x_i - \bar{x})^2$ needed to evaluate Equation 9.12 most conveniently by the formula

$$\sum (x_i - \bar{x})^2 = \sum x_i^2 - \frac{(\Sigma\, x_i)^2}{n}$$

For our example, we use data from Table 9.2 to obtain

$$\sum (x_i - \bar{x})^2 = 68.25 - \frac{(31.5)^2}{18} = 13.125$$

From Table 9.15, we obtain MSE = 56.37. We may now compute, by Equation 9.11,

$$t = \frac{18.8571 - 0}{\sqrt{56.37/13.125}} = 9.099$$

Reference to Appendix Table F indicates that we can reject H_0 at the 0.001 level of significance, and we conclude that X and Y are linearly related. This conclusion, as will always be the case in simple linear regression, is identical to that reached through analysis of variance. In other words, the two procedures are equivalent for testing $H_0{:}\beta_1 = 0$ against $H_1{:}\beta_1 \neq 0$. Note that the computed VR of 82.79 is equal to $(9.099)^2$, the square of the computed t. Also note that all tabulated values of F for 1 and $n-2$ degrees of freedom and for the α significance level are the squares of tabulated t values for $1 - \alpha/2$ and $n - 2$ degrees of freedom. For example,

$$F_{0.95,1,10} = 4.96 = [t_{0.975,10}]^2 = (2.2281)^2$$

Although the t test and analysis of variance are equivalent, for testing $H_0\colon \beta_1 = 0$ against $H_1\colon \beta_1 \neq 0$, the t test has the advantage that it can be used for the one-sided alternatives $\beta_1 < 0$ and $\beta_1 > 0$, and analysis of variance cannot.

A confidence interval for β_1

As we have already noted, b, the slope of the sample regression line, indicates the amount by which y_c changes for each unit change in x. When b_1 is positive (that is, when the observed relationship between sample values of X and Y is direct), a unit increase in x will cause y_c to be increased by an amount equal to b, and a unit decrease in x will cause y_c to decrease by an amount equal to b_1. When the observed relationship between sample values of X and Y is inverse, b_1 is negative, so a unit increase in x is accompanied by a decrease in y_c, and a unit decrease in x results in an increase in y_c—each by an amount equal to the numerical value of b_1. It is frequently worthwhile to obtain an interval estimate of β_1, the parameter of which b_1 is a point estimate. As the slope of the population regression line, β_1 tells us the amount by which $\mu_{Y|x}$ changes for a unit increase in X. Confidence intervals for β_1 allow us to draw conclusions about the amount by which, on the average, the dependent variable changes for each unit change in the independent variable. In the experiment in Example 9.1, for instance, a confidence interval for β_1 will provide us with information on the amount by which, on the average, reaction time increases for each unit increase in drug dose.

To construct a $100(1-\alpha)\%$ confidence interval for β_1, we apply the general formula for a confidence interval:

Estimator $\pm$ (reliability factor)(standard error of the estimator)

The estimator is b_1, and its standard error is given by the square root of Equation 9.11. In most cases σ_e^2 will be unknown, and we may estimate the standard error of b_1 by

$$s_{b_1} = \sqrt{\frac{\text{MSE}}{\Sigma\,(x_i - \bar{x})^2}}$$

which we have already encountered as the denominator of Equation 9.12. When the previously stated assumptions of the model are met, the reliability factor will be $t_{1-\alpha/2,n-2}$. We may then write the $100(1-\alpha)\%$ confidence interval for β_1 as

$$b_1 \pm t_{1-\alpha/2,n-2}\sqrt{\frac{\text{MSE}}{\Sigma\,(x_i - \bar{x})^2}}$$

or, alternatively,

$$C\left(b_1 - t_{1-\alpha/2,n-2}\sqrt{\frac{\text{MSE}}{\Sigma\,(x_i - \bar{x})^2}} \le \beta_1 \le b_1 + t_{1-\alpha/2,n-2}\sqrt{\frac{\text{MSE}}{\Sigma\,(x_i - \bar{x})^2}}\right) = 1 - \alpha$$

Let us now construct a 95% confidence interval for β_1 for Example 9.1. From quantities already computed, $t_{0.975,16} = 2.1199$, and we have

$$C\left(18.8571 - 2.1199\sqrt{\frac{56.37}{13.125}} \le \beta_1 \le 18.8571 + 2.1199\sqrt{\frac{56.37}{13.125}}\right) = 0.95$$

$$C(18.8571 - 4.3933 \le \beta_1 \le 18.8571 + 4.3933) = 0.95$$

$$C(14.4638 \le \beta_1 \le 23.2504) = 0.95$$

Thus we are 95% confident that the mean amount of increase in reaction time for each increase of 1 mg in the drug dose is somewhere between 14.4638 and 23.2504 msec.

The confidence interval for β_1 as a hypothesis test As we have seen in previous chapters, we can use a confidence interval for a population parameter to test hypotheses about that parameter. In the present example, we may use our 95% confidence interval for β_1 to test H_0: $\beta_1 = 0$ against H_1: $\beta_1 \neq 0$ at the 0.05 level of significance. Since the interval 14.4638–23.2504 does not contain 0, we conclude that 0 is not a candidate for the parameter being estimated, and we reject H_0. This, of course, is the same decision that we reached by means of the hypothesis-testing procedure used earlier.

EXERCISES

12 Referring to Exercise 1, (a) compute r^2 and perform the analysis of variance, (b) test H_0: $\beta_1 \neq 0$ using the t statistic and $\alpha = 0.05$, (c) construct the 95% confidence interval for β_1.

13 Refer to Exercise 2 and proceed as in Exercise 12.
14 Refer to Exercise 3 and proceed as in Exercise 12.
15 Refer to Exercise 4 and proceed as in Exercise 12.
16 Refer to Exercise 5 and proceed as in Exercise 12.
17 Refer to Exercise 6 and proceed as in Exercise 12.
18 Refer to Exercise 7 and proceed as in Exercise 12.
19 Refer to Exercise 8 and proceed as in Exercise 12.
20 Refer to Exercise 9 and proceed as in Exercise 12.
21 Refer to Exercise 10 and proceed as in Exercise 12.
22 Refer to Exercise 11 and proceed as in Exercise 12.

9.4 USING THE SIMPLE LINEAR REGRESSION EQUATION

If, after following the procedures of Section 9.3, we are unable to reject H_0: $\beta_1 = 0$, and therefore cannot conclude that there is a linear relationship between X and Y, we cannot use the sample regression equation with assurance. If, on the other hand, we reject H_0: $\beta_1 = 0$, we can use the sample regression equation with assurance for the following two purposes.

1 To predict what value Y is likely to assume for a particular value of X
2 To estimate the mean of the subpopulation of Y values presumed to exist at a particular value of X

When, for the model, the assumptions given in Section 9.1 are met, we may construct $100(1 - \alpha)\%$ *prediction intervals* and $100(1 - \alpha)\%$ *confidence intervals* for Y and the mean of the subpopulation of Y values, respectively.

Predicting Y for a particular X

Once we have deemed a sample regression equation useful on the basis of the analysis described in Section 9.3, we can use it to answer the following question. Given a new subject whose score on the independent variable X is some particular value, say x_p, what value is Y likely to assume? Consider, for example, the reaction time–drug dose example. Suppose a new schizophrenic patient admitted to the hospital is given 2 mg of the drug. What is this person's reaction time likely to be? We may obtain a point prediction by substituting $x_p = 2$ into our simple linear regression equation as follows.

$$y_c = 9.7779 + 18.8571(2) = 47.4921$$

When the previously stated assumptions are met, and when σ_e^2 is unknown, we may obtain a $100(1-\alpha)\%$ prediction interval by evaluating the following formula.

$$y_c \pm t_{1-\alpha/2, n-2} \sqrt{\text{MSE}} \sqrt{1 + \frac{1}{n} + \frac{(x_p - \bar{x})^2}{\Sigma (x_i - \bar{x})^2}} \tag{9.13}$$

where y_c is the point prediction, $t_{1-\alpha/2, n-2}$ is the tabulated value of t appropriate for the selected α and the available degrees of freedom, and

$$\sqrt{\text{MSE}} \sqrt{1 + \frac{1}{n} + \frac{(x_p - \bar{x})^2}{\Sigma (x_i - \bar{x})^2}}$$

is the standard error of the prediction.

To construct a 95% prediction interval for Y when $x_p = 2$, we make the following substitutions into Equation 9.13.

$$47.4921 \pm (2.1199)\sqrt{56.37}\sqrt{1 + \frac{1}{18} + \frac{(2-1.75)^2}{13.125}}$$

$$= 47.4921 \pm (2.1199)(7.51)(1.0297)$$

$$= 47.4921 \pm 16.3933 = 31.099 \quad \text{and} \quad 63.885$$

We are 95% confident that the reaction time of a schizophrenic person receiving 2 mg of the drug will be somewhere between 31 and 64 msec. We are able to make this statement because of the fact that if, in repetitions of the experiment, a line is fitted in the manner described above, about 95% of the prediction intervals that could be constructed for a person for whom $x_p = 2$ would include that person's true reaction time.

You should realize that here we are predicting the person's reaction time prior to administration of the drug. To find out for certain a person's reaction time relative to a certain dose of the drug, we would have to give the patient the drug and then measure his or her reaction time.

Estimating subpopulation means

When a particular value of the independent variable prevails, we can use the sample regression equation to answer a different question regarding the dependent variable. We may state this question as follows. Given a population of subjects, all of whom have a particular X score of, say x_p, what is the mean value of the dependent variable Y for this population likely to be?

To obtain a point estimate of the subpopulation of Y values for a particular value of X, we proceed in the same manner as we do to obtain a point prediction, and substitute x_p into the sample regression equation.

Suppose, for example, we have a population of schizophrenic persons, all of whom are given 2 mg of the drug discussed in Example 9.1. What can we expect the mean reaction time of this population to be? We may state the question in perhaps a more realistic manner as follows. Of all schizophrenic persons who will ever receive 2 mg of the drug, what is the mean reaction time likely to be?

Substituting 2 into the sample regression equation gives

$$y_c = 9.7779 + 18.8571(2) = 47.4921$$

Although the point prediction of Y and the point estimate of the mean of a subpopulation of Y are identical, we shall soon see that the prediction interval is not the same as the confidence interval for the subpopulation mean.

When the assumptions mentioned earlier are met, and when σ_e^2 is unknown, a $100(1 - \alpha)\%$ confidence interval for $\mu_{Y|x_p}$, the mean of the subpopulation of Y values when $X = x_p$, is given by

$$y_c \pm t_{1-\alpha/2,n-2} \sqrt{\text{MSE}} \sqrt{\frac{1}{n} + \frac{(x_p - \bar{x})^2}{\Sigma (x_i - \bar{x})^2}} \tag{9.14}$$

To construct a 95% confidence interval for $\mu_{Y|2}$ in the reaction time–drug dose example, we make the following substitutions into Equation 9.14.

$$47.4921 \pm (2.1199)\sqrt{56.37}\sqrt{\frac{1}{18} + \frac{(2 - 1.75)^2}{13.125}}$$

$$= 47.4921 \pm (2.1199)(7.51)(0.2456)$$

$$= 47.4921 \pm 3.9101 = 43.582 \quad \text{and} \quad 51.402$$

We are 95% confident that the mean reaction time of a population of schizophrenic persons, each receiving 2 mg of the drug, will be somewhere between 44 and 51 msec. We are able to make this statement because we know that if the experiment described in the example were repeated many times, about 95% of the confidence intervals computed in the manner just described would include $\mu_{Y|2}$.

You should reflect on the difference between the two situations described here. In the first case, we are interested in predicting the reaction time of a single individual. In the second case, we are estimating the mean of a whole population of individuals. The difference between the two is reflected in the formulas for the standard errors.

EXERCISES

For each of the following exercises, construct a prediction interval for Y and a confidence interval for $\mu_{Y|x_p}$. Let $\alpha = 0.05$.

23 Refer to Exercise 1 and let $x_p = 70$.
24 Refer to Exercise 2 and let $x_p = 70$.
25 Refer to Exercise 3 and let $x_p = 10$.
26 Refer to Exercise 4 and let $x_p = 10$.
27 Refer to Exercise 5 and let $x_p = 10$.
28 Refer to Exercise 6 and let $x_p = 10$.
29 Refer to Exercise 7 and let $x_p = 50$.
30 Refer to Exercise 8 and let $x_p = 5$.
31 Refer to Exercise 9 and let $x_p = 60$.
32 Refer to Exercise 10 and let $x_p = 20$.
33 Refer to Exercise 11 and let $x_p = 70$.

9.5 THE BIVARIATE MODEL

Frequently interest in the relationship between two variables X and Y centers on determining whether or not they are related, and, if so, how strongly. The investigator may or may not be interested in prediction and estimation. The appropriate analytic technique in this situation is *correlation analysis*. In contrast to regression analysis, correlation analysis requires that X, as well as Y, be a random variable. You will recall that in regression analysis X may be either random or fixed.

In the typical situation in which correlation analysis is used, the investigator draws a random sample of units of association (which may be humans, animals, places, things, points in time, and so on) from the population of interest, and takes two measurements, an X measurement and a Y measurement, on each unit of association in the sample. The investigator takes whatever values of X (as well as Y) that the sample yields. No attempt is made to limit the analysis to preselected values of X, as may be the case in regression analysis. The following example illustrates the type of problem that is amenable to correlation analysis.

Example 9.2 In a study designed to investigate the relationship between creativity and other variables, a team of psychologists gave a random sample of 20 tenth-

Figure 9.8 Scatter diagram for Example 9.2

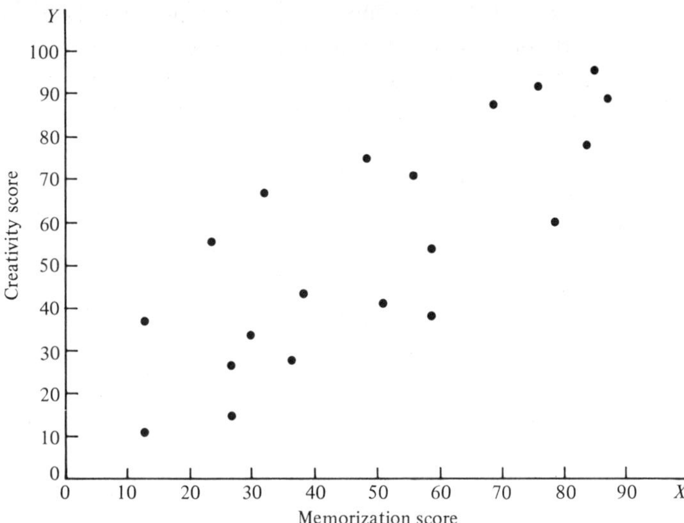

grade students a test to measure level of creativity. The investigators then asked each student to memorize a short poem. Table 9.16 shows the scores made by the students on the creativity test and scores indicating their ability to memorize the poem. Figure 9.8 shows the scatter diagram for these data.

We call a population of X and Y values, in which both X and Y are random variables, a *bivariate distribution.* A bivariate distribution is a special case (in that only two variables are involved) of a *joint distribution.* A joint distribution is a distribution in which two or more variables, say X and Y, vary together. The distribution of a single variable, say Y, is completely described by two parameters, the mean μ_Y and the variance σ_Y^2. A bivariate

Table 9.16 Creativity and memorization scores for Example 9.2

Student	Creativity score, Y	Memorization score, X	Student	Creativity score, Y	Memorization score, X
1	11	13	11	78	84
2	96	85	12	27	27
3	15	27	13	71	56
4	88	69	14	75	49
5	92	76	15	89	88
6	34	30	16	60	79
7	44	39	17	41	51
8	67	32	18	28	37
9	37	13	19	56	24
10	38	58	20	54	59

distribution, say of the variables X and Y, requires five parameters for complete characterization: μ_X, μ_Y, σ_X^2, σ_Y^2, and ρ. Only the last of these parameters, ρ, is new to you. We call the parameter ρ the population *correlation coefficient* and define it as

$$\rho = \sqrt{\frac{\sigma_{XY}^2}{\sigma_X^2 \sigma_Y^2}} = \frac{\sigma_{XY}}{\sigma_X \sigma_Y}$$

where σ_{XY}^2, called the *covariance* between X and Y, is a measure of the covariability between the two variables. We shall have more to say about the correlation coefficient, but let us first consider an important model and the assumptions on which it is based.

The correlation model

We wish now to consider a model of a bivariate distribution. Since certain correlation techniques are appropriate in analyzing data conforming to this model, we shall refer to it as the *correlation model*. More specifically, we may call it the *simple linear correlation model*, since only two variables are involved and we assume that they are related in a linear fashion. We may specify the model symbolically as follows.

$$y_i = \beta_0 + \beta_1 x_i + \xi_i \tag{9.15}$$

where y_i is a typical value of Y, β_0 and β_1 are population parameters, x_i is the ith value of the random variable X, and ξ_i is a random error term.

The parameters β_0 and β_1 have the same interpretations they had in the discussion of regression analysis. When analysis is confined to correlation, however, one does not emphasize these interpretations, since in correlation analysis ρ provides the desired information.

In correlation analysis, we do not distinguish the variables X and Y on the basis of dependence and independence, as we did in regression analysis. The two variables have equal status in this regard. In fact, under the correlation model, we may interchange the positions of X and Y in Equation 9.15 to obtain the following alternative model for describing the relationship between these two variables.

$$x_i = \beta_0' + \beta_1' y_i + \xi_i' \tag{9.16}$$

where the primes (′) indicate that the coefficients in Equation 9.16 are not, in general, numerically equal to those in Equation 9.15.

We may therefore use sample data from a bivariate population to obtain a sample regression line either of Y on X or of X on Y. In general, these two lines will not coincide. Figure 9.9 shows a scatter diagram of sample data from a bivariate population and the two regression lines that one can obtain from the data.

Figure 9.9 Scatter diagram of sample data from a bivariate distribution, showing regression lines of Y on X and X on Y

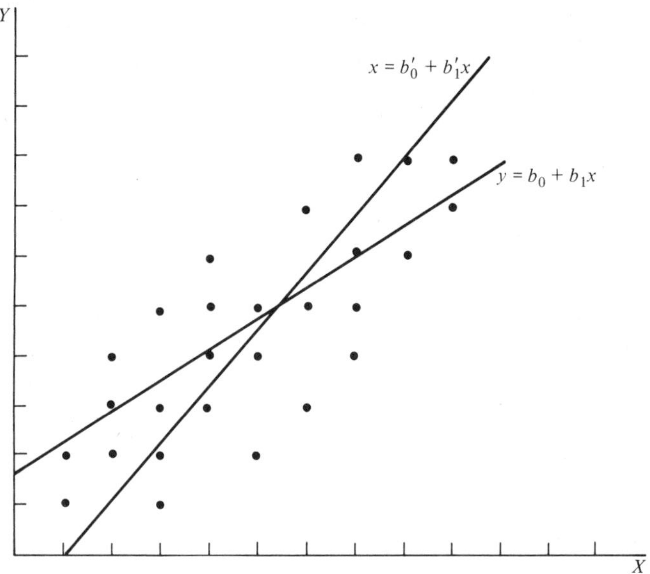

ınderlying the correlation model

Typically, it will be either impossible or impractical to examine, in its entirety, a bivariate population of interest in order to study the relationship between the two variables X and Y. One may, however, infer conclusions regarding the population from a sample drawn from the population. In order for these inferences to be valid, the following assumptions must hold.

1 The joint distribution of X and Y is normal. We call this distribution a *bivariate normal distribution*.

2 For each value of X, there is a normally distributed subpopulation of Y values.

3 The subpopulations of Y values all have the same variance.

4 The means of the subpopulations all lie on the same straight line.

5 For each value of Y, there is a subpopulation of X values that is normally distributed.

6 The subpopulations of X values all have the same variance.

7 The means of the subpopulations of X values all lie on the same straight line.

Figure 9.10 Three views of a bivariate normal distribution

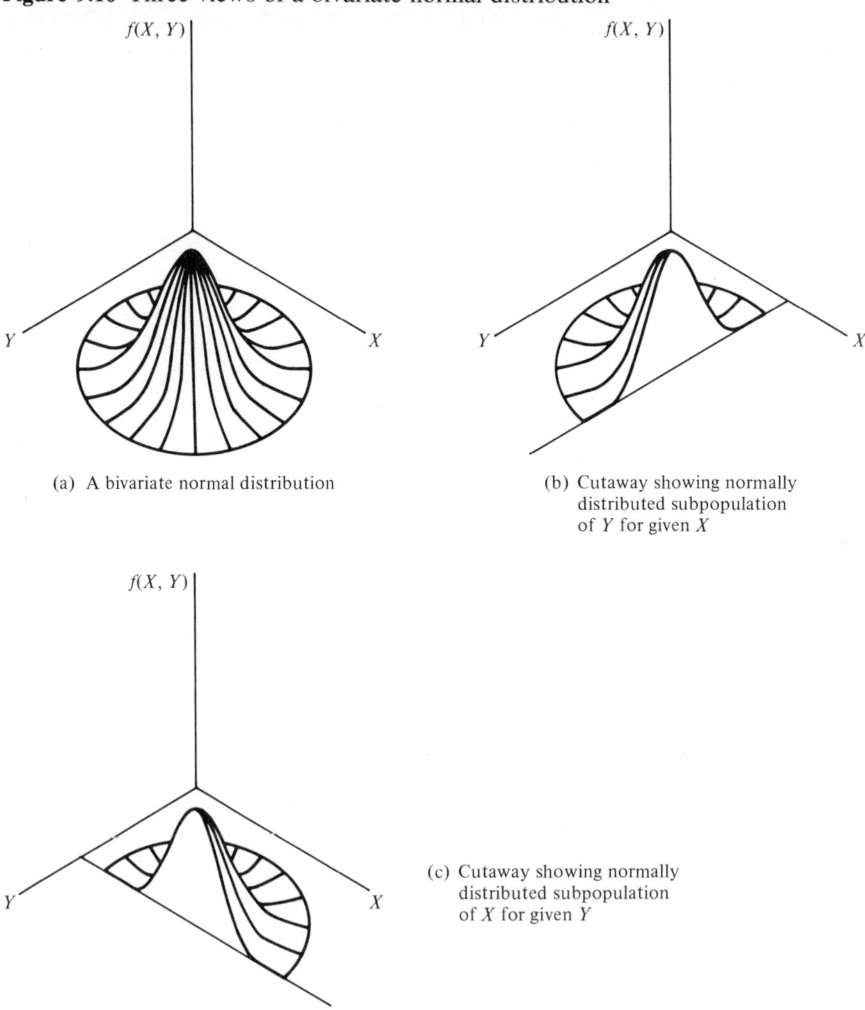

Figure 9.10(a) is a graphic representation of a bivariate normal distribution. In Figure 9.10(b) we see that a slice through the bivariate normal distribution parallel to the Y axis reveals the normally distributed subpopulation of Y values for a given value of X. In Figure 9.10(c) we see that a slice parallel to the X axis reveals the normally distributed subpopulation of X values for a given Y.

The correlation coefficient

The correlation coefficient ρ, mentioned earlier, is the parameter of primary interest in correlation. This parameter is a measure of the correlation, or linear relationship between two variables, that exists in a bivariate population. It can assume values between -1 and $+1$. If the relationship between two variables is perfectly linear and inverse, $\rho = -1$. The scatter diagram of a set of data in which the correlation coefficient is -1 would show all observations falling on a straight line, as in Figure 9.6(b). Figure 9.6(a) is a scatter diagram of data in which $\rho = +1$. Here we note that the relationship between the two variables is direct. When two variables are not correlated, $\rho = 0$. Figure 9.5(d) is a scatter diagram of data in which ρ is equal to or near 0. The correlation coefficient between two variables X and Y will always have the same sign as the slope of the regression equation of Y on X. We must surmise from these facts that the closer the numerical value of the correlation coefficient is to $+1$, or -1, the stronger the relationship between X and Y. The magnitude of ρ, then, gives us an indication of the strength of the relationship between two variables.

Estimating ρ from sample data Rarely do we know ρ. We may, however, estimate ρ from the data of a random sample of the population of interest. The point estimate of ρ is r, the sample correlation coefficient, given by

$$r = \frac{\Sigma (x_i - \bar{x})(y_i - \bar{y})}{\sqrt{\Sigma (x_i - \bar{x})^2}\sqrt{\Sigma (y_i - \bar{y})^2}} \tag{9.17}$$

Appropriate algebraic manipulation of Equation 9.17 leads to the following formula, which is computationally more convenient.

$$r = \frac{n \Sigma x_i y_i - (\Sigma x_i)(\Sigma y_i)}{\sqrt{n \Sigma x_i^2 - (\Sigma x_i)^2}\sqrt{n \Sigma y_i^2 - (\Sigma y_i)^2}} \tag{9.18}$$

To illustrate the calculation of the sample correlation coefficient, we again consider Example 9.2.

From the data in Table 9.16, we compute

$$\sum x_i y_i = 64{,}498$$

$$\sum x_i = 996$$

$$\sum y_i = 1101$$

$$\sum x_i^2 = 60{,}972$$

$$\sum y_i^2 = 73{,}681$$

When we substitute these quantities into Equation 9.18, we have

$$r = \frac{(20)(64{,}498) - (996)(1101)}{\sqrt{(20)(60{,}972) - (996)^2}\sqrt{(20)(73{,}681) - (1101)^2}} = 0.79$$

A confidence interval for ρ If the assumptions for a bivariate normal distribution are met, we may construct a confidence interval for ρ. We accomplish this by means of the following statistic, attributable to Fisher (1).

$$z_r = \frac{1}{2} \log_e \left(\frac{1+r}{1-r}\right)$$

This equation is known as *Fisher's z transformation*. One can show that z_r is approximately normally distributed with a mean of approximately $z_\rho = \frac{1}{2} \log_e [(1+\rho)/(1-\rho)]$ and an approximate variance of $1/\sqrt{n-3}$, regardless of the value of ρ. Appendix Table L gives values of z_r.

To construct a $100(1 - \alpha)\%$ confidence interval for ρ, proceed as follows.

1 Convert the observed r to z_r by means of Appendix Table L.

2 Construct a confidence interval for z_ρ by the following formula.

$$z_r \pm z_{1-\alpha/2} \frac{1}{\sqrt{n-3}}$$

3 Reconvert the resulting limits to values of r to obtain a $100(1-\alpha)\%$ confidence interval for ρ.

Let us illustrate the procedure by referring again to Example 9.2, for which $n = 20$ and $r = 0.79$. Suppose we construct a 95% confidence interval for ρ.

1 Reference to Appendix Table L shows that when $r = 0.79$, $z_r = 1.07143$

2 The 95% confidence interval for z_ρ, then, is

$$1.07143 \pm 1.96 \frac{1}{\sqrt{20-3}} = 1.07143 \pm 0.47537 = 0.59606 \text{ and } 1.54680$$

3 Again referring to Appendix Table L, we find that when $z_r = 0.59606$, r is about 0.534, and when $z_r = 1.54680$, r is approximately 0.913.

We are therefore 95% confident that ρ is somewhere between 0.53 and 0.91.

Testing H_0: $\rho = 0$ As we have noted, if two variables are not correlated, ρ will equal 0. It will be of interest, then, to be able to decide whether or not sample data are likely to have come from a population in which $\rho = 0$.

If the assumptions stated earlier are met, and if $\rho = 0$, we can show that

$$t = \frac{r\sqrt{n-2}}{\sqrt{1-r^2}}$$

follows the distribution of Student's t statistic with $n-2$ degrees of freedom. Consequently, if the absolute value of t exceeds the tabulated t for α and $n-2$, we may reject H_0: $\rho = 0$, at significance level α, in favor of H_1: $\rho \neq 0$. We may also perform one-sided tests.

For Example 9.2, let us test H_0: $\rho = 0$ against H_1: $\rho \neq 0$ at the 0.05 level of significance. We compute

$$t = \frac{0.79\sqrt{20-2}}{\sqrt{1-0.6241}} = 5.467$$

Since 5.467 is greater than $t_{0.975,18} = 2.1009$, we reject H_0, and conclude that $\rho \neq 0$, which indicates that creativity and memorization ability are correlated. We may draw this same conclusion by noting that the confidence interval for ρ computed earlier does not include 0.

Testing H_0: $\rho = \rho_0$, where $\rho_0 \neq 0$ The test that we have just demonstrated is valid only for testing H_0: $\rho = 0$. We may, however, test H_0: $\rho = \rho_0$ for any value of ρ_0 other than zero by means of the Fisher z transformation mentioned earlier. The test statistic is

$$Z = \frac{z_r - z_{\rho_0}}{1/\sqrt{n-3}}$$

which is distributed approximately as the standard normal. If the computed value of Z either is less than or equal to $-z$ or is greater than or equal to z, where z is the standard normal variable corresponding to an α significance level, we may reject H_0: $\rho = \rho_0$ in favor of H_1: $\rho \neq \rho_0$.

For Example 9.2, let us test H_0: $\rho = 0.90$ against H_1: $\rho \neq 0.90$ at the 0.05 level of significance. Referring to Appendix Table L, we have

$$Z = \frac{1.07143 - 1.47222}{1/\sqrt{20-3}} = -1.65$$

Since -1.65 is greater than -1.96, we cannot reject H_0 at the 0.05 level of significance, and we conclude that ρ may be equal to 0.90.

The use of Fisher's z transformation is not generally recommended unless n is 25 or larger. We have used the procedure here with a sample of size 20 for illustrative purposes. For small samples greater than 10, we may use a transformation proposed by Hotelling (2). This technique requires the following transformation of r.

$$z^* = z_r - \frac{3z_r + r}{4n}$$

The standard deviation of z^* is

$$\sigma_z^* = \frac{1}{\sqrt{n-1}}$$

The test statistic is

$$Z^* = \frac{z^* - \zeta^*}{1/\sqrt{n-1}} = (z^* - \zeta^*)\sqrt{n-1} \qquad \text{where } \zeta^* = z_\rho - \frac{3z_\rho + \rho}{4n}$$

(The symbol ζ is the Greek letter zeta.) Computed values of Z^* are compared for significance against tabulated values of the standard normal variable.

Using the Hotelling transformation with the data of Example 9.2 to test H_0: $\rho = 0.90$ against H_1: $\rho \neq 0.90$, we have

$$z^* = 1.07143 - \frac{3(1.07143) + 0.79}{4(20)} = 1.02138$$

and

$$\zeta^* = 1.47222 - \frac{3(1.47222) + 0.90}{4(20)} = 1.40576$$

Finally we compute

$$Z^* = (1.02138 - 1.40576)\sqrt{19} = -1.675$$

Since -1.675 is greater than -1.96, we cannot reject H_0 at the 0.05 level of significance. We note that this is the same conclusion we reached when we used Fisher's z transformation.

EXERCISES

In each of the following exercises, (a) prepare a scatter diagram, (b) compute the sample correlation coefficient, (c) test H_0: $\rho=0$ at the 0.05 level of significance and state your conclusions, (d) construct the 95% confidence interval for ρ.

34 In a study designed to investigate what variables are correlated with reading performance in the second grade, researchers collected the data in Table 9.17 on a random sample of 20 second-grade students.

Table 9.17 Data for Exercise 34

Score on emotional maturity test	10	10	18	15	20	16	22	20	25	30	25	32	28	30	36	40	40	40	42	45
Reading performance score	10	12	15	17	18	20	20	22	24	25	28	30	31	32	34	35	36	38	38	40

35 A random sample of 15 eighth-grade students were given a test to measure level of self-concept and another test to measure level of social maturity. Table 9.18 shows the results.

Table 9.18 Data for Exercise 35

Self-concept score	5	10	15	15	20	20	25	25	25	32	40	37	45	35	50
Social-maturity score	5	5	8	20	15	25	20	35	30	30	30	35	35	40	40

36 A botanist randomly selected ten plants of a species growing in a certain geographic area. The investigator took the largest basal leaf from each plant and measured it. Table 9.19 shows the dimensions, in millimeters, of the leaves.

Table 9.19 Data for Exercise 36

Width	5	15	15	30	35	35	40	45	55	60
Length	50	60	65	70	75	80	85	85	90	95

37 An experiment was conducted to investigate variables thought to be related to initiative in problem-solving situations. Subjects were a random sample of 14 juniors at a certain state university. Table 9.20 shows data obtained in the course of the experiment.

Table 9.20 Data for Exercise 37

Self-concept score	5	6	6	7	8	8	8	9	9	9	10	10	11	12
Initiative score	5	6	8	7	9	11	12	11	12	14	14	16	15	17

38 Table 9.21 shows the IQs and job-satisfaction scores of a random sample of 16 assembly-line employees with a large manufacturing firm. (The data have been coded.)

Table 9.21 Data for Exercise 38

IQ (coded)	−15	−10	−5	0	3	5	5	10	10	12	12	14	15	15	16	16
Job-satisfaction	50	55	35	40	20	13	26	8	30	10	15	5	1	8	0	5

39 Table 9.22 gives the verbal and nonverbal IQ scores (coded) for a random sample of 12 ninth-grade students.

Table 9.22 Data for Exercise 39

Verbal IQ	−5	0	5	10	15	20	0	−5	10	20	15	5
Nonverbal IQ	0	−5	6	12	20	22	5	0	15	25	20	3

40 Table 9.23 shows the vocabulary test scores and reading comprehension test scores made by a random sample of 17 eighth-grade students.

Table 9.23 Data for Exercise 40

Vocabulary	1	3	3	6	4	5	6	7	6	7	9	8	10	13	12	13	14
Reading comprehension	1	1	2	2	3	3	4	4	5	6	6	7	8	8	9	10	10

9.6 SOME PRECAUTIONS

When properly used, regression and correlation analysis provide useful and powerful tools for the analysis of data. One should avoid, however, their misuse, and misinterpretation of the results of their proper use.

One should not interpret the presence, in sample data, of an indication of a strong linear relationship between two variables as indicating a cause-and-effect relationship. A significant sample correlation coefficient between X and Y may be a reflection of any one of the following possibilities.

1 X causes Y.

2 Y causes X.

3 Some third variable causes both X and Y, either directly or indirectly.

Figure 9.11 Example of appropriate and inappropriate extrapolation

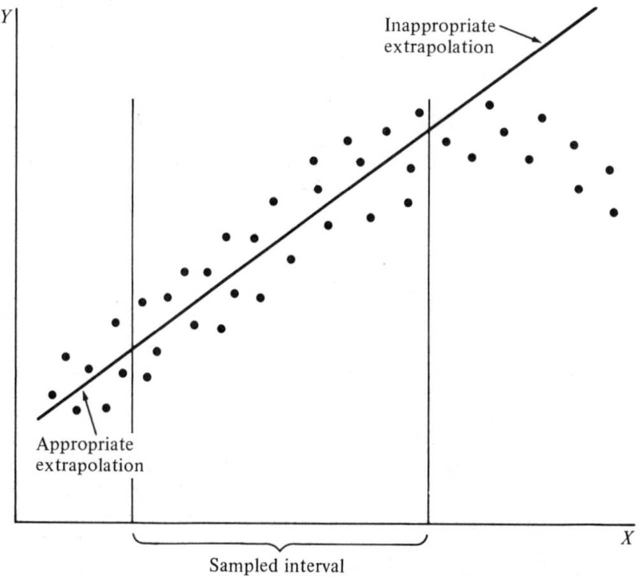

4 An unlikely event has occurred, and a sample with a significant correlation coefficient has been obtained, as a result of chance alone, from a population in which X and Y are not correlated.

5 The correlation is purely spurious.

In regression analysis, one should proceed with caution when considering the prediction of Y or the estimation of subpopulation means for values of X outside the limits of that variable represented in the sample. This practice, which we call *extrapolation*, can yield misleading results. When one uses a simple linear regression equation to predict Y for a value of X beyond the upper limit of X values in the sample, one is assuming that the relationship between X and Y continues to be linear in this region. If this assumption is not met, the prediction will be in error. The same reasoning applies to the estimation of subpopulations of Y values of X smaller than the smallest value of X in the sample. One should use extrapolation only when one knows that the relationship is linear in the area in which extrapolation takes place.

Figure 9.11 illustrates a situation in which extrapolation below the lower limit of sample X values would be appropriate, but extrapolation above the upper limit of sample X's would not.

CHAPTER SUMMARY

In this chapter we have been concerned with simple linear regression and correlation analysis, and have emphasized the use of these techniques in drawing inferences about populations.

We have suggested the following procedure for the use of regression analysis.

1 Identify the model and consider the assumptions.

2 Obtain the regression equation.

3 Evaluate the regression equation.

4 Use the regression equation.

We have discussed the use of the sample regression equation to predict the value that Y is likely to assume for a given X and to estimate the mean of a subpopulation of Y for a given X.

Our treatment of correlation analysis has focused on the correlation coefficient as a measure of the strength of the relationship between two variables. We have seen how sample data can be used to test hypotheses about, and to construct confidence intervals for, the population correlation coefficient.

Finally, we have presented some precautions that one should consider when using regression and correlation analysis.

For a more complete treatment of regression and correlation analysis, including a discussion of multiple regression and correlation, see the books by Acton (3), Draper and Smith (4), Ezekiel and Fox (5), and Neter and Wasserman (6).

SOME IMPORTANT CONCEPTS AND TERMINOLOGY

Regression
Correlation
Independent variable
Dependent variable
Assumptions underlying regression analysis
Unit of association
Method of least squares
Explained sum of squares
Unexplained sum of squares
Coefficient of determination
Prediction interval
Correlation coefficient
Assumptions underlying correlation analysis
Extrapolation

REVIEW EXERCISES

For Exercises 41 through 43, (a) find the regression equation, (b) compute r^2 and perform the analysis of variance, (c) construct a 95% confidence interval for β_1, (d) test H_0: $\beta_1 = 0$, and (e) select a value of the independent variable and construct 95% confidence and prediction intervals.

41 Table 9.24 shows the sales volumes and buying incomes for a sample of 15 trade areas in which a firm now does effective business.

Table 9.24 Data for Exercise 41

Amount of sales, Y (× \$100,000)	1.0	2.3	9.4	1.1	2.9	2.5	3.0	3.4	5.8	6.1	6.8	6.9	7.2	11.4	14.3
Effective buying income, X (× \$1,000,000)	10	70	168	22	38	25	51	60	83	90	100	124	159	176	200

42 An experiment was conducted to study the effect of a certain drug in lowering blood pressure in hypertensive adults. The independent variable is dose level, in milligrams, of the drug. The dependent variable is the difference between the lowest reading following administration of the drug and the predrug control. Table 9.25 gives the data.

Table 9.25 Data for Exercise 42

Dose, X (mg)	0.50	0.75	1.00	1.25	1.50	1.75	2.00	2.25	2.50	2.75	3.00	3.25	3.50
Reduction in blood pressure, Y	10	8	12	14	14	10	16	20	17	20	15	20	21

43 Table 9.26 shows the yield of a certain grain and the amount of fertilizer applied, as reported by ten farmers.

Table 9.26 Data for Exercise 43

Yield, Y (bushels/acre)	50	52	56	59	62	64	68	69	70	71
Fertilizer application, X (pounds/acre)	38	40	39	35	44	40	43	45	48	50

For Exercises 44 and 45, (a) compute r, (b) test H_0: $\rho = 0$ at the 0.05 level of significance, and (c) construct the 95% confidence interval for ρ.

44 Table 9.27 shows the annual sales and advertising expenditures of a certain product in ten market areas. (The data are coded for ease of calculation.)

Table 9.27 Data for Exercise 44

Sales, Y	20	25	24	30	32	40	28	50	40	50
Advertising expenditures, X	0.2	0.2	0.2	0.3	0.3	0.4	0.3	0.5	0.4	0.5

45 The measurements shown in Table 9.28 were taken on 15 apparently normal males between the ages of 16 and 25.

Table 9.28 Data for Exercise 45

Serum cholesterol, Y (mg/100 cc)	162.2	158.0	157.0	155.0	156.0	154.1	169.1	181.0	174.9	180.2	174.0
Systolic blood pressure, X	108	111	115	116	117	120	124	127	122	121	125
Serum cholesterol, Y (mg/100 cc)	160.0	180.0	155.0	190.0							
Systolic blood pressure, X	110	130	120	135							

REFERENCES

1 R. A. Fisher, "On the Probable Error of a Coefficient of Correlation Deduced from a Small Sample," *Metron*, 1 (1921), 3–21

2 H. Hotelling, "New Light on the Correlation Coefficient and Its Transforms," *Journal of the Royal Statistical Society*, Series B, 15 (1953), 193–232

3 F. S. Acton, *Analysis of Straight Line Data*, New York: Wiley, 1959
4 N. R. Draper and H. Smith, *Applied Regression Analysis*, New York: Wiley, 1966
5 Mordecai Ezekiel and Karl A. Fox, *Methods of Correlation and Regression Analysis*, third edition, New York: Wiley, 1959
6 John Neter and William Wasserman, *Applied Linear Statistical Models*, Homewood, Ill.: Richard D. Irwin, 1974

10

Some Uses of Chi-Square

We often want to test the null hypothesis that observed sample data are a result of the fact that two characteristics, as observed in a population, are unrelated. For example, we may hypothesize that people's attitudes toward a particular minority group are unrelated to their level of education. In another situation, we may wish to test the null hypothesis that in several populations the proportion of subjects having a certain characteristic is the same. We may, for example, formulate the null hypothesis that in three groups—say low-, middle-, and upper-income groups—the proportions with high, average, and poor knowledge of national politics are the same.

In each case the analysis is based on counts, or frequencies, rather than measurements, such as inches, pounds, or test scores. Investigators define mutually exclusive categories and record the observed frequency with which people fall into the categories. They then compare these observed frequencies with the frequencies that would be expected if the hypothesized conditions were in fact true.

If the discrepancy between the observed frequencies and the expected frequencies is too great to be attributed to chance when H_0 is true, we reject the null hypothesis that the particular condition of interest is true. In the sections that follow, we explain how, in various situations, one determines expected frequencies, and how to decide when the discrepancy between observed and expected frequencies is large enough to warrant rejection of the null hypothesis.

In Chapters 5 and 6, we saw how we can use the chi-square distribution to construct confidence intervals for, and to test hypotheses about, population variances. In this chapter, we again make use of this distribution in testing hypotheses of the type just described. In the following sections, we cover two types of chi-square tests: *chi-square tests of independence* and *chi-square tests of homogeneity*.

The concepts and techniques on which the tests discussed in this chapter are based were introduced in 1900 by Karl Pearson (1857–1936), who has been called the founder of the science of statistics.

10.1 THE CHI-SQUARE TEST OF INDEPENDENCE

As we have noted, an investigator may wish to know, for some population, whether or not it is likely that two criteria of classification are related. For example, we may be interested in knowing whether or not there is an association between religious involvement and the political interests of some group of people. If we conclude that two criteria of classification are not associated, we shall say that they are *independent*. Two criteria of classification are independent if the distribution of one criterion in no way depends on the distribution of the other. In the example just cited, we shall say that religious involvement and political interest are independent if knowledge of the extent of a person's religious involvement is of no help in predicting the nature of her or his political interest.

To reach a decision about the independence of two criteria in a population, as a rule we analyze sample data and make inferences about the population on the basis of the sample results. A use of the chi-square distribution called the *chi-square test of independence* is the technique most frequently employed for this type of problem.

Example 10.1 A sociologist wished to know whether it was possible to conclude that there is an association between the degree of liberalism and college class standing in a certain population of college students. The sociologist selected a sample of 500 students for the study. The null hypothesis to be tested in the chi-square test of independence is:

H_0: The two criteria of classification are independent

The alternative is:

H_1: The two criteria of classification are not independent

The appropriate null and alternative hypotheses for our example are:

H_0: Degree of liberalism and class standing are independent

H_1: The two criteria are not independent

The following steps are involved in the application of the chi-square test of independence.

1 We draw a random sample of subjects from the population of interest.

2 We classify each subject according to each of the two criteria of classification. For example, if the criteria of interest are extent of religious involvement and nature of political interest, we might classify each subject as either low, moderate, or high with respect to religious involvement, and as either leftist, centrist, or rightist with respect to politics. We may refer to the different categories into which a criterion is divided as the *levels* of that criterion.

3 Sample data are typically displayed in a table called a *contingency table*, in which the levels of one criterion of classification form the rows of the table and the levels of the other criterion form the columns. The cells formed by the intersection of rows and columns contain counts or frequencies of subjects who have been cross-classified on the basis of the two criteria.

Table 10.1 is a generalized contingency table in which a sample of n subjects have been cross-classified. The table shows that the number of subjects falling into the ith level of the first criterion and the jth level of the second criterion is n_{ij}. A total of $n_{r.}$ subjects fall into the rth category of the first criterion, and $n_{.c}$ subjects fall into the cth category of the second criterion. Table 10.2 is the contingency table for Example 10.1.

Table 10.1 Two-way classification of a sample of subjects

First criterion of classification	**Second criterion of classification**						
	LEVEL						
LEVEL	1	2	...	j	...	c	**Total**
1	n_{11}	n_{12}	...	n_{1j}	...	n_{1c}	$n_{1.}$
2	n_{21}	n_{22}	...	n_{2j}	...	n_{2c}	$n_{2.}$
⋮	⋮	⋮		⋮		⋮	⋮
i	n_{i1}	n_{i2}	...	n_{ij}	...	n_{ic}	$n_{i.}$
⋮	⋮	⋮		⋮		⋮	⋮
r	n_{r1}	n_{r2}	...	n_{rj}	...	n_{rc}	n_r
	$n_{.1}$	$n_{.2}$	...	$n_{.j}$	...	$n_{.c}$	n

Table 10.2 A sample of 500 college students classified by degree of liberalism and class standing

Class	Degree of liberalism: SLIGHT	MODERATE	HIGH	Total
Freshman	30	83	37	150
Sophomore	19	56	50	125
Junior	16	46	63	125
Senior	10	38	52	100
	75	223	202	500

4 We compute expected frequencies for each cell of the contingency table. We do this under the assumption that the null hypothesis is true; that is, that the two criteria of classification are independent.

5 We compare observed and expected frequencies by computing

$$X^2 = \sum_{i=1}^{r} \sum_{j=1}^{c} \left[\frac{(O_{ij} - E_{ij})^2}{E_{ij}}\right] \tag{10.1}$$

where O_{ij} is the observed frequency and E_{ij} is the expected frequency for the ijth cell. If the null hypothesis is true—that is, if the two criteria of classification are in fact independent—X^2 will be distributed approximately like a chi-square distribution.

6 We compare the computed value of X^2 for significance with tabulated values of χ^2. If X^2 is equal to or greater than χ^2 for the appropriate degrees of freedom and the chosen level of significance α, we reject the null hypothesis of independence, and conclude that the two criteria of classification are not independent.

Determination of expected frequencies In determining the expected frequencies for the cells of the contingency table in a chi-square test for independence, we make use of our knowledge of probability. Let us refer to the contingency table in Table 10.1. Suppose we wish to estimate the probability that a subject picked at random from the population will be characterized by level 1 of the first criterion. Our estimator of this probability is $n_{1.}/n$, the number of subjects in the sample characterized by level 1 of the first criterion divided by the total number of subjects in the sample. Similarly, to estimate the probability that a subject picked at random from the population will be categorized according to level 1 of the second criterion of classification, we compute $n_{.1}/n$. Since these probabilities are computed from marginal totals, we call them *marginal probabilities*.

To obtain an expected frequency for a particular cell, let us determine the probability that a subject picked at random from the population will fall in that cell. For example, let us determine the probability that a subject picked at random from the population will fall in the first cell of Table 10.1. That is, the probability that a subject picked at random from the population will be characterized according to level 1 of both criteria of classification. Without any knowledge of the dependence or independence of the two criteria, we would be inclined to estimate this probability by computing the joint probability n_{11}/n. Under the hypothesis that the two criteria of classification are independent, however, we follow another line of reasoning. In Chapter 2 we learned that if two events are independent, the probability of their joint occurrence is equal to the product of their individual probabilities. If we restate this rule so that it applies specifically to a contingency table, we can make the following statement.

In a contingency table, if two criteria of classification are independent, each of the joint probabilities associated with the cells of the table is equal to the product of the corresponding marginal probabilities.

Under the hypothesis of independence, then, the probability that a subject picked at random from the population is characterized by level 1 of both criteria is estimated by $(n_{1.}/n)(n_{.1}/n)$.

To obtain the expected frequencies for the cells of the contingency table, we multiply each of the joint probabilities by the total sample size n. For example, the expected frequency E_{11} of the first cell of Table 10.1 is given by

$$E_{11} = \left(\frac{n_{1.}}{n}\right)\left(\frac{n_{.1}}{n}\right)n$$

If we divide an n in the denominator into the n in the numerator, we have

$$E_{11} = \frac{(n_{1.})(n_{.1})}{n} \tag{10.2}$$

which suggests a convenient short-cut method of finding the expected frequencies. To find the expected frequency for a particular cell, we simply divide the product of the corresponding marginal totals by the grand total.

Let us illustrate the calculation of expected frequencies by means of the data of Example 10.1, which are displayed in Table 10.2.

Applying the short-cut procedure suggested by Equation 10.2, we compute the following expected frequencies.

$$E_{11} = \frac{(75)(150)}{500} = 22.50 \qquad E_{12} = \frac{(223)(150)}{500} = 66.90 \qquad E_{13} = \frac{(202)(150)}{500} = 60.60$$

$$E_{21} = \frac{(75)(125)}{500} = 18.75 \qquad E_{22} = \frac{(223)(125)}{500} = 55.75 \qquad E_{23} = \frac{(202)(125)}{500} = 50.50$$

$$E_{31} = \frac{(75)(125)}{500} = 18.75 \qquad E_{32} = \frac{(223)(125)}{500} = 55.75 \qquad E_{33} = \frac{(202)(125)}{500} = 50.50$$

$$E_{41} = \frac{(75)(100)}{500} = 15.00 \qquad E_{42} = \frac{(223)(100)}{500} = 44.60 \qquad E_{43} = \frac{(202)(100)}{500} = 40.40$$

Table 10.3 Observed and expected frequencies for Example 10.1

Class	Degree of liberalism SLIGHT	MODERATE	HIGH	Total
Freshman	30 (22.50)	83 (66.90)	37 (60.60)	150
Sophomore	19 (18.75)	56 (55.75)	50 (50.50)	125
Junior	16 (18.75)	46 (55.75)	63 (50.50)	125
Senior	10 (15.00)	38 (44.60)	52 (40.40)	100
	75	223	202	500

Table 10.3 shows the expected frequencies (in parentheses) and the observed frequencies.

We use the data in Table 10.1 and Equation 10.1 to compute the following value of the test statistic.

$$X^2 = \frac{(30 - 22.50)^2}{22.50} + \frac{(83 - 66.90)^2}{66.90} + \cdots + \frac{(52 - 40.40)^2}{40.40} = 26.752$$

Determination of degrees of freedom The number of degrees of freedom associated with X^2 computed from the data of a contingency table is equal to the number of cells that can be filled, arbitrarily, given that the marginal totals are established.

For example, consider the contingency table in Table 10.4, which has three rows and three columns. Once we have established the marginal totals, we see that it is possible to arbitrarily fill only four of the cells, since the numbers in the rows and columns have to add up to the row and column totals. A contingency table containing three rows and three columns, therefore, yields four degrees of freedom. In general, the number of degrees of freedom associated with X^2 computed from a contingency table is equal to the number of rows (r) minus 1, times the number of columns (c) minus 1. That is, $\text{df} = (r - 1)(c - 1)$.

The number of degrees of freedom for the X^2 of Example 10.1 is thus $(4 - 1)(3 - 1) = 6$.

Reference to Appendix Table G reveals that the probability of observing a value of X^2 as large as or larger than 26.752, when H_0 is true, is less than 0.005. We therefore reject H_0, and conclude that degree of liberalism and college class are not independent in the population from which the sample was drawn.

Table 10.4 A contingency table with three rows and three columns

First criterion of classification	Second criterion of classification 1	2	3	Total
1	n_{11}	n_{12}	n_{13}	$n_{1.}$
2	n_{21}	n_{22}	n_{23}	$n_{2.}$
3	n_{31}	n_{32}	n_{33}	$n_{3.}$
	$n_{.1}$	$n_{.2}$	$n_{.3}$	n

Table 10.5 A 2 × 2 contingency table

First criterion of classification	Second criterion of classification 1	2	
1	a	b	$a + b$
2	c	d	$c + d$
	$a + c$	$b + d$	n

Small expected frequencies Sometimes one may find that the expected frequencies in some of the cells of a contingency table are quite small. When this is the case, the chi-square distribution may not provide a good approximation of the distribution of X^2. Although there is not general agreement on what constitutes a small expected frequency, Cochran (1) recommends that, for contingency tables with more than one degree of freedom, a minimum expectation of one observation per cell is permissible if no more than 20% of the cells have expected frequencies less than five. One may combine adjacent cells in order to achieve the minimum expectation, provided that this does not violate the logic of the classification scheme.

The 2 × 2 contingency table A contingency table with two rows and two columns results when subjects are categorized on the basis of two criteria, each of which occurs at two levels. Such a table is called a 2 × 2 (read "two by two") contingency table. Table 10.5 is a typical table of this type. In this table a, b, c, and d are the observed frequencies, respectively, for cells 1 1, 1 2, 2 1, and 2 2.

We may use to advantage the following special formula for computing X^2 from the data of a 2 × 2 contingency table.

$$X^2 = \frac{n(ad - bc)^2}{(a + c)(b + d)(a + b)(c + d)} \tag{10.3}$$

A 2 × 2 contingency table has one degree of freedom.

Example 10.2 In a random sample of 300 high school students, a sociologist collected data regarding their fathers' occupations and the prevailing disciplinary atmosphere in their homes. Table 10.6 displays the results.

Table 10.6 A sample of 300 high school students classified by father's occupation and prevailing disciplinary atmosphere in home

Father's occupation	Disciplinary atmosphere PERMISSIVE	AUTHORITARIAN	Total
White collar	98	42	140
Blue collar	101	59	160
	199	101	300

The null and alternative hypotheses are:

H_0: Home disciplinary atmosphere and father's occupation are independent

H_1: The two criteria are not independent

We use Equation 10.3 to compute

$$X^2 = \frac{300[(98)(59) - (42)(101)]^2}{(199)(101)(140)(160)} = 1.580$$

Reference to Appendix Table G reveals that the probability of observing a value of X^2 as large as or larger than 1.580, when H_0 is true, is greater than 0.10. Therefore, we do not reject the null hypothesis, and conclude that father's occupation and disciplinary atmosphere in the home may be independent.

The problem of small expected frequencies is also liable to arise in the case of 2 × 2 contingency tables. Cochran (1) recommends that one not use chi-square analysis with 2 × 2 tables if n is less than 20 or if n is between 20 and 40 and any expected frequency is less than 5.

Yates's correction In 1934 Yates (2) proposed the use of what has come to be known as *Yates's correction for continuity* when one is computing X^2 from the data of a 2 × 2 contingency table. The purpose of the correction was to improve the approximation of the chi-square distribution to that of X^2. The correction consists of subtracting $0.5n$ from the absolute value of $ad - bc$ in the numerator of Equation 10.3. This yields

$$X_c^2 = \frac{n(|ad - bc| - 0.5n)^2}{(a + c)(b + d)(a + b)(c + d)} \tag{10.4}$$

For many years practitioners have routinely used Yates's correction for continuity, and most textbook writers have recommended its use. However, it has not gone unchallenged. Those who have raised questions include Grizzle (3), who showed that the application of Yates's correction usually results in too conservative a test (that is, the null hypothesis is not rejected as often as it should be). Consequently, some practitioners and writers have abandoned its use.

If we apply Yates's correction to the data of Example 10.2, we have, by Equation 10.4,

$$X_c^2 = \frac{300[|(98)(59) - (42)(101)| - (0.5)(300)]^2}{(199)(101)(140)(160)} = 1.287$$

Thus we see that the corrected X^2 is quite close to the uncorrected value.

EXERCISES

1 Each pupil in a random sample of 200 seventh-grade students was rated by his or her teachers as above average, average, or below average in degree of motivation. The students were also categorized according to whether their arithmetic skills scores were more than five points below the school-system median for their grade, within five points (inclusive) of the median, or more than five points above the median. Table 10.7 shows the results. Do these data provide sufficient evidence, at the 0.05 level of significance, to indicate that motivation and arithmetic skills are related?

Table 10.7 Data for Exercise 1

Arithmetic skills	Motivation			
	BELOW AVERAGE	AVERAGE	ABOVE AVERAGE	**Total**
Below median	20	35	5	60
Within 5 points of median	15	40	20	75
Above median	5	35	25	65
	40	110	50	200

2 Table 10.8 shows the results of a survey in which 250 respondents were categorized according to level of education and attitude toward student demonstrations at a certain college. Test the null hypothesis that the two criteria of classification are independent. Let $\alpha = 0.05$.

Table 10.8 Data for Exercise 2

Education	Attitude			
	AGAINST	NEUTRAL	FOR	**Total**
Less than high school	40	25	5	70
High school	40	20	5	65
Some college	30	15	30	75
College graduate	15	15	10	40
	125	75	50	250

3 Table 10.9 shows how a random sample of 150 college students was cross-classified according to grade-point average and score on a self-concept test. Can one conclude, on the basis of these data, that the two variables are related? Let $\alpha = 0.05$.

Table 10.9 Data for Exercise 3

Self-concept score	Grade point average		
	HIGH	LOW	**Total**
High	55	20	75
Low	30	45	75
	85	65	150

4 A survey was conducted to elicit respondents' attitudes toward a certain minority group. Investigators also recorded the socioeconomic status of respondents. Table 10.10 summarizes the results. Do these data provide sufficient evidence to indicate that the two variables are related? Let $\alpha = 0.01$.

Table 10.10 Data for Exercise 4

	Socioeconomic status			
Attitude	LOW	MEDIUM	HIGH	**Total**
Undecided	25	20	30	75
Unfavorable	30	45	10	85
Favorable	20	60	60	140
	75	125	100	300

5 Table 10.11 shows the results of a study in which 100 boys of junior high school age were cross-classified by delinquency status and extent of leisure-time contact with parents. Do these data provide sufficient evidence to indicate that the two variables are related? Let $\alpha = 0.05$.

Table 10.11 Data for Exercise 5

Leisure-time contact with parents	**Delinquents**	**Nondelinquents**	**Total**
High	10	29	39
Low	20	41	61
	30	70	100

6 Each of a random sample of 400 young married couples was categorized by social class of parents and religious involvement. Table 10.12 shows the results. Do these data provide sufficient evidence to indicate that the two variables are interdependent? Let $\alpha = 0.05$.

Table 10.12 Data for Exercise 6

	Social class of parents				
Religious involvement	I	II	III	IV	**Total**
Attends church	30	47	97	77	251
Active in church	10	18	35	30	93
Does not attend church	10	10	18	18	56
	50	75	150	125	400

7 Each of a random sample of 300 female patients at a large psychiatric hospital was allowed to choose from among a wide variety of magazines for leisure reading. Table 10.13 shows a breakdown of these patients by type of magazine selected and diagnosis. Test for independence. Let $\alpha = 0.05$.

Table 10.13 Data for Exercise 7

Diagnosis	Type of magazine selected NEWS	ROMANCE	CRIME	MOVIE	Total
A	35	10	5	8	58
B	10	55	12	40	117
C	10	15	10	12	47
D	15	20	33	10	78
	70	100	60	70	300

8 Each of a random sample of 500 police records of persons charged with a certain offense was categorized by social class of offender and disposition of the case. Table 10.14 shows the results. Can one conclude, on the basis of these data, that social class and disposition of case are related? Let $\alpha = 0.05$?

Table 10.14 Data for Exercise 8

Disposition of case	Social class UPPER	MIDDLE	LOW	Total
Incarceration	10	45	143	198
Probation	37	70	38	145
Fine or warning	68	60	29	157
	115	175	210	500

Table 10.15 Data for Exercise 9

Grade point average	Ever smoked marijuana? YES	NO	Total
≤ 3.0	45	7	52
>3.0	30	18	48
	75	25	100

9 Each of a random sample of 100 college students was categorized on the basis of (a) whether or not she or he had ever smoked marijuana, and (b) grade point average. Table 10.15 displays the results. Do these data provide sufficient evidence to indicate that there is a relationship between the two variables? Let $\alpha = 0.05$.

10 Each of a random sample of 300 college students was categorized according to political inclination and degree of religious commitment. Table 10.16 shows the results. Do these data provide sufficient evidence, at the 0.05 level of significance, to indicate that the two variables are interdependent?

Table 10.16 Data for Exercise 10

Religious commitment	Political inclination RADICAL	MODERATE	CONSERVATIVE	Total
Very religious	15	25	35	75
Moderately religious	38	60	75	173
Not religious	22	15	15	52
	75	100	125	300

10.2 THE CHI-SQUARE TEST OF HOMOGENEITY

In Chapter 6 we learned to test the null hypothesis that two population proportions are equal by means of the following test statistic.

$$z = \frac{(\hat{p}_1 - \hat{p}_2) - 0}{\sqrt{\dfrac{\bar{p}(1-\bar{p})}{n_1} + \dfrac{\bar{p}(1-\bar{p})}{n_2}}} \tag{10.5}$$

where $\hat{p}_1$ and $\hat{p}_2$ are sample proportions computed from samples of size n_1 and n_2, respectively, and $\bar{p}$ is the pooled estimate of the assumed common population proportion p. If the population proportions are equal, and if np and $n(1-p)$ are both greater than 5, the z of Equation 10.5 will follow approximately the standard normal distribution.

Alternatively, we may test the null hypothesis that two population proportions are equal by means of the *chi-square test of homogeneity*. We may cast the sample data in a 2×2 contingency table by using the two populations as one criterion of classification and the characteristic of interest as the other criterion, with presence and absence of the criterion being the two categories of classification. We compute the test statistic X^2 from the data by Equation 10.3 (or by Equation 10.4 if we use Yates's correction for continuity) and compare it for significance with tabulated χ^2 with one degree of freedom.

Note that the null hypothesis that two population proportions are equal, which we write symbolically as

$$H_0: p_1 = p_2$$

may be stated in words as: "The two populations are homogeneous with respect to the characteristic of interest."

Let us illustrate these ideas with an example.

Example 10.3 A sociologist, studying differences in the characteristics of college students who participate actively in campus protest activities and those who do not, wished to know whether it was possible to conclude that the proportions who perceive the existence of a generation gap between themselves and their parents are different in the two groups.

We may state the appropriate null and alternative hypotheses in words in either of two ways. First, as:

H_0: The proportion of students who perceive the existence of a generation gap between themselves and their parents is the same in the two groups

H_1: The two proportions are not the same

And second, as:

H_0: The two groups are homogeneous with respect to the existence of a perceived generation gap between individuals and their parents

H_1: The groups are not homogeneous with respect to the characteristic of interest

We state the null and alternative hypotheses symbolically as follows.

$$H_0: p_1 = p_2, \qquad H_1: p_1 \neq p_2$$

The sociologist selected a random sample of 200 students identified as active participants in campus protest activities and a random sample of 250 nonparticipants. When interviewed, 116 of the participants and 113 of the nonparticipants said that a generation gap existed between themselves and their parents. We may display the results of the interviews in a 2 × 2 contingency table, as shown in Table 10.17.

By Equation 10.3, we compute from these data

$$X^2 = \frac{450[(116)(137) - (113)(84)]^2}{(200)(250)(229)(221)} = 7.284$$

Reference to Appendix Table G, with one degree of freedom, reveals that the computed X^2 is significant at the 0.01 level. Consequently, we reject H_0, and conclude that the proportion of students who perceive the existence of a generation gap between themselves and their parents is not the same in the population of participants in campus protest activities as it is in the population of nonparticipants. In other words, we conclude that the two populations are not homogeneous with respect to the existence of a perceived generation gap between the students and their parents.

Alternatively, we may test $H_0: p_1 = p_2$ by using the test statistic given by Equation 10.5. First we compute

$$\hat{p}_1 = \frac{116}{200} = 0.580, \qquad \hat{p}_2 = \frac{113}{250} = 0.452$$

$$\bar{p} = \frac{116 + 113}{200 + 250} = \frac{229}{450} = 0.509$$

Table 10.17 Participants and nonparticipants in campus protest activities classified as to presence or absence of perceived generation gap between themselves and their parents

Perceived generation gap	Group		
	PARTICIPANTS	NONPARTICIPANTS	**Total**
Present	116	113	229
Absent	84	137	221
	200	250	450

Substituting these values into Equation 10.5 gives

$$z = \frac{0.580 - 0.452}{\sqrt{\frac{(0.509)(0.491)}{200} + \frac{(0.509)(0.491)}{250}}} = 2.699$$

We find that the computed value of z is also significant at the 0.01 level, and we reach the same conclusion as before.

We can see the relationship between the two testing procedures more clearly by noting that $z^2 = (2.699)^2 = 7.285$, which, except for rounding error, is the same as the computed X^2 of 7.284. It is also enlightening to note that the entries in Appendix Table G for one degree of freedom are the squares of the corresponding two-sided critical values of z. For example, $\chi^2_{0.95} = 3.841$ is equal to $z^2 = (1.96)^2$.

Testing for homogeneity among several populations

We may extend the chi-square test of homogeneity to test hypotheses about more than two populations. That is, we may test H_0: $p_1 = p_2 = p_3 = \cdots = p_c$, where c is the number of populations under consideration. To proceed, we draw a random sample from each population, and note in each sample the number of subjects with the characteristic of interest. We cast the results in a $2 \times c$ contingency table, and obtain expected frequencies for each resulting cell. We compute the test statistic by Equation 10.1, and compare it for significance with the appropriate tabulated value of χ^2 in Appendix Table G.

Let us illustrate by means of an example.

Example 10.4 A large company wished to know whether its blue-collar, clerical, sales, and other white-collar employees differ with respect to their rating of the desirability of the company as an employer. A random sample of employees was selected from each of the four groups (populations), and each employee was asked to rank the company as a place to work compared with other companies with which he or she was familiar. The responses were categorized as either "above average" or "average or below." Table 10.18 shows the results. We may state the appropriate null and alternative hypotheses symbolically as

$$H_0\colon p_1 = p_2 = p_3 = p_4, \qquad H_1\colon \text{Not all } p_i \text{ are equal}$$

Table 10.18 Employee ratings of their company as a place to work

	Group				
Rating	BLUE COLLAR	CLERICAL	SALES	OTHER WHITE COLLAR	**Total**
Above average	68	45	70	60	243
Average or below	57	30	30	15	132
	125	75	100	75	375

We obtain mathematically the expected frequencies for the cells of a contingency table in the chi-square test of homogeneity in the same manner as in the test of independence described in Section 10.1. That is, we obtain the expected frequency of a given cell by dividing the product of corresponding marginal totals by the grand total. The rationale underlying the procedure, however, is different.

If the c populations are homogeneous with respect to the characteristic of interest—that is, if the proportion with the characteristic of interest is the same in each population—we would expect an estimate of the common population proportion computed from the pooled sample data to provide the best estimate obtainable from the available data. For this example, we would estimate the assumed common population proportion by

$$\bar{p} = \frac{68 + 45 + 70 + 60}{125 + 75 + 100 + 75} = \frac{243}{375} = 0.648$$

We would then compute expected frequencies with the characteristic of interest for each population by multiplying each population total by the estimate of the common population proportion. For example, for the 125 blue-collar workers in our example, we would compute the expected number who would rate the company as above average by multiplying 125 by 243/375. We obtain

$$(125)\frac{243}{375} = 81.0$$

Thus we see that we obtain this expected frequency by dividing the product of appropriate marginal totals by the grand total. We obtain expected frequencies for the other cells in a similar manner. Alternatively, we may find the expected number in a given group not having the characteristic of interest by subtracting the expected number having the characteristic from the given sample size. For example, we may compute the expected number of blue-collar workers who would rate the company as average or below by subtracting 81 from 125 and we get 44.

Table 10.19 shows the observed and expected frequencies for our example. From the data in Table 10.19, we compute

$$X^2 = \frac{(68 - 81.0)^2}{81.0} + \frac{(45 - 48.6)^2}{48.6} + \cdots + \frac{(15 - 26.4)^2}{26.4} = 15.467$$

Table 10.19 Observed and expected frequencies for Example 10.4 (Expected frequencies are in parentheses)

	Group				
Rating	BLUE COLLAR	CLERICAL	SALES	OTHER WHITE COLLAR	**Total**
Above average	68 (81.0)	45 (48.6)	70 (64.8)	60 (48.6)	243
Below average	57 (44.0)	30 (26.4)	30 (35.2)	15 (26.4)	132
	125	75	100	75	375

Entering Appendix Table G with three degrees of freedom, we find that the computed value of X^2 is significant at the 0.005 level. We therefore reject the null hypothesis, and conclude that the four populations of employees are not homogeneous with respect to rating of the company.

The $r \times c$ contingency table in tests of homogeneity Frequently, when the question of the homogeneity of two or more populations arises, there will be more than two levels of the variable of interest. In other words, the variable of interest may be multinomial rather than binomial. Instead of classifying subjects simply according to whether or not they have some characteristic of interest, we may wish to specify the degree to which a characteristic is present in a subject. For example, in Example 10.4, employees might have been asked to rate their company on, say, a five-point scale consisting of such categories as "one of the best," "above average," "average," "below average," and "one of the worst." The null hypothesis to be tested, then, would be, "The populations are homogeneous with respect to the way in which subjects are distributed over the various categories." To test the hypothesis, we would draw random samples from each population, make a count of subject responses in each sample, and compute X^2 as described earlier in this section. We could display the data in an $r \times c$ contingency table in which r is the number of categories of the variable of interest and c is the number of populations sampled. We would compare the statistic X^2 for significance with the tabulated value of χ^2 corresponding to $(r-1)(c-1)$ degrees of freedom and the chosen level of significance.

The rationale for computing expected frequencies is the same as that discussed earlier in connection with the $2 \times c$ contingency table. Under the assumption of homogeneity, we pool the data to obtain expected relative frequencies for each category of the variable of interest. Wc multiply these expected relative frequencies by sample totals to obtain expected cell frequencies. In practice, one obtains expected cell frequencies by dividing the product of corresponding marginal totals by the grand total, as described previously.

Example 10.5 A random sample of young married couples was drawn from each of three communities. Each couple was asked to specify the minimum amount of education they hoped their children would obtain. Table 10.20 shows the results of the study, including expected cell frequencies.

Table 10.20 Young married couples from three communities categorized by their educational aspirations for their children

Minimum educational level desired for children	Community *A*	*B*	*C*	Total
High school	30 (25.89)	28 (21.58)	24 (34.53)	82
Trade school	30 (30.00)	19 (25.00)	46 (40.00)	95
College	90 (94.11)	78 (78.42)	130 (125.47)	298
	150	125	200	475

We may state the appropriate null and alternative hypotheses as follows.

H_0: The three populations are homogeneous with respect to the subjects' educational aspirations for their children
H_1: The three populations are not homogeneous with respect to the variable of interest

Alternatively, we may state the null and alternative hypotheses as follows.

H_0: All the proportions in the same row are the same
H_1: At least two proportions in the same row are not equal

From the data in Table 10.20, we compute

$$X^2 = \frac{(30 - 25.89)^2}{25.89} + \frac{(30 - 30.00)^2}{30.00} + \cdots + \frac{(130 - 125.47)^2}{125.47} = 8.458$$

Reference to Appendix Table G with four degrees of freedom reveals that the computed value of X^2 is not significant at the 0.05 level. Consequently, we shall not reject the hypothesis. Our conclusion is that the populations may be homogeneous with respect to the variable of interest.

Small expected frequencies The problem of small expected frequencies may arise in tests of homogeneity just as it does in tests for independence. Our conclusions about what constitutes small expected frequencies and how to handle the problem, which we discussed in connection with tests of independence in Section 10.1, also apply in tests of homogeneity.

The difference between tests of independence and tests of homogeneity As we have already pointed out, the mathematical computations in chi-square tests of independence and tests of homogeneity are the same. The two procedures do differ, however, in three respects: (1) the method by which the data are collected, (2) the rationale underlying the calculation of expected frequencies, and (3) the interpretation of the results.

In tests of independence, the investigator selects a single sample from a single population, and cross-classifies subjects on the basis of two criteria of interest. In tests of homogeneity, the researcher identifies two or more populations prior to the collection of data. He or she then selects a sample from each of the identified populations, and places subjects in each sample into one of two or more categories of the variable of interest.

The two criteria of classification in tests of independence are assumed to be independent. If this assumption is correct, the expected proportion of subjects falling into a particular cell (a joint probability) of the relevant contingency table is equal to the product of the corresponding marginal probabilities. One multiplies joint probabilities computed from sample data by the sample size to obtain expected frequencies. In tests of homogeneity, we reason that, if the populations are homogeneous with respect to the variable of interest, the use of pooled data should yield the best estimates of the proportions of subjects in each population falling into the various categories of this variable. Multiplication of these estimates by sample totals yields the expected cell frequencies.

The interpretation of the results reflects the manner in which the data are gathered and the rationale underlying the calculation of expected frequencies. In tests of independence, we think in terms of a single population and the dependence or independence of two variables. In tests of homogeneity, we think in terms of two or more populations and whether or not the subjects in these populations are distributed identically over two or more categories of a single variable.

EXERCISES

11 In a certain community, a random sample of 150 persons was selected from those who did not vote in the last primary election, and a random sample of 100 was selected from those who did vote. Table 10.21 shows the intensity of party identification of each group. Do these data provide sufficient evidence to indicate a lack of homogeneity between the two groups with respect to intensity of party identification? Let $\alpha = 0.05$.

Table 10.21 Data for Exercise 11

Intensity of party identification	Voted in last primary		Total
	YES	NO	
Strong	75	20	95
Weak	25	130	155
	100	150	250

12 In a certain state, juvenile offenders during a particular year were assigned by the courts to one of three treatment programs. A random sample of the records of juveniles in each group was selected and classified according to size of family income. Table 10.22 shows the results. On the basis of these data, can one conclude, at the 0.05 level of significance, that there is a lack of homogeneity among the treatment program groups?

13 A psychiatrist drew a random sample of the records of 150 women and a random sample of the records of 100 men from the files of a county

Table 10.22 Data for Exercise 12

Annual family income	Treatment program			Total
	CONVENTIONAL PROBATION	SUPERVISED WORK AND GROUP COUNSELING	REFORMATORY	
Welfare	30	16	38	84
< \$5,000	100	20	24	144
\$5,000–\$9,999	48	30	20	98
\$10,000 and over	22	34	18	74
	200	100	100	400

mental health clinic. The psychiatrist then classified the subjects according to whether or not they had ever suffered from depression. Table 10.23 shows the results. Do these data suggest a difference between men and women in the incidence of depression? Let $\alpha = 0.05$.

Table 10.23 Data for Exercise 13

History of depression	Sex		
	MALE	FEMALE	Total
Yes	30	90	120
No	70	60	130
	100	150	250

14 Random samples of members of four religious denominations were asked whether or not they attended church regularly. Table 10.24 shows the results. Do these data suggest a difference among denominations with respect to regular church attendance? Let $\alpha = 0.05$.

Table 10.24 Data for Exercise 14

Regular church attendance	Denomination				
	A	*B*	*C*	*D*	Total
Yes	20	30	45	32	127
No	80	90	105	48	323
	100	120	150	80	450

15 A random sample of 100 persons were selected from the rolls of five professional organizations and queried regarding the time of life at which they first decided to follow their chosen profession. Table 10.25 gives the results. From these data, can one conclude that there is a lack of homogeneity among the professions with respect to the variable of interest? Let $\alpha = 0.05$.

Table 10.25 Data for Exercise 15

Time of choice of profession	Profession					
	A	*B*	*C*	*D*	*E*	Total
Before high school	30	20	30	10	10	100
During high school	40	55	46	20	30	191
During college	30	25	24	70	60	209
	100	100	100	100	100	500

16 A random sample of 100 senior high school boys was selected from each of three athletic performance groups: high, average, and low. The boys were classified according to intelligence, as shown in Table 10.26.

Do these data suggest a difference in the distribution of intelligence among the three groups? Let $\alpha = 0.05$.

17 Psychiatrists studying alcoholism selected a random sample of patients who were receiving treatment at a county alcoholism clinic from each of three social classes. They then classified the patients according to the level of their communication with their chief therapist. Table 10.27 displays the results. Do these data suggest a lack of homogeneity among the social classes with respect to level of communication with chief therapist? Let $\alpha = 0.05$.

Table 10.26 Data for Exercise 16

	Athletic performance			
Intelligence	HIGH	AVERAGE	LOW	**Total**
High	28	30	34	92
Average	40	38	42	120
Low	32	32	24	88
	100	100	100	300

Table 10.27 Data for Exercise 17

	Social class			
Level of communication	I	II	III	**Total**
Good	45	60	68	173
Fair	10	15	25	50
Poor	5	15	32	52
	60	90	125	275

18 A random sample of 100 high school dropouts and a random sample of 150 students of the same age who did not drop out of school were classified according to whether or not they lived in crowded households. Table 10.28 gives the results. Can one conclude, on the basis of these data, that there is a difference in the two groups with respect to crowded home conditions? Let $\alpha = 0.05$.

Table 10.28 Data for Exercise 18

	High school dropout		
Member of crowded household	YES	NO	**Total**
Yes	30	60	90
No	70	90	160
	100	150	250

19 A random sample was selected from each of the four classes at a state university to participate in a survey on attitudes toward various social issues. One of the questions asked was, "Are you in favor of the legalization of marijuana?" The responses were categorized as shown in Table 10.29. Do these data suggest a lack of homogeneity among the classes with respect to the variable of interest? Let $\alpha = 0.05$.

Table 10.29 Data for Exercise 19

	Class standing				
Response	FRESHMAN	SOPHOMORE	JUNIOR	SENIOR	**Total**
Against	20	24	23	20	87
Don't care	58	52	55	30	195
For	72	64	62	70	268
	150	140	140	120	550

CHAPTER SUMMARY

In this chapter we have introduced one of the most widely used statistical techniques: the chi-square test. Specifically, we have dealt with two chi-square tests: the test of independence and the test of homogeneity.

Some of the reasons for the extensive use of chi-square tests are that the calculations are simple and that these tests do not require rigid assumptions about the sampled populations.

SOME IMPORTANT CONCEPTS AND TERMINOLOGY

Independence	Degrees of freedom for chi-square tests
Contingency table	Homogeneity
Observed frequencies	Difference between chi-square tests of independence and homogeneity
Expected frequencies	

REVIEW EXERCISES

20 A sociologist wished to know whether it was possible to conclude that the educational levels of adults in a certain city were associated with their areas of residence. The sociologist interviewed a random sample of 500 adults for the study. Table 10.30 shows the results. Test for independence.

Table 10.30 Data for Exercise 20

	Educational level			
Area of residence	ELEMENTARY SCHOOL	HIGH SCHOOL	COLLEGE	**Total**
1	52	64	24	140
2	60	59	52	171
3	50	65	74	189
	162	188	150	500

21 A researcher cross-classified 500 elementary school children by socioeconomic group and the presence or absence of a certain congenital defect. Table 10.31 gives the results. Can one conclude from these data that the defect is related to socioeconomic status?

Table 10.31 Data for Exercise 21

	Socioeconomic group				
Congenital defect	UPPER	UPPER MIDDLE	LOWER MIDDLE	LOWER	**Total**
Present	4	24	32	35	95
Absent	46	121	138	100	405
	50	145	170	135	500

22 A sample of 200 defective items of furniture, produced in two factories operated by the same company, were classified according to type of defect. Table 10.32 shows the data. Test at the 0.05 level the null hypothesis that cause of defect and factory of production are independent.

Table 10.32 Data for Exercise 22

	Factory		
Type of defect	*A*	*B*	**Total**
Scratch	35	90	125
Dent	50	25	75
	85	115	200

23 A sample of 100 employees from each shift of a large manufacturing firm was randomly selected, and each subject was asked to rate the degree of her or his satisfaction with current working conditions. Table 10.33 shows the results. Do these data provide sufficient evidence, at the 0.05 level of significance, to suggest that the employees on the different shifts are not homogeneous with respect to satisfaction with current working conditions?

Table 10.33 Data for Exercise 23

	Degree of satisfaction				
Shift	VERY SATISFIED	SATISFIED	DISSATISFIED	VERY DISSATISFIED	**Total**
3	20	30	30	20	100
2	70	10	10	10	100
1	60	10	20	10	100
	150	50	60	40	300

24 A safety director with a large industrial firm wished to evaluate two methods of teaching safety facts and rules to assembly-line employees. A random sample of 50 employees received instruction by method *A*; an independent random sample of 50 employees received instruction by method *B*. At the end of the course, the director gave each of the 100 employees a test, and rated his or her knowledge of the subject as satisfactory or unsatisfactory. Table 10.34 shows the results. Do these data provide sufficient evidence to indicate a lack of homogeneity in the two populations represented by the samples? Let $\alpha = 0.05$.

25 In a survey of personnel directors with a certain type of business firm, five subpopulations were identified on the basis of firm size. A random sample of 200 personnel directors was selected from each subpopulation, and questionnaires were sent to each of the 1000 potential respondents. Table 10.35 gives the numbers responding by size of firm. Do these data provide sufficient evidence to indicate a lack of homogeneity among the subpopulations with respect to response to the questionnaire? Let $\alpha = 0.05$.

Table 10.34 Data for Exercise 24

Test results	Method *A*	*B*	Total
Satisfactory	35	40	75
Unsatisfactory	15	10	25
	50	50	100

Table 10.35 Data for Exercise 25

Size of firm	Responded YES	NO	Total
Very small	80	120	200
Small	100	100	200
Medium	110	90	200
Large	120	80	200
Very large	140	60	200
	550	450	1000

REFERENCES

1 William G. Cochran, "Some Methods for Strengthening the Common χ^2 Tests," *Biometrics*, 10 (1954), 417–451

2 F. Yates, "Contingency Tables Involving Small Numbers and the χ^2 Tests," *Journal of the Royal Statistical Society*, Supplement I, Series B (1934), 217–235

3 J. E. Grizzle, "Continuity Correction in the χ^2 Test for 2×2 Tables," *The American Statistician*, 21 (October 1967), 28–32

11 Nonparametric Statistics

With the exception of the chi-square tests discussed in Chapter 10, the inferential procedures presented in this book have been concerned with the estimation and testing of hypotheses about population parameters. Since they are concerned with parameters, we call these procedures *parametric* procedures. Inferential procedures that are not concerned with population parameters, such as the chi-square test of independence, discussed in Chapter 10, are called *nonparametric* procedures. We also refer to these procedures as *nonparametric statistics* or *nonparametric techniques*. There is another type of procedure that we call a nonparametric procedure. These procedures are those that do not depend for validity on the nature of the population from which the data have been drawn. You will no doubt recall that many of the inferential procedures discussed so far are based on the assumption that the populations giving rise to the data are normally distributed. Such is the case for the t test, analysis of variance, and inferential procedures in regression and correlation analysis. We call statistical procedures that do not depend for validity on the functional form of the parent population distribution *distribution-free* procedures. It will be convenient, however, if we follow convention and treat both the truly nonparametric procedures and the distribution-free procedures under the heading of nonparametric procedures.

Advantages and disadvantages of nonparametric statistics The nonparametric and distribution-free procedures now available offer a number of advantages to the researcher and data analyst. Among them are the following.

1 Most of the nonparametric procedures are based on a minimum set of assumptions, and this tends to reduce the chance of their being improperly used.

2 The arithmetic computations necessary for the application of many nonparametric procedures are short and easy, so one may save time by using them.

3 Nonparametric procedures are usually easily understood by the mathematically and statistically unsophisticated.

4 One may apply nonparametric procedures when the data to be analyzed consist of only ranks or frequency counts, rather than measurements such as height, weight, length, test score, and so on. As we have already seen, the chi-square test is one nonparametric procedure that utilizes count data.

As you might expect, nonparametric procedures are not without some disadvantages. We should mention the following.

1 Because they are so easy to apply, nonparametric procedures are sometimes used in cases in which the use of a parametric procedure would be appropriate. In general, this is an undesirable practice, since, when they are appropriate, parametric procedures make more efficient use of the data.

2 The calculations necessary for the application of some of the nonparametric procedures are tedious and laborious. This is especially true of some of the nonparametric interval-estimation procedures.

When to use nonparametric statistics Nonparametric procedures provide useful, and sometimes the only, alternatives in many situations, including the following.

1 *When the hypothesis to be tested does not involve a population parameter* As we have seen, the chi-square test of independence tests null hypotheses that are not statements about population parameters. Later in this chapter we shall discuss a procedure for testing the null hypothesis that a sample drawn from some population is a random sample. Procedures are also available for testing the null hypothesis that a sample has been drawn from a symmetric distribution.

2 *When the data consist of frequency counts or ranks, rather than measurements such as height, weight, test score, and so on* Frequency

counts and ranks are examples of what we call *weak* measurement scales. Measurements in inches or pounds, most test scores, and so on are considered to constitute *strong* measurement scales. We shall discuss measurement scales in more detail in Section 11.1.

3 *When the assumptions necessary for the valid application of a parametric procedure are not met* Frequently the design and other aspects of an experiment may suggest a certain parametric analytical procedure, but close scrutiny of the data may reveal that one or more of the assumptions underlying the parametric procedure are not met. In many such instances, a nonparametric procedure is available as an alternative.

4 *When results are needed in a hurry* As already noted, one may apply many of the nonparametric procedures easily and quickly, thereby obtaining in a short time information available only after a relatively long time when one uses parametric procedures.

11.1 MEASUREMENT AND MEASUREMENT SCALES

We have already mentioned "weak" and "strong" measurement scales, with the implication that nonparametric statistical procedures are more appropriate for data based on such weak measurement scales as ranks and frequency counts. Ideas regarding measurement scales have had a tremendous impact on statistical thought in terms of the proper domains of parametric and nonparametric techniques. We may distinguish the following four scales of measurement: nominal, ordinal, interval, and ratio.

The nominal scale The weakest of the four measurement scales is the *nominal scale*. We use this scale when one object or event is distinguished from another on the basis of the names by which they are known. For example, we use the nominal scale to distinguish one mental disorder from another. Thus we have the diagnostic categories of schizophrenia, manic-depressive psychosis, psychoneurosis, and so on. The "names" employed in the application of the nominal scale of measurement need not be names in the usual sense. Numerals may also be used. The most familiar example of the use of numerals as names is the practice of assigning numbers to the members of athletic teams. The use of the nominal scale enables us to specify that one object or event is *different* from another.

The ordinal scale When we can distinguish one object or event from another on the basis of whether or not it has more or less of some characteristic than another object or event does, we can use the *ordinal scale*. We can order or rank several objects or events having varying amounts of some characteristic on the basis of the characteristic. For example, we may rank families according to socioeconomic status, students according to the order in which they complete an examination, and contestants in a beauty contest according to attractiveness. When objects or events are ranked on the basis of some characteristic, it is possible for us to determine which object or event has more or less of the characteristic than any other, but it is not possible for us to tell from the ordering alone by how much they differ. Consider, for example, three objects that are ranked first, second, and third on the basis of some characteristic. The amount by which the object ranked second differs from the object ranked first is not necessarily equal to the amount by which it differs from the object ranked third.

The interval scale When we use the *interval scale*, we are able to indicate the amount by which one object or event differs from another. Suppose, for example, that objects A and B are assigned scores of 30 and 40 on an interval scale, and that two other objects F and G are assigned scores of 60 and 70. Knowing that we are using an interval scale enables us to say that objects A and B differ by an amount equal to that by which objects F and G differ. A familiar example of an interval scale is temperature as measured in degrees. Thus, if the high temperature yesterday was 50°F, and today it was 55°F, we can say not only that today's high exceeded yesterday's, but also that the difference was 5°F.

The ratio scale The strongest, or most sophisticated, of the four measurement scales is the *ratio scale*. Using this scale enables us to indicate how many times as great one object or event is as another, in addition to indicating the amount by which they differ. Suppose, for example, that the annual incomes of families A and B are $5000 and $10,000, respectively. We can say that the annual income of family B is $5000 more than that of family A. We can also say that the annual income of family B is two times as large as that of family A. The ratio scale, unlike the other scales, is characterized by the presence of a true zero point. For example, 0°F is not a true zero point, since it does not indicate the absence of temperature. Devices customarily used to measure weight, say, use a ratio scale, since a reading of zero indicates an absence of weight.

11.2 THE BINOMIAL TEST

The first nonparametric test we discuss is the *binomial test*, which is based on the binomial distribution. We covered some of the fundamental concepts underlying the binomial test in the discussion of the binomial distribution in Chapter 3, and you may wish to reread that discussion now.

Many situations arise in which an investigator wants to test the null hypothesis that, in some population of interest, the proportion of subjects having a certain characteristic is equal to some value p. For example, an investigator may wish to test a null hypothesis regarding the proportion of college students who consider themselves right-wing in their political views, or the proportion of high school students who smoke, or the proportion of prison inmates who are repeaters, or the proportion of cancer victims who survive for 5 years or longer, or the proportion of college graduates who take jobs in fields different from their major field of concentration in college. The list could go on and on.

The null hypothesis may have a two-sided alternative or one of two possible one-sided alternatives. That is, the hypotheses may take one of the following forms.

(a) H_0: $p = p_0$
H_1: $p \neq p_0$ (Two-sided test)

(b) H_0: $p \leq p_0$
H_1: $p > p_0$ (One-sided test)

(c) H_0: $p \geq p_0$
H_1: $p < p_0$ (One-sided test)

where p_0 is the hypothesized value of the population proportion.

Assumptions underlying the binomial test The following assumptions underlie the binomial test.

1 Each of the n observations can be classified as either having or not having the characteristic of interest.

2 The n observations are independent.

3 The probability p of having the characteristic of interest remains constant throughout the sampling procedure.

Procedure Draw a random sample of size n from the population of interest, and note the number x in the sample possessing the characteristic of interest. Compute the probability of observing a number with the characteristic of interest as extreme as or more extreme than x. If this probability is equal to or less than some significance level α, reject the null hypothesis at the α level of significance. If the computed probability is greater than α, do not reject H_0, and conclude that the null hypothesis may be true. In two-sided tests, α, as usual, is divided equally between the two tails of the distribution, since either extremely large or extremely small values of x cause rejection of the null hypothesis. For small samples, one may obtain the probability of observing a number with the characteristic of interest as extreme as or more extreme than x by referring to a table of binomial probabilities, such as Appendix Table D. For samples larger than the largest n of available tables, a large-sample approximation is available, as explained, on page 357.

Figure 11.1 Rejection region for Example 11.1

X:	0	1	2	3	4	5	6	···	13	14	15	16
$P(X \leq x \mid 16, 0.40)$:	0.0003	0.0033	0.0183	0.0651	0.1665	0.3288	0.5271	···	0.9999	1.0000	1.0000	1.0000
	REJECTION REGION			ACCEPTANCE REGION								

Example 11.1 A high school physical education director wishes to know whether it is possible to conclude that fewer than 40% of high school athletes participate in sports primarily because of the benefits to health and physical fitness which come from participation.

The hypotheses are

$$H_0: p \geq 0.40, \qquad H_1: p < 0.40$$

Suppose we choose a significance level of $\alpha = 0.05$. To reach a decision, we select a random sample of 16 high school athletes, and interview each to determine her or his primary reason for participating in sports. We determine that two of the athletes participate in sports primarily for reasons of health and physical fitness.

The question to be answered is: Is the fact that, out of a random sample of 16, two had the characteristic of interest incompatible with the hypothesis that 40% or more of the subjects in the population have the characteristic of interest? To answer this question, we compute the probability of observing two or fewer with the characteristic of interest when $n = 16$ and $p = 0.40$. In other words, we wish to determine

$$P(X \leq 2 \mid 16, 0.40)$$

Reference to Appendix Table D reveals this probability to be 0.0183, and we can reject H_0 at the 0.05 level of significance, since $0.0183 < 0.05$. We conclude that the proportion of high school athletes constituting the sampled population who participate in sports primarily for reasons of health and physical fitness is less than 0.40. Figure 11.1 shows the rejection region for this example.

Large-sample approximation We may use the normal approximation of the binomial to test hypotheses (a), (b), and (c) given above when n is large and neither p nor $1-p$ is too close to zero or one. As we pointed out in Chapter 3, a rule of thumb frequently followed is that the normal approximation of the binomial is appropriate when np and $n(1-p)$ are both greater than five. Since we covered the use of the normal approximation of the binomial for testing hypotheses in considerable detail in Chapter 6, we shall not give additional examples here. You may find it helpful to reread Section 6.7 at this time.

EXERCISES

1 A sociologist believes that fewer than 20% of the heads of household in a certain area are blue-collar workers. In a random sample of 25 of the households in the area, one of the heads of household was a blue-collar worker. Do these data support the sociologist's hypothesis at the 0.05 level of significance?

2 It is hypothesized that, in a certain population of young people contemplating marriage, the proportion of couples preferring a civil marriage ceremony is the same as the proportion preferring a religious ceremony. A random sample of 20 couples revealed that five preferred a civil ceremony. On the basis of these data, can one conclude that the two proportions are not equal? Let $\alpha = 0.05$.

3 A sample of 25 residences in a certain area revealed that in 11 the average number of occupants per room was greater than 1.5 persons. Can one conclude, on the basis of these data, that the proportion of residences in the area with occupancy densities this high is greater than 0.40? Let $\alpha = 0.05$.

4 A social worker with a state board of corrections believes that fewer than 40% of the persons referred for confinement have an adequate reading ability. In a sample of 25 referrals, five had an adequate reading ability. Do these data support the social worker's hypothesis? Let $\alpha = 0.05$.

5 When a sample of 20 voters in a certain district were questioned, 15 said that they voted in the last national election. Are these data compatible with the contention that more than 50% of the voters in the district voted in the last national election? Let $\alpha = 0.05$.

11.3 THE ONE-SAMPLE RUNS TEST

The next test we discuss is based on the concept of a run as observed in a sequence of data items which we may classify as being one of two possible types. We define a *run* as a sequence of like observations preceded and followed by a different type of observation or by no observation at all. Suppose, for example, that on a certain Monday morning, male and female applicants for drivers' licenses presented themselves to the issuer of licenses in the following order.

FF M FF M FF MM F MMM FFFF

There are a total of nine runs in this sequence. We have a run consisting of two females, followed by a run of one male, followed by another run of two females, and so on.

We may use the runs test to test the null hypothesis that a sequence of observations is random. For example, we may use the runs test to test the null hypothesis that a sample drawn from some population is a random sample. The presence of either too few or too many runs in a set of data casts suspicion on the randomness of the data. Suppose that the above sequence of 18 males and females had occurred as follows.

FFFFFFFFFFF MMMMMMM

We would tend to doubt the randomness of this sequence, because there are so few runs. On the other hand, if the following sequence had occurred, we might doubt its randomness because of too many runs.

FF M FF M FF M F M F M F M F M F

The one-sample runs test enables us to decide whether or not the number of runs in a sample of data is small enough or large enough to cause rejection of the null hypothesis that the observed sequence is random.

Assumption underlying the one-sample runs test The only assumption underlying the one-sample runs test is that each observation be classifiable as belonging to one or the other of two types of observation.

Procedure The test procedure consists of the following steps.

1 Designate the number of observations of one type as n_1 and the number of observations of the other type as n_2. (The total sample size is equal to $n_1 + n_2 = n$.)

2 Determine the number of runs, r.

3 Consult Table M of the Appendix. If the observed number of runs is less than or equal to the appropriate critical value of r in Appendix Table M(a) or greater than or equal to the appropriate critical value of r in Appendix Table M(b), reject the null hypothesis of randomness at the 0.05 level of significance.

Example 11.2 Let us refer to the sequence, mentioned above, of males and females applying for drivers' licenses. We test

H_0: The sequence is random

against the alternative

H_1: The sequence is not random

We let n_1 = number of females = 11 and n_2 = number of males = 7. The number of runs r is 9, and $n = 18$. Reference to Appendix Tables M(a) and M(b) shows that the critical values are 5 and 14. Since $5 < 9 < 14$, we cannot reject H_0, and we conclude that the sequence may be random. Figure 11.2 shows the rejection region for this example.

Large-sample approximation When either n_1 or n_2 is larger than 20, we cannot use Appendix Table M to determine whether or not an observed number of runs is significant. For large samples, the quantity

Figure 11.2 Rejection region for Example 11.2

Possible number of runs		
2 3 4 5	6 7 8 9 10 11 12 13	14 15
REJECTION REGION	ACCEPTANCE REGION	REJECTION REGION

$$z = \frac{r - \left[\dfrac{2n_1n_2}{n_1 + n_2} + 1\right]}{\sqrt{\dfrac{2n_1n_2(2n_1n_2 - n_1 - n_2)}{(n_1 + n_2)^2(n_1 + n_2 - 1)}}} \tag{11.1}$$

is distributed approximately as the standard normal distribution. In Equation 11.1.

$$\frac{2n_1n_2}{n_1 + n_2} + 1$$

is the mean and

$$\sqrt{\frac{2n_1n_2(2n_1n_2 - n_1 - n_2)}{(n_1 + n_2)^2(n_1 + n_2 - 1)}}$$

is the standard deviation of the sampling distribution of r. To test a computed r for significance, we compare the r of Equation 11.1 with the z from the standard normal distribution corresponding to the chosen level of significance.

Example 11.3 The win-lose record of a certain high school basketball team, for their last 60 consecutive games, was as follows.

WWWWWW L WWWWWWW L WWWWWWWW LLL WW LLLL W
LLLL W LLLL WW LL WWWW L WWWW L WWWWWW

The null and alternative hypotheses are

H_0: The sequence of wins and losses is random

H_1: The sequence is not random

Suppose we let $\alpha = 0.05$. For the above sequence, we let n_1 = number of wins = 39 and n_2 = the number of losses = 21. The number of runs is $r = 19$.

Using Equation 11.1, we compute

$$z = \frac{19 - \left[\frac{2(39)(21)}{39 + 21} + 1\right]}{\sqrt{\frac{2(39)(21)[2(39)(21) - 39 - 21]}{(39 + 21)^2(39 + 21 - 1)}}} = -2.67$$

Since -2.67 is less than -1.96, we reject H_0, and conclude that the sequence of wins and losses is not random.

EXERCISES

6 A psychologist wished to know whether people who arrive earlier at a social function tend to make higher or lower scores on a social-skills test than those who arrive later. At a social function attended by 26 adults, the psychologist kept a record of the order in which the guests arrived. Following the affair, each guest took a social-skills test. Table 11.1 shows results. Replace each score with a plus sign (+) if it falls above the median and with a minus sign (−) if below the median. Then use the runs test to test the null hypothesis that the scores are in random order.

Table 11.1 Data for Exercise 6

Order of arrival	Score	Order of arrival	Score	Order of arrival	Score
1	52	10	53	19	29
2	42	11	42	20	53
3	26	12	59	21	59
4	39	13	36	22	47
5	41	14	54	23	52
6	54	15	21	24	36
7	36	16	24	25	43
8	55	17	49	26	48
9	56	18	57		

7 A psychiatrist in a psychiatric hospital observed a patient for 21 days. Each day the psychiatrist noted whether the patient was depressed (D) or not depressed (ND). Table 11.2 gives the results. Do these data suggest a lack of randomness in the occurrence of depression?

8 Table 11.3 displays the ranks of a random sample of 15 ninth-grade students according to their scores on an intelligence test. The rank of 1 indicates the highest score. In addition, the notation (C) indicates that the student was rated by his or her teacher as exhibiting an above-average degree of creativity. Can one conclude that in the population there is a lack of randomness in the distribution of creativity among students ranked according to intelligence?

Table 11.2 Data for Exercise 7

Day	Condition	Day	Condition
1	D	11	ND
2	ND	12	ND
3	ND	13	D
4	ND	14	D
5	D	15	D
6	D	16	D
7	ND	17	ND
8	D	18	ND
9	ND	19	D
10	ND	20	D
		21	D

Table 11.3 Data for Exercise 8

Rank according to intelligence	Rank according to intelligence
1 C	9
2 C	10 C
3	11
4 C	12 C
5 C	13
6 C	14
7	15
8 C	

9 Table 11.4 shows the ranks of 20 members of a certain youth organization according to popularity with their peers and according to their leadership ability, as perceived by the adult leaders of the organization. Consider persons with leadership ranks of 1 through 10 as leaders and the remainder as nonleaders. Do these data suggest a difference between leaders and nonleaders with respect to popularity?

Table 11.4 Data for Exercise 9

Person	Popularity rank	Leadership rank	Person	Popularity rank	Leadership rank
Ann	1	3	Karl	11	6
Bill	2	1	Larry	12	10
Carol	3	5	Matt	13	15
David	4	2	Nancy	14	13
Ed	5	7	Opal	15	20
Frank	6	12	Pat	16	17
Gail	7	11	Ray	17	19
Heather	8	9	Sandra	18	14
Iris	9	8	Ted	19	18
Jane	10	4	Veronica	20	16

11.4 THE SPEARMAN RANK CORRELATION COEFFICIENT

We often find that the nature of an investigation suggests the calculation of a measure of correlation as part of the analysis, but because the assumptions underlying the correlation methods of Chapters 9 are not met, we cannot apply these methods. An alternative in such a situation is to employ a nonparametric measure of correlation. One of the simplest and most widely used nonparametric measures of correlation for the two-variable case is the *Spearman rank correlation coefficient*.

We compute the Spearman rank correlation coefficient, designated by r_S, from the ranks assigned to the sample values of the variables X and Y of a bivariate distribution. That is, the Spearman rank correlation coefficient makes use of ranks rather than original observations. Sometimes the original observations are themselves ranks, in which case we use them. Otherwise, we convert the original observations to ranks.

We may use the Spearman rank correlation coefficient as a test statistic to test for independence between X and Y. The null hypotheses that may be tested and their alternatives are as follows.

(a) H_0: X and Y are mutually independent
H_1: Either large values of X tend to be paired with large values of Y or large values of X tend to be paired with small values of Y

(b) H_0: X and Y are mutually independent
H_1: Large values of X tend to be paired with large values of Y

(c) H_0: X and Y are mutually independent
H_1: Large values of X tend to be paired with small values of Y

The first of the above hypotheses leads to a two-sided test; the other two lead to one-sided tests.

Assumptions underlying the Spearman rank correlation coefficient When we use r_S for testing the hypotheses given above, we assume that X and Y are independent and continuous.

Procedure We calculate r_S and perform the accompanying hypothesis-testing procedure as follows.

1 Given n pairs of measurements on X and Y, rank the values of X from 1 (assigned to the smallest value of X) to n, and rank the values of Y from 1 (assigned to the smallest value of Y) to n.

2 For each pair of observations, compute $d_i = (\text{rank of } X_i) - (\text{rank of } Y_i)$.

3 Square each d_i and compute $\Sigma\, d_i^2$.

4 Compute

$$r_S = 1 - \frac{6\,\Sigma\, d_i^2}{n(n^2 - 1)} \tag{11.2}$$

5 When n is between 4 and 30, compare the computed value of r_S for significance with the appropriate critical value of r_S^* given in Appendix Table N. For the two-sided test [hypothesis (a)], reject H_0 at the α level of significance if r_S is greater than r_S^* or less than $-r_S^*$, where r_S^* is located at the intersection of the column labeled $\alpha/2$ and the row corresponding to n. For hypothesis (b), reject H_0 if r_S is greater than r_S^* for α and n. And for hypothesis (c), reject H_0 if r_S is less than $-r_S^*$ corresponding to α and n.

6 When n is equal to or greater than 10, you may compare the statistic

$$t = r_S \sqrt{\frac{n-2}{1-r_S^2}} \tag{11.3}$$

for significance with appropriate values of Student's t distribution with $n-2$ degrees of freedom. You may also compute the statistic

$$z = r_S \sqrt{n-1} \tag{11.4}$$

and compare it for significance with appropriate values of the standard normal distribution. This approximation is good for values of n as small as 10.

7 Tied observations may present a problem. The use of Appendix Table N is strictly valid only when there are no ties or when, in the event of ties, one uses some random procedure for breaking them. In practice, investigators sometimes use the table after they have used some other method for handling ties.

When the number of ties is small, one may assign to the tied observations the mean of the ranks for which they are tied, and proceed with steps 2 through 6 as given above.

If there are a large number of ties, one may use the following procedure.

1 Compute

$$T = \frac{t^3 - t}{12}$$

where t is the number of observations tied for a given rank in the X's or Y's.

2 Compute

$$r_S = \frac{\Sigma\, x^2 + \Sigma\, y^2 - \Sigma\, d_i^2}{2\sqrt{\Sigma\, x^2\, \Sigma\, y^2}} \tag{11.5}$$

where

$$\sum x^2 = \frac{n^3 - n}{12} - \sum T_x$$

$$\sum y^2 = \frac{n^3 - n}{12} - \sum T_y$$

$\sum T_x$ = the sum of the values of T for the tied ranks occurring in X

$\sum T_y$ = the sum of the values of T for the tied ranks occurring in Y

Unless the number of ties is quite large, using Equation 11.5 will make very little difference in the value of r_S.

Example 11.4 In a study of metabolism in a certain species of wild animal, a biologist obtained an activity index and data on metabolic rates for ten animals observed in captivity. Table 11.5 shows the results. The investigator wished to test

H_0: Metabolic rate and activity index are independent

against

H_1: Metabolic rate tends to increase as the activity index increases

Suppose we choose a significance level of 0.05. Table 11.6 gives the intermediate calculations necessary for computing r_S.

In assigning ranks to the values of Y, we note that animals 3 and 6 are tied. Since these values of Y occupy positions 4 and 5 in the ranked sequence, we assign to each a rank equal to the mean of the ranks 4 and 5, that is, $(4 + 5)/2 = 4.5$.

Table 11.5 Metabolic rate and activity index in ten wild animals observed in captivity

Animal	Activity index, X	Metabolic rate, Y
1	50	0.16
2	300	0.20
3	225	0.19
4	600	0.25
5	450	0.23
6	275	0.19
7	200	0.21
8	150	0.18
9	500	0.24
10	100	0.17

Table 11.6 Intermediate calculations for Example 11.4

Animal	X	Y	Rank of X	Rank of Y	d_i	d_i^2
1	50	0.16	1	1	0	0
2	300	0.20	7	6	1	1
3	225	0.19	5	4.5	0.5	0.25
4	600	0.25	10	10	0	0
5	450	0.23	8	8	0	0
6	275	0.19	6	4.5	1.5	2.25
7	200	0.21	4	7	−3	9
8	150	0.18	3	3	0	0
9	500	0.24	9	9	0	0
10	100	0.17	2	2	0	0
						$11.50 = \sum d_i^2$

Using the data in Table 11.6 and Equation 11.2, we compute

$$r_S = 1 - \frac{6(11.50)}{10(10^2 - 1)} = 0.93$$

Reference to Appendix Table N reveals that when $n = 10$, $\alpha = 0.05$, and we have a one-sided test, the critical value of r_S^* is 0.5515. Since 0.93 is greater than 0.5515, we reject H_0, and conclude that metabolic rate increases as the activity index increases.

Now let us illustrate the calculation of r_S and test it for significance for a case in which there are several ties.

Example 11.5 Each of 20 new employees of a manufacturing plant took a test designed to measure her/his aptitude for the job she/he was hired to fill. At the end of six months, each of these employees was assigned, by her or his supervisor, a score measuring how well the supervisor thought the employee was performing the job. A psychologist with the firm wished to test

H_0: Aptitude scores and supervisor ratings are independent

H_1: High aptitude scores tend to be accompanied by high rating by supervisor

at the 0.05 level of significance. Table 11.7 shows the original scores, the ranks, and the intermediate calculations for r_S.

Table 11.7 Original scores and intermediate calculations for Example 11.5

Employee	Aptitude score X	Rank of X	Supervisor's rating, Y	Rank of Y	d_i	d_i^2
1	79	12	86	13	−1	1
2	80	14	96	20	−6	36
3	51	1	53	1	0	0
4	76	9	77	8.5	0.5	0.25
5	99	20	90	17.5	2.5	6.25
6	78	11	85	11.5	−0.5	0.25
7	76	9	88	14.5	−5.5	30.25
8	80	14	85	11.5	2.5	6.25
9	52	2.5	55	3	−0.5	0.25
10	55	4	56	4	0	0
11	90	18	88	14.5	3.5	12.25
12	62	6	66	6	0	0
13	81	16	90	17.5	−1.5	2.25
14	52	2.5	54	2	0.5	0.25
15	69	7	58	5	2	4
16	76	9	77	8.5	0.5	0.25
17	58	5	68	7	−2	4
18	90	18	89	16	2	4
19	80	14	84	10	4	16
20	90	18	95	19	−1	1
						$124.50 = \sum d_i^2$

First, let us compute r_S without making an adjustment for ties. By Equation 11.2, we have

$$r_S = 1 - \frac{6(124.50)}{20(20^2 - 1)} = 0.9064$$

Reference to Appendix Table N reveals that the critical value of r_S^* for this example is 0.3789. Since 0.9064 is greater than 0.3789, we reject H_0, and conclude that high aptitude scores tend to be accompanied by high ratings by supervisor.

Now let us compute r_S by adjusting for ties. First we note that, when we assign the ranks to the X values, there are three sets of ties in which three observations are tied for a given rank, and one set of two tied ranks. Thus we compute

$$T_x = \frac{3^3 - 3}{12} + \frac{3^3 - 3}{12} + \frac{2^3 - 2}{12} = 4.5$$

In assigning ranks to the Y observations, we find four sets of tied ranks with $t = 2$, so

$$T_y = 4\left(\frac{2^3 - 2}{12}\right) = 2$$

We now compute

$$\sum x^2 = \frac{20^3 - 20}{12} - 4.5 = 660.5 \qquad \text{and} \qquad \sum y^2 = \frac{20^3 - 20}{12} - 2 = 663$$

Finally, by Equation 11.5, we have

$$r_S = \frac{660.5 + 663 - 124.50}{2\sqrt{(660.5)(663)}} = 0.9059$$

which is very close to that obtained without the adjustment for ties. Since there are a number of ties in this example, the use of Appendix Table N is not strictly valid. Alternatively, we can use Equation 11.3 and compute (using the last computed value of r_S)

$$t = 0.9059\sqrt{\frac{20 - 2}{1 - (0.9059)^2}} = 9.08$$

which is significant at the 0.05 level. For comparative purposes, let us also compute, by Equation 11.4,

$$z = 0.9059\sqrt{20 - 1} = 3.95$$

which is also significant at the 0.05 level.

10 A test designed to measure anxiety was given to a random sample of 12 eighth-grade students. Following the test, each subject was asked to memorize a short poem. A record was kept of the number of minutes required before each subject was able to recite the poem without error. Table 11.8 shows the results of the experiment. Compute r_S. Can one conclude from these data that high anxiety scores tend to be paired with the requirement of a large number of trials to learn the poem?

Table 11.8 Data for Exercise 10

Subject	Anxiety score, X	Trials, Y	Subject	Anxiety score, X	Trials, Y
1	94	6	7	23	1
2	76	4	8	47	3
3	88	5	9	48	2
4	94	6	10	52	3
5	87	5	11	68	3
6	87	6	12	50	2

11 A random sample of 16 short stories written by students in a creative writing class were ranked by the teacher on the basis of originality and craftsmanship. Table 11.9 gives the results. Compute r_S. Test the null hypothesis that originality and craftsmanship are independent.

Table 11.9 Data for Exercise 11

Story	Originality rank	Craftsmanship rank	Story	Originality rank	Craftsmanship rank
1	14	13	9	1	3
2	8	7	10	2	4
3	16	15	11	7	6
4	9	11	12	3	2
5	10.5	12	13	15	16
6	5	8	14	12	9
7	4	1	15	6	5
8	10.5	10	16	13	14

12 Table 11.10 shows the scores made by 20 seventh-grade students on a test designed to measure level of self-concept. The students' teachers ranked them on the basis of degree of initiative generally exhibited in problem-solving situations. The table also shows these ranks. Compute r_S. On the basis of these data, can one conclude that high self-concept scores tend to be paired with high initiative scores?

Table 11.10 Data for Exercise 12

Student	Self-concept score	Rank on initiative	Student	Self-concept score	Rank on initiative
1	70	12	11	73	11
2	80	18	12	76	9
3	61	8	13	71	7
4	84	20	14	68	10
5	56	5	15	84	14
6	84	17	16	76	6
7	57	4	17	73	15
8	85	19	18	82	13
9	61	3	19	83	16
10	63	2	20	51	1

13 Table 11.11 shows the ranks of 15 high school students according to (a) popularity with peers and (b) academic performance. Compute r_S. Test the null hypothesis that popularity and academic performance are independent.

Table 11.11 Data for Exercise 13

Student	Popularity rank	Academic performance rank	Student	Popularity rank	Academic performance rank
1	14	15	9	9.5	5
2	8	6	10	1	3
3	12.5	7	11	6	10
4	11	1	12	5	11
5	9.5	14	13	7	2
6	12.5	4	14	2	9
7	15	13	15	3	8
8	4	12			

14 Table 11.12 shows ten occupations ranked on the basis of prestige by a panel of high school seniors and by a panel of high school counselors. Compute r_S. Test the null hypothesis that students' ratings and counselors' ratings are independent.

Table 11.12 Data for Exercise 14

Occupation	Students' ranking	Counselors' ranking	Occupation	Students' ranking	Counselors' ranking
1	9	5	6	4	6
2	8	4	7	1	2
3	7	3	8	6	7
4	10	8	9	3	9
5	2	1	10	5	10

15 Referring to Exercise 34 of Chapter 9, compute the coefficient of rank correlation, and test for significance. Compare the results with those obtained by parametric methods.

16 Refer to Exercise 35 of Chapter 9, and proceed as in Exercise 15.

17 As part of a larger study, a medical research team collected aldosterone-secretion values and body-surface-area values on 18 infants and children. Table 11.13 gives the results. Do these data provide sufficient evidence to indicate that aldosterone-secretion levels and body-surface-area values in infants and children are not independent?

Table 11.13 Data for Exercise 17

Subject	Body surface area, X	Aldosterone level, Y	Subject	Body surface area, X	Aldosterone level, Y
1	0.20	22	10	0.60	130
2	0.25	41	11	0.30	62
3	0.50	80	12	0.28	36
4	0.60	50	13	0.30	100
5	0.70	70	14	0.40	100
6	0.55	110	15	0.25	70
7	0.40	32	16	0.72	122
8	0.55	69	17	0.70	90
9	0.75	120	18	0.50	50

18 Table 11.14 shows the scores made by 20 high school seniors on two psychological tests. Do these data provide sufficient evidence to indicate that these measures of hostility and self-concept are not independent?

Table 11.14 Data for Exercise 18

Student	Self-concept score, X	Hostility score, Y	Student	Self-concept score, X	Hostility score, Y
1	20	48	11	60	40
2	20	64	12	60	64
3	20	80	13	70	20
4	30	70	14	70	40
5	40	48	15	80	30
6	40	70	16	80	48
7	50	20	17	90	10
8	50	40	18	90	20
9	50	56	19	90	40
10	60	10	20	100	30

19 Table 11.15 shows the scores made by mothers and their daughters on a test designed to measure level of anxiety. Can one conclude, on the basis of these data, that anxiety scores in mothers and their daughters are not independent?

Table 11.15 Data for Exercise 19

Mother's score, X	Daughter's score, Y	Mother's score, X	Daughter's score, Y	Mother's score, X	Daughter's score, Y
10	20	50	40	80	60
20	20	60	50	80	70
30	10	60	60	80	92
30	30	60	80	92	80
30	40	70	40	100	70
40	30	70	50	100	100
40	50	70	80	110	80
50	20	70	100	80	81
50	30	80	30		

20 Thirty predatory fish were placed in individual tanks. A uniform number of smaller fish, all of equal size, were placed in each tank. Table 11.16 shows a measure of the size of each predatory fish and the number of smaller fish remaining in the corresponding tank at the end of six weeks. Do these data provide sufficient evidence to indicate that size of predatory fish and number of smaller fish remaining are not independent?

Table 11.16 Data for Exercise 20

Tank	Size of predatory fish, X	Number of smaller fish, Y	Tank	Size of predatory fish, X	Number of smaller fish, Y
1	10	18	16	80	10
2	20	12	17	80	12
3	20	15	18	90	9
4	30	13	19	90	11
5	30	17	20	90	13
6	40	11	21	100	6
7	40	14	22	100	7
8	50	10	23	100	8
9	50	12	24	100	9
10	60	10	25	110	6
11	60	13	26	110	8
12	60	15	27	110	10
13	70	9	28	120	5
14	70	12	29	120	7
15	80	8	30	120	10

11.5 THE MEDIAN TEST

As we have seen, a question investigators often ask is whether or not the means of two populations are equal. In Chapter 6 we learned to test H_0: $\mu_1 = \mu_2$ either by means of the t test or by application of the central-limit theorem, depending on the circumstances. If the assumptions underlying these procedures are not met, we must seek an alternative approach. One alternative is the nonparametric procedure called the *median test.*

The median test focuses on the median, rather than the mean, as the measure of central tendency. In symmetrical distributions, the median and mean coincide. Thus, if the two populations under consideration are symmetric, inferences about the median apply to the mean also.

Assumptions underlying the median test The use of the median test is based on the assumptions that the samples are random and that they are independent of each other. The data must be measured on at least an ordinal scale.

Procedure To test the null hypothesis that two population medians are equal, we draw an independent random sample from each population under consideration. We designate the size of the first sample by n_1, and the size of the second sample, which need not be equal in size to the first, by n_2. We designate the total number of observations in the two samples by N.

We then determine the median of the N observations obtained by merging the two samples, and make a count to determine how many observations from each of the original samples are greater than and how many are less than or equal to the median of the combined observations. We may display the resulting frequencies in a 2×2 contingency table such as Table 11.17.

If the minimum expected cell frequencies, as specified in Chapter 10, are met, we may compute X^2 by Equation 10.3 and compare it for significance with χ^2 with one degree of freedom and the desired value of α.

Table 11.17 Data display for median test

Number of observations	Sample 1	Sample 2	Total
Greater than median	a	b	$a + b$
Less than or equal to median	c	d	$c + d$
	$a + c = n_1$	$b + d = n_2$	$N = n_1 + n_2$

Example 11.6 An educator wished to know whether it was possible to conclude that freshman students of a given high school who come from rural elementary schools and those who come from urban elementary schools differ with respect to median verbal reasoning scores. A random sample of 15 students from rural and 18 from urban elementary schools made the scores on a verbal reasoning test shown in Table 11.18.

The appropriate null and alternative hypotheses are

H_0: The medians of the two populations are equal
H_1: The medians of the two populations are not equal

We shall let $\alpha = 0.05$.

Table 11.18 Verbal reasoning test scores made by 15 high school freshmen from rural elementary schools and 18 freshmen from urban elementary schools

Students from rural schools	66	73	63	65	80	74	69	80	63	54	42	80	57	79	52			
Students from urban schools	68	91	89	60	73	90	79	97	64	62	91	69	64	89	99	76	77	96

The median of the 33 observations combined is 73. When we count the number in each sample greater than and the number less than or equal to the median, we have the results shown in Table 11.19.

By Equation 10.3, we compute

$$X^2 = \frac{33\,[(5)(7) - (11)(10)]^2}{(15)(18)(17)(16)} = 2.528$$

Reference to Appendix Table G reveals that $X^2 = 2.528$ is not significant at the 0.05 level. Consequently, we do not reject H_0, and we conclude that the two population medians may be equal.

Table 11.19 Contingency table for Example 11.6

Number of observations	**Sample** RURAL	URBAN	**Total**
Greater than 73	5	11	16
Less than or equal to 73	10	7	17
	15	18	33

EXERCISES

21 A test designed to measure level of self-esteem was given to two samples of prepubescent children who were outpatients at a children's psychiatric clinic. Sample A consisted of 10 children who were experiencing difficulties with gender identification. Sample B consisted of 12 children who were being seen at the clinic for other problems. Table 11.20 shows the test scores for these two groups. Can one conclude, on the basis of these data, that the populations represented by the samples have different medians? Let $\alpha = 0.05$.

Table 11.20 Data for Exercise 21

Sample *A*	54	39	35	58	42	35	31	59	31	59		
Sample *B*	72	78	67	70	80	68	57	52	54	64	55	72

22 Two random samples were drawn from the inmates of a state prison system. One sample was drawn from those prisoners who were reared in

rural environments, and the other from those reared in urban environments. Table 11.21 gives the ages at time of first arrest for the two samples. On the basis of these data, can one conclude that the median age at time of first arrest is different in the two populations? Let $\alpha = 0.05$.

Table 11.21 Data for Exercise 22

Rural environment	41	20	38	25	39	40	34	35	27	25	39	38	32	18	23	38
Urban environment	16	19	20	30	31	24	18	16	22	24	21	37	24	19	13	15

23 Table 11.22 shows the reading comprehension scores of a random sample of students who dropped out of high school at age 16. Also shown are the scores of a random sample of their classmates who did not drop out. All tests were administered within 3 months of the time at which the dropouts left school. Do these data provide sufficient evidence to indicate that the two populations have different medians? Let $\alpha = 0.05$.

Table 11.22 Data for Exercise 23

Dropouts	53	36	58	52	46	60	44	69	49	44	42	73	67	59	60	61	53	80	54	80
	62																			
Classmates	84	92	62	81	51	67	74	73	78	57	58	91	89	57	64	52	80	62	51	72
	64	66	59	74	77															

11.6 THE SIGN TEST

The median test, discussed in Section 11.5, is appropriate when the data consist of two independent samples. Frequently the data available for analysis consist of observations on two related samples. Related samples arise when an investigator tries to eliminate one or more extraneous sources of variation in order to detect a difference between two groups. For example, a researcher may take measurements of the same subjects before and after some intervening experience, such as the application of some treatment. Or an investigator may match subjects on the basis of one or more variables in order to yield related samples. We discussed these techniques in Chapters 5 and 6 under the heading of paired comparisons. We learned in Chapter 6 that, if certain assumptions are met, we may use the t test to test hypotheses about the population mean difference on the basis of the mean difference observed between two related samples.

When the assumptions underlying the parametric t test are not met, an alternative approach is to use a nonparametric test. One such test that is appropriate for use with two related samples is the *sign test*. The use of the sign test has been traced as far back as 1710, and hence it is perhaps the oldest of the nonparametric procedures. Like the median test, this test also focuses on the median as the measure of central tendency. In particular, one uses the sign test to test hypotheses about the median difference, and one obtains differences by comparing pairs of observations from the two related samples.

Assumptions underlying the sign test The valid use of the sign test rests on the assumptions that the data represent measurement on at least an ordinal scale and that the individual pairs of observations are mutually independent.

Procedure Perhaps the most frequent application of the sign test is in testing the null hypothesis that the median difference is zero. Suppose we designate one set of scores by X and the set of scores comprising the related population by Y. Samples of size n from each set of scores will yield pairs of observations, which may be designated $(x_1, y_1), (x_2, y_2), \ldots, (x_i, Y_i), \ldots, (x_n, y_n)$. The test requires the calculation of the differences $x_1 - y_1, x_2 - y_2, \ldots, x_i - y_i, \ldots, x_n - y_n$. If x_i is larger than y_i, the difference is recorded as a plus (+). If x_i is smaller than y_i, the difference is recorded as a minus (–). The test utilizes only the sample of pluses and minuses, and hence the name sign test.

If, in fact, the median difference is zero, we would expect that a random sample of differences, $x_i - y_i$, would yield about as many pluses as minuses. A larger number of either sign than can reasonably be explained on the basis of chance casts doubt on the hypothesis that the median difference is zero. We may state the null hypothesis in compact form, then, as

$$H_0: P(+) = P(-) = 0.5$$

and we see that the sign test is a special case of the binomial test. The observed sequence of pluses and minuses constitutes a sample from a binomial population with parameter $p = 0.5$. Figure 11.3 shows the applicable sampling distribution for the sign test when $n = 10$. The alternative hypothesis may be either one-sided or two-sided.

Figure 11.3 Graphic representation of the sampling distribution for the sign test when $n = 10$

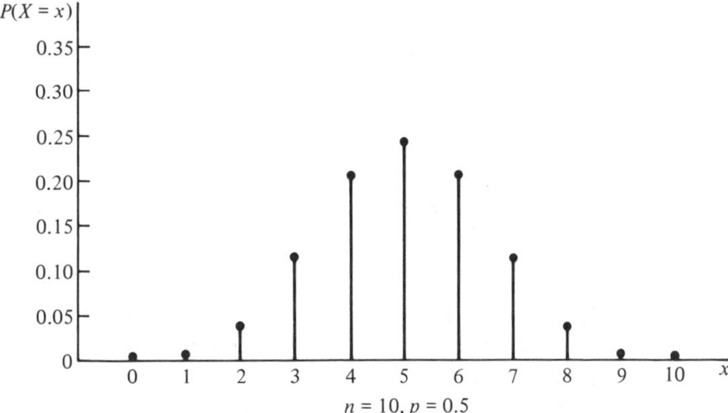

In applying the sign test, we focus on the sign that occurs less frequently, and determine the probability of obtaining, when the null hypothesis is true, as few of that sign as actually observed, or fewer. If this probability is equal to or less than some chosen level of significance α, we reject H_0. We may use Appendix Table D for determining probabilities, when the sample sizes are represented in the table.

If a difference $x_i - y_i$ is equal to zero, we eliminate the pair from the sample, and reduce the sample size accordingly.

Example 11.7 In a study of the general mental health status of retired persons, a team of psychologists selected a random sample of persons due to retire within the next two years. Each subject was given a test designed to measure his or her level of mental health. One year after retirement, each subject was again given the test. Table 11.23 gives the results.

Table 11.23 Mental health status of 15 subjects before and after retirement

Before, X	After, Y	Sign of $x_i - y_i$	Before, X	After, Y	Sign of $x_i - y_i$
76	70	+	87	92	−
80	75	+	96	85	+
86	84	+	98	88	+
87	90	−	77	76	+
85	81	+	80	85	−
95	95	0	87	81	+
97	87	+	89	84	+
75	72	+			

Investigators wished to know whether they could conclude that the level of mental health in the population from which the sample was drawn tends to be lower after retirement than before. If the subjects' mental health levels do tend to be lower after retirement than before, the population median difference computed from before (X) and after (Y) scores will be positive. That is, the population of pluses and minuses based on the differences, $x_i - y_i$, will contain more pluses than minuses. In a random sample from this population, then, we would expect fewer minuses than pluses. We may state the null and alternative hypotheses as follows.

$$H_0\colon P(+) = P(-) = 0.5, \qquad H_1\colon P(+) > P(-)$$

Suppose we let $\alpha = 0.05$.

We note in Table 11.23 that there is one difference of zero. We eliminate this observation from the analysis, and have an effective n of 14. We also note that three differences yielded minuses. We wish to know the probability of observing three or fewer minus signs in a sample of size 14 when the probability of a minus sign is 0.5. By entering Appendix Table D with $n = 14$ and $p = 0.5$, we find the probability we seek to be 0.0288. Since 0.0288 is less than the chosen significance level of 0.05, we reject H_0, and conclude

Figure 11.4 Rejection region for Example 11.7

Possible number of minus signs:	0	1	2	3	4	…	11	12	13	14
Probability of this number or fewer:	0.0001	0.0010	0.0066	0.0288	0.0899	…	0.9937	0.9993	1.0000	1.0000
	REJECTION REGION				ACCEPTANCE REGION					

that the median difference is positive; that is, that the level of mental health in the sampled population tends to be lower after retirement than before. Figure 11.4 shows the rejection region for this example.

Large-sample approximation When n is greater than or equal to 12, we may use the normal approximation to the binomial. The test statistic is

$$z = \frac{(k + 0.5) - 0.5n}{0.5\sqrt{n}} \tag{11.6}$$

where $k =$ the number of occurrences of the less frequent sign. We compare the computed value of z for significance with the appropriate value from the standard normal distribution.

Example 11.8 Twenty seventh-grade students participated in an experiment designed to evaluate the effectiveness of a certain program for increasing reading comprehension. Each subject was given a reading comprehension test before and after participating in the program. Table 11.24 shows the results.

Table 11.24 Reading comprehension scores of 20 seventh-grade students before and after participation in a program designed to improve reading comprehension

Subject	1	2	3	4	5	6	7	8	9	10	11	12	13	14	15	16	17	18	19	20
Before, X	80	65	73	60	60	73	76	53	57	65	61	54	77	66	59	74	52	56	63	67
After, Y	85	75	84	55	72	81	84	50	65	75	79	62	85	78	69	88	60	52	75	80
Sign of difference	+	+	+	−	+	+	+	−	+	+	+	+	+	+	+	+	+	−	+	+

The investigator wished to test

H_0: $P(+) = P(-) = 0.5$ (The program has no effect.)

against the alternative

H_1: $P(+) > P(-)$ (The program improves reading comprehension.)

Suppose we let $\alpha = 0.05$. Using the data in Table 11.24, we compute, by Equation 11.6,

$$z = \frac{(4 + 0.5) - 0.5(20)}{0.5\sqrt{20}} = -2.46$$

Since -2.46 is less than -1.96, we reject H_0, and conclude that the program is effective in increasing reading comprehension.

EXERCISES

24 In an experiment designed to compare two methods of teaching swimming, college freshmen girls who did not know how to swim were divided into two groups of ten girls each. Each student was matched with a student in the other group on the basis of age, physical condition, athletic aptitude, and other relevant variables. At the end of the instructional period, a swimming coach, who was unaware of the method of instruction by which each girl had been taught, evaluated each girl's swimming ability. Table 11.25 shows the results. Do these data provide sufficient evidence to indicate that method A is superior to method B? Let $\alpha = 0.05$.

25 An experiment was designed to evaluate the effectiveness of a certain treatment regimen in the relief of depression in psychiatric patients. Prior to being treated, each of the 60 patients who participated in the experiment was evaluated by a psychiatrist to determine level of depression. An evaluation following treatment revealed that 42 patients had improved. Do these data provide sufficient evidence to indicate that the treatment is effective in the relief of depression in psychiatric patients? Let $\alpha = 0.05$.

26 A manufacturer of pocket calculators can place one of two models on the market. In order to get some idea of consumer preference, the manufacturer gives each of 15 students, randomly drawn from the students at a local engineering school, each of the two models to use for one month. At the end of the period, 11 students say they prefer model A; three prefer model B; and one has no preference. Do these data provide sufficient evidence to indicate that model A will be preferred by the population represented? Let $\alpha = 0.05$.

Table 11.25 Data for Exercise 24

Swimming ability score, by method of instruction										
Pair	1	2	3	4	5	6	7	8	9	10
Method *A*	8	8	8	9	8	7	6	9	6	7
Method *B*	6	7	5	9	6	5	4	7	8	6

11.7 THE WILCOXON MATCHED-PAIRS SIGNED-RANKS TEST

Another test that is appropriate for use with the data of two related samples is one known as the *Wilcoxon matched-pairs signed-ranks test.* We saw in Section 11.6 that the sign test utilizes only the sign of the difference between the members of pairs of observations. The test we discuss in this section makes use of the relative magnitudes of the differences, as well as their signs.

Thus we see that the Wilcoxon matched-pairs signed-ranks test uses more of the information contained in the data than the sign test. When the investigator can choose between the sign test and the Wilcoxon test, the latter is preferred, since, because it uses more information, it is usually the more powerful test.

Assumptions underlying the Wilcoxon matched-pairs signed-ranks test The assumptions underlying the Wilcoxon matched-pairs signed-ranks test are as follows.

1 The observed matched pairs constitute a random (bivariate) sample of the variables X and Y.

2 The differences between matched pairs are mutually independent values of a continuous random variable.

3 The distribution of differences is symmetric.

4 The differences are measured on at least an interval scale.

We may use the Wilcoxon matched-pairs signed-ranks test to test the following null hypotheses and their alternatives.

(a) H_0: $E(X) \geq E(Y)$ (or $\mu_X \geq \mu_Y$)
H_1: $E(X) < E(Y)$ (or $\mu_X < \mu_Y$)

(b) H_0: $E(X) \leq E(Y)$ (or $\mu_X \leq \mu_Y$)
H_1: $E(X) > E(Y)$ (or $\mu_X > \mu_Y$)

(c) H_0: $E(X) = E(Y)$ (or $\mu_X = \mu_Y$)
H_1: $E(X) \neq E(Y)$ (or $\mu_X \neq \mu_Y$)

Procedure To apply the Wilcoxon matched-pairs signed-ranks test, carry out the following steps.

1 For n matched pairs of observations, designate the observations on X from the one group by $x_1, x_2, \ldots, x_i, \ldots, x_n$, and the corresponding matched observations on Y from the other group by $y_1, y_2, \ldots, y_i, \ldots, y_n$. This yields the pairs $(x_1, y_1), (x_2, y_2), \ldots, (x_i, y_i), \ldots, (x_n, y_n)$.

2 For each pair of observations, determine the signed difference $d_i = (y_i - x_i)$. Omit from the analysis any pair for which $x_i = y_i$, and reduce n accordingly.

3 Rank the d_i, disregarding their signs. When two or more d_i are tied for the same rank, give each the mean of the rank positions they would otherwise occupy.

4 Assign to each rank the sign (+ or −) of the d_i giving rise to that rank.

5 Compute

$$T = \text{sum of the positive signed ranks} \tag{11.7}$$

6 To determine whether or not to reject H_0, compare the computed T with critical values given in Appendix Table O, as described below.

To determine whether or not to reject H_0 at the α level of significance, follow one of the procedures given below, depending on which pair of hypotheses is under consideration.

(a) For H_0: $E(X) \geq E(Y)$, a sufficiently large value of T will cause rejection. This is true because T is equal to the sum of the positive ranks, and a large T indicates the occurrence of large differences in which y_i is greater than x_i. In this case, reject H_0 if T is greater than

$$w_{1-\alpha} = \frac{n(n+1)}{2} - w_\alpha$$

where w_α is the critical value of T, corresponding to the level of significance, given in Appendix Table O.

(b) For H_0: $E(X) \leq E(Y)$, a sufficiently small value of T will cause rejection, since a small T indicates the occurrence of numerically large differences in which y_i is less than x_i. Reject H_0 if computed T is less than w_α, where w_α is the critical value of T, corresponding to the α level of significance, given in Appendix Table O.

(c) For H_0: $E(X) = E(Y)$, either sufficiently large or sufficiently small values of T will cause rejection. In this case, reject H_0 if computed T exceeds

$$w_{1-\alpha/2} = \frac{n(n+1)}{2} - w_{\alpha/2}$$

or is less than $w_{\alpha/2}$, where $w_{\alpha/2}$ is given in Appendix Table O.

Example 11.9 To illustrate the application of the Wilcoxon matched-pairs signed-ranks test, let us use the data of Example 11.7, given in Table 11.23.

Table 11.26 shows the intermediate computations necessary for computing T. The null and alternative hypotheses are

$$H_0: E(X) \leq E(Y), \qquad H_1: E(X) > E(Y)$$

Table 11.26 Calculations for Wilcoxon matched-pairs signed-ranks test for Example 11.9

Subject	Mental health status score BEFORE, X	AFTER, Y	$d_i = y_i - x_i$	Rank of $\|d_i\|$	Signed ranks
1	76	70	−6	10.5	−10.5
2	80	75	−5	7.5	−7.5
3	86	84	−2	2	−2
4	87	90	+3	3.5	+3.5
5	85	81	−4	5	−5
6	95	95	0		
7	97	87	−10	12.5	−12.5
8	75	72	−3	3.5	−3.5
9	87	92	+5	7.5	+7.5
10	96	85	−11	14	−14
11	98	88	−10	12.5	−12.5
12	77	76	−1	1	−1
13	80	85	+5	7.5	+7.5
14	87	81	−6	10.5	−10.5
15	89	84	−5	7.5	−7.5
					$T = 18.5$

Suppose we let $\alpha = 0.05$.

Entering Appendix Table O with $n = 14$, we find the critical value of T when $\alpha = 0.05$ (one-sided test) to be 26. Since the computed T of 18.5 is less than 26, we reject H_0, and conclude that, in the sampled population, level of mental health tends to be lower after retirement than before. You will recall that this is the same conclusion we reached with the sign test. Figure 11.5 shows the rejection region for Exercise 11.9.

Large-sample approximation For samples larger than 20, we cannot use Appendix Table O. For large samples, however,

$$z = \frac{T - [n(n+1)]/4}{\sqrt{[n(n+1)(2n+1)]/24}} \tag{11.8}$$

is distributed approximately as the standard normal distribution.

Figure 11.5 Rejection region for Example 11.9

Some possible critical values (w_p) of the test statistic:	13	16	22	26	32	39	44	48
p:	0.005	0.01	0.025	0.05	0.10	0.20	0.30	0.40

REJECTION REGION (0.005–0.025) | ACCEPTANCE REGION (0.05–0.40)

Table 11.27 Intermediate calculations for Wilcoxon matched-pairs signed-ranks test for Example 11.10

Subject	Reading comprehension scores BEFORE, X	AFTER, Y	$d_i = y_i - x_i$	Rank of $\lvert d_i \rvert$	Signed ranks
1	80	85	+5	3.5	+3.5
2	65	75	+10	12	+12
3	73	84	+11	14	+14
4	60	55	−5	3.5	−3.5
5	60	72	+12	16	+16
6	73	81	+8	7.5	+7.5
7	76	84	+8	7.5	+7.5
8	53	50	−3	1	−1
9	57	65	+8	7.5	+7.5
10	65	75	+10	12	+12
11	61	79	+18	20	+20
12	54	62	+8	7.5	+7.5
13	77	85	+8	7.5	+7.5
14	66	78	+12	16	+16
15	59	69	+10	12	+12
16	74	88	+14	19	+19
17	52	60	+8	7.5	+7.5
18	56	52	−4	2	−2
19	63	75	+12	16	+16
20	67	80	+13	18	+18
					$T = 203.5$

Example 11.10 To illustrate the use of the normal approximation, let us refer to Example 11.8 and the data displayed in Table 11.24. Table 11.27 shows the intermediate calculations.

The null and alternative hypotheses are

$$H_0\colon E(X) \geq E(Y), \qquad H_1\colon E(X) < E(Y)$$

Suppose we let $\alpha = 0.05$. By Equation 11.8, we compute

$$z = \frac{203.5 - [20(20 + 1)]/4}{\sqrt{[20(20 + 1)(40 + 1)]/24}} = 3.68$$

Since 3.68 exceeds the critical z of 1.645, we reject H_0, and conclude that the program does improve reading comprehension.

EXERCISES

27 Table 11.28 shows the level-of-severity scores of ten stutterers before and after a certain treatment. Do these data provide sufficient evidence to indicate that the treatment is effective? Let $\alpha = 0.05$.

28 A team of mental health professionals evaluated a random sample of 15 executives six months prior to their retirement, and assigned to each a

Table 11.28 Data for Exercise 27

Subject	Level-of-severity score BEFORE	AFTER	Subject	Level-of-severity score BEFORE	AFTER
1	49	32	6	29	31
2	28	15	7	46	34
3	25	28	8	31	18
4	28	20	9	32	15
5	41	28	10	37	21

level-of-adjustment score. The team evaluated the former executives again in one year after retirement, and again gave each a level-of-adjustment score. Table 11.29 shows the results. Do these data provide sufficient evidence to indicate that, in the population, level-of-adjustment scores, on the average, are lower after retirement than before? Let $\alpha = 0.05$.

Table 11.29 Data for Exercise 28

Subject	Level-of-adjustment score BEFORE	AFTER	Subject	Level-of-adjustment score BEFORE	AFTER
1	85	80	9	94	80
2	86	72	10	73	85
3	83	90	11	80	75
4	93	85	12	92	85
5	81	91	13	87	80
6	76	66	14	76	71
7	99	92	15	76	65
8	93	88			

29 Refer to Exercise 24 and apply the Wilcoxon matched-pairs signed-ranks test. Compare the results with those obtained previously.

11.8 THE MANN-WHITNEY TEST

The sign test and the Wilcoxon matched-pairs signed-ranks test discussed in Sections 11.6 and 11.7, respectively, were designed for use with related samples. In this section we present a test known as the *Mann-Whitney test*, which is appropriate when the inferential procedure is based on two independent samples.

If certain assumptions are tenable, we may use the Mann-Whitney test to test the null hypothesis that the means of two populations are equal. The test, therefore, provides a nonparametric alternative to the two-sample t test when the assumptions of the latter are not met.

Assumptions underlying the Mann-Whitney test The following assumptions are necessary for the valid use of the Mann-Whitney test in testing the null hypothesis that two population means are equal.

1 Each of the samples has been drawn at random from its population.

2 There is independence among observations within each sample, as well as between the two samples.

3 The random variable under consideration is continuous in both populations.

4 The data represent measurement on at least an ordinal scale.

5 The two population-distribution functions differ only with respect to location, if they differ at all.

Procedure Designate the random variable of interest as it occurs in population 1 by X and as it occurs in population 2 by Y. Designate the observations in a random sample of size n from population 1 by $x_1, x_2, \ldots, x_i, \ldots, x_n$, and the observations in a sample of size m from population 2 by $y_1, y_2, \ldots, y_i, \ldots, y_m$.

When the assumptions stated earlier are met, the following null hypotheses, along with their alternatives, may be tested.

(a) H_0: $E(X) = E(Y)$, H_1: $E(X) \neq E(Y)$

(b) H_0: $E(X) \geq E(Y)$, H_1: $E(X) < E(y)$

(c) H_0: $E(X) \leq E(Y)$, H_1: $E(X) > E(Y)$

To obtain a value of the test statistic, begin by assigning ranks to the $n + m$ observations of the combined samples. Assign the rank of 1 to the smallest of the $n + m$ observations, the rank of 2 to the next smallest, and so on, to the rank $n + m$, which is assigned to the largest. Assign to tied observations the mean of the rank positions they would have occupied had there been no ties.

The test statistic is

$$T = S - \frac{n(n+1)}{2} \tag{11.9}$$

where S is the sum of the ranks assigned to the x_i.

The decision to reject or not to reject H_0 at the α level of significance depends on the magnitude of T and on which of the hypotheses—A, B, or C—is being tested. The following criteria are observed.

(a) For H_0: $E(X) = E(Y)$, either sufficiently small or sufficiently large values of T will cause rejection. Therefore, reject H_0 if computed T is less than $w_{\alpha/2}$ or greater than $w_{1-\alpha/2}$, where $w_{\alpha/2}$ is the critical value of T given in Appendix Table P, and $w_{1-\alpha/2}$ is given by

$$w_{1-\alpha/2} = nm - w_{\alpha/2} \tag{11.10}$$

(b) For H_0: $E(X) \geq E(Y)$, sufficiently small values of T will cause rejection. Therefore, reject H_0 if computed T is less than w_α, where w_α is the critical value of T obtained by entering Appendix Table P with n, m, and α.

(c) For H_0: $E(X) \leq E(Y)$, sufficiently large values of T will cause rejection. Therefore reject H_0 if computed T is greater than $w_{1-\alpha}$, where

$$w_{1-\alpha} = nm - w_\alpha \qquad (11.11)$$

Example 11.11 To illustrate the application of the Mann-Whitney test, let us refer to Example 11.6 and the data displayed in Table 11.18. For demonstration purposes, let us assume that these data satisfy the assumptions of the Mann-Whitney test given earlier. Table 11.30 shows assignment of ranks.

The null and alternative hypotheses are

$$H_0: E(X) = E(Y), \qquad H_1: E(X) \neq E(Y)$$

Suppose we let $\alpha = 0.05$. From Table 11.30, we find the sum of the ranks of the X values to be $S = 190.5$. By Equation 11.9, then,

$$T = 190.5 - \frac{15(16)}{2} = 70.5$$

For this two-sided test, the critical values of T are 81, which we obtain by entering Appendix Table P with $m = 18$, $n = 15$, and $\alpha/2 = 0.025$. And, by Equation 11.10, $15(18) - 81 = 189$. Since 70.5 is less than 81, we reject H_0, and conclude on the basis of this test that, at this particular high school, freshman students who come from rural elementary schools and those who come from urban elementary schools do differ with respect to mean verbal reasoning scores. Note that this is the opposite of the conclusion we reached when we applied the median test in Section 11.5. These results illustrate that the Mann-Whitney test is more powerful than the median test. Figure 11.6 shows the rejection region for Example 11.11.

Table 11.30 Verbal reasoning test scores made by 15 freshmen from rural and 18 freshmen from urban elementary schools, with ranks assigned for use in the Mann-Whitney test

Rural, X	Urban, Y	Rank	Rural, X	Urban, Y	Rank	Rural, X	Urban, Y	Rank
42		1	66		12	80		24
52		2		68	13	80		24
54		3	69		14.5	80		24
57		4		69	14.5		89	26.5
	60	5	73		16.5		89	26.5
	62	6		73	16.5		90	28
63		7.5	74		18		91	29.5
63		7.5		76	19		91	29.5
	64	9.5		77	20		96	31
	64	9.5	79		21.5		97	32
65		11		79	21.5		99	33

Figure 11.6 Rejection region for Example 11.1

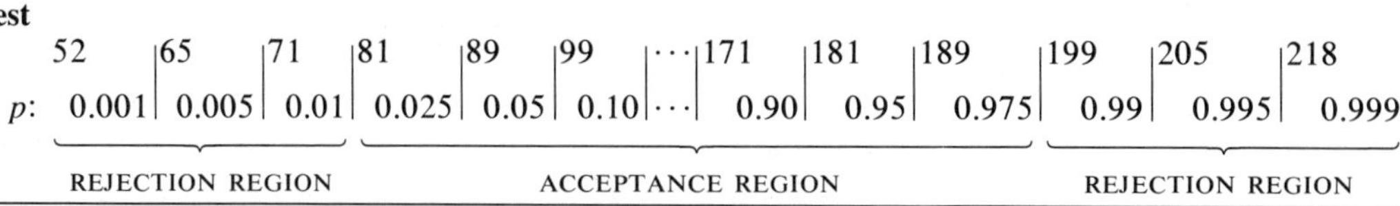

For values of n or m greater than 20, we cannot use Appendix Table P. Fortunately, in this case we may use a normal approximation in the hypothesis-testing procedure. The statistic

$$z = \frac{T - mn/2}{\sqrt{[(n)(m)(n + m + 1)]/12}}$$

is distributed approximately as the standard normal distribution when n and m are large.

EXERCISES

30 A psychologist conducted a study based on a random sample of eight high school seniors who scored high and eight who scored low on a test designed to measure an individual's level of authoritarianism. The psychologist also gave each of the 16 subjects a test to measure the extent of his or her need to achieve. Table 11.31 shows these scores by level of authoritarianism. Can one conclude from these data that, in the population, subjects who have low authoritarianism levels score higher on the need-to-achieve test? Let $\alpha = 0.05$.

Table 11.31 Data for Exercise 30

Authoritarianism level								
HIGH	19	9	13	8	6	17	14	18
LOW	16	19	11	18	9	13	20	18

31 A random sample of 15 elderly persons was selected from a population of those classified as being in the high-income group. An independent random sample of 12 subjects was selected from a population of low-income elderly persons. Each subject was given a test to measure her or his lack of adjustment to old age. Table 11.32 gives the scores. Do these data provide sufficient evidence to indicate that elderly persons in the low-income group score higher on the test to measure lack of adjustment to old age than those in the high-income group? ($\alpha = 0.05$.)

Table 11.32 Data for Exercise 31

High-income group	21	26	25	27	14	10	17	23	17	15	16	23	25	22	14
Low-income group	28	25	30	35	26	27	22	26	31	31	16	35			

32 Table 11.33 shows the scores made by random samples of smokers and nonsmokers on a test designed to measure level of neurosis. Can one conclude, on the basis of these data, that nonsmokers tend to have lower neurosis scores than smokers? ($\alpha = 0.05$.)

Table 11.33 Data for Exercise 32

Smokers	21	33	15	31	22	15	22	29	33	18	21	32	33	16	23	20	30	17	29	19
Nonsmokers	13	16	17	26	19	27	22	28	20	13	18	23	20	28	15	10	23	26	13	14
Smokers	17	34	23	17	23	25														
Nonsmokers	21	21	21	11	14															

CHAPTER SUMMARY

In this chapter we have introduced a number of useful nonparametric statistical techniques, some of which are convenient alternatives that we may use when, for some reason, we cannot use an analogous parametric technique. In order to put the topic into proper perspective, we have also introduced the concepts of distribution-free statistics and the different measurement scales.

Only some of the most widely used techniques can be presented in a general introductory book such as this. Those who need to learn more about available nonparametric procedures should consult the elementary textbooks by Maxwell (1), Mosteller and Rourke (2), Pierce (3), Noether (4), Tate and and Clelland (5), Conover (6), and Siegel (7).

SOME IMPORTANT CONCEPTS AND TERMINOLOGY

Nonparametric statistics
Distribution-free statistics
Nominal scale
Ordinal scale
Interval scale
Ratio scale
Runs

REVIEW EXERCISES

33 Table 11.34 shows the ranks of ten small kitchen appliances on the basis of appeal to homemakers and durability as determined by laboratory tests. Do these data provide sufficient evidence to indicate a lack of independence between the two variables? Let $\alpha = 0.05$.

Table 11.34 Data for Exercise 33

Appeal-to-homemakers rank	1	2	3	4	5	6	7	8	9	10
Durability rank	3	4	2	1	5	9	8	6	10	7

34 The quality-control department of a factory periodically takes samples of items coming off the assembly line, computes the mean for each sample, and plots the results on a control chart. The quality-control engineer is particularly interested in knowing whether the mean of a given sample is larger or smaller than the desired population mean of all items manufactured. The following sequence shows which of the sample means computed from the last 40 consecutive samples fell above (a) and below (b) the desired population mean.

bbbbbbb aaaa bbb aa bbb aa
bbbb a bbb aa bb aaaaaaa

Does the sequence appear to be random?

35 Table 11.35 shows the supine systolic blood pressures of 15 adult males taken before and after a 30-minute period of rest. Use the sign test to determine whether or not these data provide sufficient evidence to indicate that rest lowers supine blood pressure in adult males. Let $\alpha = 0.05$.

Table 11.35 Data for Exercise 35

Before rest	156	152	132	148	154	130	174	162	144	142	140	128	152	156	172
After rest	144	154	148	136	142	126	170	160	136	130	132	124	144	150	170

36 Use the Wilcoxon matched-pairs signed-ranks test to reach a decision regarding the data of Exercise 35.

37 A psychologist studying aggression in human subjects administered a test designed to measure that attribute to a random sample of adolescents who were the youngest children in families of three or more children. The psychologist also administered the test to an independent random sample of adolescents who came from single-child families. Table 11.36 shows the results. Do these data provide sufficient evidence to indicate that the two populations differ with respect to average aggression level, as measured by the test? Let $\alpha = 0.05$ and use the median test.

Table 11.36 Data for Exercise 37

Youngest children	84	84	86	96	93	93	87	92	90	87	94	99	96	97	97	96
Only children	93	73	92	76	89	77	87	81	87	84	70	71	73	73	71	

38 Use the Mann-Whitney test to answer the question posed in Exercise 37.

39 In a random sample of 15 college freshmen at a state university, two stated that they were attending the college of their first choice. Do these data provide sufficient evidence to indicate that, in the population of students from which the sample was drawn, fewer than 40% are attending the college of their first choice? Let $\alpha = 0.05$.

REFERENCES

1 A. E. Maxwell, *Analyzing Qualitative Data*, New York: Wiley, 1961

2 Frederick Mosteller and Robert E. K. Rourke, *Sturdy Statistics: Nonparametric and Order Statistics*, Reading, Mass.: Addison-Wesley, 1973

3 Albert Pierce, *Fundamentals of Nonparametric Statistics*, Belmont, Calif.: Dickensen, 1970

4 G. E. Noether, *Introduction to Statistics: A Fresh Approach*, Boston: Houghton Mifflin, 1971

5 Merle W. Tate and Richard C. Clelland, *Nonparametric and Shortcut Statistics in the Social, Biological, and Medical Sciences*, Danville, Ill.: Interstate Printers and Publishers, 1957

6 W. J. Conover, *Practical Nonparametric Statistics*, New York: Wiley, 1971

7 Sidney Siegel, *Nonparametric Statistics for the Behaviorial Sciences*, New York: McGraw-Hill, 1956

Answers to Odd-Numbered Exercises

Some of the following answers were obtained by means of computer programs. Electronic desk and pocket calculators were used to obtain the remaining answers. In the latter case, unless otherwise indicated, intermediate calculations employed the significant digits to the full capacity of the calculator used, to avoid unnecessary copying of intermediate results. The answers given here may differ from your own because of rounding, unless you use the same method of calculation.

Chapter 1

1.

Class interval	Frequency	Cumulative frequency	Relative frequency	Cumulative relative frequency
155–159	3	3	0.020	0.020
160–164	9	12	0.060	0.080
165–169	17	29	0.113	0.193
170–174	22	51	0.147	0.340
175–179	53	104	0.353	0.693
180–184	23	127	0.153	0.846
185–189	20	147	0.133	0.979
190–194	3	150	0.020	0.999
	150		0.999	

3.

Class interval	Frequency	Cumulative frequency	Relative frequency	Cumulative relative frequency
30– 44	6	6	0.06	0.06
45– 59	9	15	0.09	0.15
60– 74	10	25	0.10	0.25
75– 89	17	42	0.17	0.42
90–104	21	63	0.21	0.63
105–119	17	80	0.17	0.80
120–134	11	91	0.11	0.91
135–149	9	100	0.09	1.00
	100		1.00	

11. $\bar{x} = 2.5$ Range $= 0.4$
Median $= 2.5$ $s^2 = 0.0229$
Modes $= 2.3$ and 2.5 $s = 0.15$

13. $\bar{x} = 4.7$ Range $= 8$
Median $= 4$ $s^2 = 8.0952$
Mode $= 2$ $s = 2.8$

15. $\bar{x} = 7$ Range $= 6$
Median $= 8$ $s^2 = 4.4$
Mode $= 8$ $s = 2.1$

17. $\bar{x} = 11.30$ $s^2 = 51.0202$
Median $= 10.0$ $s = 7.14$
Modal class: $5 - 9$

19. $\bar{x} = 119.10$ $s^2 = 83.9293$
Median $= 119.50$ $s = 9.16$
Modal class: $120 - 124$

21. $\bar{x} = 4.67$ $s^2 = 3.3810$
Median $= 4$ $s = 1.84$

23. $\bar{x} = 64$ $s^2 = 63.3333$
Median $= 62.5$ $s = 7.96$

25.

Class interval	Frequency (f_i)	Cumulative frequency	Relative frequency	Cumulative relative frequency
30–39	33	33	0.33	0.33
40–49	28	61	0.28	0.61
50–59	20	81	0.20	0.81
60–69	10	91	0.10	0.91
70–79	4	95	0.04	0.95
80–89	3	98	0.03	0.98
90–99	2	100	0.02	1.00
	100		1.00	

$\bar{x} = 48.6$ $s^2 = 212.3131$
Median $= 45.6$ $s = 14.57$

Chapter 2

1. (a) $A \cup B = \{t, u, v, w, x, y, z\}$
(b) $A \cap B = \{w, x, y, z\}$
(c) A and B are not equal since they do not contain exactly the same elements.
(d) A and B are not disjoint since they have in common the elements w, x, y, and z.

3. (a) A (b) ϕ (c) A (d) U

5.

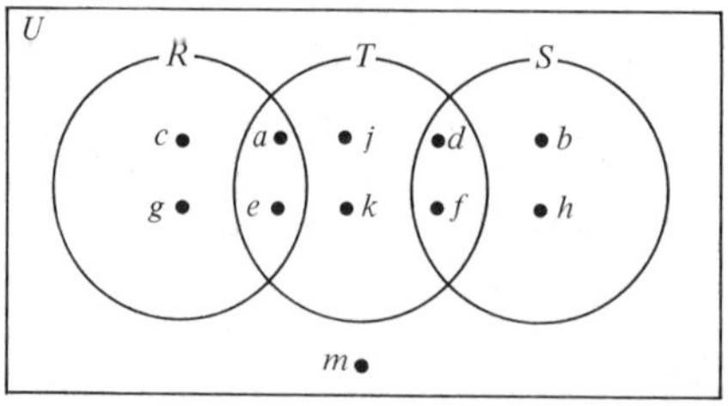

U
R T S
c g a e j k d f b h
m

(a) $R' = \{b, d, f, h, j, k, m\}$
(b) $S' = \{a, c, e, g, j, k, m\}$
(c) $T' = \{b, c, g, h, m\}$
(d) $R \cap S = \{\phi\}$
(e) $R \cap T = \{a, e\}$
(f) $T \cap S = \{d, f\}$
(g) $R \cup S = \{a, b, c, d, e, f, g, h\}$
(h) $S \cup T = \{a, b, d, e, f, h, j, k\}$
(i) $T \cup R = \{a, c, d, e, f, g, j, k\}$

7. (a) 30 (b) $25 + 44 - 15 = 54$ (c) 33 (d) $100 - 15 = 85$
(e) 100 (f) $100 - 23 = 77$ (g) 7 (h) $100 - 20 = 80$

9. (a) $A \cup B = \{a, b, c, d, e, f, g, h, i, j, n, o, p, q, r\}$
(b) $A \cap B = \{d, e, f, g, h\}$
(c) $A \cup C = \{a, b, c, d, e, f, g, h, i, j, k, l, m, n, o\}$
(d) $C \cup B = \{d, e, f, g, h, i, j, k, l, m, n, o, p, q, r\}$
(e) $A \cap C = \{g, h, i, j\}$
(f) $C \cap B = \{g, h, n, o\}$
(g) $A \cap B \cap C = \{g, h\}$
(h) $A \cup B \cup C = \{a, b, c, d, e, f, g, h, i, j, k, l, m, n, o, p, q, r\}$
(i) $(A \cup B \cup C)' = \{\phi\}$
(j) $B' = \{a, b, c, i, j, k, l, m\}$
(k) $(A \cup B)' = \{k, l, m\}$

11. $\binom{12}{4} = \frac{12!}{4!\,8!} = 495$

13. ${}_{16}P_9 = \frac{16!}{7!} = 4{,}151{,}347{,}200$

15. ${}_{10}P_{10} = \frac{10!}{0!} = 3{,}628{,}800$

17. ${}_{10}P_5 = \frac{10!}{5!} = 30{,}240$

19. $\binom{14}{9} = \frac{14!}{9!\,5!} = 2{,}002$

21. $(2)(3)(3) = 18$

23. $\binom{8}{4}\binom{6}{3} = \frac{8!}{4!\,4!} \cdot \frac{6!}{3!\,3!} = (70)(20) = 1400$

25. No. For example, $P(A \cap L) \neq P(A)P(L)$. That is, $50/1000 \neq (300/1000)(200/1000)$, or $0.05 \neq 0.06$.

27. (a) $P(B|A) = \frac{P(A \cap B)}{P(A)} = \frac{0.25}{0.6} = 0.4167$

(b) $P(A \cap B) \neq P(B)P(A)$
$P(A \cap B) \neq [1 - P(B')][P(A)]$
$0.25 \neq (1 - 0.7)(0.6)$
$0.25 \neq 0.18$
Therefore A and B are not independent.
(c) $P(A') = 1 - P(A) = 1 - 0.6 = 0.4$

29. $(0.9)(0.8)(0.7) = 0.504$

31. (a) (1) $300/500 = 0.60$
(2) $135/500 = 0.27$
(3) $125/500 = 0.25$ in each case
(4) $115/125 = 0.92$
(5) $(65/500) + (125/500) - (15/500) = 175/500 = 0.35$

(b) (1) $P(A \cap R) = 5/500 = 0.01$
(2) $P(B \cup C) = (125/500) + (300/500) - (115/500) = 310/500 = 0.62$
(3) $P(D') = 1 - P(D) = 1 - (125/500) = 0.75$
(4) $P(N|D) = 60/125 = 0.48$
(5) $P(B|R) = 5/65 = 0.08$
(6) $P(C) = 300/500 = 0.60$

33.

	A	A'	
B	0.1575	0.2925	0.45
B'	0.1925	0.3575	0.55
	0.35	0.65	1.00

35.

Event	**$P(C_i)$, prior probability**	**$P(E\|C_i)$, likelihood**	**$P(E \cap C_i)$, joint probability**	**$P(C_i\|E)$, posterior probability**
C_1	0.45	0.03	0.0135	0.4030
C_2	0.30	0.05	0.0150	0.4478
C_3	0.25	0.02	0.0050	0.1493
	1.00		0.0335	1.0000

37.

Event	**$P(D_i)$ prior probability**	**$P(S\|D_i)$, likelihood**	**$P(S \cap D_i)$, joint probability**	**$(PD_i\|S)$, posterior probability**
D_1	0.20	0.25	0.05	0.0794
D_2	0.30	0.60	0.18	0.2857
D_3	0.50	0.80	0.40	0.6349
	1.00		0.63	1.0000

39.

Event	**P(area), prior probability**	**$P(W\|$area$)$, likelihood**	**$P(W \cap$ area$)$, joint probability**	**$P($area$\|W)$, posterior probability**
A	0.45	0.30	0.135	0.6136
B	0.30	0.20	0.060	0.2727
C	0.25	0.10	0.025	0.1136
	1.00		0.220	1.0000

41.

Event	$P(A_i)$, prior probability	$P(M\|A_i)$, likelihood	$P(M \cap A_i)$, joint probability	$P(A_i\|M)$, posterior probability
A_1	0.20	0.05	0.010	0.0141
A_2	0.40	0.80	0.320	0.4507
A_3	0.40	0.95	0.380	0.5352
	1.00		0.710	1.0000

43. (a) $50/1147 = 0.0436$ (b) $\dfrac{329 + 554 - 160}{1147} = \dfrac{723}{1147} = 0.6303$

(c) $\dfrac{1147 - 23}{1147} = \dfrac{1124}{1147} = 0.9799$ (d) $14/35 = 0.4000$

(e) $74/1147 = 0.0645$ (f) $5/23 = 0.2174$

(g) $554/1147 = 0.4830$ (h) $140/259 = 0.5405$

(i) $140/1147 = 0.1221$

45. $\dbinom{10}{4} = 10!/6!\,4! = 210$

47. (a) $2300/5000 = 0.4600$ (b) $125/5000 = 0.0250$

(c) $325/1825 = 0.1781$ (d) $\dfrac{1425 + 2300 - 700}{5000} = \dfrac{3025}{5000} = 0.6050$

(e) $500/1225 = 0.4082$ (f) $400/5000 = 0.0800$

(g) $\dfrac{2300 + 1825}{5000} = \dfrac{4125}{5000} = 0.8250$

49.

Event	Prior probability	Likelihood	Joint probability	Posterior probability
A	0.3500	0.7143	0.2500	0.4274
B	0.2500	0.6000	0.1500	0.2564
C	0.2250	0.5111	0.1150	0.1966
D	0.1750	0.4000	0.0700	0.1197
			0.5850	1.0000

Chapter 3

1. (a) 0.0577 (b) 0.5881
3. (a) 0.1406 (b) 0.1106 (c) 0.8894
5. 0.0819
7. 0.0593
9. (a) 0.0016 (b) 0.6822 (c) 0.3178 (d) 0.5848 (e) 0.2484
11. (a) 0.5328 (b) 0.0228 (c) 0.0228
13. (a) 0.1587 (b) 0.0062

15. (a) 0.7938 (b) 0.2514 (c) 0.2514

17. (a) 0.1587 (b) 0.1587 (c) 0.0228

19. 0.0806

21. 0.1319

23. (a) 0.6826 (b) 0.3085 (c) 0.8413

25. (a) 0.6826 (b) 0.0228 (c) 0.0228 (d) 0.8164

27. (a) 0.1276 (b) 0.0509 (c) 0.9487

29. 0.2131

Chapter 4

1. 0.9270 3. 0.0228 5. 0.1359 7. 0.4778 9. 0.0526

11. Yes. $0.005 > P(t \leq -3) < 0.0005$

13. 0.7479 15. (a) 0.8764 (b) 0.0618

17. 0.0475 19. 0.2033 21. 30.578

23. 13.120 and 40.646

25. Yes. $P(\chi^2 > 42) > 0.01$

27. Greater than 0.05, but less than 0.10

29. 0.01 31. 0.05 33. 0.0475

35. (a) $\mu_{\bar{x}} = 24.5$, $\sigma_{\bar{x}} = 0.5$ (b) 0.1587

37. Assume that populations are normally distributed with equal variances; 0.0071

39. 0.5411

Chapter 5

1. $C(\$12{,}098 \leq \mu \leq \$12{,}902) = 0.95$ 3. 249 5. 246

7. $C(43 \leq \mu \leq 47) = 0.95$

9. $C(24 \leq \mu \leq 28) = 0.90$. Assume that the population from which the sample was drawn is normally distributed.

11. $C(18 \leq \mu \leq 22) = 0.95$ 13. $C(86 \leq \mu \leq 94) = 0.95$

15. $C(46 \leq \mu \leq 54) = 0.95$ 17. $C(2.9 \leq \mu_A - \mu_B \leq 7.1) = 0.95$

19. $C[2.0 \leq (\mu_I - \mu_{II}) \leq 18.0] = 0.95$

21. Assume that the samples constitute independent simple random samples from normally distributed populations with equal variances. $C[1.2 \leq (\mu_A - \mu_B) \leq 14.8] = 0.90$.

23. $C[7.8 \leq (\mu_A - \mu_B) \leq 20.2] = 0.95$

25. Assume that the sample of differences constitutes a random sample from a normally distributed population of differences. $C(0.92 \leq \mu_d \leq 3.08) = 0.99$

27. $C(-11.05 \leq \mu_d \leq -8.79) = 0.95$ 29. $C(0.16 \leq p \leq 0.24) = 0.90$

31. $C(0.20 \leq p \leq 0.30) = 0.90$ 33. $C(0.71 \leq p \leq 0.79) = 0.95$

35. 385 37. 2401 39. 87

41. $C[0.004 \leq (p_A - p_B) \leq 0.22] = 0.90$

43. $C[0.01 \leq (p_1 - p_2) \leq 0.17] = 0.95$

45. $C(2.20 \leq \sigma^2 \leq 8.11 = 0.95,$ $C(2.13 \leq \sigma^2 \leq 7.80 = 0.95.$

47. Assume that the sample was drawn from a normally distributed population. $C(26.83 \leq \sigma^2 \leq 85.15) = 0.95,$ $C(26.15 \leq \sigma^2 \leq 82.56) = 0.95$

49. $C(57.80 \leq \sigma^2 \leq 325.83) = 0.99,$ $C(65.74 \leq \sigma^2 \leq 240.64) = 0.99$

51. $C(0.0488 \leq \sigma^2 \leq 0.1548) = 0.95,$ $C(0.0475 \leq \sigma^2 \leq 0.1501) = 0.95$

53. $C[0.60 \leq \left(\sigma_1^2 / \sigma_2^2\right) \leq 3.95] = 0.90$. Assume that the two sample variances are computed from independent random samples from normally distributed populations.

55. Assume that the samples are independent and random and that the heights are normally distributed in both populations. $C[0.74 \leq (\sigma_G^2 / \sigma_B^2) \leq 4.13] = 0.95$

57. $C(5903.96 \leq \mu \leq 6096.04) = 0.95$

59. (a) $C(116.71 \leq \mu \leq 123.29) = 0.90,$
(b) $C(116.08 \leq \mu \leq 123.92) = 0.95$

61. (a) $C(97.26 \leq \mu \leq 102.74) = 0.90,$
(b) $C(96.73 \leq \mu \leq 103.27) = 0.95$

63. $C(91.61 \leq \mu \leq 104.39) = 0.95$ 65. $C(0.201 \leq p \leq 0.299) = 0.95$

Chapter 6

1. Since $520 < 529.16$, do not reject H_0. Alternatively, since computed $z = (520 - 516)/8 = 0.5 < 1.645$, do not reject H_0. $p = 0.5 - 0.1915 = 0.3085$

3. Since $z = 2.5 > 1.96$, reject H_0 at 0.05 level of significance. $p = 0.0062$

5. Since $5 < 5.4867$, reject H_0. Alternatively, since

$$z = \frac{5 - 6}{1.5/\sqrt{25}} = -3.33 < -1.7109,$$

reject H_0. $0.0005 < p < 0.005$

7. Since $4.6 > 4.51$, H_0 cannot be rejected. Alternatively, since

$$z = \frac{4.6 - 5}{3.9/\sqrt{169}} = -1.33 > -1.645,$$

H_0 cannot be rejected. $p = 0.0918$

9. Assume that the sample constitutes a random sample from a normally distributed population. Since 94 > 86.259, reject H_0. Alternatively,

$$t = \frac{94 - 81}{12/\sqrt{16}} = 4.33$$

Since 4.33 > 1.7530, reject H_0. $p < 0.0005$

11. Assume that the sample constitutes a simple random sample from a normally distributed population. Since 30 < 30.58 H_0 can be rejected. Alternatively, computed

$$t = \frac{30 - 34}{10/\sqrt{25}} = -2$$

Since $-2 < -1.7109$, H_0 can be rejected. $0.025 < p < 0.05$

13. Since 5 < 5.2, reject H_0. Alternatively, since

$$z = \frac{5 - 6}{4/\sqrt{64}} = -2 < -1.645,$$

reject H_0.

15.

Alternative value of μ	β	Value of power function $(1 - \beta)$
516	0.9500	0.0500
521	0.8461	0.1539
528	0.5596	0.4404
533	0.3156	0.6844
539	0.1093	0.8907
544	0.0314	0.9686
547	0.0129	0.9871

17.

Alternative value of μ	β	Value of power function $(1 - \beta)$
4.25	0.9900	0.0100
4.50	0.8599	0.1401
4.75	0.4325	0.5675
5.00	0.0778	0.9222
5.25	0.0038	0.9962

19.

Alternative value of μ	β	Value of power function $(1 - \beta)$
5	0.9500	0.0500
4.9	0.9032	0.0968
4.7	0.7357	0.2643
4.5	0.4880	0.5120
4.3	0.2420	0.7580
4.1	0.0853	0.9147
4.0	0.0446	0.9554
3.8	0.0089	0.9911

21.

Alternative value of μ	β	Value of power function $(1 - \beta)$
6	0.9500	0.0500
5.8	0.8849	0.1151
5.6	0.7881	0.2119
5.4	0.6554	0.3446
5.2	0.5000	0.5000
5.0	0.3446	0.6554
4.8	0.2119	0.7881
4.6	0.1151	0.8849
4.4	0.0548	0.9452
4.2	0.0228	0.9772

23. Since $-10 < -6.45$, reject H_0. Alternatively, since

$$t = \frac{-10 - 0}{\sqrt{(72.95/11) + (72.95/10)}} = -2.68 < -1.7291$$

reject H_0.

25. Since $5 > 3.32$, reject H_0. Alternatively, since

$$t = \frac{5 - 0}{1.92} = 2.60 > 1.7291$$

reject H_0.

27. Since $10 > 3.29$, reject H_0. Alternatively, since

$$z = \frac{10 - 0}{\sqrt{(100/50) + (90/45)}} = 5 > 1.645$$

reject H_0.

29. Since $9 > 5.43$, reject H_0. Alternatively, since

$$\frac{9 - 0}{3.16} = 2.85 > 1.7171$$

reject H_0.

31. Since

$$t = \frac{-11.08 - 0}{5.78/\sqrt{12}} = -6.64 < -1.7959,$$

reject H_0.

33. Since

$$t = \frac{8.87 - 0}{12.66/\sqrt{15}} = 2.71 > 1.7613$$

reject H_0.

35. Since $z = -1.00 > -2.58$, you cannot reject H_0.

37. Since $z = -3.27 < -1.645$, reject H_0.

39. Since $z = 2.5 > 1.645$, reject H_0.

41. Since $z = 0.67 < 1.645$, you cannot reject H_0.

43. Since $z = -2.24 < -1.645$, reject H_0.

45. Since $z = 1 < 1.96$, do not reject H_0.

47. Since $z = 3 > 1.645$, reject H_0.

49. Since $-0.83 < -2.58$, do not reject H_0.

51. Since $z = -2.5 < -1.645$, reject H_0.

53. $\chi^2 = 19.6 < 20.483$, and you cannot reject H_0.

55. $\chi^2 = 45.36 > 23.685$, and you should reject H_0.

57. Assume that the samples are independent random samples from normally distributed populations.

$$\text{VR} = \frac{2916}{4624} = 0.63$$

Critical values of F are 4.99 and $1/4.99 = 0.20$. Since $0.20 < 0.63 < 4.99$, you cannot reject H_0.

59. Assume that the two groups are independent random samples from normally distributed populations. Since VR = 81/36 = 2.25 > 1.89, reject H_0.

61. Assume that the samples are independent random samples from normally distributed populations. Since VR = 12/4 = 3 > 2.75, reject H_0.

63. Reject H_0, since 160 < 168.60

65. Reject H_0, since 190 < 200.06.

67. Reject H_0, since 81 < 88.74.

69. Reject H_0, since 0 > −3.55.

71. 0.42, 0.54. Reject H_0, since 0.40 is not in the interval.

73. Since 2.95 > 1.645, reject H_0.

75. No, since −1.2 > −1.645.

77. Since $t = 4.03 > 1.8595$, reject H_0.

79. $t = -1.58 > -1.7459$. Do not reject H_0. Assumptions: Data are computed from two independent random samples drawn from normally distributed populations with equal variances.

81. $z = -1.88 > -1.96$. Therefore you cannot reject H_0.

83. $t = -0.4444 > -1.7530$. Therefore do not reject H_0.

Chapter 8

1. Since 19.56 > 3.35, reject H_0.

3. Since 4.68 > 2.92, reject H_0.

5. Since 16.66 > 3.01, reject H_0.

7. Since 29.82 > 4.31, reject H_0

9. Since 14.27 > 3.47, reject H_0.

11. Since 5.92 > 5.46, reject H_0 at the 0.025 level of significance.

13. Since 22.09 > 3.98, reject H_0.

15. Since 22.43 > 7.21, reject H_0 at the 0.005 level of significance.

17. HSD = 5.53

	40.00	25.25	16.38
40.00	—	14.75*	23.62*
25.25		—	8.87*
16.38			—

*Significant differences at 0.05 level.

19. HSD = 3.26

	19.27	15.36	10.09	9.27
19.27	—	3.91*	9.18*	10.00*
15.36		—	5.27*	6.09*
10.09			—	0.82
9.27				—

*Significant differences at 0.05 level.

21. $n_j = 14$

23. $n_j \approx 53$

25. (a) ≈ 0.60 (b) ≈ 0.32

27. Since $6.00 > 5.14$, reject H_0.

29. Since $6.05 > 3.49$, reject H_0.

31. Since $10.65 > 5.12$, reject H_0.

33. Since $21.41 > 3.26$, reject H_0.

35. Since $4.55 > 3.26$, reject H_0.

37. Since $392.09 > 6.94$, reject H_0.

39. Since $30.33 > 6.49$, you can reject H_0 at the 0.005 level of significance.

41. Since $57.41 > 3.86$, reject H_0.

Chapter 9

1. $y_c = 25.839 + 0.6036x$

3. $y_c = 7.71 + 0.4543x$

5. $y_c = 97.98 - 4.33x$

7. $y_c = 56.09 - 0.6465x$

9. $y_c = -90.07 + 2x$

11. $y_c = 31.72 + 0.6873x$

13. (a) $r^2 = 0.74$, VR $= 26.00$ (b) Since $5.10 > 2.2622$, reject H_0.
(c) $C(0.3224 \leq \beta_1 \leq 0.8366) = 0.95$

15. (a) $r^2 = 0.65$, VR $= 15.21$ (b) Since $3.90 > 2.3060$, reject H_0.
(c) $C(0.0656 \leq \beta_1 \leq 0.2556) = 0.95$

17. (a) $r^2 = 0.71$, VR $= 19.69$ (b) Since $-4.44 < -2.3060$, reject H_0.
(c) $C(-5.12 \leq \beta_1 \leq -1.62) = 0.95$

19. (a) $r^2 = 0.29$, VR $= 5.72$ (b) Since $2.39 > 2.1448$, reject H_0.
(c) $C(0.0403 \leq \beta_1 \leq 0.7427) = 0.95$

21. (a) $r^2 = 0.87$, VR $= 52.64$ (b) Since $7.26 > 2.3060$, reject H_0.
(c) $C(0.4187 \leq \beta_1 \leq 0.8089) = 0.95$

23. Prediction interval: 56.511, 79.671 Confidence interval: 64.552, 71.630

25. Prediction interval: 9.703, 14.803 Confidence interval: 11.527, 12.979

27. Prediction interval: 45.95, 63.41 Confidence interval: 50.94, 58.42

29. Prediction interval: 7.43, 40.09 Confidence interval: 19.04, 28.48

31. Prediction interval: 17.55, 42.31 Confidence interval: 26.34, 33.52

33. Prediction interval: 69.73, 89.93 Confidence interval: 76.78, 82.88

35. (b) $r = 0.87$ (c) Since $6.36 > 2.1604$, reject H_0. (d) 0.65, 0.96

37. (b) $r = 0.95$ (c) Since $10.54 > 2.1788$, reject H_0. (d) 0.85, 0.98

39. (b) $r = 0.95$ (c) Since $9.62 > 2.2281$, reject H_0. (d) 0.83, 0.99

41. (a) $y_c = 0.0568 + 0.0605x$ (b) $r^2 = 0.91$, VR $= 131.21$
(c) 0.05, 0.07 (d) $t = 11.46$

43. (a) $y_c = 8.2064 + 1.2771x$ (b) $r^2 = 0.61$, VR $= 12.52$
(c) 0.4446, 2.1096 (d) $t = 3.54$

45. (a) $r = 0.779$ (b) Since $t = 4.48 > 2.1604$, reject H_0. (c) 0.44, 0.92

Chapter 10

1. Since $21.8928 > \chi^2_{(4)} = 9.488$, reject H_0.
3. Since $16.968 > \chi^2_{(1)} = 3.841$, reject H_0.
5. Since $0.58 < \chi^2_{(1)} = 3.841$, you cannot reject H_0.
7. Since $98.86 > \chi^2_{(9)} = 16.919$, reject H_0.
9. Since $7.69 > \chi^2_{(1)} = 3.841$, reject H_0.
11. Since $96.84 > \chi^2_{(1)} = 3.841$, reject H_0.
13. Since $21.63 > \chi^2_{(1)} = 3.841$, reject H_0.
15. Since $84.11 > \chi^2_{(8)} = 15.507$, reject H_0.
17. Since $10.1994 > \chi^2_{(4)} = 9.488$, reject H_0.
19. Since $8.77 < 12.592$, do not reject H_0.
21. Since $8.71 > \chi^2_{(3)} = 7.815$, you can reject H_0 at the 0.05 level of significance.
23. Since $59 > \chi^2_{(6)} = 12.592$, reject H_0.
25. Since $X^2 = 40.40 > \chi^2_{(4)} = 9.488$, reject H_0.

Chapter 11

1. $P(X \leq 1|25, 0.20) = 0.0274$ 3. $P(X \geq 11|25, 0.40) = 0.4141$
5. $P(X \geq 15|20, 0.50) = 0.0207$
7. $r = 9$. Since $6 < 9 < 17$, you cannot reject H_0.
9. $r = 4$. Since $4 < 6$, reject H_0. 11. $r_S = 0.93$
13. Adjusted for ties, $r_S = 0.15742$. Unadjusted for ties, $r_S = 0.15893$.
15. r_S (adjusted for ties) $= 0.9691$ 17. Adjusted for ties, $r_S = 0.62$
19. Adjusted for ties, $r_S = 0.75$
21. $X^2 = 2.93$. Since $2.93 < \chi^2_{(1)} = 3.841$, do not reject H_0.
23. $X^2 = 7.430$. Since $7.430 > \chi^2_{(1)} = 3.841$, reject H_0.
25. $z = -2.97$. Since $-2.97 < -1.645$, reject H_0.
27. Since $T = 3 < w_{0.05} = 11$, reject H_0.
29. Since $T = 5.5 < w_{0.05} = 9$, reject H_0.
31. Since $T = 23.5 < w_{0.05} = 56$, reject H_0. 33. $r_S = 0.75$
35. $P(x \leq 2|15, 0.5) = 0.0037$. Do not reject H_0.
37. Since $X^2 = 7.429 > 3.841$, reject H_0.
39. $P(x \leq 2|15, 0.40) = 0.0271$. Since $0.0271 < 0.05$, reject H_0.

Appendix

SUMMATION NOTATION

We frequently use the symbol Σ in this text. It is mathematical shorthand notation used to indicate that the items following it are to be added. When necessary for clarity, we include an index of summation, usually *i*, as part of the notation. For example,

$$\sum_{i=1}^{4} x_i$$

instructs us to add the values of x from x_1 through x_4. That is,

$$\sum_{i=1}^{4} x_i = x_1 + x_2 + x_3 + x_4$$

Similarly, Σx or Σx_i instructs us to add *all* values of x, where the meaning of "all" is apparent from the context.

The following are some useful algebraic properties of summation notation that you will encounter in the text.

1 The summation of a constant c is n times the constant, when n is the number of values of the index of summation. That is,

$$\sum_{i=1}^{n} c = nc$$

For example,

$$\sum_{i=1}^{4} 5 = 4(5) = 20$$

2 The summation of a constant times a variable is equal to the constant times the summation of the variable. That is, given that c is a constant, then

$$\sum_{i=1}^{n} cx_i = c \sum_{i=1}^{n} x_i$$

For example,

$$\sum_{i=1}^{4} 5(x_i) = 5 \sum_{i=1}^{4} x_i$$

If $x_1 = 2$, $x_2 = 3$, $x_3 = 6$, and $x_4 = 10$, we have

$$5(2 + 3 + 6 + 10) = 5(21) = 105$$

3 The summation of a sum (or difference) is the sum (or difference) of the individual sums. In symbols,

$$\sum_{i=1}^{n} (x_i \pm y_i) = \sum_{i=1}^{n} x_i \pm \sum_{i=1}^{n} y_i$$

This last property extends to more than two components. For example,

$$\sum_{i=1}^{n} (x_i + y_i - z_i) = \sum_{i=1}^{n} x_i + \sum_{i=1}^{n} y_i - \sum_{i=1}^{n} z_i$$

Table A Random digits

42916	50199	26435	97117	77100	62919	74498	14252	11052	70038
49019	02101	14580	14421	58592	30885	60248	29783	39125	97534
04421	62261	52644	36493	53146	31906	00208	98915	27613	58180
74606	07765	21788	03093	69158	44498	51540	61267	70550	90599
76288	24031	13826	61989	54283	95614	20378	35853	86644	68259
42866	46273	43621	93636	23582	59351	29828	53006	06004	00427
99017	74447	14581	32223	89571	38437	43037	17654	32705	02726
67245	80759	07378	06307	51311	52458	57898	15213	72105	18792
29317	02377	60654	51918	97109	38972	71750	81431	69776	00892
56457	56692	88071	93055	31559	77054	33921	24189	47537	18470
66908	96815	00106	47915	72072	34460	04085	74036	99640	88672
83966	92418	68500	70046	30009	99166	49224	68804	34733	69265
53196	82252	58476	40657	09612	15380	70717	33052	93954	14642
63291	73919	67613	81329	27561	97499	79346	28385	20829	73829
62205	91166	04127	19669	17699	31072	16918	81168	72908	00561
83502	34546	70327	79999	26659	68085	43541	69983	09041	05677
20293	65765	45954	12799	49028	44691	19957	40928	81503	07030
72932	94622	89404	69024	73518	29828	35482	83798	92363	13918
69803	06247	23872	32055	36776	77634	01444	88377	50827	83716
01155	81380	11691	18090	13236	34313	13390	31223	64796	40116
44290	82296	81987	09423	44272	24414	43248	50536	52161	18884
16980	43552	32970	87214	99340	79058	70912	03514	87351	05102
73249	52463	51467	18602	28336	41484	49543	74121	04575	78007
06050	29975	60715	02040	12974	02831	52032	69726	67679	13772
34216	50564	74588	70102	62585	25511	38134	13802	98334	76947
04607	52269	21767	98347	69224	44987	31255	00344	60841	53970
92738	66714	58465	83216	95109	31032	99817	18844	31514	44004
76152	98002	84257	47518	53932	46337	96349	17004	81135	26247
24405	52117	41434	82281	02756	40000	26893	71507	55783	78195
99046	61444	59911	58255	45299	60971	72833	61883	52645	60945
26312	73154	21070	90104	42013	27302	55283	13166	14051	81929
36315	59502	91215	86654	44578	04159	63389	43516	48971	40922
52467	19775	71391	63601	84377	63350	59557	74397	06289	74426
66790	72193	63999	20307	47423	55164	93870	43783	06851	90065
16427	71681	64661	59249	74118	46257	69308	31035	64498	19592
63988	01319	15012	95770	82029	99778	81793	73836	11528	81863
67468	22553	71756	30281	28244	58696	72161	46240	63452	56485
60477	14463	49722	95808	73193	37865	84147	46004	43753	92444
95384	28822	12047	59393	14588	22723	64262	93653	00284	05594
51396	45671	08283	96848	27039	20852	38008	65531	65322	51775
70321	26394	01403	77390	52111	27816	33570	28064	41906	81867
98710	50639	43559	34442	25514	32178	83688	31018	11232	70459
61664	16238	04228	33224	18550	02255	34597	64773	97872	28450
12906	19628	77265	38578	00958	67476	92199	70519	32591	80452
07633	02489	78236	70986	74294	29591	31175	20817	64727	70957
35933	31203	16796	66581	55006	90733	07198	65126	54346	42214
57652	46065	59420	33920	44589	70899	41795	86683	27317	74817
86860	69306	49382	48964	92022	98252	47414	05190	66648	35104
54447	02332	11406	27021	60064	70307	42155	15810	08324	36194
69865	39302	09057	46982	14177	94534	90536	44442	43337	16371

Source: The Rand Corporation, *A Million Random Digits With* 100,000 *Normal Deviates*, Glencoe, Ill.: The Free Press, 1955. Used by permission of The Rand Corporation.

Table B Squares and square roots

n	n^2	$\sqrt{n}$	$\sqrt{10n}$	$(10n)^2$
1.0	1.00	1.00000	3.16228	100
1.1	1.21	1.04881	3.31662	121
1.2	1.44	1.09545	3.46410	144
1.3	1.69	1.14018	3.60555	169
1.4	1.96	1.18322	3.74166	196
1.5	2.25	1.22474	3.87298	225
1.6	2.56	1.26491	4.00000	256
1.7	2.89	1.30384	4.12311	289
1.8	3.24	1.34164	4.24264	324
1.9	3.61	1.37840	4.35890	361
2.0	4.00	1.41421	4.47214	400
2.1	4.41	1.44914	4.58258	441
2.2	4.84	1.48324	4.69042	484
2.3	5.29	1.51658	4.79583	529
2.4	5.76	1.54919	4.89898	576
2.5	6.25	1.58114	5.00000	625
2.6	6.76	1.61245	5.09902	676
2.7	7.29	1.64317	5.19615	729
2.8	7.84	1.67332	5.29150	784
2.9	8.41	1.70294	5.38516	841
3.0	9.00	1.73205	5.47723	900
3.1	9.61	1.76068	5.56776	961
3.2	10.24	1.78885	5.65685	1024
3.3	10.89	1.81659	5.74456	1089
3.4	11.56	1.84391	5.83095	1156
3.5	12.25	1.87083	5.91608	1225
3.6	12.96	1.89737	6.00000	1296
3.7	13.69	1.92354	6.08276	1369
3.8	14.44	1.94936	6.16441	1444
3.9	15.21	1.97484	6.24500	1521
4.0	16.00	2.00000	6.32456	1600
4.1	16.81	2.02485	6.40312	1681
4.2	17.64	2.04939	6.48074	1764
4.3	18.49	2.07364	6.55744	1849
4.4	19.36	2.09762	6.63325	1936
4.5	20.25	2.12132	6.70820	2025
4.6	21.16	2.14476	6.78233	2116
4.7	22.09	2.16795	6.85565	2209
4.8	23.04	2.19089	6.92820	2304
4.9	24.01	2.21359	7.00000	2401
5.0	25.00	2.23607	7.07107	2500
5.1	26.01	2.25832	7.14143	2601
5.2	27.04	2.28035	7.21110	2704
5.3	28.09	2.30217	7.28011	2809
5.4	29.16	2.32379	7.34847	2916
5.5	30.25	2.34521	7.41620	3025
5.6	31.36	2.36643	7.48331	3136

Table B Squares and square roots (continued)

n	n^2	$\sqrt{n}$	$\sqrt{10n}$	$(10n)^2$
5.7	32.49	2.38747	7.54983	3249
5.8	33.64	2.40832	7.61577	3364
5.9	34.81	2.42899	7.68115	3481
6.0	36.00	2.44949	7.74597	3600
6.1	37.21	2.46982	7.81025	3721
6.2	38.44	2.48998	7.87401	3844
6.3	39.69	2.50998	7.93725	3969
6.4	40.96	2.52982	8.00000	4096
6.5	42.25	2.54951	8.06226	4225
6.6	43.56	2.56905	8.12404	4356
6.7	44.89	2.58844	8.18535	4489
6.8	46.24	2.60768	8.24621	4624
6.9	47.61	2.62679	8.30662	4761
7.0	49.00	2.64575	8.36660	4900
7.1	50.41	2.66458	8.42615	5041
7.2	51.84	2.68328	8.48528	5184
7.3	53.29	2.70185	8.54400	5329
7.4	54.76	2.72029	8.60233	5476
7.5	56.25	2.73861	8.66025	5625
7.6	57.76	2.75681	8.71780	5776
7.7	59.29	2.77489	8.77496	5929
7.8	60.84	2.79285	8.83176	6084
7.9	62.41	2.81069	8.88819	6241
8.0	64.00	2.82843	8.94427	6400
8.1	65.61	2.84605	9.00000	6561
8.2	67.24	2.86356	9.05539	6724
8.3	68.89	2.88097	9.11043	6889
8.4	70.56	2.89828	9.16515	7056
8.5	72.25	2.91548	9.21954	7225
8.6	73.96	2.93258	9.27362	7396
8.7	75.69	2.94958	9.32738	7569
8.8	77.44	2.96648	9.38083	7744
8.9	79.21	2.98329	9.43398	7921
9.0	81.00	3.00000	9.48683	8100
9.1	82.81	3.01662	9.53939	8281
9.2	84.64	3.03315	9.59166	8464
9.3	86.49	3.04959	9.64365	8649
9.4	88.36	3.06594	9.69536	8836
9.5	90.25	3.08221	9.74679	9025
9.6	92.16	3.09839	9.79796	9216
9.7	94.09	3.11448	9.84886	9409
9.8	96.04	3.13050	9.89949	9604
9.9	98.01	3.14643	9.94987	9801

Table C Logarithms

N	0	1	2	3	4	5	6	7	8	9
10	0000	0043	0086	0128	0170	0212	0253	0294	0334	0374
11	0414	0453	0492	0531	0569	0607	0645	0682	0719	0755
12	0792	0828	0864	0899	0934	0969	1004	1038	1072	1106
13	1139	1173	1206	1239	1271	1303	1335	1367	1399	1430
14	1461	1492	1523	1553	1584	1614	1644	1673	1703	1732
15	1761	1790	1818	1847	1875	1903	1931	1959	1987	2014
16	2041	2068	2095	2122	2148	2175	2201	2227	2253	2279
17	2304	2330	2355	2380	2405	2430	2455	2480	2504	2529
18	2553	2577	2601	2625	2648	2672	2695	2718	2742	2765
19	2788	2810	2833	2856	2878	2900	2923	2945	2967	2989
20	3010	3032	3054	3075	3096	3118	3139	3160	3181	3201
21	3222	3243	3263	3284	3304	3324	3345	3365	3385	3404
22	3424	3444	3464	3483	3502	3522	3541	3560	3579	3598
23	3617	3636	3655	3674	3692	3711	3729	3747	3766	3784
24	3802	3820	3838	3856	3874	3892	3909	3927	3945	3962
25	3979	3997	4014	4031	4048	4065	4082	4099	4116	4133
26	4150	4166	4183	4200	4216	4232	4249	4265	4281	4298
27	4314	4330	4346	4362	4378	4393	4409	4425	4440	4456
28	4472	4487	4502	4518	4533	4548	4564	4579	4594	4609
29	4624	4639	4654	4669	4683	4698	4713	4728	4742	4757
30	4771	4786	4800	4814	4829	4843	4857	4871	4886	4900
31	4914	4928	4942	4955	4969	4983	4997	5011	5024	5038
32	5051	5065	5079	5092	5105	5119	5132	5145	5159	5172
33	5185	5198	5211	5224	5237	5250	5263	5276	5289	5302
34	5315	5328	5340	5353	5366	5378	5391	5403	5416	5428
35	5441	5453	5465	5478	5490	5502	5514	5527	5539	5551
36	5563	5575	5587	5599	5611	5623	5635	5647	5658	5670
37	5682	5694	5705	5717	5729	5740	5752	5763	5775	5786
38	5798	5809	5821	5832	5843	5855	5866	5877	5888	5899
39	5911	5922	5933	5944	5955	5966	5977	5988	5999	6010
40	6021	6031	6042	6053	6064	6075	6085	6096	6107	6117
41	6128	6138	6149	6160	6170	6180	6191	6201	6212	6222
42	6232	6243	6253	6263	6274	6284	6294	6304	6314	6325
43	6335	6345	6355	6365	6375	6385	6395	6405	6415	6425
44	6435	6444	6454	6464	6474	6484	6493	6503	6513	6522
45	6532	6542	6551	6561	6571	6580	6590	6599	6609	6618
46	6628	6637	6646	6656	6665	6675	6684	6693	6702	6712
47	6721	6730	6739	6749	6758	6767	6776	6785	6794	6803
48	6812	6821	6830	6839	6848	6857	6866	6875	6884	6893
49	6902	6911	6920	6928	6937	6946	6955	6964	6972	6981
50	6990	6998	7007	7016	7024	7033	7042	7050	7059	7067
51	7076	7084	7093	7101	7110	7118	7126	7135	7143	7152
52	7160	7168	7177	7185	7193	7202	7210	7218	7226	7235
53	7243	7251	7259	7267	7275	7284	7292	7300	7308	7316

Source: John E. Freund, *Modern Elementary Statistics,* third edition, Englewood Cliffs, N.J.: Prentice-Hall, 1967

Table C Logarithms (continued)

N	0	1	2	3	4	5	6	7	8	9
54	7324	7332	7340	7348	7356	7364	7372	7380	7388	7396
55	7404	7412	7419	7427	7435	7443	7451	7459	7466	7474
56	7482	7490	7497	7505	7513	7520	7528	7536	7543	7551
57	7559	7566	7574	7582	7589	7597	7604	7612	7619	7627
58	7634	7642	7649	7657	7664	7672	7679	7686	7694	7701
59	7709	7716	7723	7731	7738	7745	7752	7760	7767	7774
60	7782	7789	7796	7803	7810	7818	7825	7832	7839	7846
61	7853	7860	7868	7875	7882	7889	7896	7903	7910	7917
62	7924	7931	7938	7945	7952	7959	7966	7973	7980	7987
63	7993	8000	8007	8014	8021	8028	8035	8041	8048	8055
64	8062	8069	8075	8082	8089	8096	8102	8109	8116	8122
65	8129	8136	8142	8149	8156	8162	8169	8176	8182	8189
66	8195	8202	8209	8215	8222	8228	8235	8241	8248	8254
67	8261	8267	8274	8280	8287	8293	8299	8306	8312	8319
68	8325	8331	8338	8344	8351	8357	8363	8370	8376	8382
69	8388	8395	8401	8407	8414	8420	8426	8432	8439	8445
70	8451	8457	8463	8470	8476	8482	8488	8494	8500	8506
71	8513	8519	8525	8531	8537	8543	8549	8555	8561	8567
72	8573	8579	8585	8591	8597	8603	8609	8615	8621	8627
73	8633	8639	8645	8651	8657	8663	8669	8675	8681	8686
74	8692	8698	8704	8710	8716	8722	8727	8733	8739	8745
75	8751	8756	8762	8768	8774	8779	8785	8791	8797	8802
76	8808	8814	8820	8825	8831	8837	8842	8848	8854	8859
77	8865	8871	8876	8882	8887	8893	8899	8904	8910	8915
78	8921	8927	8932	8938	8943	8949	8954	8960	8965	8971
79	8976	8982	8987	8993	8998	9004	9009	9015	9020	9025
80	9031	9036	9042	9047	9053	9058	9063	9069	9074	9079
81	9085	9090	9096	9101	9106	9112	9117	9122	9128	9133
82	9138	9143	9149	9154	9159	9165	9170	9175	9180	9186
83	9191	9196	9201	9206	9212	9217	9222	9227	9232	9238
84	9243	9248	9253	9258	9263	9269	9274	9279	9284	9289
85	9294	9299	9304	9309	9315	9320	9325	9330	9335	9340
86	9345	9350	9355	9360	9365	9370	9375	9380	9385	9390
87	9395	9400	9405	9410	9415	9420	9425	9430	9435	9440
88	9445	9450	9455	9460	9465	9469	9474	9479	9484	9489
89	9494	9499	9504	9509	9513	9518	9523	9528	9533	9538
90	9542	9547	9552	9557	9562	9566	9571	9576	9581	9586
91	9590	9595	9600	9605	9609	9614	9619	9624	9628	9633
92	9638	9643	9647	9652	9657	9661	9666	9671	9675	9680
93	9685	9689	9694	9699	9703	9708	9713	9717	9722	9727
94	9731	9736	9741	9745	9750	9754	9759	9763	9768	9773
95	9777	9782	9786	9791	9795	9800	9805	9809	9814	9818
96	9823	9827	9832	9836	9841	9845	9850	9854	9859	9863
97	9868	9872	9877	9881	9886	9890	9894	9899	9903	9908
98	9912	9917	9921	9926	9930	9934	9939	9943	9948	9952
99	9956	9961	9965	9969	9974	9978	9983	9987	9991	9996

Table D Binomial probability distribution

$P(x|n, p) = \binom{n}{x} p^x q^{n-x}$

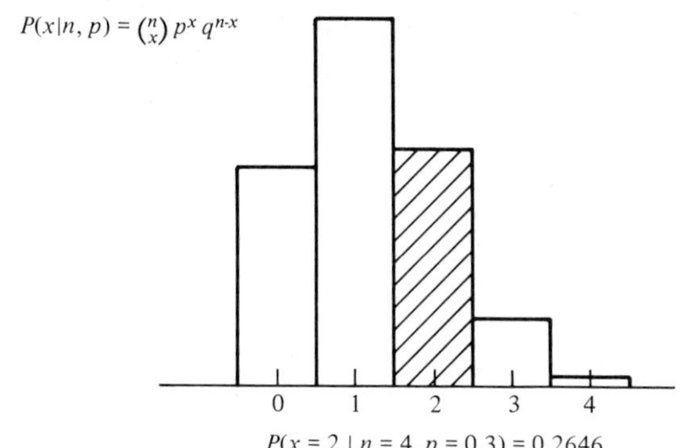

$P(x = 2 \mid n = 4, p = 0.3) = 0.2646$

n = 1

x \ p	.01	.02	.03	.04	.05	.06	.07	.08	.09	.10
0	.9900	.9800	.9700	.9600	.9500	.9400	.9300	.9200	.9100	.9000
1	.0100	.0200	.0300	.0400	.0500	.0600	.0700	.0800	.0900	.1000
	.11	.12	.13	.14	.15	.16	.17	.18	.19	.20
0	.8900	.8800	.8700	.8600	.8500	.8400	.8300	.8200	.8100	.8000
1	.1100	.1200	.1300	.1400	.1500	.1600	.1700	.1800	.1900	.2000
	.21	.22	.23	.24	.25	.26	.27	.28	.29	.30
0	.7900	.7800	.7700	.7600	.7500	.7400	.7300	.7200	.7100	.7000
1	.2100	.2200	.2300	.2400	.2500	.2600	.2700	.2800	.2900	.3000
	.31	.32	.33	.34	.35	.36	.37	.38	.39	.40
0	.6900	.6800	.6700	.6600	.6500	.6400	.6300	.6200	.6100	.6000
1	.3100	.3200	.3300	.3400	.3500	.3600	.3700	.3800	.3900	.4000
	.41	.42	.43	.44	.45	.46	.47	.48	.49	.50
0	.5900	.5800	.5700	.5600	.5500	.5400	.5300	.5200	.5100	.5000
1	.4100	.4200	.4300	.4400	.4500	.4600	.4700	.4800	.4900	.5000

n = 2

x \ p	.01	.02	.03	.04	.05	.06	.07	.08	.09	.10
0	.9801	.9604	.9409	.9216	.9025	.8836	.8649	.8464	.8281	.8100
1	.0198	.0392	.0582	.0768	.0950	.1128	.1302	.1472	.1638	.1800
2	.0001	.0004	.0009	.0016	.0025	.0036	.0049	.0064	.0081	.0100
	.11	.12	.13	.14	.15	.16	.17	.18	.19	.20
0	.7921	.7744	.7569	.7396	.7225	.7056	.6889	.6724	.6561	.6400
1	.1958	.2112	.2262	.2408	.2550	.2688	.2822	.2952	.3078	.3200
2	.0121	.0144	.0169	.0196	.0225	.0256	.0289	.0324	.0361	.0400
	.21	.22	.23	.24	.25	.26	.27	.28	.29	.30
0	.6241	.6084	.5929	.5776	.5625	.5476	.5329	.5184	.5041	.4900
1	.3318	.3432	.3542	.3648	.3750	.3848	.3942	.4032	.4118	.4200
2	.0441	.0484	.0529	.0576	.0625	.0676	.0729	.0784	.0841	.0900
	.31	.32	.33	.34	.35	.36	.37	.38	.39	.40
0	.4761	.4624	.4489	.4356	.4225	.4096	.3969	.3844	.3721	.3600
1	.4278	.4352	.4422	.4488	.4550	.4608	.4662	.4712	.4758	.4800
2	.0961	.1024	.1089	.1156	.1225	.1296	.1369	.1444	.1521	.1600
	.41	.42	.43	.44	.45	.46	.47	.48	.49	.50
0	.3481	.3364	.3249	.3136	.3025	.2916	.2809	.2704	.2601	.2500
1	.4838	.4872	.4902	.4928	.4950	.4968	.4982	.4992	.4998	.5000
2	.1681	.1764	.1849	.1936	.2025	.2116	.2209	.2304	.2401	.2500

Source: Charles T. Clark and Lawrence L. Schkade, *Statistical Analysis for Administrative Decisions,* second edition, Cincinnati, Ohio: South-Western Publishing Co., 1974. Used by permission of the publisher.

Table D Binomial probability distribution (continued)

$n = 3$

x \ p	.01	.02	.03	.04	.05	.06	.07	.08	.09	.10
0	.9704	.9412	.9127	.8847	.8574	.8306	.8044	.7787	.7536	.7290
1	.0294	.0576	.0847	.1106	.1354	.1590	.1816	.2031	.2236	.2430
2	.0003	.0012	.0026	.0046	.0071	.0102	.0137	.0177	.0221	.0270
3	.0000	.0000	.0000	.0001	.0001	.0002	.0003	.0005	.0007	.0010

x \ p	.11	.12	.13	.14	.15	.16	.17	.18	.19	.20
0	.7050	.6815	.6585	.6361	.6141	.5927	.5718	.5514	.5314	.5120
1	.2614	.2788	.2952	.3106	.3251	.3387	.3513	.3631	.3740	.3840
2	.0323	.0380	.0441	.0506	.0574	.0645	.0720	.0797	.0877	.0960
3	.0013	.0017	.0022	.0027	.0034	.0041	.0049	.0058	.0069	.0080

x \ p	.21	.22	.23	.24	.25	.26	.27	.28	.29	.30
0	.4930	.4746	.4565	.4390	.4219	.4052	.3890	.3732	.3579	.3430
1	.3932	.4015	.4091	.4159	.4219	.4271	.4316	.4355	.4386	.4410
2	.1045	.1133	.1222	.1313	.1406	.1501	.1597	.1693	.1791	.1890
3	.0093	.0106	.0122	.0138	.0156	.0176	.0197	.0220	.0244	.0270

x \ p	.31	.32	.33	.34	.35	.36	.37	.38	.39	.40
0	.3285	.3144	.3008	.2875	.2746	.2621	.2500	.2383	.2270	.2160
1	.4428	.4439	.4444	.4443	.4436	.4424	.4406	.4382	.4354	.4320
2	.1989	.2089	.2189	.2289	.2389	.2488	.2587	.2686	.2783	.2880
3	.0298	.0328	.0359	.0393	.0429	.0467	.0507	.0549	.0593	.0640

x \ p	.41	.42	.43	.44	.45	.46	.47	.48	.49	.50
0	.2054	.1951	.1852	.1756	.1664	.1575	.1489	.1406	.1327	.1250
1	.4282	.4239	.4191	.4140	.4084	.4024	.3961	.3894	.3823	.3750
2	.2975	.3069	.3162	.3252	.3341	.3428	.3512	.3594	.3674	.3750
3	.0689	.0741	.0795	.0852	.0911	.0973	.1038	.1106	.1176	.1250

$n = 4$

x \ p	.01	.02	.03	.04	.05	.06	.07	.08	.09	.10
0	.9606	.9224	.8853	.8493	.8145	.7807	.7481	.7164	.6857	.6561
1	.0388	.0753	.1095	.1416	.1715	.1993	.2252	.2492	.2713	.2916
2	.0006	.0023	.0051	.0088	.0135	.0191	.0254	.0325	.0402	.0486
3	.0000	.0000	.0001	.0002	.0005	.0008	.0013	.0019	.0027	.0036
4	.0000	.0000	.0000	.0000	.0000	.0000	.0000	.0000	.0001	.0001

x \ p	.11	.12	.13	.14	.15	.16	.17	.18	.19	.20
0	.6274	.5997	.5729	.5470	.5220	.4979	.4746	.4521	.4305	.4096
1	.3102	.3271	.3424	.3562	.3685	.3793	.3888	.3970	.4039	.4096
2	.0575	.0669	.0767	.0870	.0975	.1084	.1195	.1307	.1421	.1536
3	.0047	.0061	.0076	.0094	.0115	.0138	.0163	.0191	.0222	.0256
4	.0001	.0002	.0003	.0004	.0005	.0007	.0008	.0010	.0013	.0016

x \ p	.21	.22	.23	.24	.25	.26	.27	.28	.29	.30
0	.3895	.3702	.3515	.3336	.3164	.2999	.2840	.2687	.2541	.2401
1	.4142	.4176	.4200	.4214	.4219	.4214	.4201	.4180	.4152	.4116
2	.1651	.1767	.1882	.1996	.2109	.2221	.2331	.2439	.2544	.2646
3	.0293	.0332	.0375	.0420	.0469	.0520	.0575	.0632	.0693	.0756
4	.0019	.0023	.0028	.0033	.0039	.0046	.0053	.0061	.0071	.0081

x \ p	.31	.32	.33	.34	.35	.36	.37	.38	.39	.40
0	.2267	.2138	.2015	.1897	.1785	.1678	.1575	.1478	.1385	.1296
1	.4074	.4025	.3970	.3910	.3845	.3775	.3701	.3623	.3541	.3456
2	.2745	.2841	.2933	.3021	.3105	.3185	.3260	.3330	.3396	.3456
3	.0822	.0891	.0963	.1038	.1115	.1194	.1276	.1361	.1447	.1536
4	.0092	.0105	.0119	.0134	.0150	.0168	.0187	.0209	.0231	.0256

x \ p	.41	.42	.43	.44	.45	.46	.47	.48	.49	.50
0	.1212	.1132	.1056	.0983	.0915	.0850	.0789	.0731	.0677	.0625
1	.3368	.3278	.3185	.3091	.2995	.2897	.2799	.2700	.2600	.2500
2	.3511	.3560	.3604	.3643	.3675	.3702	.3723	.3738	.3747	.3750
3	.1627	.1719	.1813	.1908	.2005	.2102	.2201	.2300	.2400	.2500
4	.0283	.0311	.0342	.0375	.0410	.0448	.0488	.0531	.0576	.0625

Table D Binomial probability distribution (continued)

$n = 5$

x \ p	.01	.02	.03	.04	.05	.06	.07	.08	.09	.10
0	.9510	.9039	.8587	.8154	.7738	.7339	.6957	.6591	.6240	.5905
1	.0480	.0922	.1328	.1699	.2036	.2342	.2618	.2866	.3086	.3280
2	.0010	.0038	.0082	.0142	.0214	.0299	.0394	.0498	.0610	.0729
3	.0000	.0001	.0003	.0006	.0011	.0019	.0030	.0043	.0060	.0081
4	.0000	.0000	.0000	.0000	.0000	.0001	.0001	.0002	.0003	.0004

x \ p	.11	.12	.13	.14	.15	.16	.17	.18	.19	.20
0	.5584	.5277	.4984	.4704	.4437	.4182	.3939	.3707	.3487	.3277
1	.3451	.3598	.3724	.3829	.3915	.3983	.4034	.4069	.4089	.4096
2	.0853	.0981	.1113	.1247	.1382	.1517	.1652	.1786	.1919	.2048
3	.0105	.0134	.0166	.0203	.0244	.0289	.0338	.0392	.0450	.0512
4	.0007	.0009	.0012	.0017	.0022	.0028	.0035	.0043	.0053	.0064
5	.0000	.0000	.0000	.0001	.0001	.0001	.0001	.0002	.0002	.0003

x \ p	.21	.22	.23	.24	.25	.26	.27	.28	.29	.30
0	.3077	.2887	.2707	.2536	.2373	.2219	.2073	.1935	.1804	.1681
1	.4090	.4072	.4043	.4003	.3955	.3898	.3834	.3762	.3685	.3602
2	.2174	.2297	.2415	.2529	.2637	.2739	.2836	.2926	.3010	.3087
3	.0578	.0648	.0721	.0798	.0879	.0962	.1049	.1138	.1229	.1323
4	.0077	.0091	.0108	.0126	.0146	.0169	.0194	.0221	.0251	.0284
5	.0004	.0005	.0006	.0008	.0010	.0012	.0014	.0017	.0021	.0024

x \ p	.31	.32	.33	.34	.35	.36	.37	.38	.39	.40
0	.1564	.1454	.1350	.1252	.1160	.1074	.0992	.0916	.0845	.0778
1	.3513	.3421	.3325	.3226	.3124	.3020	.2914	.2808	.2700	.2592
2	.3157	.3220	.3275	.3323	.3364	.3397	.3423	.3441	.3452	.3456
3	.1418	.1515	.1613	.1712	.1811	.1911	.2010	.2109	.2207	.2304
4	.0319	.0357	.0397	.0441	.0488	.0537	.0590	.0646	.0706	.0768
5	.0029	.0034	.0039	.0045	.0053	.0060	.0069	.0079	.0090	.0102

x \ p	.41	.42	.43	.44	.45	.46	.47	.48	.49	.50
0	.0715	.0656	.0602	.0551	.0503	.0459	.0418	.0380	.0345	.0312
1	.2484	.2376	.2270	.2164	.2059	.1956	.1854	.1755	.1657	.1562
2	.3452	.3442	.3424	.3400	.3369	.3332	.3289	.3240	.3185	.3125
3	.2399	.2492	.2583	.2671	.2757	.2838	.2916	.2990	.3060	.3125
4	.0834	.0902	.0974	.1049	.1128	.1209	.1293	.1380	.1470	.1562
5	.0116	.0131	.0147	.0165	.0185	.0206	.0229	.0255	.0282	.0312

$n = 6$

x \ p	.01	.02	.03	.04	.05	.06	.07	.08	.09	.10
0	.9415	.8858	.8330	.7828	.7351	.6899	.6470	.6064	.5679	.5314
1	.0571	.1085	.1546	.1957	.2321	.2642	.2922	.3164	.3370	.3543
2	.0014	.0055	.0120	.0204	.0305	.0422	.0550	.0688	.0833	.0984
3	.0000	.0002	.0005	.0011	.0021	.0036	.0055	.0080	.0110	.0146
4	.0000	.0000	.0000	.0000	.0001	.0002	.0003	.0005	.0008	.0012
5	.0000	.0000	.0000	.0000	.0000	.0000	.0000	.0000	.0000	.0001

x \ p	.11	.12	.13	.14	.15	.16	.17	.18	.19	.20
0	.4970	.4644	.4336	.4046	.3771	.3513	.3269	.3040	.2824	.2621
1	.3685	.3800	.3888	.3952	.3993	.4015	.4018	.4004	.3975	.3932
2	.1139	.1295	.1452	.1608	.1762	.1912	.2057	.2197	.2331	.2458
3	.0188	.0236	.0289	.0349	.0415	.0486	.0562	.0643	.0729	.0819
4	.0017	.0024	.0032	.0043	.0055	.0069	.0086	.0106	.0128	.0154
5	.0001	.0001	.0002	.0003	.0004	.0005	.0007	.0009	.0012	.0015
6	.0000	.0000	.0000	.0000	.0000	.0000	.0000	.0000	.0000	.0001

x \ p	.21	.22	.23	.24	.25	.26	.27	.28	.29	.30
0	.2431	.2252	.2084	.1927	.1780	.1642	.1513	.1393	.1281	.1176
1	.3877	.3811	.3735	.3651	.3560	.3462	.3358	.3251	.3139	.3025
2	.2577	.2687	.2789	.2882	.2966	.3041	.3105	.3160	.3206	.3241
3	.0913	.1011	.1111	.1214	.1318	.1424	.1531	.1639	.1746	.1852
4	.0182	.0214	.0249	.0287	.0330	.0375	.0425	.0478	.0535	.0595
5	.0019	.0024	.0030	.0036	.0044	.0053	.0063	.0074	.0087	.0102
6	.0001	.0001	.0001	.0002	.0002	.0003	.0004	.0005	.0006	.0007

Table D Binomial probability distribution (continued)

$n = 6$ (continued)

x \ p	.31	.32	.33	.34	.35	.36	.37	.38	.39	.40
0	.1079	.0989	.0905	.0827	.0754	.0687	.0625	.0568	.0515	.0467
1	.2909	.2792	.2673	.2555	.2437	.2319	.2203	.2089	.1976	.1866
2	.3267	.3284	.3292	.3290	.3280	.3261	.3235	.3201	.3159	.3110
3	.1957	.2061	.2162	.2260	.2355	.2446	.2533	.2616	.2693	.2765
4	.0660	.0727	.0799	.0873	.0951	.1032	.1116	.1202	.1291	.1382
5	.0119	.0137	.0157	.0180	.0205	.0232	.0262	.0295	.0330	.0369
6	.0009	.0011	.0013	.0015	.0018	.0022	.0026	.0030	.0035	.0041

x \ p	.41	.42	.43	.44	.45	.46	.47	.48	.49	.50
0	.0422	.0381	.0343	.0308	.0277	.0248	.0222	.0198	.0176	.0156
1	.1759	.1654	.1552	.1454	.1359	.1267	.1179	.1095	.1014	.0938
2	.3055	.2994	.2928	.2856	.2780	.2699	.2615	.2527	.2436	.2344
3	.2831	.2891	.2945	.2992	.3032	.3065	.3091	.3110	.3121	.3125
4	.1475	.1570	.1666	.1763	.1861	.1958	.2056	.2153	.2249	.2344
5	.0410	.0455	.0503	.0554	.0609	.0667	.0729	.0795	.0864	.0938
6	.0048	.0055	.0063	.0073	.0083	.0095	.0108	.0122	.0138	.0156

$n = 7$

x \ p	.01	.02	.03	.04	.05	.06	.07	.08	.09	.10
0	.9321	.8681	.8080	.7514	.6983	.6485	.6017	.5578	.5168	.4783
1	.0659	.1240	.1749	.2192	.2573	.2897	.3170	.3396	.3578	.3720
2	.0020	.0076	.0162	.0274	.0406	.0555	.0716	.0886	.1061	.1240
3	.0000	.0003	.0008	.0019	.0036	.0059	.0090	.0128	.0175	.0230
4	.0000	.0000	.0000	.0001	.0002	.0004	.0007	.0011	.0017	.0026
5	.0000	.0000	.0000	.0000	.0000	.0000	.0000	.0001	.0001	.0002

x \ p	.11	.12	.13	.14	.15	.16	.17	.18	.19	.20
0	.4423	.4087	.3773	.3479	.3206	.2951	.2714	.2493	.2288	.2097
1	.3827	.3901	.3946	.3965	.3960	.3935	.3891	.3830	.3756	.3670
2	.1419	.1596	.1769	.1936	.2097	.2248	.2391	.2523	.2643	.2753
3	.0292	.0363	.0441	.0525	.0617	.0714	.0816	.0923	.1033	.1147
4	.0036	.0049	.0066	.0086	.0109	.0136	.0167	.0203	.0242	.0287
5	.0003	.0004	.0006	.0008	.0012	.0016	.0021	.0027	.0034	.0043
6	.0000	.0000	.0000	.0000	.0001	.0001	.0001	.0002	.0003	.0004

x \ p	.21	.22	.23	.24	.25	.26	.27	.28	.29	.30
0	.1920	.1757	.1605	.1465	.1335	.1215	.1105	.1003	.0910	.0824
1	.3573	.3468	.3356	.3237	.3115	.2989	.2860	.2731	.2600	.2471
2	.2850	.2935	.3007	.3067	.3115	.3150	.3174	.3186	.3186	.3177
3	.1263	.1379	.1497	.1614	.1730	.1845	.1956	.2065	.2169	.2269
4	.0336	.0389	.0447	.0510	.0577	.0648	.0724	.0803	.0886	.0972
5	.0054	.0066	.0080	.0097	.0115	.0137	.0161	.0187	.0217	.0250
6	.0005	.0006	.0008	.0010	.0013	.0016	.0020	.0024	.0030	.0036
7	.0000	.0000	.0000	.0000	.0001	.0001	.0001	.0001	.0002	.0002

x \ p	.31	.32	.33	.34	.35	.36	.37	.38	.39	.40
0	.0745	.0672	.0606	.0546	.0490	.0440	.0394	.0352	.0314	.0280
1	.2342	.2215	.2090	.1967	.1848	.1732	.1619	.1511	.1407	.1306
2	.3156	.3127	.3088	.3040	.2985	.2922	.2853	.2778	.2698	.2613
3	.2363	.2452	.2535	.2610	.2679	.2740	.2793	.2838	.2875	.2903
4	.1062	.1154	.1248	.1345	.1442	.1541	.1640	.1739	.1838	.1935
5	.0286	.0326	.0369	.0416	.0466	.0520	.0578	.0640	.0705	.0774
6	.0043	.0051	.0061	.0071	.0084	.0098	.0113	.0131	.0150	.0172
7	.0003	.0003	.0004	.0005	.0006	.0008	.0009	.0011	.0014	.0016

x \ p	.41	.42	.43	.44	.45	.46	.47	.48	.49	.50
0	.0249	.0221	.0195	.0173	.0152	.0134	.0117	.0103	.0090	.0078
1	.1211	.1119	.1032	.0950	.0872	.0798	.0729	.0664	.0604	.0547
2	.2524	.2431	.2336	.2239	.2140	.2040	.1940	.1840	.1740	.1641
3	.2923	.2934	.2937	.2932	.2918	.2897	.2867	.2830	.2786	.2734
4	.2031	.2125	.2216	.2304	.2388	.2468	.2543	.2612	.2676	.2734
5	.0847	.0923	.1003	.1086	.1172	.1261	.1353	.1447	.1543	.1641
6	.0196	.0223	.0252	.0284	.0320	.0358	.0400	.0445	.0494	.0547
7	.0019	.0023	.0027	.0032	.0037	.0044	.0051	.0059	.0068	.0078

Table D Binomial probability distribution (continued)

$n = 8$

x \ p	.01	.02	.03	.04	.05	.06	.07	.08	.09	.10
0	.9227	.8508	.7837	.7214	.6634	.6096	.5596	.5132	.4703	.4305
1	.0746	.1389	.1939	.2405	.2793	.3113	.3370	.3570	.3721	.3826
2	.0026	.0099	.0210	.0351	.0515	.0695	.0888	.1087	.1288	.1488
3	.0001	.0004	.0013	.0029	.0054	.0089	.0134	.0189	.0255	.0331
4	.0000	.0000	.0001	.0002	.0004	.0007	.0013	.0021	.0031	.0046
5	.0000	.0000	.0000	.0000	.0000	.0000	.0001	.0001	.0002	.0004

x \ p	.11	.12	.13	.14	.15	.16	.17	.18	.19	.20
0	.3937	.3596	.3282	.2992	.2725	.2479	.2252	.2044	.1853	.1678
1	.3892	.3923	.3923	.3897	.3847	.3777	.3691	.3590	.3477	.3355
2	.1684	.1872	.2052	.2220	.2376	.2518	.2646	.2758	.2855	.2936
3	.0416	.0511	.0613	.0723	.0839	.0959	.1084	.1211	.1339	.1468
4	.0064	.0087	.0115	.0147	.0185	.0228	.0277	.0332	.0393	.0459
5	.0006	.0009	.0014	.0019	.0026	.0035	.0045	.0058	.0074	.0092
6	.0000	.0001	.0001	.0002	.0002	.0003	.0005	.0006	.0009	.0011
7	.0000	.0000	.0000	.0000	.0000	.0000	.0000	.0000	.0001	.0001

x \ p	.21	.22	.23	.24	.25	.26	.27	.28	.29	.30
0	.1517	.1370	.1236	.1113	.1001	.0899	.0806	.0722	.0646	.0576
1	.3226	.3092	.2953	.2812	.2670	.2527	.2386	.2247	.2110	.1977
2	.3002	.3052	.3087	.3108	.3115	.3108	.3089	.3058	.3017	.2965
3	.1596	.1722	.1844	.1963	.2076	.2184	.2285	.2379	.2464	.2541
4	.0530	.0607	.0689	.0775	.0865	.0959	.1056	.1156	.1258	.1361
5	.0113	.0137	.0165	.0196	.0231	.0270	.0313	.0360	.0411	.0467
6	.0015	.0019	.0025	.0031	.0038	.0047	.0058	.0070	.0084	.0100
7	.0001	.0002	.0002	.0003	.0004	.0005	.0006	.0008	.0010	.0012
8	.0000	.0000	.0000	.0000	.0000	.0000	.0000	.0000	.0001	.0001

x \ p	.31	.32	.33	.34	.35	.36	.37	.38	.39	.40
0	.0514	.0457	.0406	.0360	.0319	.0281	.0248	.0218	.0192	.0168
1	.1847	.1721	.1600	.1484	.1373	.1267	.1166	.1071	.0981	.0896
2	.2904	.2835	.2758	.2675	.2587	.2494	.2397	.2297	.2194	.2090
3	.2609	.2668	.2717	.2756	.2786	.2805	.2815	.2815	.2806	.2787
4	.1465	.1569	.1673	.1775	.1875	.1973	.2067	.2157	.2242	.2322
5	.0527	.0591	.0659	.0732	.0808	.0888	.0971	.1058	.1147	.1239
6	.0118	.0139	.0162	.0188	.0217	.0250	.0285	.0324	.0367	.0413
7	.0015	.0019	.0023	.0028	.0033	.0040	.0048	.0057	.0067	.0079
8	.0001	.0001	.0001	.0002	.0002	.0003	.0004	.0004	.0005	.0007

x \ p	.41	.42	.43	.44	.45	.46	.47	.48	.49	.50
0	.0147	.0128	.0111	.0097	.0084	.0072	.0062	.0053	.0046	.0039
1	.0816	.0742	.0672	.0608	.0548	.0493	.0442	.0395	.0352	.0312
2	.1985	.1880	.1776	.1672	.1569	.1469	.1371	.1275	.1183	.1094
3	.2759	.2723	.2679	.2627	.2568	.2503	.2431	.2355	.2273	.2188
4	.2397	.2465	.2526	.2580	.2627	.2665	.2695	.2717	.2730	.2734
5	.1332	.1428	.1525	.1622	.1719	.1816	.1912	.2006	.2098	.2188
6	.0463	.0517	.0575	.0637	.0703	.0774	.0848	.0926	.1008	.1094
7	.0092	.0107	.0124	.0143	.0164	.0188	.0215	.0244	.0277	.0312
8	.0008	.0010	.0012	.0014	.0017	.0020	.0024	.0028	.0033	.0039

$n = 9$

x \ p	.01	.02	.03	.04	.05	.06	.07	.08	.09	.10
0	.9135	.8337	.7602	.6925	.6302	.5730	.5204	.4722	.4279	.3874
1	.0830	.1531	.2116	.2597	.2985	.3292	.3525	.3695	.3809	.3874
2	.0034	.0125	.0262	.0433	.0629	.0840	.1061	.1285	.1507	.1722
3	.0001	.0006	.0019	.0042	.0077	.0125	.0186	.0261	.0348	.0446
4	.0000	.0000	.0001	.0003	.0006	.0012	.0021	.0034	.0052	.0074
5	.0000	.0000	.0000	.0000	.0000	.0001	.0002	.0003	.0005	.0008
6	.0000	.0000	.0000	.0000	.0000	.0000	.0000	.0000	.0000	.0001

Table D Binomial probability distribution (continued)

$n = 9$ (continued)

x \ p	.11	.12	.13	.14	.15	.16	.17	.18	.19	.20
0	.3504	.3165	.2855	.2573	.2316	.2082	.1869	.1676	.1501	.1342
1	.3897	.3884	.3840	.3770	.3679	.3569	.3446	.3312	.3169	.3020
2	.1927	.2119	.2295	.2455	.2597	.2720	.2823	.2908	.2973	.3020
3	.0556	.0674	.0800	.0933	.1069	.1209	.1349	.1489	.1627	.1762
4	.0103	.0138	.0179	.0228	.0283	.0345	.0415	.0490	.0573	.0661
5	.0013	.0019	.0027	.0037	.0050	.0066	.0085	.0108	.0134	.0165
6	.0001	.0002	.0003	.0004	.0006	.0008	.0012	.0016	.0021	.0028
7	.0000	.0000	.0000	.0000	.0000	.0001	.0001	.0001	.0002	.0003

x \ p	.21	.22	.23	.24	.25	.26	.27	.28	.29	.30
0	.1199	.1069	.0952	.0846	.0751	.0665	.0589	.0520	.0458	.0404
1	.2867	.2713	.2558	.2404	.2253	.2104	.1960	.1820	.1685	.1556
2	.3049	.3061	.3056	.3037	.3003	.2957	.2899	.2831	.2754	.2668
3	.1891	.2014	.2130	.2238	.2336	.2424	.2502	.2569	.2624	.2668
4	.0754	.0852	.0954	.1060	.1168	.1278	.1388	.1499	.1608	.1715
5	.0200	.0240	.0285	.0335	.0389	.0449	.0513	.0583	.0657	.0735
6	.0036	.0045	.0057	.0070	.0087	.0105	.0127	.0151	.0179	.0210
7	.0004	.0005	.0007	.0010	.0012	.0016	.0020	.0025	.0031	.0039
8	.0000	.0000	.0001	.0001	.0001	.0001	.0002	.0002	.0003	.0004

x \ p	.31	.32	.33	.34	.35	.36	.37	.38	.39	.40
0	.0355	.0311	.0272	.0238	.0207	.0180	.0156	.0135	.0117	.0101
1	.1433	.1317	.1206	.1102	.1004	.0912	.0826	.0747	.0673	.0605
2	.2576	.2478	.2376	.2270	.2162	.2052	.1941	.1831	.1721	.1612
3	.2701	.2721	.2731	.2729	.2716	.2693	.2660	.2618	.2567	.2508
4	.1820	.1921	.2017	.2109	.2194	.2272	.2344	.2407	.2462	.2508
5	.0818	.0904	.0994	.1086	.1181	.1278	.1376	.1475	.1574	.1672
6	.0245	.0284	.0326	.0373	.0424	.0479	.0539	.0603	.0671	.0743
7	.0047	.0057	.0069	.0082	.0098	.0116	.0136	.0158	.0184	.0212
8	.0005	.0007	.0008	.0011	.0013	.0016	.0020	.0024	.0029	.0035
9	.0000	.0000	.0000	.0001	.0001	.0001	.0001	.0002	.0002	.0003

x \ p	.41	.42	.43	.44	.45	.46	.47	.48	.49	.50
0	.0087	.0074	.0064	.0054	.0046	.0039	.0033	.0028	.0023	.0020
1	.0542	.0484	.0431	.0383	.0339	.0299	.0263	.0231	.0202	.0176
2	.1506	.1402	.1301	.1204	.1110	.1020	.0934	.0853	.0776	.0703
3	.2442	.2369	.2291	.2207	.2119	.2027	.1933	.1837	.1739	.1641
4	.2545	.2573	.2592	.2601	.2600	.2590	.2571	.2543	.2506	.2461
5	.1769	.1863	.1955	.2044	.2128	.2207	.2280	.2347	.2408	.2461
6	.0819	.0900	.0983	.1070	.1160	.1253	.1348	.1445	.1542	.1641
7	.0244	.0279	.0318	.0360	.0407	.0458	.0512	.0571	.0635	.0703
8	.0042	.0051	.0060	.0071	.0083	.0097	.0114	.0132	.0153	.0176
9	.0003	.0004	.0005	.0006	.0008	.0009	.0011	.0014	.0016	.0020

$n = 10$

x \ p	.01	.02	.03	.04	.05	.06	.07	.08	.09	.10
0	.9044	.8171	.7374	.6648	.5987	.5386	.4840	.4344	.3894	.3487
1	.0914	.1667	.2281	.2770	.3151	.3438	.3643	.3777	.3851	.3874
2	.0042	.0153	.0317	.0519	.0746	.0988	.1234	.1478	.1714	.1937
3	.0001	.0008	.0026	.0058	.0105	.0168	.0248	.0343	.0452	.0574
4	.0000	.0000	.0001	.0004	.0010	.0019	.0033	.0052	.0078	.0112
5	.0000	.0000	.0000	.0000	.0001	.0001	.0003	.0005	.0009	.0015
6	.0000	.0000	.0000	.0000	.0000	.0000	.0000	.0000	.0001	.0001

x \ p	.11	.12	.13	.14	.15	.16	.17	.18	.19	.20
0	.3118	.2785	.2484	.2213	.1969	.1749	.1552	.1374	.1216	.1074
1	.3854	.3798	.3712	.3603	.3474	.3331	.3178	.3017	.2852	.2684
2	.2143	.2330	.2496	.2639	.2759	.2856	.2929	.2980	.3010	.3020
3	.0706	.0847	.0995	.1146	.1298	.1450	.1600	.1745	.1883	.2013
4	.0153	.0202	.0260	.0326	.0401	.0483	.0573	.0670	.0773	.0881
5	.0023	.0033	.0047	.0064	.0085	.0111	.0141	.0177	.0218	.0264
6	.0002	.0004	.0006	.0009	.0012	.0018	.0024	.0032	.0043	.0055
7	.0000	.0000	.0000	.0001	.0001	.0002	.0003	.0004	.0006	.0008
8	.0000	.0000	.0000	.0000	.0000	.0000	.0000	.0000	.0001	.0001

Table D Binomial probability distribution (continued)

$n = 10$ (continued)

x \ p	.21	.22	.23	.24	.25	.26	.27	.28	.29	.30
0	.0947	.0834	.0733	.0643	.0563	.0492	.0430	.0374	.0326	.0282
1	.2517	.2351	.2188	.2030	.1877	.1730	.1590	.1456	.1330	.1211
2	.3011	.2984	.2942	.2885	.2816	.2735	.2646	.2548	.2444	.2335
3	.2134	.2244	.2343	.2429	.2503	.2563	.2609	.2642	.2662	.2668
4	.0993	.1108	.1225	.1343	.1460	.1576	.1689	.1798	.1903	.2001
5	.0317	.0375	.0439	.0509	.0584	.0664	.0750	.0839	.0933	.1029
6	.0070	.0088	.0109	.0134	.0162	.0195	.0231	.0272	.0317	.0368
7	.0011	.0014	.0019	.0024	.0031	.0039	.0049	.0060	.0074	.0090
8	.0001	.0002	.0002	.0003	.0004	.0005	.0007	.0009	.0011	.0014
9	.0000	.0000	.0000	.0000	.0000	.0000	.0001	.0001	.0001	.0001

x \ p	.31	.32	.33	.34	.35	.36	.37	.38	.39	.40
0	.0245	.0211	.0182	.0157	.0135	.0115	.0098	.0084	.0071	.0060
1	.1099	.0995	.0898	.0808	.0725	.0649	0578	.0514	.0456	.0403
2	.2222	.2107	.1990	.0873	.1757	.1642	.1529	.1419	.1312	.1209
3	.2662	.2644	.2614	.2573	.2522	.2462	.2394	.2319	.2237	.2150
4	.2093	.2177	.2253	.2320	.2377	.2424	.2461	.2487	.2503	.2508
5	.1128	.1229	.1332	.1434	.1536	.1636	.1734	.1829	.1920	.2007
6	.0422	.0482	.0547	.0616	.0689	.0767	.0849	.0934	.1023	.1115
7	.0108	.0130	.0154	.0181	.0212	.0247	.0285	.0327	.0374	.0425
8	.0018	.0023	.0028	.0035	.0043	.0052	.0063	.0075	.0090	.0106
9	.0002	.0002	.0003	.0004	.0005	.0006	.0008	.0010	.0013	.0016
10	.0000	.0000	.0000	.0000	.0000	.0000	.0000	.0001	.0001	.0001

x \ p	.41	.42	.43	.44	.45	.46	.47	.48	.49	.50
0	.0051	.0043	.0036	.0030	.0025	.0021	.0017	.0014	.0012	.0010
1	.0355	.0312	.0273	.0238	.0207	.0180	.0155	.0133	.0114	.0098
2	.1111	.1017	.0927	.0843	.0763	.0688	.0619	.0554	.0494	.0439
3	.2058	.1963	.1865	.1765	.1665	.1564	.1464	.1364	.1267	.1172
4	.2503	.2488	.2462	.2427	.2384	.2331	.2271	.2204	.2130	.2051
5	.2087	.2162	.2229	.2289	.2340	.2383	.2417	.2441	.2456	.2461
6	.1209	.1304	.1401	.1499	.1596	.1692	.1786	.1878	.1966	.2051
7	.0480	.0540	.0604	.0673	.0746	.0824	.0905	.0991	.1080	.1172
8	.0125	.0147	.0171	.0198	.0229	.0263	.0301	.0343	.0389	.0439
9	.0019	.0024	.0029	.0035	.0042	.0050	.0059	.0070	.0083	.0098
10	.0001	.0002	.0002	.0003	.0003	.0004	.0005	.0006	.0008	.0010

$n = 11$

x \ p	.01	.02	.03	.04	.05	.06	.07	.08	.09	.10
0	.8953	.8007	.7153	.6382	.5688	.5063	.4501	.3996	.3544	.3138
1	.0995	.1798	.2433	.2925	.3293	.3555	.3727	.3823	.3855	.3835
2	.0050	.0183	.0376	.0609	.0867	.1135	.1403	.1662	.1906	.2131
3	.0002	.0011	.0035	.0076	.0137	.0217	.0317	.0434	.0566	.0710
4	.0000	.0000	.0002	.0006	.0014	.0028	.0048	.0075	.0112	.0158
5	.0000	.0000	.0000	.0000	.0001	.0002	.0005	.0009	.0015	.0025
6	.0000	.0000	.0000	.0000	.0000	.0000	.0000	.0001	.0002	.0003

x \ p	.11	.12	.13	.14	.15	.16	.17	.18	.19	.20
0	.2775	.2451	.2161	.1903	.1673	.1469	.1288	.1127	.0985	.0859
1	.3773	.3676	.3552	.3408	.3248	.3078	.2901	.2721	.2541	.2362
2	.2332	.2507	.2654	.2774	.2866	.2932	.2971	.2987	.2980	.2953
3	.0865	.1025	.1190	.1355	.1517	.1675	.1826	.1967	.2097	.2215
4	.0214	.0280	.0356	.0441	.0536	.0638	.0748	.0864	.0984	.1107
5	.0037	.0053	.0074	.0101	.0132	.0170	.0214	.0265	.0323	.0388
6	.0005	.0007	.0011	.0016	.0023	.0032	.0044	.0058	.0076	.0097
7	.0000	.0001	.0001	.0002	.0003	.0004	.0006	.0009	.0013	.0017
8	.0000	.0000	.0000	.0000	.0000	.0000	.0001	.0001	.0001	.0002

Table D Binomial probability distribution (continued)

$n = 11$ (continued)

x \ p	.21	.22	.23	.24	.25	.26	.27	.28	.29	.30
0	.0748	.0650	.0564	.0489	.0422	.0364	.0314	.0270	.0231	.0198
1	.2187	.2017	.1854	.1697	.1549	.1408	.1276	.1153	.1038	.0932
2	.2907	.2845	.2768	.2680	.2581	.2474	.2360	.2242	.2121	.1998
3	.2318	.2407	.2481	.2539	.2581	.2608	.2619	.2616	.2599	.2568
4	.1232	.1358	.1482	.1603	.1721	.1832	.1937	.2035	.2123	.2201
5	.0459	.0536	.0620	.0709	.0803	.0901	.1003	.1108	.1214	.1321
6	.0122	.0151	.0185	.0224	.0268	.0317	.0371	.0431	.0496	.0566
7	.0023	.0030	.0039	.0050	.0064	.0079	.0098	.0120	.0145	.0173
8	.0003	.0004	.0006	.0008	.0011	.0014	.0018	.0023	.0030	.0037
9	.0000	.0000	.0001	.0001	.0001	.0002	.0002	.0003	.0004	.0005

x \ p	.31	.32	.33	.34	.35	.36	.37	.38	.39	.40
0	.0169	.0144	.0122	.0104	.0088	.0074	.0062	.0052	.0044	.0036
1	.0834	.0744	.0662	.0587	.0518	.0457	.0401	.0351	.0306	.0266
2	.1874	.1751	.1630	.1511	.1395	.1284	.1177	.1075	.0978	.0887
3	.2526	.2472	.2408	.2335	.2254	.2167	.2074	.1977	.1876	.1774
4	.2269	.2326	.2372	.2406	.2428	.2438	.2436	.2423	.2399	.2365
5	.1427	.1533	.1636	.1735	.1830	.1920	.2003	.2079	.2148	.2207
6	.0641	.0721	.0806	.0894	.0985	.1080	.1176	.1274	.1373	.1471
7	.0206	.0242	.0283	.0329	.0379	.0434	.0494	.0558	.0627	.0701
8	.0046	.0057	.0070	.0085	.0102	.0122	.0145	.0171	.0200	.0234
9	.0007	.0009	.0011	.0015	.0018	.0023	.0028	.0035	.0043	.0052
10	.0001	.0001	.0001	.0001	.0002	.0003	.0003	.0004	.0005	.0007

x \ p	.41	.42	.43	.44	.45	.46	.47	.48	.49	.50
0	.0030	.0025	.0021	.0017	.0014	.0011	.0009	.0008	.0006	.0005
1	.0231	.0199	.0171	.0147	.0125	.0107	.0090	.0076	.0064	.0054
2	.0801	.0721	.0646	.0577	.0513	.0454	.0401	.0352	.0308	.0269
3	.1670	.1566	.1462	.1359	.1259	.1161	.1067	.0976	.0888	.0806
4	.2321	.2267	.2206	.2136	.2060	.1978	.1892	.1801	.1707	.1611
5	.2258	.2299	.2329	.2350	.2360	.2359	.2348	.2327	.2296	.2256
6	.1569	.1664	.1757	.1846	.1931	.2010	.2083	.2148	.2206	.2256
7	.0779	.0861	.0947	.1036	.1128	.1223	.1319	.1416	.1514	.1611
8	.0271	.0312	.0357	.0407	.0462	.0521	.0585	.0654	.0727	.0806
9	.0063	.0075	.0090	.0107	.0126	.0148	.0173	.0201	.0233	.0269
10	.0009	.0011	.0014	.0017	.0021	.0025	.0031	.0037	.0045	.0054
11	.0001	.0001	.0001	.0001	.0002	.0002	.0002	.0003	.0004	.0005

$n = 12$

x \ p	.01	.02	.03	.04	.05	.06	.07	.08	.09	.10
0	.8864	.7847	.6938	.6127	.5404	.4759	.4186	.3677	.3225	.2824
1	.1074	.1922	.2575	.3064	.3413	.3645	.3781	.3837	.3827	.3766
2	.0060	.0216	.0438	.0702	.0988	.1280	.1565	.1835	.2082	.2301
3	.0002	.0015	.0045	.0098	.0173	.0272	.0393	.0532	.0686	.0852
4	.0000	.0001	.0003	.0009	.0021	.0039	.0067	.0104	.0153	.0213
5	.0000	.0000	.0000	.0001	.0002	.0004	.0008	.0014	.0024	.0038
6	.0000	.0000	.0000	.0000	.0000	.0000	.0001	.0001	.0003	.0005

x \ p	.11	.12	.13	.14	.15	.16	.17	.18	.19	.20
0	.2470	.2157	.1880	.1637	.1422	.1234	.1069	.0924	.0798	.0687
1	.3663	.3529	.3372	.3197	.3012	.2821	.2627	.2434	.2245	.2062
2	.2490	.2647	.2771	.2863	.2924	.2955	.2960	.2939	.2897	.2835
3	.1026	.1203	.1380	.1553	.1720	.1876	.2021	.2151	.2265	.2362
4	.0285	.0369	.0464	.0569	.0683	.0804	.0931	.1062	.1195	.1329
5	.0056	.0081	.0111	.0148	.0193	.0245	.0305	.0373	.0449	.0532
6	.0008	.0013	.0019	.0028	.0040	.0054	.0073	.0096	.0123	.0155
7	.0001	.0001	.0002	.0004	.0006	.0009	.0013	.0018	.0025	.0033
8	.0000	.0000	.0000	.0000	.0001	.0001	.0002	.0002	.0004	.0005
9	.0000	.0000	.0000	.0000	.0000	.00000	.0000	.0000	.0000	.0001

Table D Binomial probability distribution (continued)

$n = 12$ (continued)

$x \backslash p$	.21	.22	.23	.24	.25	.26	.27	.28	.29	.30
0	.0591	.0507	.0434	.0371	.0317	.0270	.0229	.0194	.0164	.0138
1	.1885	.1717	.1557	.1407	.1267	.1137	.1016	.0906	.0804	.0712
2	.2756	.2663	.2558	.2444	.2323	.2197	.2068	.1937	.1807	.1678
3	.2442	.2503	.2547	.2573	.2581	.2573	.2549	.2511	.2460	.2397
4	.1460	.1589	.1712	.1828	.1936	.2034	.2122	.2197	.2261	.2311
5	.0621	.0717	.0818	.0924	.1032	.1143	.1255	.1367	.1477	.1585
6	.0193	.0236	.0285	.0340	.0401	.0469	.0542	.0620	.0704	.0792
7	.0044	.0057	.0073	.0092	.0115	.0141	.0172	.0207	.0246	.0291
8	.0007	.0010	.0014	.0018	.0024	.0031	.0040	.0050	.0063	.0078
9	.0001	.0001	.0002	.0003	.0004	.0005	.0007	.0009	.0011	.0015
10	.0000	.0000	.0000	.0000	.0000	.0001	.0001	.0001	.0001	.0002

$x \backslash p$	.31	.32	.33	.34	.35	.36	.37	.38	.39	.40
0	.0116	.0098	.0082	.0068	.0057	.0047	.0039	.0032	.0027	.0022
1	.0628	.0552	.0484	.0422	.0368	.0319	.0276	.0237	.0204	.0174
2	.1552	.1429	.1310	.1197	.1088	.0986	.0890	.0800	.0716	.0639
3	.2324	.2241	.2151	.2055	.1954	.1849	.1742	.1634	.1526	.1419
4	.2349	.2373	.2384	.2382	.2367	.2340	.2302	.2254	.2195	.2128
5	.1688	.1787	.1879	.1963	.2039	.2106	.2163	.2210	.2246	.2270
6	.0885	.0981	.1079	.1180	.1281	.1382	.1482	.1580	.1675	.1766
7	.0341	.0396	.0456	.0521	.0591	.0666	.0746	.0830	.0918	.1009
8	.0096	.0116	.0140	.0168	.0199	.0234	.0274	.0318	.0367	.0420
9	.0019	.0024	.0031	.0038	.0048	.0059	.0071	.0087	.0104	.0125
10	.0003	.0003	.0005	.0006	.0008	.0010	.0013	.0016	.0020	.0025
11	.0000	.0000	.0000	.0001	.0001	.0001	.0001	.0002	.0002	.0003

$x \backslash p$	.41	.42	.43	.44	.45	.46	.47	.48	.49	.50
0	.0018	.0014	.0012	.0010	.0008	.0006	.0005	.0004	.0003	.0002
1	.0148	.0126	.0106	.0090	.0075	.0063	.0052	.0043	.0036	.0029
2	.0567	.0502	.0442	.0388	.0339	.0294	.0255	.0220	.0189	.0161
3	.1314	.1211	.1111	.1015	.0923	.0836	.0754	.0676	.0604	.0537
4	.2054	.1973	.1886	.1794	.1700	.1602	.1504	.1405	.1306	.1208
5	.2284	.2285	.2276	.2256	.2225	.2184	.2134	.2075	.2008	.1934
6	.1851	.1931	.2003	.2068	.2124	.2171	.2208	.2234	.2250	.2256
7	.1103	.1198	.1295	.1393	.1489	.1585	.1678	.1768	.1853	.1934
8	.0479	.0542	.0611	.0684	.0762	.0844	.0930	.1020	.1113	.1208
9	.0148	.0175	.0205	.0239	.0277	.0319	.0367	.0418	.0475	.0537
10	.0031	.0038	.0046	.0056	.0068	.0082	.0098	.0116	.0137	.0161
11	.0004	.0005	.0006	.0008	.0010	.0013	.0016	.0019	.0024	.0029
12	.0000	.0000	.0000	.0001	.0001	.0001	.0001	.0001	.0002	.0002

$n = 13$

$x \backslash p$	.01	.02	.03	.04	.05	.06	.07	.08	.09	.10
0	.8775	.7690	.6730	.5882	.5133	.4474	.3893	.3383	.2935	.2542
1	.1152	.2040	.2706	.3186	.3512	.3712	.3809	.3824	.3773	.3672
2	.0070	.0250	.0502	.0797	.1109	.1422	.1720	.1995	.2239	.2448
3	.0003	.0019	.0057	.0122	.0214	.0333	.0475	.0636	.0812	.0997
4	.0000	.0001	.0004	.0013	.0028	.0053	.0089	.0138	.0201	.0277
5	.0000	.0000	.0000	.0001	.0003	.0006	.0012	.0022	.0036	.0055
6	.0000	.0000	.0000	.0000	.0000	.0001	.0001	.0003	.0005	.0008
7	.0000	.0000	.0000	.0000	.0000	.0000	.0000	.0000	.0000	.0001

$x \backslash p$	.11	.12	.13	.14	.15	.16	.17	.18	.19	.20
0	.2198	.1898	.1636	.1408	.1209	.1037	.0887	.0758	.0646	.0550
1	.3532	.3364	.3178	.2979	.2774	.2567	.2362	.2163	.1970	.1787
2	.2619	.2753	.2849	.2910	.2937	.2934	.2903	.2848	.2773	.2680
3	.1187	.1376	.1561	.1737	.1900	.2049	.2180	.2293	.2385	.2457
4	.0367	.0469	.0583	.0707	.0838	.0976	.1116	.1258	.1399	.1535
5	.0082	.0115	.0157	.0207	.0266	.0335	.0412	.0497	.0591	.0691
6	.0013	.0021	.0031	.0045	.0063	.0085	.0112	.0145	.0185	.0230
7	.0002	.0003	.0005	.0007	.0011	.0016	.0023	.0032	.0043	.0058
8	.0000	.0000	.0001	.0001	.0001	.0002	.0004	.0005	.0008	.0011
9	.0000	.0000	.0000	.0000	.0000	.0000	.0000	.0001	.0001	.0001

Table D Binomial probability distribution (continued)

$n = 13$ (continued)

$x \backslash p$	.21	.22	.23	.24	.25	.26	.27	.28	.29	.30
0	.0467	.0396	.0334	.0282	.0238	.0200	.0167	.0140	.0117	.0097
1	.1613	.1450	.1299	.1159	.1029	.0911	.0804	.0706	.0619	.0540
2	.2573	.2455	.2328	.2195	.2059	.1921	.1784	.1648	.1516	.1388
3	.2508	.2539	.2550	.2542	.2517	.2475	.2419	.2351	.2271	.2181
4	.1667	.1790	.1904	.2007	.2097	.2174	.2237	.2285	.2319	.2337
5	.0797	.0909	.1024	.1141	.1258	.1375	.1489	.1600	.1705	.1803
6	.0283	.0342	.0408	.0480	.0559	.0644	.0734	.0829	.0928	.1030
7	.0075	.0096	.0122	.0152	.0186	.0226	.0272	.0323	.0379	.0442
8	.0015	.0020	.0027	.0036	.0047	.0060	.0075	.0094	.0116	.0142
9	.0002	.0003	.0005	.0006	.0009	.0012	.0015	.0020	.0026	.0034
10	.0000	.0000	.0001	.0001	.0001	.0002	.0002	.0003	.0004	.0006
11	.0000	.0000	.0000	.0000	.0000	.0000	.0000	.0000	.0000	.0001

$x \backslash p$	.31	.32	.33	.34	.35	.36	.37	.38	.39	.40
0	.0080	.0066	.0055	.0045	.0037	.0030	.0025	.0020	.0016	.0013
1	.0469	.0407	.0351	.0302	.0259	.0221	.0188	.0159	.0135	.0113
2	.1265	.1148	.1037	.0933	.0836	.0746	.0663	.0586	.0516	.0453
3	.2084	.1981	.1874	.1763	.1651	.1538	.1427	.1317	.1210	.1107
4	.2341	.2331	.2307	.2270	.2222	.2163	.2095	.2018	.1934	.1845
5	.1893	.1974	.2045	.2105	.2154	.2190	.2215	.2227	.2226	.2214
6	.1134	.1239	.1343	.1446	.1546	.1643	.1734	.1820	.1898	.1968
7	.0509	.0583	.0662	.0745	.0833	.0924	.1019	.1115	.1213	.1312
8	.0172	.0206	.0244	.0288	.0336	.0390	.0449	.0513	.0582	.0656
9	.0043	.0054	.0067	.0082	.0101	.0122	.0146	.0175	.0207	.0243
10	.0008	.0010	.0013	.0017	.0022	.0027	.0034	.0043	.0053	.0065
11	.0001	.0001	.0002	.0002	.0003	.0004	.0006	.0007	.0009	.0012
12	.0000	.0000	.0000	.0000	.0000	.0000	.0001	.0001	.0001	.0001

$x \backslash p$	.41	.42	.43	.44	.45	.46	.47	.48	.49	.50
0	.0010	.0008	.0007	.0005	.0004	.0003	.0003	.0002	.0002	.0001
1	.0095	.0079	.0066	.0054	.0045	.0037	.0030	.0024	.0020	.0016
2	.0395	.0344	.0298	.0256	.0220	.0188	.0160	.0135	.0114	.0095
3	.1007	.0913	.0823	.0739	.0660	.0587	.0519	.0457	.0401	.0349
4	.1750	.1653	.1553	.1451	.1350	.1250	.1151	.1055	.0962	.0873
5	.2189	.2154	.2108	.2053	.1989	.1917	.1838	.1753	.1664	.1571
6	.2029	.2080	.2121	.2151	.2169	.2177	.2173	.2158	.2131	.2095
7	.1410	.1506	.1600	.1690	.1775	.1854	.1927	.1992	.2048	.2095
8	.0735	.0818	.0905	.0996	.1089	.1185	.1282	.1379	.1476	.1571
9	.0284	.0329	.0379	.0435	.0495	.0561	.0631	.0707	.0788	.0873
10	.0079	.0095	.0114	.0137	.0162	.0191	.0224	.0261	.0303	.0349
11	.0015	.0019	.0024	.0029	.0036	.0044	.0054	.0066	.0079	.0095
12	.0002	.0002	.0003	.0004	.0005	.0006	.0008	.0010	.0013	.0016
13	.0000	.0000	.0000	.0000	.0000	.0000	.0001	.0001	.0001	.0001

$n = 14$

$x \backslash p$	.01	.02	.03	.04	.05	.06	.07	.08	.09	.10
0	.8687	.7536	.6528	.5647	.4877	.4205	.3620	.3112	.2670	.2288
1	.1229	.2153	.2827	.3294	.3593	.3758	.3815	.3788	.3698	.3559
2	.0081	.0286	.0568	.0892	.1229	.1559	.1867	.2141	.2377	.2570
3	.0003	.0023	.0070	.0149	.0259	.0398	.0562	.0745	.0940	.1142
4	.0000	.0001	.0006	.0017	.0037	.0070	.0116	.0178	.0256	.0349
5	.0000	.0000	.0000	.0001	.0004	.0009	.0018	.0031	.0051	.0078
6	.0000	.0000	.0000	.0000	.0000	.0001	.0002	.0004	.0008	.0013
7	.0000	.0000	.0000	.0000	.0000	.0000	.0000	.0000	.0001	.0002

$x \backslash p$	.11	.12	.13	.14	.15	.16	.17	.18	.19	.20
0	.1956	.1670	.1423	.1211	.1028	.0871	.0736	.0621	.0523	.0440
1	.3385	.3188	.2977	.2759	.2539	.2322	.2112	.1910	.1719	.1539
2	.2720	.2826	.2892	.2919	.2912	.2875	.2811	.2725	.2620	.2501
3	.1345	.1542	.1728	.1901	.2056	.2190	.2303	.2393	.2459	.2501
4	.0457	.0578	.0710	.0851	.0998	.1147	.1297	.1444	.1586	.1720
5	.0113	.0158	.0212	.0277	.0352	.0437	.0531	.0634	.0744	.0860
6	.0021	.0032	.0048	.0068	.0093	.0125	.0163	.0209	.0262	.0322
7	.0003	.0005	.0008	.0013	.0019	.0027	.0038	.0052	.0070	.0092
8	.0000	.0001	.0001	.0002	.0003	.0005	.0007	.0010	.0014	.0020
9	.0000	.0000	.0000	.0000	.0000	.0001	.0001	.0001	.0002	.0003

Table D Binomial probability distribution (continued)

$n = 14$ (continued)

$x \backslash p$	.21	.22	.23	.24	.25	.26	.27	.28	.29	.30
0	.0369	.0309	.0258	.0214	.0178	.0148	.0122	.0101	.0083	.0068
1	.1372	.1218	.1077	.0948	.0832	.0726	.0632	.0548	.0473	.0407
2	.2371	.2234	.2091	.1946	.1802	.1659	.1519	.1385	.1256	.1134
3	.2521	.2520	.2499	.2459	.2402	.2331	.2248	.2154	.2052	.1943
4	.1843	.1955	.2052	.2135	.2202	.2252	.2286	.2304	.2305	.2290
5	.0980	.1103	.1226	.1348	.1468	.1583	.1691	.1792	.1883	.1963
6	.0391	.0466	.0549	.0639	.0734	.0834	.0938	.1045	.1153	.1262
7	.0119	.0150	.0188	.0231	.0280	.0335	.0397	.0464	.0538	.0618
8	.0028	.0037	.0049	.0064	.0082	.0103	.0128	.0158	.0192	.0232
9	.0005	.0007	.0010	.0013	.0018	.0024	.0032	.0041	.0052	.0066
10	.0001	.0001	.0001	.0002	.0003	.0004	.0006	.0008	.0011	.0014
11	.0000	.0000	.0000	.0000	.0000	.0001	.0001	.0001	.0002	.0002

$x \backslash p$	.31	.32	.33	.34	.35	.36	.37	.38	.39	.40
0	.0055	.0045	.0037	.0030	.0024	.0019	.0016	.0012	.0010	.0008
1	.0349	.0298	.0253	.0215	.0181	.0152	.0128	.0106	.0088	.0073
2	.1018	.0911	.0811	.0719	.0634	.0557	.0487	.0424	.0367	.0317
3	.1830	.1715	.1598	.1481	.1366	.1253	.1144	.1039	.0940	.0845
4	.2261	.2219	.2164	.2098	.2022	.1938	.1848	.1752	.1652	.1549
5	.2032	.2088	.2132	.2161	.2178	.2181	.2170	.2147	.2112	.2066
6	.1369	.1474	.1575	.1670	.1759	.1840	.1912	.1974	.2026	.2066
7	.0703	.0793	.0886	.0983	.1082	.1183	.1283	.1383	.1480	.1574
8	.0276	.0326	.0382	.0443	.0510	.0582	.0659	.0742	.0828	.0918
9	.0083	.0102	.0125	.0152	.0183	.0218	.0258	.0303	.0353	.0408
10	.0019	.0024	.0031	.0039	.0049	.0061	.0076	.0093	.0113	.0136
11	.0003	.0004	.0006	.0007	.0010	.0013	.0016	.0021	.0026	.0033
12	.0000	.0000	.0001	.0001	.0001	.0002	.0002	.0003	.0004	.0005
13	.0000	.0000	.0000	.0000	.0000	.0000	.0000	.0000	.0000	.0001

$x \backslash p$	.41	.42	.43	.44	.45	.46	.47	.48	.49	.50
0	.0006	.0005	.0004	.0003	.0002	.0002	.0001	.0001	.0001	.0001
1	.0060	.0049	.0040	.0033	.0027	.0021	.0017	.0014	.0011	.0009
2	.0272	.0233	.0198	.0168	.0141	.0118	.0099	.0082	.0068	.0056
3	.0757	.0674	.0597	.0527	.0462	.0403	.0350	.0303	.0260	.0222
4	.1446	.1342	.1239	.1138	.1040	.0945	.0854	.0768	.0687	.0611
5	.2009	.1943	.1869	.1788	.1701	.1610	.1515	.1418	.1320	.1222
6	.2094	.2111	.2115	.2108	.2088	.2057	.2015	.1963	.1902	.1833
7	.1663	.1747	.1824	.1892	.1952	.2003	.2043	.2071	.2089	.2095
8	.1011	.1107	.1204	.1301	.1398	.1493	.1585	.1673	.1756	.1833
9	.0469	.0534	.0605	.0682	.0762	.0848	.0937	.1030	.1125	.1222
10	.0163	.0193	.0228	.0268	.0312	.0361	.0415	.0475	.0540	.0611
11	.0041	.0051	.0063	.0076	.0093	.0112	.0134	.0160	.0189	.0222
12	.0007	.0009	.0012	.0015	.0019	.0024	.0030	.0037	.0045	.0056
13	.0001	.0001	.0001	.0002	.0002	.0003	.0004	.0005	.0007	.0009
14	.0000	.0000	.0000	.0000	.0000	.0000	.0000	.0000	.0000	.0001

$n = 15$

$x \backslash p$	.01	.02	.03	.04	.05	.06	.07	.08	.09	.10
0	.8601	.7386	.6333	.5421	.4633	.3953	.3367	.2863	.2430	.2059
1	.1303	.2261	.2938	.3388	.3658	.3785	.3801	.3734	.3605	.3432
2	.0092	.0323	.0636	.0988	.1348	.1691	.2003	.2273	.2496	.2669
3	.0004	.0029	.0085	.0178	.0307	.0468	.0653	.0857	.1070	.1285
4	.0000	.0002	.0008	.0022	.0049	.0090	.0148	.0223	.0317	.0428
5	.0000	.0000	.0001	.0002	.0006	.0013	.0024	.0043	.0069	.0105
6	.0000	.0000	.0000	.0000	.0000	.0001	.0003	.0006	.0011	.0019
7	.0000	.0000	.0000	.0000	.0000	.0000	.0000	.0001	.0001	.0003

Table D Binomial probability distribution (continued)

n = 15 (continued)

x \ p	.11	.12	.13	.14	.15	.16	.17	.18	.19	.20
0	.1741	.1470	.1238	.1041	.0874	.0731	.0611	.0510	.0424	.0352
1	.3228	.3006	.2775	.2542	.2312	.2090	.1878	.1678	.1492	.1319
2	.2793	.2870	.2903	.2897	.2856	.2787	.2692	.2578	.2449	.2309
3	.1496	.1696	.1880	.2044	.2184	.2300	.2389	.2452	.2489	.2501
4	.0555	.0694	.0843	.0998	.1156	.1314	.1468	.1615	.1752	.1876
5	.0151	.0208	.0277	.0357	.0449	.0551	.0662	.0780	.0904	.1032
6	.0031	.0047	.0069	.0097	.0132	.0175	.0226	.0285	.0353	.0430
7	.0005	.0008	.0013	.0020	.0030	.0043	.0059	.0081	.0107	.0138
8	.0001	.0001	.0002	.0003	.0005	.0008	.0012	.0018	.0025	.0035
9	.0000	.0000	.0000	.0000	.0001	.0001	.0002	.0003	.0005	.0007
10	.0000	.0000	.0000	.0000	.0000	.0000	.0000	.0000	.0001	.0001
	.21	.22	.23	.24	.25	.26	.27	.28	.29	.30
0	.0291	.0241	.0198	.0163	.0134	.0109	.0089	.0072	.0059	.0047
1	.1162	.1018	.0889	.0772	.0668	.0576	.0494	.0423	.0360	.0305
2	.2162	.2010	.1858	.1707	.1559	.1416	.1280	.1150	.1029	.0916
3	.2490	.2457	.2405	.2336	.2252	.2156	.2051	.1939	.1821	.1700
4	.1986	.2079	.2155	.2213	.2252	.2273	.2276	.2262	.2231	.2186
5	.1161	.1290	.1416	.1537	.1651	.1757	.1852	.1935	.2005	.2061
6	.0514	.0606	.0705	.0809	.0917	.1029	.1142	.1254	.1365	.1472
7	.0176	.0220	.0271	.0329	.0393	.0465	.0543	.0627	.0717	.0811
8	.0047	.0062	.0081	.0104	.0131	.0163	.0201	.0244	.0293	.0348
9	.0010	.0014	.0019	.0025	.0034	.0045	.0058	.0074	.0093	.0116
10	.0002	.0002	.0003	.0005	.0007	.0009	.0013	.0017	.0023	.0030
11	.0000	.0000	.0000	.0001	.0001	.0002	.0002	.0003	.0004	.0006
12	.0000	.0000	.0000	.0000	.0000	.0000	.0000	.0000	.0001	.0001
	.31	.32	.33	.34	.35	.36	.37	.38	.39	.40
0	.0038	.0031	.0025	.0020	.0016	.0012	.0010	.0008	.0006	.0005
1	.0258	.0217	.0182	.0152	.0126	.0104	.0086	.0071	.0058	.0047
2	.0811	.0715	.0627	.0547	.0476	.0411	.0354	.0303	.0259	.0219
3	.1579	.1457	.1338	.1222	.1110	.1002	.0901	.0805	.0716	.0634
4	.2128	.2057	.1977	.1888	.1792	.1692	.1587	.1481	.1374	.1268
5	.210	.2130	.2142	.2140	.2123	.2093	.2051	.1997	.1933	.1859
6	.1575	.1671	.1759	.1837	.1906	.1963	.2008	.2040	.2059	.2066
7	.0910	.1011	.1114	.1217	.1319	.1419	.1516	.1608	.1693	.1771
8	.0409	.0476	.0549	.0627	.0710	.0798	.0890	.0985	.1082	.1181
9	.0143	.0174	.0210	.0251	.0298	.0349	.0407	.0470	.0538	.0612
10	.0038	.0049	.0062	.0078	.0096	.0118	.0143	.0173	.0206	.0245
11	.0008	.0011	.0014	.0018	.0024	.0030	.0038	.0048	.0060	.0074
12	.0001	.0002	.0002	.0003	.0004	.0006	.0007	.0010	.0013	.0016
13	.0000	.0000	.0000	.0000	.0001	.0001	.0001	.0001	.0002	.0003
	.41	.42	.43	.44	.45	.46	.47	.48	.49	.50
0	.0004	.0003	.0002	.0002	.0001	.0001	.0001	.0001	.0000	.0000
1	.0038	.0031	.0025	.0020	.0016	.0012	.0010	.0008	.0006	.0005
2	.0185	.0156	.0130	.0108	.0090	.0074	.0060	.0049	.0040	.0032
3	.0558	.0489	.0426	.0369	.0318	.0272	.0232	.0197	.0166	.0139
4	.1163	.1061	.0963	.0869	.0780	.0696	.0617	.0545	.0478	.0417
5	.1778	.1691	.1598	.1502	.1404	.1304	.1204	.1106	.1010	.0916
6	.2060	.2041	.2010	.1967	.1914	.1851	.1780	.1702	.1617	.1527
7	.1840	.1900	.1949	.1987	.2013	.2028	.2030	.2020	.1997	.1964
8	.1279	.1376	.1470	.1561	.1647	.1727	.1800	.1864	.1919	.1964
9	.0691	.0775	.0863	.0954	.1048	.1144	.1241	.1338	.1434	.1527
10	.0288	.0337	.0390	.0450	.0515	.0585	.0661	.0741	.0827	.0916
11	.0091	.0111	.0134	.0161	.0191	.0226	.0266	.0311	.0361	.0417
12	.0021	.0027	.0034	.0042	.0052	.0064	.0079	.0096	.0116	.0139
13	.0003	.0004	.0006	.0008	.0010	.0013	.0016	.0020	.0026	.0032
14	.0000	.0000	.0001	.0001	.0001	.0002	.0002	.0003	.0004	.0005

Table D Binomial probability distribution (continued)

$n = 16$

x \ p	.01	.02	.03	.04	.05	.06	.07	.08	.09	.10
0	.8515	.7238	.6143	.5204	.4401	.3716	.3131	.2634	.2211	.1853
1	.1376	.2363	.3040	.3469	.3706	.3795	.3771	.3665	.3499	.3294
2	.0104	.0362	.0705	.1084	.1463	.1817	.2129	.2390	.2596	.2745
3	.0005	.0034	.0102	.0211	.0359	.0541	.0748	.0970	.1198	.1423
4	.0000	.0002	.0010	.0029	.0061	.0112	.0183	.0274	.0385	.0514
5	.0000	.0000	.0001	.0003	.0008	.0017	.0033	.0057	.0091	.0137
6	.0000	.0000	.0000	.0000	.0001	.0002	.0005	.0009	.0017	.0028
7	.0000	.0000	.0000	.0000	.0000	.0000	.0000	.0001	.0002	.0004
8	.0000	.0000	.0000	.0000	.0000	.0000	.0000	.0000	.0000	.0001

x \ p	.11	.12	.13	.14	.15	.16	.17	.18	.19	.20
0	.1550	.1293	.1077	.0895	.0743	.0614	.0507	.0418	.0343	.0281
1	.3065	.2822	.2575	.2332	.2097	.1873	.1662	.1468	.1289	.1126
2	.2841	.2886	.2886	.2847	.2775	.2675	.2554	.2416	.2267	.2111
3	.1638	.1837	.2013	.2163	.2285	.2378	.2441	.2475	.2482	.2463
4	.0658	.0814	.0977	.1144	.1311	.1472	.1625	.1766	.1892	.2001
5	.0195	.0266	.0351	.0447	.0555	.0673	.0799	.0930	.1065	.1201
6	.0044	.0067	.0096	.0133	.0180	.0235	.0300	.0374	.0458	.0550
7	.0008	.0013	.0020	.0031	.0045	.0064	.0088	.0117	.0153	.0197
8	.0001	.0002	.0003	.0006	.0009	.0014	.0020	.0029	.0041	.0055
9	.0000	.0000	.0000	.0001	.0001	.0002	.0004	.0006	.0008	.0012
10	.0000	.0000	.0000	.0000	.0000	.0000	.0001	.0001	.0001	.0002

x \ p	.21	.22	.23	.24	.25	.26	.27	.28	.29	.30
0	.0230	.0188	.0153	.0124	.0100	.0081	.0065	.0052	.0042	.0033
1	.0979	.0847	.0730	.0626	.0535	.0455	.0385	.0325	.0273	.0228
2	.1952	.1792	.1635	.1482	.1336	.1198	.1068	.0947	.0835	.0732
3	.2421	.2359	.2279	.2185	.2079	.1964	.1843	.1718	.1591	.1465
4	.2092	.2162	.2212	.2242	.2252	.2243	.2215	.2171	.2112	.2040
5	.1334	.1464	.1586	.1699	.1802	.1891	.1966	.2026	.2071	.2099
6	.0650	.0757	.0869	.0984	.1101	.1218	.1333	.1445	.1551	.1649
7	.0247	.0305	.0371	.0444	.0524	.0611	.0704	.0803	.0905	.1010
8	.0074	.0097	.0125	.0158	.0197	.0242	.0293	.0351	.0416	.0487
9	.0017	.0024	.0033	.0044	.0058	.0075	.0096	.0121	.0151	.0185
10	.0003	.0005	.0007	.0010	.0014	.0019	.0025	.0033	.0043	.0056
11	.0000	.0001	.0001	.0002	.0002	.0004	.0005	.0007	.0010	.0013
12	.0000	.0000	.0000	.0000	.0000	.0001	.0001	.0001	.0002	.0002

x \ p	.31	.32	.33	.34	.35	.36	.37	.38	.39	.40
0	.0026	.0021	.0016	.0013	.0010	.0008	.0006	.0005	.0004	.0003
1	.0190	.0157	.0130	.0107	.0087	.0071	.0058	.0047	.0038	.0030
2	.0639	.0555	.0480	.0413	.0353	.0301	.0255	.0215	.0180	.0150
3	.1341	.1220	.1103	.0992	.0888	.0790	.0699	.0615	.0538	.0468
4	.1958	.1865	.1766	.1662	.1553	.1444	.1333	.1224	.1118	.1014
5	.2111	.2107	.2088	.2054	.2008	.1949	.1879	.1801	.1715	.1623
6	.1739	.1818	.1885	.1940	.1982	.2010	.2024	.2024	.2010	.1983
7	.1116	.1222	.1326	.1428	.1524	.1615	.1698	.1772	.1836	.1889
8	.0564	.0647	.0735	.0827	.0923	.1022	.1122	.1222	.1320	.1417
9	.0225	.0271	.0322	.0379	.0442	.1511	.0586	.0666	.0750	.0840
10	.0071	.0089	.0111	.0137	.0167	.0201	.0241	.0286	.0336	.0392
11	.0017	.0023	.0030	.0038	.0049	.0062	.0077	.0095	.0117	.0142
12	.0003	.0004	.0006	.0008	.0011	.0014	.0019	.0024	.0031	.0040
13	.0000	.0001	.0001	.0001	.0002	.0003	.0003	.0005	.0006	.0008
14	.0000	.0000	.0000	.0000	.0000	.0000	.0000	.0001	.0001	.0001

x \ p	.41	.42	.43	.44	.45	.46	.47	.48	.49	.50
0	.0002	.0002	.0001	.0001	.0001	.0001	.0000	.0000	.0000	.0000
1	.0024	.0019	.0015	.0012	.0009	.0007	.0005	.0004	.0003	.0002
2	.0125	.0103	.0085	.0069	.0056	.0046	.0037	.0029	.0023	.0018
3	.0405	.0349	.0299	.0254	.0215	.0181	.0151	.0126	.0104	.0085
4	.0915	.0821	.0732	.0649	.0572	.0501	.0436	.0378	.0325	.0278
5	.1526	.1426	.1325	.1224	.1123	.1024	.0929	.0837	.0749	.0667
6	.1944	.1894	.1833	.1762	.1684	.1600	.1510	.1416	.1319	.1222
7	.1930	.1959	.1975	.1978	.1969	.1947	.1912	.1867	.1811	.1746
8	.1509	.1596	.1676	.1749	.1812	.1865	.1908	.1939	.1958	.1964
9	.0932	.1027	.1124	.1221	.1318	.1413	.1504	.1591	.1672	.1746
10	.0453	.0521	.0594	.0672	.0755	.0842	.0934	.1028	.1124	.1222
11	.0172	.0206	.0244	.0288	.0337	.0391	.0452	.0518	.0589	.0667
12	.0050	.0062	.0077	.0094	.0115	.0139	.0167	.0199	.0236	.0278
13	.0011	.0014	.0018	.0023	.0029	.0036	.0046	.0057	.0070	.0085
14	.0002	.0002	.0003	.0004	.0005	.0007	.0009	.0011	.0014	.0018
15	.0000	.0000	.0000	.0000	.0001	.0001	.0001	.0001	.0002	.0002

Table D Binomial probability distribution (continued)

$n = 17$

x \ p	.01	.02	.03	.04	.05	.06	.07	.08	.09	.10
0	.8429	.7093	.5958	.4996	.4181	.3493	.2912	.2423	.2012	.1668
1	.1447	.2461	.3133	.3539	.3741	.3790	.3726	.3582	.3383	.3150
2	.0117	.0402	.0775	.1180	.1575	.1935	.2244	.2492	.2677	.2800
3	.0006	.0041	.0120	.0246	.0415	.0618	.0844	.1083	.1324	.1556
4	.0000	.0003	.0013	.0036	.0076	.0138	.0222	.0330	.0458	.0605
5	.0000	.0000	.0001	.0004	.0010	.0023	.0044	.0075	.0118	.0175
6	.0000	.0000	.0000	.0000	.0001	.0003	.0007	.0013	.0023	.0039
7	.0000	.0000	.0000	.0000	.0000	.0000	.0001	.0002	.0004	.0007
8	.0000	.0000	.0000	.0000	.0000	.0000	.0000	.0000	.0000	.0001
	.11	.12	.13	.14	.15	.16	.17	.18	.19	.20
0	.1379	.1138	.0937	.0770	.0631	.0516	.0421	.0343	.0278	.0225
1	.2898	.2638	.2381	.2131	.1893	.1671	.1466	.1279	.1109	.0957
2	.2865	.2878	.2846	.2775	.2673	.2547	.2402	.2245	.2081	.1914
3	.1771	.1963	.2126	.2259	.2359	.2425	.2460	.2464	.2441	.2393
4	.0766	.0937	.1112	.1287	.1457	.1617	.1764	.1893	.2004	.2093
5	.0246	.0332	.0432	.0545	.0668	.0801	.0939	.1081	.1222	.1361
6	.0061	.0091	.0129	.0177	.0236	.0305	.0385	.0474	.0573	.0680
7	.0012	.0019	.0030	.0045	.0065	.0091	.0124	.0164	.0211	.0267
8	.0002	.0003	.0006	.0009	.0014	.0022	.0032	.0045	.0062	.0084
9	.0000	.0000	.0001	.0002	.0003	.0004	.0006	.0010	.0015	.0021
10	.0000	.0000	.0000	.0000	.0000	.0001	.0001	.0002	.0003	.0004
11	.0000	.0000	.0000	.0000	.0000	.0000	.0000	.0000	.0000	.0001
	.21	.22	.23	.24	.25	.26	.27	.28	.29	.30
0	.0182	.0146	.0118	.0094	.0075	.0060	.0047	.0038	.0030	.0023
1	.0822	.0702	.0597	.0505	.0426	.0357	.0299	.0248	.0206	.0169
2	.1747	.1584	.1427	.1277	.1136	.1005	.0883	.0772	.0672	.0581
3	.2322	.2234	.2131	.2016	.1893	.1765	.1634	.1502	.1372	.1245
4	.2161	.2205	.2228	.2228	.2209	.2170	.2115	.2044	.1961	.1868
5	.1493	.1617	.1730	.1830	.1914	.1982	.2033	.2067	.2083	.2081
6	.0794	.0912	.1034	.1156	.1276	.1393	.1504	.1608	.1701	.1784
7	.0332	.0404	.0485	.0573	.0668	.0769	.0874	.0982	.1092	.1201
8	.0110	.0143	.0181	.0226	.0279	.0338	.0404	.0478	.0558	.0644
9	.0029	.0040	.0054	.0071	.0093	.0119	.0150	.0186	.0228	.0276
10	.0006	.0009	.0013	.0018	.0025	.0033	.0044	.0058	.0074	.0095
11	.0001	.0002	.0002	.0004	.0005	.0007	.0010	.0014	.0019	.0026
12	.0000	.0000	.0000	.0001	.0001	.0001	.0002	.0003	.0004	.0006
13	.0000	.0000	.0000	.0000	.0000	.0000	.0000	.0000	.0001	.0001
	.31	.32	.33	.34	.35	.36	.37	.38	.39	.40
0	.0018	.0014	.0011	.0009	.0007	.0005	.0004	.0003	.0002	.0002
1	.0139	.0114	.0093	.0075	.0060	.0048	.0039	.0031	.0024	.0019
2	.0500	.0428	.0364	.0309	.0260	.0218	.0182	.0151	.0125	.0102
3	.1123	.1007	.0898	.0795	.0701	.0614	.0534	.0463	.0398	.0341
4	.1766	.1659	.1547	.1434	.1320	.1208	.1099	.0993	.0892	.0796
5	.2063	.2030	.1982	.1921	.1849	.1767	.1677	.1582	.1482	.1379
6	.1854	.1910	.1952	.1979	.1991	.1988	.1970	.1939	.1895	.1839
7	.1309	.1413	.1511	.1602	.1685	.1757	.1818	.1868	.1904	.1927
8	.0735	.0831	.0930	.1032	.1134	.1235	.1335	.1431	.1521	.1606
9	.0330	.0391	.0458	.0531	.0611	.0695	.0784	.0877	.0973	.1070
10	.0119	.0147	.0181	.0219	.0263	.0313	.0368	.0430	.0498	.0571
11	.0034	.0044	.0057	.0072	.0090	.0112	.0138	.0168	.0202	.0242
12	.0008	.0010	.0014	.0018	.0024	.0031	.0040	.0051	.0065	.0081
13	.0001	.0002	.0003	.0004	.0005	.0007	.0009	.0012	.0016	.0021
14	.0000	.0000	.0000	.0001	.0001	.0001	.0002	.0002	.0003	.0004
15	.0000	.0000	.0000	.0000	.0000	.0000	.0000	.0000	.0000	.0001

Table D Binomial probability distribution (continued)

$n = 17$ (continued)

x \ p	.41	.42	.43	.44	.45	.46	.47	.48	.49	.50
0	.0001	.0001	.0001	.0001	.0000	.0000	.0000	.0000	.0000	.0000
1	.0015	.0012	.0009	.0007	.0005	.0004	.0003	.0002	.0002	.0001
2	.0084	.0068	.0055	.0044	.0035	.0028	.0022	.0017	.0013	.0010
3	.0290	.0246	.0207	.0173	.0144	.0119	.0097	.0079	.0064	.0052
4	.0706	.0622	.0546	.0475	.0411	.0354	.0302	.0257	.0217	.0182
5	.1276	.1172	.1070	.0971	.0875	.0784	.0697	.0616	.0541	.0472
6	.1773	.1697	.1614	.1525	.1432	.1335	.1237	.1138	.1040	.0944
7	.1936	.1932	.1914	.1883	.1841	.1787	.1723	.1650	.1570	.1484
8	.1682	.1748	.1805	.1850	.1883	.1903	.1910	.1904	.1886	.1855
9	.1169	.1266	.1361	.1453	.1540	.1621	.1694	.1758	.1812	.1855
10	.0650	.0733	.0822	.0914	.1008	.1105	.1202	.1298	.1393	.1484
11	.0287	.0338	.0394	.0457	.0525	.0599	.0678	.0763	.0851	.0944
12	.0100	.0122	.0149	.0179	.0215	.0255	.0301	.0352	.0409	.0472
13	.0027	.0034	.0043	.0054	.0068	.0084	.0103	.0125	.0151	.0182
14	.0005	.0007	.0009	.0012	.0016	.0020	.0026	.0033	.0041	.0052
15	.0001	.0001	.0001	.0002	.0003	.0003	.0005	.0006	.0008	.0010
16	.0000	.0000	.0000	.0000	.0000	.0000	.0001	.0001	.0001	.0001

$n = 18$

x \ p	.01	.02	.03	.04	.05	.06	.07	.08	.09	.10
0	.8345	.6951	.5780	.4796	.3972	.3283	.2708	.2229	.1831	.1501
1	.1517	.2554	.3217	.3597	.3763	.3772	.3669	.3489	.3260	.3002
2	.0130	.0443	.0846	.1274	.1683	.2047	.2348	.2579	.2741	.2835
3	.0007	.0048	.0140	.0283	.0473	.0697	.0942	.1196	.1446	.1680
4	.0000	.0004	.0016	.0044	.0093	.0167	.0266	.0390	.0536	.0700
5	.0000	.0000	.0001	.0005	.0014	.0030	.0056	.0095	.0148	.0218
6	.0000	.0000	.0000	.0000	.0002	.0004	.0009	.0018	.0032	.0052
7	.0000	.0000	.0000	.0000	.0000	.0000	.0001	.0003	.0005	.0010
8	.0000	.0000	.0000	.0000	.0000	.0000	.0000	.0000	.0001	.0002

x \ p	.11	.12	.13	.14	.15	.16	.17	.18	.19	.20
0	.1227	.1002	.0815	.0662	.0536	.0434	.0349	.0281	.0225	.0180
1	.2731	.2458	.2193	.1940	.1704	.1486	.1288	.1110	.0951	.0811
2	.2869	.2850	.2785	.2685	.2556	.2407	.2243	.2071	.1897	.1723
3	.1891	.2072	.2220	.2331	.2406	.2445	.2450	.2425	.2373	.2297
4	.0877	.1060	.1244	.1423	.1592	.1746	.1882	.1996	.2087	.2153
5	.0303	.0405	.0520	.0649	.0787	.0931	.1079	.1227	.1371	.1507
6	.0081	.0120	.0168	.0229	.0301	.0384	.0479	.0584	.0697	.0816
7	.0017	.0028	.0043	.0064	.0091	.0126	.0168	.0220	.0280	.0350
8	.0003	.0005	.0009	.0014	.0022	.0033	.0047	.0066	.0090	.0120
9	.0000	.0001	.0001	.0003	.0004	.0007	.0011	.0016	.0024	.0033
10	.0000	.0000	.0000	.0000	.0001	.0001	.0002	.0003	.0005	.0008
11	.0000	.0000	.0000	.0000	.0000	.0000	.0000	.0001	.0001	.0001

x \ p	.21	.22	.23	.24	.25	.26	.27	.28	.29	.30
0	.0144	.0114	.0091	.0072	.0056	.0044	.0035	.0027	.0021	.0016
1	.0687	.0580	.0487	.0407	.0338	.0280	.0231	.0189	.0155	.0126
2	.1553	.1390	.1236	.1092	.0958	.0836	.0725	.0626	.0537	.0458
3	.2202	.2091	.1969	.1839	.1704	.1567	.1431	.1298	.1169	.1046
4	.2195	.2212	.2205	.2177	.2130	.2065	.1985	.1892	.1790	.1681
5	.1634	.1747	.1845	.1925	.1988	.2031	.2055	.2061	.2048	.2017
6	.0941	.1067	.1194	.1317	.1436	.1546	.1647	.1736	.1812	.1873
7	.0429	.0516	.0611	.0713	.0820	.0931	.1044	.1157	.1269	.1376
8	.0157	.0200	.0251	.0310	.0376	.0450	.0531	.0619	.0713	.0811
9	.0046	.0063	.0083	.0109	.0139	.0176	.0218	.0267	.0323	.0386
10	.0011	.0016	.0022	.0031	.0042	.0056	.0073	.0094	.0119	.0149
11	.0002	.0003	.0005	.0007	.0010	.0014	.0020	.0026	.0035	.0046
12	.0000	.0001	.0001	.0001	.0002	.0003	.0004	.0006	.0008	.0012
13	.0000	.0000	.0000	.0000	.0000	.0000	.0001	.0001	.0002	.0002

Table D Binomial probability distribution (continued)

n = 18 (continued)

x \ p	.31	.32	.33	.34	.35	.36	.37	.38	.39	.40
0	.0013	.0010	.0007	.0006	.0004	.0003	.0002	.0002	.0001	.0001
1	.0102	.0082	.0066	.0052	.0042	.0033	.0026	.0020	.0016	.0012
2	.0388	.0327	.0275	.0229	.0190	.0157	.0129	.0105	.0086	.0069
3	.0930	.0822	.0722	.0630	.0547	.0471	.0404	.0344	.0292	.0246
4	.1567	.1450	.1333	.1217	.1104	.0994	.0890	.0791	.0699	.0614
5	.1971	.1911	.1838	.1755	.1664	.1566	.1463	.1358	.1252	.1146
6	.1919	.1948	.1962	.1959	.1941	.1908	.1862	.1803	.1734	.1655
7	.1478	.1572	.1656	.1730	.1792	.1840	.1875	.1895	.1900	.1892
8	.0913	.1017	.1122	.1226	.1327	.1423	.1514	.1597	.1671	.1734
9	.0456	.0532	.0614	.0701	.0794	.0890	.0988	.1087	.1187	.1284
10	.0184	.0225	.0272	.0325	.0385	.0450	.0522	.0600	.0683	.0771
11	.0060	.0077	.0097	.0122	.0151	.0184	.0223	.0267	.0318	.0374
12	.0016	.0021	.0028	.0037	.0047	.0060	.0076	.0096	.0118	.0145
13	.0003	.0005	.0006	.0009	.0012	.0016	.0021	.0027	.0035	.0045
14	.0001	.0001	.0001	.0002	.0002	.0003	.0004	.0006	.0008	.0011
15	.0000	.0000	.0000	.0000	.0000	.0000	.0001	.0001	.0001	.0002

x \ p	.41	.42	.43	.44	.45	.46	.47	.48	.49	.50
0	.0001	.0001	.0000	.0000	.0000	.0000	.0000	.0000	.0000	.0000
1	.0009	.0007	.0005	.0004	.0003	.0002	.0002	.0001	.0001	.0001
2	.0055	.0044	.0035	.0028	.0022	.0017	.0013	.0010	.0008	.0006
3	.0206	.0171	.0141	.0116	.0095	.0077	.0062	.0050	.0039	.0031
4	.0536	.0464	.0400	.0342	.0291	.0246	.0206	.0172	.0142	.0117
5	.1042	.0941	.0844	.0753	.0666	.0586	.0512	.0444	.0382	.0327
6	.1569	.1477	.1380	.1281	.1181	.1081	.0983	.0887	.0796	.0708
7	.1869	.1833	.1785	.1726	.1657	.1579	.1494	.1404	.1310	.1214
8	.1786	.1825	.1852	.1864	.1864	.1850	.1822	.1782	.1731	.1669
9	.1379	.1469	.1552	.1628	.1694	.1751	.1795	.1828	.1848	.1855
10	.0862	.0957	.1054	.1151	.1248	.1342	.1433	.1519	.1598	.1669
11	.0436	.0504	.0578	.0658	.0742	.0831	.0924	.1020	.1117	.1214
12	.0177	.0213	.0254	.0301	.0354	.0413	.0478	.1549	.0626	.0708
13	.0057	.0071	.0089	.0109	.0134	.0162	.0196	.0234	.0278	.0327
14	.0014	.0018	.0024	.0031	.0039	.0049	.0062	.0077	.0095	.0117
15	.0003	.0004	.0005	.0006	.0009	.0011	.0015	.0019	.0024	.0031
16	.0000	.0000	.0001	.0001	.0001	.0002	.0002	.0003	.0004	.0006
17	.0000	.0000	.0000	.0000	.0000	.0000	.0000	.0000	.0000	.0001

n = 19

x \ p	.01	.02	.03	.04	.05	.06	.07	.08	.09	.10
0	.8262	.6812	.5606	.4604	.3774	.3086	.2519	.2051	.1666	.1351
1	.1586	.2642	.3294	.3645	.3774	.3743	.3602	.3389	.3131	.2852
2	.0144	.0485	.0917	.1367	.1787	.2150	.2440	.2652	.2787	.2852
3	.0008	.0056	.0161	.0323	.0533	.0778	.1041	.1307	.1562	.1796
4	.0000	.0005	.0020	.0054	.0112	.0199	.0313	.0455	.0618	.0798
5	.0000	.0000	.0002	.0007	.0018	.0038	.0071	.0119	.0183	.0266
6	.0000	.0000	.0000	.0001	.0002	.0006	.0012	.0024	.0042	.0069
7	.0000	.0000	.0000	.0000	.0000	.0001	.0002	.0004	.0008	.0014
8	.0000	.0000	.0000	.0000	.0000	.0000	.0000	.0001	.0001	.0002

x \ p	.11	.12	.13	.14	.15	.16	.17	.18	.19	.20
0	.1092	.0881	.0709	.0569	.0456	.0364	.0290	.0230	.0182	.0144
1	.2565	.2284	.2014	.1761	.1529	.1318	.1129	.0961	.0813	.0685
2	.2854	.2803	.2708	.2581	.2428	.2259	.2081	.1898	.1717	.1540
3	.1999	.2166	.2293	.2381	.2428	.2439	.2415	.2361	.2282	.2182
4	.0988	.1181	.1371	.1550	.1714	.1858	.1979	.2073	.2141	.2182
5	.0366	.0483	.0614	.0757	.0907	.1062	.1216	.1365	.1507	.1636
6	.0106	.0154	.0214	.0288	.0374	.0472	.0581	.0699	.0825	.0955
7	.0024	.0039	.0059	.0087	.0122	.0167	.0221	.0285	.0359	.0443
8	.0004	.0008	.0013	.0021	.0032	.0048	.0068	.0094	.0126	.0166
9	.0001	.0001	.0002	.0004	.0007	.0011	.0017	.0025	.0036	.0051
10	.0000	.0000	.0000	.0001	.0001	.0002	.0003	.0006	.0009	.0013
11	.0000	.0000	.0000	.0000	.0000	.0000	.0001	.0001	.0002	.0003

Table D Binomial probability distribution (continued)

$n = 19$ (continued)

x \ p	.21	.22	.23	.24	.25	.26	.27	.28	.29	.30
0	.0113	.0089	.0070	.0054	.0042	.0033	.0025	.0019	.0015	.0011
1	.0573	.0477	.0396	.0326	.0268	.0219	.0178	.0144	.0116	.0093
2	.1371	.1212	.1064	.0927	.0803	.0692	.0592	.0503	.0426	.0358
3	.2065	.1937	.1800	.1659	.1517	.1377	.1240	.1109	.0985	.0869
4	.2196	.2185	.2151	.2096	.2023	.1935	.1835	.1726	.1610	.1491
5	.1751	.1849	.1928	.1986	.2023	.2040	.2036	.2013	.1973	.1916
6	.1086	.1217	.1343	.1463	.1574	.1672	.1757	.1827	.1880	.1916
7	.0536	.0637	.0745	.0858	.0974	.1091	.1207	.1320	.1426	.1525
8	.0214	.0270	.0334	.0406	.0487	.0575	.0670	.0770	.0874	.0981
9	.0069	.0093	.0122	.0157	.0198	.0247	.0303	.0366	.0436	.0514
10	.0018	.0026	.0036	.0050	.0066	.0087	.0112	.0142	.0178	.0220
11	.0004	.0006	.0009	.0013	.0018	.0025	.0034	.0045	.0060	.0077
12	.0001	.0001	.0002	.0003	.0004	.0006	.0008	.0012	.0016	.0022
13	.0000	.0000	.0000	.0000	.0001	.0001	.0002	.0002	.0004	.0005
14	.0000	.0000	.0000	.0000	.0000	.0000	.0000	.0000	.0001	.0001

x \ p	.31	.32	.33	.34	.35	.36	.37	.38	.39	.40
0	.0009	.0007	.0005	.0004	.0003	.0002	.0002	.0001	.0001	.0001
1	.0074	.0059	.0046	.0036	.0029	.0022	.0017	.0013	.0010	.0008
2	.0299	.0249	.0206	.0169	.0138	.0112	.0091	.0073	.0058	.0046
3	.0762	.0664	.0574	.0494	.0422	.0358	.0302	.0253	.0211	.0175
4	.1370	.1249	.1131	.1017	.0909	.0806	.0710	.0621	.0540	.0467
5	.1846	.1764	.1672	.1572	.1468	.1360	.1251	.1143	.1036	.0933
6	.1935	.1936	.1921	.1890	.1844	.1785	.1714	.1634	.1546	.1451
7	.1615	.1692	.1757	.1808	.1844	.1865	.1870	.1860	.1835	.1797
8	.1088	.1195	.1298	.1397	.1489	.1573	.1647	.1710	.1760	.1797
9	.0597	.0687	.0782	.0880	.0980	.1082	.1182	.1281	.1375	.1464
10	.0268	.0323	.0385	.0453	.0528	.0608	.0694	.0785	.0879	.0976
11	.0099	.0124	.0155	.0191	.0233	.0280	.0334	.0394	.0460	.0532
12	.0030	.0039	.0051	.0066	.0083	.0105	.0131	.0161	.0196	.0237
13	.0007	.0010	.0014	.0018	.0024	.0032	.0041	.0053	.0067	.0085
14	.0001	.0002	.0003	.0004	.0006	.0008	.0010	.0014	.0018	.0024
15	.0000	.0000	.0000	.0001	.0001	.0001	.0002	.0003	.0004	.0005
16	.0000	.0000	.0000	.0000	.0000	.0000	.0000	.0000	.0001	.0001

x \ p	.41	.42	.43	.44	.45	.46	.47	.48	.49	.50
0	.0000	.0000	.0000	.0000	.0000	.0000	.0000	.0000	.0000	.0000
1	.0006	.0004	.0003	.0002	.0002	.0001	.0001	.0001	.0001	.0000
2	.0037	.0029	.0022	.0017	.0013	.0010	.0008	.0006	.0004	.0003
3	.0144	.0118	.0096	.0077	.0062	.0049	.0039	.0031	.0024	.0018
4	.0400	.0341	.0289	.0243	.0203	.0168	.0138	.0113	.0092	.0074
5	.0834	.0741	.0653	.0572	.0497	.0429	.0368	.0313	.0265	.0222
6	.1353	.1252	.1150	.1049	.0949	.0853	.0751	.0674	.0593	.0518
7	.1746	.1683	.1611	.1530	.1443	.1350	.1254	.1156	.1058	.0961
8	.1820	.1829	.1823	.1803	.1771	.1725	.1668	.1601	.1525	.1442
9	.1546	.1618	.1681	.1732	.1771	.1796	.1808	.1806	.1791	.1762
10	.1074	.1172	.1268	.1361	.1449	.1530	.1603	.1667	.1721	.1762
11	.0611	.0694	.0783	.0875	.0970	.1066	.1163	.1259	.1352	.1442
12	.0283	.0335	.0394	.0458	.0529	.0606	.0688	.0775	.0866	.0961
13	.0106	.0131	.0160	.0194	.0233	.0278	.0328	.0385	.0448	.0518
14	.0032	.0041	.0052	.0065	.0082	.0101	.0125	.0152	.0185	.0222
15	.0007	.0010	.0013	.0017	.0022	.0029	.0037	.0047	.0059	.0074
16	.0001	.0002	.0002	.0003	.0005	.0006	.0008	.0011	.0014	.0018
17	.0000	.0000	.0000	.0000	.0001	.0001	.0001	.0002	.0002	.0003

$n = 20$

x \ p	.01	.02	.03	.04	.05	.06	.07	.08	.09	.10
0	.8179	.6676	.5438	.4420	.3585	.2901	.2342	.1887	.1516	.1216
1	.1652	.2725	.3364	.3683	.3774	.3703	.3526	.3282	.3000	.2702
2	.0159	.0528	.0988	.1458	.1887	.2246	.2521	.2711	.2818	.2852
3	.0010	.0065	.0183	.0364	.0596	.0860	.1139	.1414	.1672	.1901
4	.0000	.0006	.0024	.0065	.0133	.0233	.0364	.0523	.0703	.0898
5	.0000	.0000	.0002	.0009	.0022	.0048	.0088	.0145	.0222	.0319
6	.0000	.0000	.0000	.0001	.0003	.0008	.0017	.0032	.0055	.0089
7	.0000	.0000	.0000	.0000	.0000	.0001	.0002	.0005	.0011	.0020
8	.0000	.0000	.0000	.0000	.0000	.0000	.0000	.0001	.0002	.0004
9	.0000	.0000	.0000	.0000	.0000	.0000	.0000	.0000	.0000	.0001

Table D Binomial probability distribution (continued)

n = 20 (continued)

x \ p	.11	.12	.13	.14	.15	.16	.17	.18	.19	.20
0	.0972	.0776	.0617	.0490	.0388	.0306	.0241	.0189	.0148	.0115
1	.2403	.2115	.1844	.1595	.1368	.1165	.0986	.0829	.0693	.0576
2	.2822	.2740	.2618	.2466	.2293	.2109	.1919	.1730	.1545	.1369
3	.2093	.2242	.2347	.2409	.2428	.2410	.2358	.2278	.2175	.2054
4	.1099	.1299	.1491	.1666	.1821	.1951	.2053	.2125	.2168	.2182
5	.0435	.0567	.0713	.1868	.1028	.1189	.1345	.1493	.1627	.1746
6	.0134	.0193	.0266	.0353	.0454	.0566	.0689	.0819	.0954	.1091
7	.0033	.0053	.0080	.0115	.0160	.0216	.0282	.0360	.0448	.0545
8	.0007	.0012	.0019	.0030	.0046	.0067	.0094	.0128	.0171	.0222
9	.0001	.0002	.0004	.0007	.0011	.0017	.0026	.0038	.0053	.0074
10	.0000	.0000	.0001	.0001	.0002	.0004	.0006	.0009	.0014	.0020
11	.0000	.0000	.0000	.0000	.0000	.0001	.0001	.0002	.0003	.0005
12	.0000	.0000	.0000	.0000	.0000	.0000	.0000	.0000	.0001	.0001

x \ p	.21	.22	.23	.24	.25	.26	.27	.28	.29	.30
0	.0090	.0069	.0054	.0041	.0032	.0024	.0018	.0014	.0011	.0008
1	.0477	.0392	.0321	.0261	.0211	.0170	.0137	.0109	.0087	.0068
2	.1204	.1050	.0910	.0783	.0669	.0569	.0480	.0403	.0336	.0278
3	.1920	.1777	.1631	.1484	.1339	.1199	.1065	.0940	.0823	.0716
4	.2169	.2131	.2070	.1991	.1897	.1790	.1675	.1553	.1429	.1304
5	.1845	.1923	.1979	.2012	.2023	.2013	.1982	.1933	.1868	.1789
6	.1226	.1356	.1478	.1589	.1686	.1768	.1833	.1879	.1907	.1916
7	.0652	.0765	.0883	.1003	.1124	.1242	.1356	.1462	.1558	.1643
8	.0282	.0351	.0429	.0515	.0609	.0709	.0815	.0924	.1034	.1144
9	.0100	.0132	.0171	.0217	.0271	.0332	.0402	.0479	.0563	.0654
10	.0029	.0041	.0056	.0075	.0099	.0128	.0163	.0205	.0253	.0308
11	.0007	.0010	.0015	.0022	.0030	.0041	.0055	.0072	.0094	.0120
12	.0001	.0002	.0003	.0005	.0008	.0011	.0015	.0021	.0029	.0039
13	.0000	.0000	.0001	.0001	.0002	.0002	.0003	.0005	.0007	.0010
14	.0000	.0000	.0000	.0000	.0000	.0000	.0001	.0001	.0001	.0002

x \ p	.31	.32	.33	.34	.35	.36	.37	.38	.39	.40
0	.0006	.0004	.0003	.0002	.0002	.0001	.0001	.0001	.0001	.0000
1	.0054	.0042	.0033	.0025	.0020	.0015	.0011	.0009	.0007	.0005
2	.0229	.0188	.0153	.0124	.0100	.0080	.0064	.0050	.0040	.0031
3	.0619	.0531	.0453	.0383	.0323	.0270	.0224	.0185	.0152	.0123
4	.1181	.1062	.0947	.0839	.0738	.0645	.0559	.0482	.0412	.0350
5	.1698	.1599	.1493	.1384	.1272	.1161	.1051	.0945	.0843	.0746
6	.1907	.1881	.1839	.1782	.1712	.1632	.1543	.1447	.1347	.1244
7	.1714	.1770	.1811	.1836	.1844	.1836	.1812	.1774	.1722	.1659
8	.1251	.1354	.1450	.1537	.1614	.1678	.1730	.1767	.1790	.1797
9	.0750	.0849	.0952	.1056	.1158	.1259	.1354	.1444	.1526	.1597
10	.0370	.0440	.0516	.0598	.0686	.0779	.0875	.0974	.1073	.1171
11	.0151	.0188	.0231	.0280	.0336	.0398	.0467	.0542	.0624	.0710
12	.0051	.0066	.0085	.0108	.0136	.0168	.0206	.0249	.0299	.0355
13	.0014	.0019	.0026	.0034	.0045	.0058	.0074	.0094	.0118	.0146
14	.0003	.0005	.0006	.0009	.0012	.0016	.0022	.0029	.0038	.0049
15	.0001	.0001	.0001	.0002	.0003	.0004	.0005	.0007	.0010	.0013
16	.0000	.0000	.0000	.0000	.0000	.0001	.0001	.0001	.0002	.0003

x \ p	.41	.42	.43	.44	.45	.46	.47	.48	.49	.50
0	.0000	.0000	.0000	.0000	.0000	.0000	.0000	.0000	.0000	.0000
1	.0004	.0003	.0002	.0001	.0001	.0001	.0001	.0000	.0000	.0000
2	.0024	.0018	.0014	.0011	.0008	.0006	.0005	.0003	.0002	.0002
3	.0100	.0080	.0064	.0051	.0040	.0031	.0024	.0019	.0014	.0011
4	.0295	.0247	.0206	.0170	.0139	.0113	.0092	.0074	.0059	.0046
5	.0656	.0573	.0496	.0427	.0365	.0309	.0260	.0217	.0180	.0148
6	.1140	.1037	.0936	.0839	.0746	.0658	.0577	.0501	.0432	.0370
7	.1585	.1502	.1413	.1318	.1221	.1122	.1023	.0925	.0830	.0739
8	.1790	.1768	.1732	.1683	.1623	.1553	.1474	.1388	.1296	.1201
9	.1658	.1707	.1742	.1763	.1771	.1763	.1742	.1708	.1661	.1602
10	.1268	.1359	.1446	.1524	.1593	.1652	.1700	.1734	.1755	.1762
11	.0801	.0895	.0991	.1089	.1185	.1280	.1370	.1455	.1533	.1602
12	.0417	.0486	.0561	.0642	.0727	.0818	.0911	.1007	.1105	.1201
13	.0178	.0217	.0260	.0310	.0366	.0429	.0497	.0572	.0653	.0739
14	.0062	.0078	.0098	.0122	.0150	.0183	.0221	.0264	.0314	.0370
15	.0017	.0023	.0030	.0038	.0049	.0062	.0078	.0098	.0121	.0148
16	.0004	.0005	.0007	.0009	.0013	.0017	.0022	.0028	.0036	.0046
17	.0001	.0001	.0001	.0002	.0002	.0003	.0005	.0006	.0008	.0011
18	.0000	.0000	.0000	.0000	.0000	.0000	.0001	.0001	.0001	.0002

Table D Binomial probability distribution (continued)

$n = 25$

x \ p	.01	.02	.03	.04	.05	.06	.07	.08	.09	.10
0	.7778	.6035	.4670	.3604	.2774	.2129	.1630	.1244	.0946	.0718
1	.1964	.3079	.3611	.3754	.3650	.3398	.3066	.2704	.2340	.1994
2	.0238	.0754	.1340	.1877	.2305	.2602	.2770	.2821	.2777	.2659
3	.0018	.0118	.0318	.0600	.0930	.1273	.1598	.1881	.2106	.2265
4	.0001	.0013	.0054	.0137	.0269	.0447	.0662	.0899	.1145	.1384
5	.0000	.0001	.0007	.0024	.0060	.0120	.0209	.0329	.0476	.0646
6	.0000	.0000	.0001	.0003	.0010	.0026	.0052	.0095	.0157	.0239
7	.0000	.0000	.0000	.0000	.0001	.0004	.0011	.0022	.0042	.0072
8	.0000	.0000	.0000	.0000	.0000	.0001	.0002	.0004	.0009	.0018
9	0	.0000	.0000	.0000	.0000	.0000	.0000	.0001	.0002	.0004
10	0	.0000	.0000	.0000	.0000	.0000	.0000	.0000	.0000	.0001

x \ p	.11	.12	.13	.14	.15	.16	.17	.18	.19	.20
0	.0543	.0409	.0308	.0230	.0172	.0128	.0095	.0070	.0052	.0038
1	.1678	.1395	.1149	.0938	.0759	.0609	.0486	.0384	.0302	.0236
2	.2488	.2283	.2060	.1832	.1607	.1392	.1193	.1012	.0851	.0708
3	.2358	.2387	.2360	.2286	.2174	.2033	.1874	.1704	.1530	.1358
4	.1603	.1790	.1940	.2047	.2110	.2130	.2111	.2057	.1974	.1867
5	.0832	.1025	.1217	.1399	.1564	.1704	.1816	.1897	.1945	.1960
6	.0343	.0466	.0606	.0759	.0920	.1082	.1240	.1388	.1520	.1633
7	.0115	.0173	.0246	.0336	.0441	.0559	.0689	.0827	.0968	.1108
8	.0032	.0053	.0083	.0123	.0175	.0240	.0318	.0408	.0511	.0623
9	.0007	.0014	.0023	.0038	.0058	.0086	.0123	.0169	.0226	.0294
10	.0001	.0003	.0006	.0010	.0016	.0026	.0040	.0059	.0085	.0118
11	.0000	.0001	.0001	.0002	.0004	.0007	.0011	.0018	.0027	.0040
12	.0000	.0000	.0000	.0000	.0001	.0002	.0003	.0005	.0007	.0012
13	.0000	.0000	.0000	.0000	.0000	.0000	.0001	.0001	.0002	.0003
14	.0000	.0000	.0000	.0000	.0000	.0000	.0000	.0000	.0000	.0001

x \ p	.21	.22	.23	.24	.25	.26	.27	.28	.29	.30
0	.0028	.0020	.0015	.0010	.0008	.0005	.0004	.0003	.0002	.0001
1	.0183	.0141	.0109	.0083	.0063	.0047	.0035	.0026	.0020	.0014
2	.0585	.0479	.0389	.0314	.0251	.0199	.0157	.0123	.0096	.0074
3	.1192	.1035	.0891	.0759	.0641	.0537	.0446	.0367	.0300	.0243
4	.1742	.1606	.1463	.1318	.1175	.1037	.0906	.0785	.0673	.0572
5	.1945	.1903	.1836	.1749	.1645	.1531	.1408	.1282	.1155	.1030
6	.1724	.1789	.1828	.1841	.1828	.1793	.1736	.1661	.1572	.1472
7	.1244	.1369	.1482	.1578	.1654	.1709	.1743	.1754	.1743	.1712
8	.0744	.0869	.0996	.1121	.1241	.1351	.1450	.1535	.1602	.1651
9	.0373	.0463	.0562	.0669	.0781	.0897	.1013	.1127	.1236	.1336
10	.0159	.0209	.0269	.0338	.0417	.0504	.0600	.0701	.0808	.0916
11	.0058	.0080	.0109	.0145	.0189	.0242	.0302	.0372	.0450	.0536
12	.0018	.0026	.0038	.0054	.0074	.0099	.0130	.0169	.0214	.0268
13	.0005	.0007	.0011	.0017	.0025	.0035	.0048	.0066	.0088	.0115
14	.0001	.0002	.0003	.0005	.0007	.0010	.0015	.0022	.0031	.0042
15	.0000	.0000	.0001	.0001	.0002	.0003	.0004	.0006	.0009	.0013
16	.0000	.0000	.0000	.0000	.0000	.0001	.0001	.0002	.0002	.0004
17	.0000	.0000	.0000	.0000	.0000	.0000	.0000	.0000	.0001	.0001

x \ p	.31	.32	.33	.34	.35	.36	.37	.38	.39	.40
0	.0001	.0001	.0000	.0000	.0000	.0000	.0000	.0000	.0000	.0000
1	.0011	.0008	.0006	.0004	.0003	.0002	.0001	.0001	.0001	.0000
2	.0057	.0043	.0033	.0025	.0018	.0014	.0010	.0007	.0005	.0004
3	.0195	.0156	.0123	.0097	.0076	.0058	.0045	.0034	.0026	.0019
4	.0482	.0403	.0334	.0274	.0224	.0181	.0145	.0115	.0091	.0071
5	.0910	.0797	.0691	.0594	.0506	.0427	.0357	.0297	.0244	.0199
6	.1363	.1250	.1134	.1020	.0908	.0801	.0700	.0606	.0520	.0442
7	.1662	.1596	.1516	.1426	.1327	.1222	.1115	.1008	.0902	.0800
8	.1680	.1690	.1681	.1652	.1607	.1547	.1474	.1390	.1298	.1200
9	.1426	.1502	.1563	.1608	.1635	.1644	.1635	.1609	.1567	.1511
10	.1025	.1131	.1232	.1325	.1409	.1479	.1536	.1578	.1603	.1612
11	.0628	.0726	.0828	.0931	.1034	.1135	.1230	.1319	.1398	.1465
12	.0329	.0399	.0476	.0560	.0650	.0745	.0843	.0943	.1043	.1140
13	.0148	.0188	.0234	.0288	.0350	.0419	.0495	.0578	.0667	.0760
14	.0057	.0076	.0099	.0127	.0161	.0202	.0249	.0304	.0365	.0434
15	.0019	.0026	.0036	.0048	.0064	.0083	.0107	.0136	.0171	.0212
16	.0005	.0008	.0011	.0015	.0021	.0029	.0039	.0052	.0068	.0088
17	.0001	.0002	.0003	.0004	.0006	.0009	.0012	.0017	.0023	.0031
18	.0000	.0000	.0001	.0001	.0001	.0002	.0003	.0005	.0007	.0009
19	.0000	.0000	.0000	.0000	.0000	.0000	.0001	.0001	.0002	.0002

Table D Binomial probability distribution (continued)

x \ p	.41	.42	.43	.44	.45	.46	.47	.48	.49	.50
					$n = 25$ (continued)					
0	.0000	.0000	.0000	.0000	.0000	.0000	.0000	.0000	.0000	.0000
1	.0000	.0000	.0000	.0000	.0000	.0000	.0000	.0000	.0000	.0000
2	.0003	.0002	.0001	.0001	.0001	.0000	.0000	.0000	.0000	.0000
3	.0014	.0011	.0008	.0006	.0004	.0003	.0002	.0001	.0001	.0001
4	.0055	.0042	.0032	.0024	.0018	.0014	.0010	.0007	.0005	.0004
5	.0161	.0129	.0102	.0081	.0063	.0049	.0037	.0028	.0021	.0016
6	.0372	.0311	.0257	.0211	.0172	.0138	.0110	.0087	.0068	.0053
7	.0703	.0611	.0527	.0450	.0381	.0319	.0265	.0218	.0178	.0143
8	.1099	.0996	.0895	.0796	.0701	.0612	.0529	.0453	.0384	.0322
9	.1442	.1363	.1275	.1181	.1084	.0985	.0886	.0790	.0697	.0609
10	.1603	.1579	.1539	.1485	.1419	.1342	.1257	.1166	.1071	.0974
11	.1519	.1559	.1583	.1591	.1583	.1559	.1521	.1468	.1404	.1328
12	.1232	.1317	.1393	.1458	.1511	.1550	.1573	.1581	.1573	.1550
13	.0856	.0954	.1051	.1146	.1236	.1320	.1395	.1460	.1512	.1550
14	.0510	.0592	.0680	.0772	.0867	.0964	.1060	.1155	.1245	.1328
15	.0260	.0314	.0376	.0445	.0520	.0602	.0690	.0782	.0877	.0974
16	.0113	.0142	.0177	.0218	.0266	.0321	.0382	.0451	.0527	.0609
17	.0042	.0055	.0071	.0091	.0115	.0145	.0179	.0220	.0268	.0322
18	.0013	.0018	.0024	.0032	.0042	.0055	.0071	.0090	.0114	.0143
19	.0003	.0005	.0007	.0009	.0013	.0017	.0023	.0031	.0040	.0053
20	.0001	.0001	.0001	.0002	.0003	.0004	.0006	.0009	.0012	.0016
21	.0000	.0000	.0000	.0000	.0001	.0001	.0001	.0002	.0003	.0004
22	.0000	.0000	.0000	.0000	.0000	.0000	.0000	.0000	.0000	.0001

x \ p	.01	.02	.03	.04	.05	.06	.07	.08	.09	.10
					$n = 50$					
0	.6050	.3642	.2181	.1299	.0769	.0453	.0266	.0155	.0090	.0052
1	.3056	.3716	.3372	.2706	.2025	.1447	.0999	.0672	.0443	.0286
2	.0756	.1858	.2555	.2762	.2611	.2262	.1843	.1433	.1073	.0779
3	.0122	.0607	.1264	.1842	.2199	.2311	.2219	.1993	.1698	.1386
4	.0015	.0145	.0459	.0902	.1360	.1733	.1963	.2037	.1973	.1809
5	.0001	.0027	.0131	.0346	.0658	.1018	.1359	.1629	.1795	.1849
6	.0000	.0004	.0030	.0108	.0260	.0487	.0767	.1063	.1332	.1541
7	.0000	.0001	.0006	.0028	.0086	.0195	.0363	.0581	.0828	.1076
8	.0000	.0000	.0001	.0006	.0024	.0067	.0147	.0271	.0440	.0643
9	.0000	.0000	.0000	.0001	.0006	.0020	.0052	.0110	.0203	.0333
10	.0000	.0000	.0000	.0000	.0001	.0005	.0016	.0039	.0082	.0152
11	0	.0000	.0000	.0000	.0000	.0001	.0004	.0012	.0030	.0061
12	0	.0000	.0000	.0000	.0000	.0000	.0001	.0004	.0010	.0022
13	0	0	.0000	.0000	.0000	.0000	.0000	.0001	.0003	.0007
14	0	0	.0000	.0000	.0000	.0000	.0000	.0000	.0001	.0002
15	0	0	0	.0000	.0000	.0000	.0000	.0000	.0000	.0001
16	0	0	0	.0000	.0000	.0000	.0000	.0000	.0000	.0000
17	0	0	0	0	.0000	.0000	.0000	.0000	.0000	.0000
18	0	0	0	0	0	.0000	.0000	.0000	.0000	.0000
19	0	0	0	0	0	.0000	.0000	.0000	.0000	.0000
20	0	0	0	0	0	0	.0000	.0000	.0000	.0000
21	0	0	0	0	0	0	0	.0000	.0000	.0000
22	0	0	0	0	0	0	0	0	.0000	.0000
23	0	0	0	0	0	0	0	0	0	.0000

Table D Binomial probability distribution (continued)

$n = 50$

x \ p	.11	.12	.13	.14	.15	.16	.17	.18	.19	.20
0	.0029	.0017	.0009	.0005	.0003	.0002	.0001	.0000	.0000	.0000
1	.0182	.0114	.0071	.0043	.0026	.0016	.0009	.0005	.0003	.0002
2	.0552	.0382	.0259	.0172	.0113	.0073	.0046	.0029	.0018	.0011
3	.1091	.0833	.0619	.0449	.0319	.0222	.0151	.0102	.0067	.0044
4	.1584	.1334	.1086	.0858	.0661	.0496	.0364	.0262	.0185	.0128
5	.1801	.1674	.1493	.1286	.1072	.0869	.0687	.0530	.0400	.0295
6	.1670	.1712	.1674	.1570	.1419	.1242	.1055	.0872	.0703	.0554
7	.1297	.1467	.1572	.1606	.1575	.1487	.1358	.1203	.1037	.0870
8	.0862	.1075	.1263	.1406	.1493	.1523	.1495	.1420	.1307	.1169
9	.0497	.0684	.0880	.1068	.1230	.1353	.1429	.1454	.1431	.1364
10	.0252	.0383	.0539	.0713	.0890	.1057	.1200	.1309	.1376	.1398
11	.0113	.0190	.0293	.0422	.0571	.0732	.0894	.1045	.1174	.1271
12	.0045	.0084	.0142	.0223	.0328	.0453	.0595	.0745	.0895	.1033
13	.0016	.0034	.0062	.0106	.0169	.0252	.0356	.0478	.0613	.0755
14	.0005	.0012	.0025	.0046	.0079	.0127	.0193	.0277	.0380	.0499
15	.0002	.0004	.0009	.0018	.0033	.0058	.0095	.0146	.0214	.0299
16	.0000	.0001	.0003	.0006	.00p3	.0024	.0042	.0070	.0110	.0164
17	.0000	.0000	.0001	.0002	.0005	.0009	.0017	.0031	.0052	.0082
18	.0000	.0000	.0000	.0001	.0001	.0003	.0007	.0012	.0022	.0037
19	.0000	.0000	.0000	.0000	.0000	.0001	.0002	.0005	.0009	.0016
20	.0000	.0000	.0000	.0000	.0000	.0000	.0001	.0002	.0003	.0006
21	.0000	.0000	.0000	.0000	.0000	.0000	.0000	.0000	.0001	.0002
22	.0000	.0000	.0000	.0000	.0000	.0000	.0000	.0000	.0000	.0001
23	.0000	.0000	.0000	.0000	.0000	.0000	.0000	.0000	.0000	.0000

x \ p	.21	.22	.23	.24	.25	.26	.27	.28	.29	.30
0	.0000	.0000	.0000	.0000	.0000	.0000	.0000	.0000	.0000	.0000
1	.0001	.0001	.0000	.0000	.0000	.0000	.0000	.0000	.0000	.0000
2	.0007	.0004	.0002	.0001	.0001	.0000	.0000	.0000	.0000	.0000
3	.0028	.0018	.0011	.0007	.0004	.0002	.0001	.0001	.0000	.0000
4	.0088	.0059	.0039	.0025	.0016	.0010	.0006	.0004	.0002	.0001
5	.0214	.0152	.0106	.0073	.0049	.0033	.0021	.0014	.0009	.0006
6	.0427	.0322	.0238	.0173	.0123	.0087	.0060	.0040	.0027	.0018
7	.0713	.0571	.0447	.0344	.0259	.0191	.0139	.0099	.0069	.0048
8	.1019	.0865	.0718	.0583	.0463	.0361	.0276	.0207	.0152	.0110
9	.1263	.1139	.1001	.0859	.0721	.0592	.0476	.0375	.0290	.0220
10	.1377	.1317	.1226	.1113	.0985	.0852	.0721	.0598	.0485	.0386
11	.1331	.1351	.1332	.1278	.1194	.1089	.0970	.0845	.0721	.0602
12	.1150	.1238	.1293	.1311	.1294	.1244	.1166	.1068	.0957	.0838
13	.0894	.1021	.1129	.1210	.1261	.1277	.1261	.1215	.1142	.1050
14	.0628	.0761	.0891	.1010	.1110	.1186	.1233	.1248	.1233	.1189
15	.0400	.0515	.0639	.0766	.0888	.1000	.1094	.1165	.1209	.1223
16	.0233	.0318	.0417	.0529	.0648	.0769	.0885	.0991	.1080	.1147
17	.0124	.0179	.0249	.0334	.0432	.0540	.0655	.0771	.0882	.0983
18	.0060	.0093	.0137	.0193	.0264	.0348	.0444	.0550	.0661	.0772
19	.0027	.0044	.0069	.0103	.0148	.0206	.0277	.0360	.0454	.0558
20	.0011	.0019	.0032	.0050	.0077	.0112	.0159	.0217	.0288	.0370
21	.0004	.0008	.0014	.0023	.0036	.0056	.0084	.0121	.0168	.0227
22	.0001	.0003	.0005	.0009	.0016	.0026	.0041	.0062	.0090	.0128
23	.0000	.0001	.0002	.0004	.0006	.0011	.0018	.0029	.0045	.0067
24	.0000	.0000	.0001	.0001	.0002	.0004	.0008	.0013	.0021	.0032
25	.0000	.0000	.0000	.0000	.0001	.0002	.0003	.0005	.0009	.0014
26	.0000	.0000	.0000	.0000	.0000	.0001	.0001	.0002	.0003	.0006
27	.0000	.0000	.0000	.0000	.0000	.0000	.0000	.0001	.0001	.0002
28	.0000	.0000	.0000	.0000	.0000	.0000	.0000	.0000	.0000	.0001

Table D Binomial probability distribution (continued)

$n = 50$ (continued)

x \ p	.31	.32	.33	.34	.35	.36	.37	.38	.39	.40
0	.0000	.0000	.0000	.0000	.0000	.0000	.0000	.0000	.0000	0
1	.0000	.0000	.0000	.0000	.0000	.0000	.0000	.0000	.0000	.0000
2	.0000	.0000	.0000	.0000	.0000	.0000	.0000	.0000	.0000	.0000
3	.0000	.0000	.0000	.0000	.0000	.0000	.0000	.0000	.0000	.0000
4	.0001	.0000	.0000	.0000	.0000	.0000	.0000	.0000	.0000	.0000
5	.0003	.0002	.0001	.0001	.0000	.0000	.0000	.0000	.0000	.0000
6	.0011	.0007	.0005	.0003	.0002	.0001	.0001	.0000	.0000	.0000
7	.0032	.0022	.0014	.0009	.0006	.0004	.0002	.0001	.0001	.0000
8	.0078	.0055	.0037	.0025	.0017	.0011	.0007	.0004	.0003	.0002
9	.0164	.0120	.0086	.0061	.0042	.0029	.0019	.0013	.0008	.0005
10	.0301	.0231	.0174	.0128	.0093	.0066	.0046	.0032	.0022	.0014
11	.0493	.0395	.0311	.0240	.0182	.0136	.0099	.0071	.0050	.0035
12	.0719	.0604	.0498	.0402	.0319	.0248	.0189	.0142	.0105	.0076
13	.0944	.0831	.0717	.0606	.0502	.0408	.0325	.0255	.0195	.0147
14	.1121	.1034	.0933	.0825	.0714	.0607	.0505	.0412	.0330	.0260
15	.1209	.1168	.1103	.1020	.0923	.0819	.0712	.0606	.0507	.0415
16	.1188	.1202	.1189	.1149	.1088	.1008	.0914	.0813	.0709	.0606
17	.1068	.1132	.1171	.1184	.1171	.1133	.1074	.0997	.0906	.0808
18	.0880	.0976	.1057	.1118	.1156	.1169	.1156	.1120	.1062	.0987
19	.0666	.0774	.0877	.0970	.1048	.1107	.1144	.1156	.1144	.1109
20	.0463	.0564	.0670	.0775	.0875	.0965	.1041	.1098	.1134	.1146
21	.0297	.0379	.0471	.0570	.0673	.0776	.0874	.0962	.1035	.1091
22	.0176	.0235	.0306	.0387	.0478	.0575	.0676	.0777	.0873	.0959
23	.0096	.0135	.0183	.0243	.0313	.0394	.0484	.0580	.0679	.0778
24	.0049	.0071	.0102	.0141	.0190	.0249	.0319	.0400	.0489	.0584
25	.0023	.0035	.0052	.0075	.0106	.0146	.0195	.0255	.0325	.0405
26	.0010	.0016	.0025	.0037	.0055	.0079	.0110	.0150	.0200	.0259
27	.0004	.0007	.0011	.0017	.0026	.0039	.0058	.0082	.0113	.0154
28	.0001	.0003	.0004	.0007	.0012	.0018	.0028	.0041	.0060	.0084
29	.0000	.0001	.0002	.0003	.0005	.0008	.0012	.0019	.0029	.0043
30	.0000	.0000	.0001	.0001	.0002	.0003	.0005	.0008	.0013	.0020
31	.0000	.0000	.0000	.0000	.0001	.0001	.0002	.0003	.0005	.0009
32	.0000	.0000	.0000	.0000	.0000	.0000	.0001	.0001	.0002	.0003
33	.0000	.0000	.0000	.0000	.0000	.0000	.0000	.0000	.0001	.0001

x \ p	.41	.42	.43	.44	.45	.46	.47	.48	.49	.50
0	0	0	0	0	0	0	0	0	0	0
1	.0000	.0000	.0000	0	0	0	0	0	0	0
2	.0000	.0000	.0000	.0000	.0000	.0000	.0000	0	0	0
3	.0000	.0000	.0000	.0000	.0000	.0000	.0000	.0000	.0000	.0000
4	.0000	.0000	.0000	.0000	.0000	.0000	.0000	.0000	.0000	.0000
5	.0000	.0000	.0000	.0000	.0000	.0000	.0000	.0000	.0000	.0000
6	.0000	.0000	.0000	.0000	.0000	.0000	.0000	.0000	.0000	.0000
7	.0000	.0000	.0000	.0000	.0000	.0000	.0000	.0000	.0000	.0000
8	.0001	.0001	.0000	.0000	.0000	.0000	.0000	.0000	.0000	.0000
9	.0003	.0002	.0001	.0001	.0000	.0000	.0000	.0000	.0000	.0000
10	.0009	.0006	.0004	.0002	.0001	.0001	.0001	.0000	.0000	.0000
11	.0024	.0016	.0010	.0007	.0004	.0003	.0002	.0001	.0001	.0000
12	.0054	.0037	.0026	.0017	.0011	.0007	.0005	.0003	.0002	.0001
13	.0109	.0079	.0057	.0040	.0027	.0018	.0012	.0008	.0005	.0003
14	.0200	.0152	.0113	.0082	.0059	.0041	.0029	.0019	.0013	.0008

Table D Binomial probability distribution (continued)

$n = 50$ (continued)

$x \backslash p$	.41	.42	.43	.44	.45	.46	.47	.48	.49	.50
15	.0334	.0264	.0204	.0155	.0116	.0085	.0061	.0043	.0030	.0020
16	.0508	.0418	.0337	.0267	.0207	.0158	.0118	.0086	.0062	.0044
17	.0706	.0605	.0508	.0419	.0339	.0269	.0209	.0159	.0119	.0087
18	.0899	.0803	.0703	.0604	.0508	.0420	.0340	.0270	.0210	.0160
19	.1053	.0979	.0893	.0799	.0700	.0602	.0507	.0419	.0340	.0270
20	.1134	.1099	.1044	.0973	.0888	.0795	.0697	.0600	.0506	.0419
21	.1126	.1137	.1126	.1092	.1038	.0967	.0884	.0791	.0695	.0598
22	.1031	.1086	.1119	.1131	.1119	.1086	.1033	.0963	.0880	.0788
23	.0872	.0957	.1028	.1082	.1115	.1126	.1115	.1082	.1029	.0960
24	.0682	.0780	.0872	.0956	.1026	.1079	.1112	.1124	.1112	.1080
25	.0493	.0587	.0684	.0781	.0873	.0956	.1026	.1079	.1112	.1123
26	.0329	.0409	.0497	.0590	.0687	.0783	.0875	.0957	.1027	.1080
27	.0203	.0263	.0333	.0412	.0500	.0593	.0690	.0786	.0877	.0960
28	.0116	.0157	.0206	.0266	.0336	.0415	.0502	.0596	.0692	.0788
29	.0061	.0086	.0118	.0159	.0208	.0268	.0338	.0417	.0504	.0598
30	.0030	.0044	.0062	.0087	.0119	.0160	.0210	.0270	.0339	.0419
31	.0013	.0020	.0030	.0044	.0063	.0088	.0120	.0161	.0210	.0270
32	.0006	.0009	.0014	.0021	.0031	.0044	.0063	.0088	.0120	.0160
33	.0002	.0003	.0006	.0009	.0014	.0021	.0031	.0044	.0063	.0087
34	.0001	.0001	.0002	.0003	.0006	.0009	.0014	.0020	.0030	.0044
35	.0000	.0000	.0001	.0001	.0002	.0003	.0005	.0009	.0013	.0020
36	.0000	.0000	.0000	.0000	.0001	.0001	.0002	.0003	.0005	.0008
37	.0000	.0000	.0000	.0000	.0000	.0000	.0001	.0001	.0002	.0003
38	.0000	.0000	.0000	.0000	.0000	.0000	.0000	.0000	.0001	.0001

$n = 75$

$x \backslash p$	.01	.02	.03	.04	.05	.06	.07	.08	.09	.10
0	.4706	.2198	.1018	.0468	.0213	.0097	.0043	.0019	.0008	.0004
1	.3565	.3364	.2362	.1463	.0843	.0462	.0244	.0125	.0063	.0031
2	.1332	.2540	.2703	.2255	.1641	.1091	.0680	.0404	.0230	.0127
3	.0327	.1261	.2034	.2287	.2101	.1695	.1246	.0854	.0554	.0343
4	.0060	.0463	.1132	.1715	.1991	.1947	.1688	.1337	.0985	.0685
5	.0009	.0134	.0497	.1015	.1488	.1765	.1804	.1651	.1384	.1081
6	.0001	.0032	.0179	.0493	.0914	.1314	.1584	.1674	.1597	.1402
7	.0000	.0006	.0055	.0203	.0474	.0827	.1176	.1435	.1557	.1535
8	.0000	.0001	.0014	.0072	.0212	.0449	.0752	.1061	.1309	.1450
9	.0000	.0000	.0003	.0022	.0083	.0213	.0421	.0687	.0964	.1199
10	.0000	.0000	.0001	.0006	.0029	.0090	.0209	.0394	.0629	.0880
11	.0000	.0000	.0000	.0002	.0009	.0034	.0093	.0203	.0368	.0578
12	0	.0000	.0000	.0000	.0003	.0012	.0037	.0094	.0194	.0342
13	0	.0000	.0000	.0000	.0001	.0004	.0014	.0040	.0093	.0184
14	0	.0000	.0000	.0000	.0000	.0001	.0005	.0015	.0041	.0091
15	0	.0000	.0000	.0000	.0000	.0000	.0001	.0005	.0016	.0041
16	0	0	.0000	.0000	.0000	.0000	.0000	.0002	.0006	.0017
17	0	0	.0000	.0000	.0000	.0000	.0000	.0001	.0002	.0007
18	0	0	0	.0000	.0000	.0000	.0000	.0000	.0001	.0002
19	0	0	0	.0000	.0000	.0000	.0000	.0000	.0000	.0001
20	0	0	0	0	.0000	.0000	.0000	.0000	.0000	.0000
21	0	0	0	0	.0000	.0000	.0000	.0000	.0000	.0000
22	0	0	0	0	0	.0000	.0000	.0000	.0000	.0000
23	0	0	0	0	0	.0000	.0000	.0000	.0000	.0000
24	0	0	0	0	0	0	.0000	.0000	.0000	.0000
25	0	0	0	0	0	0	.0000	.0000	.0000	.0000
26	0	0	0	0	0	0	0	.0000	.0000	.0000
27	0	0	0	0	0	0	0	0	.0000	.0000
28	0	0	0	0	0	0	0	0	.0000	.0000
29	0	0	0	0	0	0	0	0	0	.0000
30	0	0	0	0	0	0	0	0	0	0
31	0	0	0	0	0	0	0	0	0	0
32	0	0	0	0	0	0	0	0	0	0
33	0	0	0	0	0	0	0	0	0	0
34	0	0	0	0	0	0	0	0	0	0
35	0	0	0	0	0	0	0	0	0	0

Table D Binomial probability distribution (continued)

n = 75 (continued)

x \ p	.11	.12	.13	.14	.15	.16	.17	.18	.19	.20
0	.0002	.0001	.0000	.0000	.0000	.0000	.0000	.0000	.0000	.0000
1	.0015	.0007	.0003	.0001	.0001	.0000	.0000	.0000	.0000	.0000
2	.0068	.0035	.0018	.0009	.0004	.0002	.0001	.0000	.0000	.0000
3	.0204	.0117	.0066	.0036	.0019	.0010	.0005	.0002	.0001	.0001
4	.0454	.0288	.0176	.0104	.0060	.0034	.0018	.0010	.0005	.0003
5	.0797	.0558	.0374	.0241	.0150	.0091	.0053	.0030	.0017	.0009
6	.1149	.0888	.0652	.0458	.0309	.0201	.0127	.0077	.0046	.0027
7	.1400	.1193	.0961	.0735	.0538	.0378	.0256	.0167	.0106	.0065
8	.1470	.1383	.1220	.1018	.0807	.0612	.0446	.0313	.0212	.0139
9	.1353	.1404	.1357	.1233	.1060	.0868	.0679	.0511	.0370	.0258
10	.1104	.1264	.1339	.1325	.1235	.1091	.0919	.0740	.0572	.0426
11	.0806	.1018	.1182	.1275	.1288	.1228	.1112	.0960	.0793	.0630
12	.0531	.0741	.0942	.1107	.1212	.1248	.1214	.1124	.0992	.0840
13	.0318	.0489	.0682	.0873	.1037	.1152	.1205	.1195	.1128	.1017
14	.0174	.0296	.0451	.0629	.0810	.0971	.1093	.1162	.1172	.1126
15	.0088	.0164	.0274	.0417	.0581	.0752	.0911	.1037	.1118	.1145
16	.0041	.0084	.0154	.0254	.0385	.0537	.0699	.0854	.0983	.1073
17	.0017	.0040	.0080	.0144	.0236	.0355	.0497	.0650	.0800	.0931
18	.0007	.0017	.0038	.0075	.0134	.0218	.0328	.0460	.0605	.0750
19	.0003	.0007	.0017	.0037	.0071	.0125	.0202	.0303	.0426	.0563
20	.0001	.0003	.0007	.0017	.0035	.0066	.0116	.0186	.0280	.0394
21	.0000	.0001	.0003	.0007	.0016	.0033	.0062	.0107	.0172	.0258
22	.0000	.0000	.0001	.0003	.0007	.0016	.0031	.0058	.0099	.0158
23	.0000	.0000	.0000	.0001	.0003	.0007	.0015	.0029	.0053	.0091
24	.0000	.0000	.0000	.0000	.0001	.0003	.0007	.0014	.0027	.0049
25	.0000	.0000	.0000	.0000	.0000	.0001	.0003	.0006	.0013	.0025
26	.0000	.0000	.0000	.0000	.0000	.0000	.0001	.0003	.0006	.0012
27	.0000	.0000	.0000	.0000	.0000	.0000	.0000	.0001	.0002	.0005
28	.0000	.0000	.0000	.0000	.0000	.0000	.0000	.0000	.0001	.0002
29	.0000	.0000	.0000	.0000	.0000	.0000	.0000	.0000	.0000	.0001
30	.0000	.0000	.0000	.0000	.0000	.0000	.0000	.0000	.0000	.0000
31	0	.0000	.0000	.0000	.0000	.0000	.0000	.0000	.0000	.0000
32	0	.0000	.0000	.0000	.0000	.0000	.0000	.0000	.0000	.0000
33	0	0	.0000	.0000	.0000	.0000	.0000	.0000	.0000	.0000
34	0	0	0	.0000	.0000	.0000	.0000	.0000	.0000	.0000
35	0	0	0	0	.0000	.0000	.0000	.0000	.0000	.0000

x \ p	.21	.22	.23	.24	.25	.26	.27	.28	.29	.30
0	.0000	.0000	.0000	.0000	.0000	.0000	.0000	.0000	0	0
1	.0000	.0000	.0000	.0000	.0000	.0000	.0000	.0000	.0000	.0000
2	.0000	.0000	.0000	.0000	.0000	.0000	.0000	.0000	.0000	.0000
3	.0000	.0000	.0000	.0000	.0000	.0000	.0000	.0000	.0000	.0000
4	.0001	.0001	.0000	.0000	.0000	.0000	.0000	.0000	.0000	.0000
5	.0005	.0002	.0001	.0001	.0000	.0000	.0000	.0000	.0000	.0000
6	.0015	.0008	.0004	.0002	.0001	.0001	.0000	.0000	.0000	.0000
7	.0039	.0023	.0013	.0007	.0004	.0002	.0001	.0001	.0000	.0000
8	.0088	.0055	.0033	.0019	.0011	.0006	.0003	.0002	.0001	.0000
9	.0175	.0115	.0073	.0045	.0027	.0016	.0009	.0005	.0003	.0001
10	.0307	.0213	.0144	.0094	.0060	.0037	.0022	.0013	.0007	.0004
11	.0481	.0355	.0254	.0176	.0118	.0077	.0049	.0030	.0018	.0011
12	.0683	.0535	.0404	.0296	.0209	.0144	.0096	.0062	.0039	.0024
13	.0879	.0731	.0585	.0453	.0338	.0245	.0172	.0118	.0078	.0050
14	.1035	.0913	.0774	.0633	.0500	.0381	.0282	.0202	.0141	.0095
15	.1119	.1047	.0940	.0813	.0677	.0545	.0424	.0320	.0234	.0166
16	.1116	.1107	.1053	.0962	.0846	.0718	.0589	.0467	.0359	.0267
17	.1029	.1084	.1092	.1055	.0979	.0876	.0756	.0630	.0508	.0397
18	.0881	.0985	.1051	.1073	.1052	.0991	.0900	.0789	.0669	.0549
19	.0703	.0834	.0942	.1017	.1052	.1045	.0999	.0921	.0820	.0705
20	.0523	.0658	.0788	.0899	.0982	.1028	.1035	.1003	.0938	.0846
21	.0364	.0486	.0616	.0744	.0857	.0946	.1002	.1021	.1003	.0950
22	.0238	.0337	.0452	.0576	.0701	.0816	.0910	.0975	.1005	.1000
23	.0146	.0219	.0311	.0419	.0539	.0661	.0776	.0874	.0946	.0987
24	.0084	.0134	.0201	.0287	.0389	.0503	.0622	.0736	.0838	.0917
25	.0045	.0077	.0123	.0185	.0265	.0360	.0469	.0584	.0698	.0801
26	.0023	.0042	.0070	.0112	.0170	.0244	.0334	.0437	.0548	.0660
27	.0011	.0021	.0038	.0064	.0103	.0155	.0224	.0308	.0406	.0514
28	.0005	.0010	.0020	.0035	.0059	.0094	.0142	.0206	.0285	.0377
29	.0002	.0005	.0009	.0018	.0032	.0053	.0085	.0130	.0188	.0262

Table D Binomial probability distribution (continued)

$n = 75$ (continued)

x \ p	.21	.22	.23	.24	.25	.26	.27	.28	.29	.30
30	.0001	.0002	.0004	.0009	.0016	.0029	.0048	.0077	.0118	.0172
31	.0000	.0001	.0002	.0004	.0008	.0015	.0026	.0044	.0070	.0107
32	.0000	.0000	.0001	.0002	.0004	.0007	.0013	.0023	.0039	.0063
33	.0000	.0000	.0000	.0001	.0002	.0003	.0006	.0012	.0021	.0035
34	.0000	.0000	.0000	.0000	.0001	.0001	.0003	.0006	.0011	.0019
35	.0000	.0000	.0000	.0000	.0000	.0001	.0001	.0003	.0005	.0009
36	.0000	.0000	.0000	.0000	.0000	.0000	.0001	.0001	.0002	.0004
37	.0000	.0000	.0000	.0000	.0000	.0000	.0000	.0000	.0001	.0002
38	.0000	.0000	.0000	.0000	.0000	.0000	.0000	.0000	.0000	.0001

x \ p	.31	.32	.33	.34	.35	.36	.37	.38	.39	.40
0	0	0	0	0	0	0	0	0	0	0
1	.0000	0	0	0	0	0	0	0	0	0
2	.0000	.0000	.0000	.0000	0	0	0	0	0	0
3	.0000	.0000	.0000	.0000	.0000	.0000	0	0	0	0
4	.0000	.0000	.0000	.0000	.0000	.0000	.0000	.0000	.0000	0
5	.0000	.0000	.0000	.0000	.0000	.0000	.0000	.0000	.0000	.0000
6	.0000	.0000	.0000	.0000	.0000	.0000	.0000	.0000	.0000	.0000
7	.0000	.0000	.0000	.0000	.0000	.0000	.0000	.0000	.0000	.0000
8	.0000	.0000	.0000	.0000	.0000	.0000	.0000	.0000	.0000	.0000
9	.0001	.0000	.0000	.0000	.0000	.0000	.0000	.0000	.0000	.0000
10	.0002	.0001	.0001	.0000	.0000	.0000	.0000	.0000	.0000	.0000
11	.0006	.0003	.0002	.0001	.0001	.0000	.0000	.0000	.0000	.0000
12	.0014	.0008	.0005	.0003	.0001	.0001	.0000	.0000	.0000	.0000
13	.0032	.0019	.0011	.0007	.0004	.0002	.0001	.0001	.0000	.0000
14	.0063	.0040	.0025	.0015	.0009	.0005	.0003	.0002	.0001	.0000
15	.0115	.0077	.0050	.0032	.0020	.0012	.0007	.0004	.0002	.0001
16	.0193	.0136	.0093	.0061	.0040	.0025	.0015	.0009	.0005	.0003
17	.0301	.0222	.0158	.0110	.0074	.0049	.0031	.0019	.0012	.0007
18	.0436	.0336	.0251	.0182	.0129	.0088	.0059	.0038	.0024	.0015
19	.0588	.0474	.0371	.0282	.0208	.0149	.0104	.0070	.0046	.0030
20	.0739	.0625	.0512	.0407	.0314	.0235	.0171	.0121	.0083	.0056
21	.0870	.0770	.0660	.0549	.0442	.0346	.0263	.0194	.0139	.0097
22	.0959	.0890	.0798	.0694	.0585	.0478	.0379	.0292	.0218	.0159
23	.0993	.0965	.0906	.0824	.0725	.0619	.0513	.0412	.0321	.0244
24	.0967	.0984	.0967	.0919	.0846	.0755	.0652	.0547	.0445	.0352
25	.0886	.0944	.0972	.0966	.0930	.0866	.0782	.0684	.0581	.0479
26	.0765	.0855	.0920	.0957	.0963	.0937	.0883	.0806	.0714	.0614
27	.0624	.0730	.0823	.0895	.0941	.0956	.0941	.0897	.0829	.0742
28	.0481	.0589	.0695	.0790	.0868	.0922	.0947	.0942	.0908	.0848
29	.0350	.0449	.0554	.0660	.0758	.0841	.0902	.0936	.0941	.0917
30	.0241	.0324	.0419	.0521	.0626	.0725	.0812	.0880	.0922	.0937
31	.0157	.0221	.0299	.0390	.0489	.0592	.0692	.0783	.0856	.0907
32	.0097	.0143	.0203	.0276	.0362	.0458	.0559	.0659	.0753	.0831
33	.0057	.0088	.0130	.0185	.0254	.0336	.0428	.0527	.0627	.0722
34	.0032	.0051	.0079	.0118	.0169	.0233	.0310	.0399	.0495	.0595
35	.0017	.0028	.0046	.0071	.0107	.0154	.0214	.0286	.0371	.0464
36	.0008	.0015	.0025	.0041	.0064	.0096	.0139	.0195	.0263	.0344
37	.0004	.0007	.0013	.0022	.0036	.0057	.0086	.0126	.0178	.0242
38	.0002	.0003	.0006	.0011	.0019	.0032	.0051	.0077	.0114	.0161
39	.0001	.0002	.0003	.0006	.0010	.0017	.0028	.0045	.0069	.0102
40	.0000	.0001	.0001	.0003	.0005	.0009	.0015	.0025	.0040	.0061
41	.0000	.0000	.0001	.0001	.0002	.0004	.0007	.0013	.0022	.0035
42	.0000	.0000	.0000	.0000	.0001	.0002	.0004	.0006	.0011	.0019
43	.0000	.0000	.0000	.0000	.0000	.0001	.0002	.0003	.0005	.0010
44	.0000	.0000	.0000	.0000	.0000	.0000	.0001	.0001	.0003	.0005
45	.0000	.0000	.0000	.0000	.0000	.0000	.0000	.0001	.0001	.0002
46	.0000	.0000	.0000	.0000	.0000	.0000	.0000	.0000	.0000	.0001

x \ p	.41	.42	.43	.44	.45	.46	.47	.48	.49	.50
0	0	0	0	0	0	0	0	0	0	0
1	0	0	0	0	0	0	0	0	0	0
2	0	0	0	0	0	0	0	0	0	0
3	0	0	0	0	0	0	0	0	0	0
4	0	0	0	0	0	0	0	0	0	0
5	.0000	0	0	0	0	0	0	0	0	0
6	.0000	.0000	.0000	0	0	0	0	0	0	0
7	.0000	.0000	.0000	.0000	.0000	0	0	0	0	0
8	.0000	.0000	.0000	.0000	.0000	.0000	0	0	0	0
9	.0000	.0000	.0000	.0000	.0000	.0000	.0000	.0000	0	0

Table D Binomial probability distribution (continued)

$n = 75$ (continued)

x \ p	.41	.42	.43	.44	.45	.46	.47	.48	.49	.50
10	.0000	.0000	.0000	.0000	.0000	.0000	.0000	.0000	.0000	.0000
11	.0000	.0000	.0000	.0000	.0000	.0000	.0000	.0000	.0000	.0000
12	.0000	.0000	.0000	.0000	.0000	.0000	.0000	.0000	.0000	.0000
13	.0000	.0000	.0000	.0000	.0000	.0000	.0000	.0000	.0000	.0000
14	.0000	.0000	.0000	.0000	.0000	.0000	.0000	.0000	.0000	.0000
15	.0001	.0000	.0000	.0000	.0000	.0000	.0000	.0000	.0000	.0000
16	.0002	.0001	.0000	.0000	.0000	.0000	.0000	.0000	.0000	.0000
17	.0004	.0002	.0001	.0001	.0000	.0000	.0000	.0000	.0000	.0000
18	.0009	.0005	.0003	.0002	.0001	.0000	.0000	.0000	.0000	.0000
19	.0019	.0011	.0007	.0004	.0002	.0001	.0001	.0000	.0000	.0000
20	.0036	.0023	.0014	.0008	.0005	.0003	.0002	.0001	.0000	.0000
21	.0066	.0043	.0028	.0017	.0010	.0006	.0004	.0002	.0001	.0001
22	.0112	.0077	.0051	.0033	.0021	.0013	.0008	.0004	.0003	.0001
23	.0179	.0128	.0089	.0060	.0040	.0025	.0016	.0009	.0006	.0003
24	.0270	.0201	.0146	.0103	.0070	.0047	.0030	.0019	.0012	.0007
25	.0383	.0298	.0225	.0165	.0117	.0081	.0055	.0036	.0023	.0014
26	.0512	.0414	.0326	.0249	.0185	.0133	.0093	.0063	.0042	.0027
27	.0645	.0544	.0446	.0355	.0274	.0206	.0150	.0106	.0073	.0049
28	.0769	.0676	.0577	.0478	.0384	.0300	.0228	.0168	.0120	.0083
29	.0866	.0793	.0705	.0609	.0510	.0415	.0327	.0251	.0187	.0135
30	.0923	.0881	.0816	.0733	.0639	.0541	.0445	.0355	.0275	.0207
31	.0931	.0926	.0893	.0836	.0760	.0670	.0573	.0476	.0384	.0300
32	.0889	.0922	.0927	.0903	.0854	.0784	.0699	.0604	.0507	.0413
33	.0805	.0870	.0911	.0925	.0911	.0871	.0807	.0727	.0635	.0538
34	.0691	.0778	.0849	.0898	.0921	.0916	.0884	.0829	.0753	.0665
35	.0563	.0660	.0750	.0826	.0882	.0914	.0919	.0896	.0848	.0779
36	.0434	.0531	.0629	.0721	.0802	.0865	.0905	.0919	.0905	.0865
37	.0318	.0405	.0500	.0597	.0692	.0777	.0846	.0894	.0917	.0912
38	.0221	.0294	.0377	.0469	.0566	.0662	.0750	.0825	.0881	.0912
39	.0146	.0202	.0270	.0350	.0439	.0535	.0631	.0723	.0803	.0865
40	.0091	.0131	.0183	.0247	.0324	.0410	.0504	.0600	.0694	.0779
41	.0054	.0081	.0118	.0166	.0226	.0298	.0381	.0473	.0569	.0665
42	.0030	.0048	.0072	.0106	.0150	.0206	.0274	.0354	.0443	.0538
43	.0016	.0026	.0042	.0064	.0094	.0134	.0186	.0250	.0327	.0413
44	.0008	.0014	.0023	.0036	.0056	.0083	.0120	.0168	.0228	.0300
45	.0004	.0007	.0012	.0020	.0032	.0049	.0073	.0107	.0151	.0207
46	.0002	.0003	.0006	.0010	.0017	.0027	.0042	.0064	.0095	.0135
47	.0001	.0001	.0003	.0005	.0008	.0014	.0023	.0037	.0056	.0083
48	.0000	.0001	.0001	.0002	.0004	.0007	.0012	.0020	.0031	.0049
49	.0000	.0000	.0000	.0001	.0002	.0003	.0006	.0010	.0017	.0027
50	.0000	.0000	.0000	.0000	.0001	.0001	.0003	.0005	.0008	.0014
51	.0000	.0000	.0000	.0000	.0000	.0001	.0001	.0002	.0004	.0007
52	.0000	.0000	.0000	.0000	.0000	.0000	.0000	.0001	.0002	.0003
53	.0000	.0000	.0000	.0000	.0000	.0000	.0000	.0000	.0001	.0001
54	.0000	.0000	.0000	.0000	.0000	.0000	.0000	.0000	.0000	.0001

$n = 100$

x \ p	.01	.02	.03	.04	.05	.06	.07	.08	.09	.10
0	.3660	.1326	.0476	.0169	.0059	.0021	.0007	.0002	.0001	.0000
1	.3697	.2707	.1471	.0703	.0312	.0131	.0053	.0021	.0008	.0003
2	.1849	.2734	.2252	.1450	.0812	.0414	.0198	.0090	.0039	.0016
3	.0610	.1823	.2275	.1973	.1396	.0864	.0486	.0254	.0125	.0059
4	.0149	.0902	.1706	.1994	.1781	.1338	.0888	.0536	.0301	.0159
5	.0029	.0353	.1013	.1595	.1800	.1639	.1283	.0895	.0571	.0339
6	.0005	.0114	.0496	.1052	.1500	.1657	.1529	.1233	.0895	.0596
7	.0001	.0031	.0206	.0589	.1060	.1420	.1545	.1440	.1188	.0889
8	.0000	.0007	.0074	.0285	.0649	.1054	.1352	.1455	.1366	.1148
9	.0000	.0002	.0023	.0121	.0349	.0687	.1040	.1293	.1381	.1304
10	.0000	.0000	.0007	.0046	.0167	.0399	.0712	.1024	.1243	.1319
11	.0000	.0000	.0002	.0016	.0072	.0209	.0439	.0728	.1006	.1199
12	.0000	.0000	.0000	.0005	.0028	.0099	.0245	.0470	.0738	.0988
13	.0000	.0000	.0000	.0001	.0010	.0043	.0125	.0276	.0494	.0743
14	0	.0000	.0000	.0000	.0003	.0017	.0058	.0149	.0304	.0513
15	0	.0000	.0000	.0000	.0001	.0006	.0025	.0074	.0172	.0327
16	0	.0000	.0000	.0000	.0000	.0002	.0010	.0034	.0090	.0193
17	0	.0000	.0000	.0000	.0000	.0001	.0004	.0015	.0044	.0106
18	0	0	.0000	.0000	.0000	.0000	.0001	.0006	.0020	.0054
19	0	0	.0000	.0000	.0000	.0000	.0000	.0002	.0009	.0026

Table D Binomial probability distribution (continued)

$n = 100$ (continued)

x \ p	.01	.02	.03	.04	.05	.06	.07	.08	.09	.10
20	0	0	.0000	.0000	.0000	.0000	.0000	.0001	.0003	.0012
21	0	0	0	.0000	.0000	.0000	.0000	.0000	.0001	.0005
22	0	0	0	.0000	.0000	.0000	.0000	.0000	.0000	.0002
23	0	0	0	0	.0000	.0000	.0000	.0000	.0000	.0001
24	0	0	0	0	.0000	.0000	.0000	.0000	.0000	.0000
25	0	0	0	0	.0000	.0000	.0000	.0000	.0000	.0000
26	0	0	0	0	0	.0000	.0000	.0000	.0000	.0000
27	0	0	0	0	0	.0000	.0000	.0000	.0000	.0000
28	0	0	0	0	0	0	.0000	.0000	.0000	.0000
29	0	0	0	0	0	0	.0000	.0000	.0000	.0000
30	0	0	0	0	0	0	0	.0000	.0000	.0000
31	0	0	0	0	0	0	0	.0000	.0000	.0000
32	0	0	0	0	0	0	0	0	.0000	.0000
33	0	0	0	0	0	0	0	0	.0000	.0000
34	0	0	0	0	0	0	0	0	0	.0000
35	0	0	0	0	0	0	0	0	0	0
36	0	0	0	0	0	0	0	0	0	0
37	0	0	0	0	0	0	0	0	0	0
38	0	0	0	0	0	0	0	0	0	0
39	0	0	0	0	0	0	0	0	0	0
40	0	0	0	0	0	0	0	0	0	0
41	0	0	0	0	0	0	0	0	0	0
42	0	0	0	0	0	0	0	0	0	0
43	0	0	0	0	0	0	0	0	0	0
44	0	0	0	0	0	0	0	0	0	0
45	0	0	0	0	0	0	0	0	0	0
46	0	0	0	0	0	0	0	0	0	0
47	0	0	0	0	0	0	0	0	0	0
48	0	0	0	0	0	0	0	0	0	0

x \ p	.11	.12	.13	.14	.15	.16	.17	.18	.19	.20
0	.0000	.0000	.0000	.0000	.0000	.0000	.0000	.0000	.0000	.0000
1	.0001	.0000	.0000	.0000	.0000	.0000	.0000	.0000	.0000	.0000
2	.0007	.0003	.0001	.0000	.0000	.0000	.0000	.0000	.0000	.0000
3	.0027	.0012	.0005	.0002	.0001	.0000	.0000	.0000	.0000	.0000
4	.0080	.0038	.0018	.0008	.0003	.0001	.0001	.0000	.0000	.0000
5	.0189	.0100	.0050	.0024	.0011	.0005	.0002	.0001	.0000	.0000
6	.0369	.0215	.0119	.0063	.0031	.0015	.0007	.0003	.0001	.0001
7	.0613	.0394	.0238	.0137	.0075	.0039	.0020	.0009	.0004	.0002
8	.0881	.0625	.0414	.0259	.0153	.0086	.0047	.0024	.0012	.0006
9	.1112	.0871	.0632	.0430	.0276	.0168	.0098	.0054	.0029	.0015
10	.1251	.1080	.0860	.0637	.0444	.0292	.0182	.0108	.0062	.0034
11	.1265	.1205	.1051	.0849	.0640	.0454	.0305	.0194	.0118	.0069
12	.1160	.1219	.1165	.1025	.0838	.0642	.0463	.0316	.0206	.0128
13	.0970	.1125	.1179	.1130	.1001	.0827	.0642	.0470	.0327	.0216
14	.0745	.0954	.1094	.1143	.1098	.0979	.0817	.0641	.0476	.0335
15	.0528	.0745	.0938	.1067	.1111	.1070	.0960	.0807	.0640	.0481
16	.0347	.0540	.0744	.0922	.1041	.1082	.1044	.0941	.0798	.0638
17	.0212	.0364	.0549	.0742	.0908	.1019	.1057	.1021	.0924	.0789
18	.0121	.0229	.0379	.0557	.0739	.0895	.0998	.1033	.1000	.0909
19	.0064	.0135	.0244	.0391	.0563	.0736	.0882	.0979	.1012	.0981
20	.0032	.0074	.0148	.0258	.0402	.0567	.0732	.0870	.0962	.0993
21	.0015	.0039	.0084	.0160	.0270	.0412	.0571	.0728	.0859	.0946
22	.0007	.0019	.0045	.0094	.0171	.0282	.0420	.0574	.0724	.0849
23	.0003	.0009	.0023	.0052	.0103	.0182	.0292	.0427	.0576	.0720
24	.0001	.0004	.0011	.0027	.0058	.0111	.0192	.0301	.0433	.0577
25	.0000	.0002	.0005	.0013	.0031	.0064	.0119	.0201	.0309	.0439
26	.0000	.0001	.0002	.0006	.0016	.0035	.0071	.0127	.0209	.0316
27	.0000	.0000	.0001	.0003	.0008	.0018	.0040	.0076	.0134	.0217
28	.0000	.0000	.0000	.0001	.0004	.0009	.0021	.0044	.0082	.0141
29	.0000	.0000	.0000	.0000	.0002	.0004	.0011	.0024	.0048	.0088
30	.0000	.0000	.0000	.0000	.0001	.0002	.0005	.0012	.0027	.0052
31	.0000	.0000	.0000	.0000	.0000	.0001	.0002	.0006	.0014	.0029
32	.0000	.0000	.0000	.0000	.0000	.0000	.0001	.0003	.0007	.0016
33	.0000	.0000	.0000	.0000	.0000	.0000	.0000	.0001	.0003	.0008
34	.0000	.0000	.0000	.0000	.0000	.0000	.0000	.0001	.0002	.0004
35	.0000	.0000	.0000	.0000	.0000	.0000	.0000	.0000	.0001	.0002
36	.0000	.0000	.0000	.0000	.0000	.0000	.0000	.0000	.0000	.0001
37	0	.0000	.0000	.0000	.0000	.0000	.0000	.0000	.0000	.0000
38	0	.0000	.0000	.0000	.0000	.0000	.0000	.0000	.0000	.0000
39	0	0	.0000	.0000	.0000	.0000	.0000	.0000	.0000	.0000

Table D Binomial probability distribution (continued)

$n = 100$ (continued)

x \ p	.11	.12	.13	.14	.15	.16	.17	.18	.19	.20
40	0	0	0	.0000	.0000	.0000	.0000	.0000	.0000	.0000
41	0	0	0	.0000	.0000	.0000	.0000	.0000	.0000	.0000
42	0	0	0	. 0	.0000	.0000	.0000	.0000	.0000	.0000
43	0	0	0	0	0	.0000	.0000	.0000	.0000	.0000
44	0	0	0	0	0	.0000	.0000	.0000	.0000	.0000
45	0	0	0	0	0	0	.0000	.0000	.0000	.0000
46	0	0	0	0	0	0	0	.0000	.0000	.0000
47	0	0	0	0	0	0	0	.0000	.0000	.0000
48	0	0	0	0	0	0	0	0	.0000	.0000

x \ p	.21	.22	.23	.24	.25	.26	.27	.28	.29	.30
0	.0000	.0000	0	0	0	0	0	0	0	0
1	.0000	.0000	.0000	.0000	0	0	0	0	0	0
2	.0000	.0000	.0000	.0000	.0000	.0000	.0000	0	0	0
3	.0000	.0000	.0000	.0000	.0000	.0000	.0000	.0000	.0000	0
4	.0000	.0000	.0000	.0000	.0000	.0000	.0000	.0000	.0000	.0000
5	.0000	.0000	.0000	.0000	.0000	.0000	.0000	.0000	.0000	.0000
6	.0000	.0000	.0000	.0000	.0000	.0000	.0000	.0000	.0000	.0000
7	.0001	.0000	.0000	.0000	.0000	.0000	.0000	.0000	.0000	.0000
8	.0003	.0001	.0001	.0000	.0000	.0000	.0000	.0000	.0000	.0000
9	.0007	.0003	.0002	.0001	.0000	.0000	.0000	.0000	.0000	.0000
10	.0018	.0009	.0004	.0002	.0001	.0000	.0000	.0000	.0000	.0000
11	.0038	.0021	.0011	.0005	.0003	.0001	.0001	.0000	.0000	.0000
12	.0076	.0043	.0024	.0012	.0006	.0003	.0001	.0001	.0000	.0000
13	.0136	.0082	.0048	.0027	.0014	.0007	.0004	.0002	.0001	.0000
14	.0225	.0144	.0089	.0052	.0030	.0016	.0009	.0004	.0002	.0001
15	.0343	.0233	.0152	.0095	.0057	.0033	.0018	.0010	.0005	.0002
16	.0484	.0350	.0241	.0159	.0100	.0061	.0035	.0020	.0011	.0006
17	.0636	.0487	.0356	.0248	.0165	.0106	.0065	.0038	.0022	.0012
18	.0780	.0634	.0490	.0361	.0254	.0171	.0111	.0069	.0041	.0024
19	.0895	.0772	.0631	.0492	.0365	.0259	.0177	.0115	.0072	.0044
20	.0963	.0881	.0764	.0629	.0493	.0369	.0264	.0182	.0120	.0076
21	.0975	.0947	.0869	.0756	.0626	.0494	.0373	.0269	.0186	.0124
22	.0931	.0959	.0932	.0858	.0749	.0623	.0495	.0376	.0273	.0190
23	.0839	.0917	.0944	.0919	.0847	.0743	.0621	.0495	.0378	.0277
24	.0716	.0830	.0905	.0931	.0906	.0837	.0736	.0618	.0496	.0380
25	.0578	.0712	.0822	.0893	.0918	.0894	.0828	.0731	.0615	.0496
26	.0444	.0579	.0708	.0814	.0883	.0906	.0883	.0819	.0725	.0613
27	.0323	.0448	.0580	.0704	.0806	.0873	.0896	.0873	.0812	.0720
28	.0224	.0329	.0451	.0580	.0701	.0799	.0864	.0886	.0864	.0804
29	.0148	.0231	.0335	.0455	.0580	.0697	.0793	.0855	.0876	.0856
30	.0093	.0154	.0237	.0340	.0458	.0580	.0694	.0787	.0847	.0868
31	.0056	.0098	.0160	.0242	.0344	.0460	.0580	.0691	.0781	.0840
32	.0032	.0060	.0103	.0165	.0248	.0349	.0462	.0579	.0688	.0776
33	.0018	.0035	.0063	.0107	.0170	.0252	.0352	.0464	.0579	.0685
34	.0009	.0019	.0037	.0067	.0112	.0175	.0257	.0356	.0466	.0579
35	.0005	.0010	.0021	.0040	.0070	.0116	.0179	.0261	.0359	.0468
36	.0002	.0005	.0011	.0023	.0042	.0073	.0120	.0183	.0265	.0362
37	.0001	.0003	.0006	.0012	.0024	.0045	.0077	.0123	.0187	.0268
38	.0000	.0001	.0003	.0006	.0013	.0026	.0047	.0079	.0127	.0191
39	.0000	.0001	.0001	.0003	.0007	.0015	.0028	.0049	.0082	.0130
40	.0000	.0000	.0001	.0002	.0004	.0008	.0016	.0029	.0051	.0085
41	.0000	.0000	.0000	.0001	.0002	.0004	.0008	.0017	.0031	.0053
42	.0000	.0000	.0000	.0000	.0001	.0002	.0004	.0009	.0018	.0032
43	.0000	.0000	.0000	.0000	.0000	.0001	.0002	.0005	.0010	.0019
44	.0000	.0000	.0000	.0000	.0000	.0000	.0001	.0002	.0005	.0010
45	.0000	.0000	.0000	.0000	.0000	.0000	.0000	.0001	.0003	.0005
46	.0000	.0000	.0000	.0000	.0000	.0000	.0000	.0001	.0001	.0003
47	.0000	.0000	.0000	.0000	.0000	.0000	.0000	.0000	.0001	.0001
48	.0000	.0000	.0000	.0000	.0000	.0000	.0000	.0000	.0000	.0001

x \ p	.31	.32	.33	.34	.35	.36	.37	.38	.39	.40
0	0	0	0	0	0	0	0	0	0	0
1	0	0	0	0	0	0	0	0	0	0
2	0	0	0	0	0	0	0	0	0	0
3	0	0	0	0	0	0	0	0	0	0
4	0	0	0	0	0	0	0	0	0	0
5	.0000	.0000	0	0	0	0	0	0	0	0
6	.0000	.0000	.0000	.0000	0	0	0	0	0	0
7	.0000	.0000	.0000	.0000	.0000	0	0	0	0	0
8	.0000	.0000	.0000	.0000	.0000	.0000	.0000	0	0	0
9	.0000	.0000	.0000	.0000	.0000	.0000	.0000	.0000	0	0

Table D Binomial probability distribution (continued)

n = 100 (continued)

x \ p	.31	.32	.33	.34	.35	.36	.37	.38	.39	.40
10	.0000	.0000	.0000	.0000	.0000	.0000	.0000	.0000	.0000	.0000
11	.0000	.0000	.0000	.0000	.0000	.0000	.0000	.0000	.0000	.0000
12	.0000	.0000	.0000	.0000	.0000	.0000	.0000	.0000	.0000	.0000
13	.0000	.0000	.0000	.0000	.0000	.0000	.0000	.0000	.0000	.0000
14	.0000	.0000	.0000	.0000	.0000	.0000	.0000	.0000	.0000	.0000
15	.0001	.0001	.0000	.0000	.0000	.0000	.0000	.0000	.0000	.0000
16	.0003	.0001	.0001	.0000	.0000	.0000	.0000	.0000	.0000	.0000
17	.0006	.0003	.0002	.0001	.0000	.0000	.0000	.0000	.0000	.0000
18	.0013	.0007	.0004	.0002	.0001	.0000	.0000	.0000	.0000	.0000
19	.0025	.0014	.0008	.0004	.0002	.0001	.0000	.0000	.0000	.0000
20	.0046	.0027	.0015	.0008	.0004	.0002	.0001	.0001	.0000	.0000
21	.0079	.0049	.0029	.0016	.0009	.0005	.0002	.0001	.0001	.0000
22	.0127	.0082	.0051	.0030	.0017	.0010	.0005	.0003	.0001	.0001
23	.0194	.0131	.0085	.0053	.0032	.0018	.0010	.0006	.0003	.0001
24	.0280	.0198	.0134	.0088	.0055	.0033	.0019	.0011	.0006	.0003
25	.0382	.0283	.0201	.0137	.0090	.0057	.0035	.0020	.0012	.0006
26	.0496	.0384	.0286	.0204	.0140	.0092	.0059	.0036	.0021	.0012
27	.0610	.0495	.0386	.0288	.0207	.0143	.0095	.0060	.0037	.0022
28	.0715	.0608	.0495	.0387	.0290	.0209	.0145	.0097	.0062	.0038
29	.0797	.0710	.0605	.0495	.0388	.0292	.0211	.0147	.0098	.0063
30	.0848	.0791	.0706	.0603	.0494	.0389	.0294	.0213	.0149	.0100
31	.0860	.0840	.0785	.0702	.0601	.0494	.0389	.0295	.0215	.0151
32	.0833	.0853	.0834	.0779	.0698	.0599	.0493	.0390	.0296	.0217
33	.0771	.0827	.0846	.0827	.0774	.0694	.0597	.0493	.0390	.0297
34	.0683	.0767	.0821	.0840	.0821	.0769	.0691	.0595	.0492	.0391
35	.0578	.0680	.0763	.0816	.0834	.0816	.0765	.0688	.0593	.0491
36	.0469	.0578	.0678	.0759	.0811	.0829	.0811	.0761	.0685	.0591
37	.0365	.0471	.0578	.0676	.0755	.0806	.0824	.0807	.0757	.0682
38	.0272	.0367	.0472	.0577	.0674	.0752	.0802	.0820	.0803	.0754
39	.0194	.0275	.0369	.0473	.0577	.0672	.0749	.0799	.0816	.0799
40	.0133	.0197	.0277	.0372	.0474	.0577	.0671	.0746	.0795	.0812
41	.0087	.0136	.0200	.0280	.0373	.0475	.0577	.0670	.0744	.0792
42	.0055	.0090	.0138	.0203	.0282	.0375	.0476	.0576	.0668	.0742
43	.0033	.0057	.0092	.0141	.0205	.0285	.0377	.0477	.0576	.0667
44	.0019	.0035	.0059	.0094	.0143	.0207	.0287	.0378	.0477	.0576
45	.0011	.0020	.0036	.0060	.0096	.0145	.0210	.0289	.0380	.0478
46	.0006	.0011	.0021	.0037	.0062	.0098	.0147	.0212	.0290	.0381
47	.0003	.0006	.0012	.0022	.0038	.0063	.0099	.0149	.0213	.0292
48	.0001	.0003	.0007	.0012	.0023	.0039	.0064	.0101	.0151	.0215
49	.0001	.0002	.0003	.0007	.0013	.0023	.0040	.0066	.0102	.0152
50	.0000	.0001	.0002	.0004	.0007	.0013	.0024	.0041	.0067	.0103
51	.0000	.0000	.0001	.0002	.0004	.0007	.0014	.0025	.0042	.0068
52	.0000	.0000	.0000	.0001	.0002	.0004	.0008	.0014	.0025	.0042
53	.0000	.0000	.0000	.0000	.0001	.0002	.0004	.0008	.0015	.0026
54	.0000	.0000	.0000	.0000	.0000	.0001	.0002	.0004	.0008	.0015
55	.0000	.0000	.0000	.0000	.0000	.0000	.0001	.0002	.0004	.0008
56	.0000	.0000	.0000	.0000	.0000	.0000	.0000	.0001	.0002	.0004
57	.0000	.0000	.0000	.0000	.0000	.0000	.0000	.0001	.0001	.0002
58	.0000	.0000	.0000	.0000	.0000	.0000	.0000	.0000	.0001	.0001
59	.0000	.0000	.0000	.0000	.0000	.0000	.0000	.0000	.0000	.0001

x \ p	.41	.42	.43	.44	.45	.46	.47	.48	.49	.50
0	0	0	0	0	0	0	0	0	0	0
1	0	0	0	0	0	0	0	0	0	0
2	0	0	0	0	0	0	0	0	0	0
3	0	0	0	0	0	0	0	0	0	0
4	0	0	0	0	0	0	0	0	0	0
5	0	0	0	0	0	0	0	0	0	0
6	0	0	0	0	0	0	0	0	0	0
7	0	0	0	0	0	0	0	0	0	0
8	0	0	0	0	0	0	0	0	0	0
9	0	0	0	0	0	0	0	0	0	0
10	0	0	0	0	0	0	0	0	0	0
11	.0000	0	0	0	0	0	0	0	0	0
12	.0000	.0000	0	0	0	0	0	0	0	0
13	.0000	.0000	.0000	.0000	0	0	0	0	0	0
14	.0000	.0000	.0000	.0000	.0000	0	0	0	0	0

Table D Binomial probability distribution (continued)

x \ p	.41	.42	.43	.44	.45	.46	.47	.48	.49	.50
					$n = 100$ (continued)					
15	.0000	.0000	.0000	.0000	.0000	.0000	0	0	0	0
16	.0000	.0000	.0000	.0000	.0000	.0000	.0000	.0000	0	0
17	.0000	.0000	.0000	.0000	.0000	.0000	.0000	.0000	.0000	0
18	.0000	.0000	.0000	.0000	.0000	.0000	.0000	.0000	.0000	.0000
19	.0000	.0000	.0000	.0000	.0000	.0000	.0000	.0000	.0000	.0000
20	.0000	.0000	.0000	.0000	.0000	.0000	.0000	.0000	.0000	.0000
21	.0000	.0000	.0000	.0000	.0000	.0000	.0000	.0000	.0000	.0000
22	.0000	.0000	.0000	.0000	.0000	.0000	.0000	.0000	.0000	.0000
23	.0001	.0000	.0000	.0000	.0000	.0000	.0000	.0000	.0000	.0000
24	.0002	.0001	.0000	.0000	.0000	.0000	.0000	.0000	.0000	.0000
25	.0003	.0002	.0001	.0000	.0000	.0000	.0000	.0000	.0000	.0000
26	.0007	.0003	.0002	.0001	.0000	.0000	.0000	.0000	.0000	.0000
27	.0013	.0007	.0004	.0002	.0001	.0000	.0000	.0000	.0000	.0000
28	.0023	.0013	.0007	.0004	.0002	.0001	.0000	.0000	.0000	.0000
29	.0039	.0024	.0014	.0008	.0004	.0002	.0001	.0000	.0000	.0000
30	.0065	.0040	.0024	.0014	.0008	.0004	.0002	.0001	.0001	.0000
31	.0102	.0066	.0041	.0025	.0014	.0008	.0004	.0002	.0001	.0001
32	.0152	.0103	.0067	.0042	.0025	.0015	.0008	.0004	.0002	.0001
33	.0218	.0154	.0104	.0068	.0043	.0026	.0015	.0008	.0004	.0002
34	.0298	.0219	.0155	.0105	.0069	.0043	.0026	.0015	.0009	.0005
35	.0391	.0299	.0220	.0156	.0106	.0069	.0044	.0026	.0015	.0009
36	.0491	.0391	.0300	.0221	.0157	.0107	.0070	.0044	.0027	.0016
37	.0590	.0490	.0391	.0300	.0222	.0157	.0107	.0070	.0044	.0027
38	.0680	.0588	.0489	.0391	.0301	.0222	.0158	.0108	.0071	.0045
39	.0751	.0677	.0587	.0489	.0391	.0301	.0223	.0158	.0108	.0071
40	.0796	.0748	.0675	.0586	.0488	.0391	.0301	.0223	.0159	.0108
41	.0809	.0793	.0745	.0673	.0584	.0487	.0391	.0301	.0223	.0159
42	.0790	.0806	.0790	.0743	.0672	.0583	.0487	.0390	.0301	.0223
43	.0740	.0787	.0804	.0788	.0741	.0670	.0582	.0486	.0390	.0301
44	.0666	.0739	.0785	.0802	.0786	.0739	.0669	.0581	.0485	.0390
45	.0576	.0666	.0737	.0784	.0800	.0784	.0738	.0668	.0580	.0485
46	.0479	.0576	.0665	.0736	.0782	.0798	.0783	.0737	.0667	.0580
47	.0382	.0480	.0576	.0665	.0736	.0781	.0797	.0781	.0736	.0666
48	.0293	.0383	.0480	.0577	.0665	.0735	.0781	.0797	.0781	.0735
49	.0216	.0295	.0384	.0481	.0577	.0664	.0735	.0780	.0796	.0780
50	.0153	.0218	.0296	.0385	.0482	.0577	.0665	.0735	.0780	.0796
51	.0104	.0155	.0219	.0297	.0386	.0482	.0578	.0665	.0735	.0780
52	.0068	.0105	.0156	.0220	.0298	.0387	.0483	.0578	.0665	.0735
53	.0043	.0069	.0106	.0156	.0221	.0299	.0388	.0483	.0579	.0666
54	.0026	.0044	.0070	.0107	.0157	.0221	.0299	.0388	.0484	.0580
55	.0015	.0026	.0044	.0070	.0108	.0158	.0222	.0300	.0389	.0485
56	.0008	.0015	.0027	.0044	.0071	.0108	.0158	.0222	.0300	.0390
57	.0005	.0009	.0016	.0027	.0045	.0071	.0108	.0158	.0223	.0301
58	.0002	.0005	.0009	.0016	.0027	.0045	.0071	.0108	.0159	.0223
59	.0001	.0002	.0005	.0009	.0016	.0027	.0045	.0071	.0109	.0159
60	.0001	.0001	.0002	.0005	.0009	.0016	.0027	.0045	.0071	.0108
61	.0000	.0001	.0001	.0002	.0005	.0009	.0016	.0027	.0045	.0071
62	.0000	.0000	.0001	.0001	.0002	.0005	.0009	.0016	.0027	.0045
63	.0000	.0000	.0000	.0001	.0001	.0002	.0005	.0009	.0016	.0027
64	.0000	.0000	.0000	.0000	.0001	.0001	.0002	.0005	.0009	.0016
65	.0000	.0000	.0000	.0000	.0000	.0001	.0001	.0002	.0005	.0009
66	.0000	.0000	.0000	.0000	.0000	.0000	.0001	.0001	.0002	.0005
67	.0000	.0000	.0000	.0000	.0000	.0000	.0000	.0001	.0001	.0002
68	.0000	.0000	.0000	.0000	.0000	.0000	.0000	.0000	.0001	.0001
69	.0000	.0000	.0000	.0000	.0000	.0000	.0000	.0000	.0000	.0001

Table E Normal curve areas

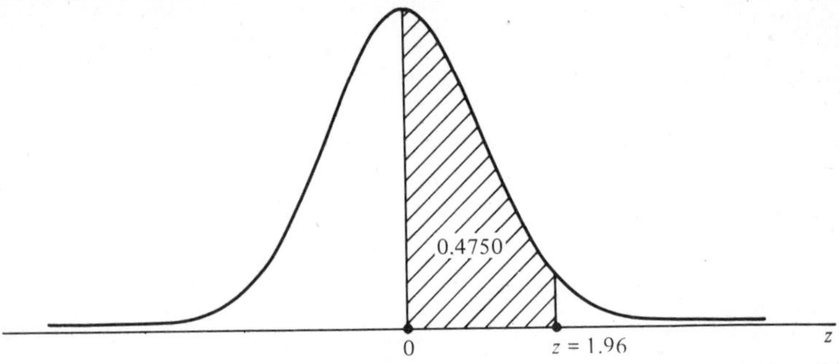

Entries in the body of the table give the area under the standard normal curve from 0 to *z*.

z	.00	.01	.02	.03	.04	.05	.06	.07	.08	.09
0.0	.0000	.0040	.0080	.0120	.0160	.0199	.0239	.0279	.0319	.0359
0.1	.0398	.0438	.0478	.0517	.0557	.0596	.0636	.0675	.0714	.0753
0.2	.0793	.0832	.0871	.0910	.0948	.0987	.1026	.1064	.1103	.1141
0.3	.1179	.1217	.1255	.1293	.1331	.1368	.1406	.1443	.1480	.1517
0.4	.1554	.1591	.1628	.1664	.1700	.1736	.1772	.1808	.1844	.1879
0.5	.1915	.1950	.1985	.2019	.2054	.2088	.2123	.2157	.2190	.2224
0.6	.2257	.2291	.2324	.2357	.2389	.2422	.2454	.2486	.2517	.2549
0.7	.2580	.2611	.2642	.2673	.2704	.2734	.2764	.2794	.2823	.2852
0.8	.2881	.2910	.2939	.2967	.2995	.3023	.3051	.3078	.3106	.3133
0.9	.3159	.3186	.3212	.3238	.3264	.3289	.3315	.3340	.3365	.3389
1.0	.3413	.3438	.3461	.3485	.3508	.3531	.3554	.3577	.3599	.3621
1.1	.3643	.3665	.3686	.3708	.3729	.3749	.3770	.3790	.3810	.3830
1.2	.3849	.3869	.3888	.3907	.3925	.3944	.3962	.3980	.3997	.4015
1.3	.4032	.4049	.4066	.4082	.4099	.4115	.4131	.4147	.4162	.4177
1.4	.4192	.4207	.4222	.4236	.4251	.4265	.4279	.4292	.4306	.4319
1.5	.4332	.4345	.4357	.4370	.4382	.4394	.4406	.4418	.4429	.4441
1.6	.4452	.4463	.4474	.4484	.4495	.4505	.4515	.4525	.4535	.4545
1.7	.4554	.4564	.4573	.4582	.4591	.4599	.4608	.4616	.4625	.4633
1.8	.4641	.4649	.4656	.4664	.4671	.4678	.4686	.4693	.4699	.4706
1.9	.4713	.4719	.4726	.4732	.4738	.4744	.4750	.4756	.4761	.4767
2.0	.4772	.4778	.4783	.4788	.4793	.4798	.4803	.4808	.4812	.4817
2.1	.4821	.4826	.4830	.4834	.4838	.4842	.4846	.4850	.4854	.4857
2.2	.4861	.4864	.4868	.4871	.4875	.4878	.4881	.4884	.4887	.4890
2.3	.4893	.4896	.4898	.4901	.4904	.4906	.4909	.4911	.4913	.4916
2.4	.4918	.4920	.4922	.4925	.4927	.4929	.4931	.4932	.4934	.4936
2.5	.4938	.4940	.4941	.4943	.4945	.4946	.4948	.4949	.4951	.4952
2.6	.4953	.4955	.4956	.4957	.4959	.4960	.4961	.4962	.4963	.4964
2.7	.4965	.4966	.4967	.4968	.4969	.4970	.4971	.4972	.4973	.4974
2.8	.4974	.4975	.4976	.4977	.4977	.4978	.4979	.4979	.4980	.4981
2.9	.4981	.4982	.4982	.4983	.4984	.4984	.4985	.4985	.4986	.4986
3.0	.4987	.4987	.4987	.4988	.4988	.4989	.4989	.4989	.4990	.4990

Source: John E. Freund and Frank J. Williams, *Elementary Business Statistics: The Modern Approach,* second edition, Englewood Cliffs, N.J.: Prentice-Hall, 1972. Reprinted by permission of the publisher.

Table F Percentiles of the t distribution

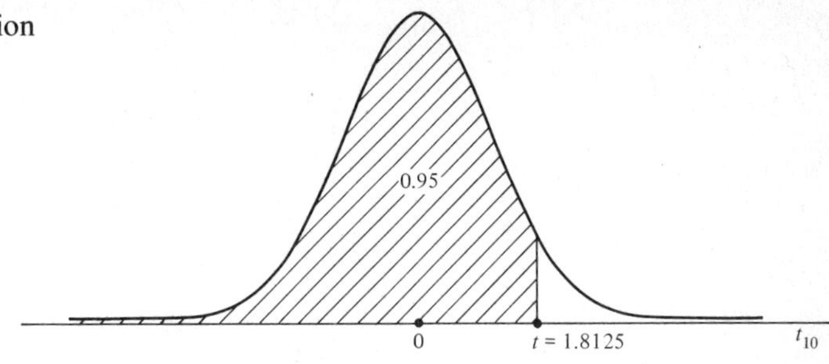

df	$t_{0.90}$	$t_{0.95}$	$t_{0.975}$	$t_{0.99}$	$t_{0.995}$	$t_{0.9995}$
1	3.078	6.3138	12.706	31.821	63.657	636.619
2	1.886	2.9200	4.3027	6.965	9.9248	31.598
3	1.638	2.3534	3.1825	4.541	5.8409	12.924
4	1.533	2.1318	2.7764	3.747	4.6041	8.610
5	1.476	2.0150	2.5706	3.365	4.0321	6.869
6	1.440	1.9432	2.4469	3.143	3.7074	5.959
7	1.415	1.8946	2.3646	2.998	3.4995	5.408
8	1.397	1.8595	2.3060	2.896	3.3554	5.041
9	1.383	1.8331	2.2622	2.821	3.2498	4.781
10	1.372	1.8125	2.2281	2.764	3.1693	4.587
11	1.363	1.7959	2.2010	2.718	3.1058	4.437
12	1.356	1.7823	2.1788	2.681	3.0545	4.318
13	1.350	1.7709	2.1604	2.650	3.0123	4.221
14	1.345	1.7613	2.1448	2.624	2.9768	4.140
15	1.341	1.7530	2.1315	2.602	2.9467	4.073
16	1.337	1.7459	2.1199	2.583	2.9208	4.015
17	1.333	1.7396	2.1098	2.567	2.8982	3.965
18	1.330	1.7341	2.1009	2.552	2.8784	3.922
19	1.328	1.7291	2.0930	2.539	2.8609	3.883
20	1.325	1.7247	2.0860	2.528	2.8453	3.850
21	1.323	1.7207	2.0796	2.518	2.8314	3.819
22	1.321	1.7171	2.0739	2.508	2.8188	3.792
23	1.319	1.7139	2.0687	2.500	2.8073	3.767
24	1.318	1.7109	2.0639	2.492	2.7969	3.745
25	1.316	1.7081	2.0595	2.485	2.7874	3.725
26	1.315	1.7056	2.0555	2.479	2.7787	3.707
27	1.314	1.7033	2.0518	2.473	2.7707	3.690
28	1.313	1.7011	2.0484	2.467	2.7633	3.674
29	1.311	1.6991	2.0452	2.462	2.7564	3.659
30	1.310	1.6973	2.0423	2.457	2.7500	3.646
35	1.3062	1.6896	2.0301	2.438	2.7239	3.5915
40	1.3031	1.6839	2.0211	2.423	2.7045	3.5511
45	1.3007	1.6794	2.0141	2.412	2.6896	3.5207
50	1.2987	1.6759	2.0086	2.403	2.6778	3.4965
60	1.2959	1.6707	2.0003	2.390	2.6603	3.4606
70	1.2938	1.6669	1.9945	2.381	2.6480	3.4355
80	1.2922	1.6641	1.9901	2.374	2.6388	3.4169
90	1.2910	1.6620	1.9867	2.368	2.6316	3.4022
100	1.2901	1.6602	1.9840	2.364	2.6260	3.3909
120	1.2887	1.6577	1.9799	2.358	2.6175	3.3736
140	1.2876	1.6558	1.9771	2.353	2.6114	3.3615
160	1.2869	1.6545	1.9749	2.350	2.6070	3.3527
180	1.2863	1.6534	1.9733	2.347	2.6035	3.3456
200	1.2858	1.6525	1.9719	2.345	2.6006	3.3400
∞	1.282	1.645	1.96	2.326	2.576	3.2905

Source: Reproduced from *Documenta Geigy, Scientific Tables,* seventh edition. Courtesy of Ciba-Geigy Ltd., Basel, Switzerland, 1970.

Table G Percentiles of the chi-square distribution

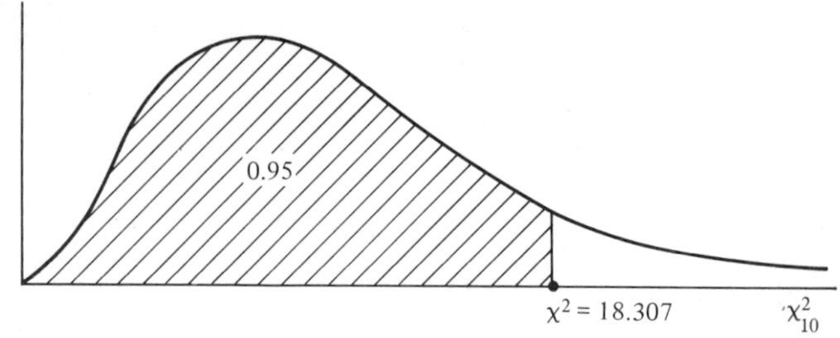

df	$\chi^2_{0.005}$	$\chi^2_{0.025}$	$\chi^2_{0.05}$	$\chi^2_{0.90}$	$\chi^2_{0.95}$	$\chi^2_{0.975}$	$\chi^2_{0.99}$	$\chi^2_{0.995}$
1	0.0000393	0.000982	0.00393	2.706	3.841	5.024	6.635	7.879
2	0.0100	0.0506	0.103	4.605	5.991	7.378	9.210	10.597
3	0.0717	0.216	0.352	6.251	7.815	9.348	11.345	12.838
4	0.207	0.484	0.711	7.779	9.488	11.143	13.277	14.860
5	0.412	0.831	1.145	9.236	11.070	12.832	15.086	16.750
6	0.676	1.237	1.635	10.645	12.592	14.449	16.812	18.548
7	0.989	1.690	2.167	12.017	14.067	16.013	18.475	20.278
8	1.344	2.180	2.733	13.362	15.507	17.535	20.090	21.955
9	1.735	2.700	3.325	14.684	16.919	19.023	21.666	23.589
10	2.156	3.247	3.940	15.987	18.307	20.483	23.209	25.188
11	2.603	3.816	4.575	17.275	19.675	21.920	24.725	26.757
12	3.074	4.404	5.226	18.549	21.026	23.336	26.217	28.300
13	3.565	5.009	5.892	19.812	22.362	24.736	27.688	29.819
14	4.075	5.629	6.571	21.064	23.685	26.119	29.141	31.319
15	4.601	6.262	7.261	22.307	24.996	27.488	30.578	32.801
16	5.142	6.908	7.962	23.542	26.296	28.845	32.000	34.267
17	5.697	7.564	8.672	24.769	27.587	30.191	33.409	35.718
18	6.265	8.231	9.390	25.989	28.869	31.526	34.805	37.156
19	6.844	8.907	10.117	27.204	30.144	32.852	36.191	38.582
20	7.434	9.591	10.851	28.412	31.410	34.170	37.566	39.997
21	8.034	10.283	11.591	29.615	32.671	35.479	38.932	41.401
22	8.643	10.982	12.338	30.813	33.924	36.781	40.289	42.796
23	9.260	11.688	13.091	32.007	35.172	38.076	41.638	44.181
24	9.886	12.401	13.848	33.196	36.415	39.364	42.980	45.558
25	10.520	13.120	14.611	34.382	37.652	40.646	44.314	46.928
26	11.160	13.844	15.379	35.563	38.885	41.923	45.642	48.290
27	11.808	14.573	16.151	36.741	40.113	43.194	46.963	49.645
28	12.461	15.308	16.928	37.916	41.337	44.461	48.278	50.993
29	13.121	16.047	17.708	39.087	42.557	45.722	49.588	52.336
30	13.787	16.791	18.493	40.256	43.773	46.979	50.892	53.672
35	17.192	20.569	22.465	46.059	49.802	53.203	57.342	60.275
40	20.707	24.433	26.509	51.805	55.758	59.342	63.691	66.766
45	24.311	28.366	30.612	57.505	61.656	65.410	69.957	73.166
50	27.991	32.357	34.764	63.167	67.505	71.420	76.154	79.490
60	35.535	40.482	43.188	74.397	79.082	83.298	88.379	91.952
70	43.275	48.758	51.739	85.527	90.531	95.023	100.425	104.215
80	51.172	57.153	60.391	96.578	101.879	106.629	112.329	116.321
90	59.196	65.647	69.126	107.565	113.145	118.136	124.116	128.299
100	67.328	74.222	77.929	118.498	124.342	129.561	135.807	140.169

Source: A. Hald and S. A. Sinkbaek, "A Table of Percentage Points of the χ^2 Distribution," *Skandinavisk Aktuarietidskrift,* 33 (1950), 168–175. Used by permission.

Table H Percentiles of the *F* distribution

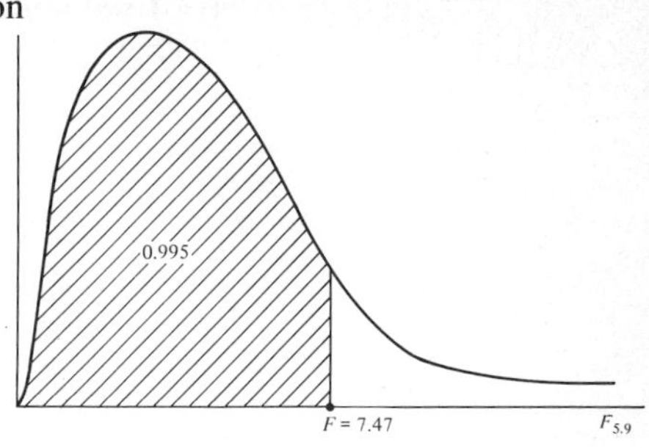

$F_{0.995}$

Denominator degrees of freedom	1	2	3	4	5	6	7	8	9
1	16211	20000	21615	22500	23056	23437	23715	23925	24091
2	198.5	199.0	199.2	199.2	199.3	199.3	199.4	199.4	199.4
3	55.55	49.80	47.47	46.19	45.39	44.84	44.43	44.13	43.88
4	31.33	26.28	24.26	23.15	22.46	21.97	21.62	21.35	21.14
5	22.78	18.31	16.53	15.56	14.94	14.51	14.20	13.96	13.77
6	18.63	14.54	12.92	12.03	11.46	11.07	10.79	10.57	10.39
7	16.24	12.40	10.88	10.05	9.52	9.16	8.89	8.68	8.51
8	14.69	11.04	9.60	8.81	8.30	7.95	7.69	7.50	7.34
9	13.61	10.11	8.72	7.96	7.47	7.13	6.88	6.69	6.54
10	12.83	9.43	8.08	7.34	6.87	6.54	6.30	6.12	5.97
11	12.23	8.91	7.60	6.88	6.42	6.10	5.86	5.68	5.54
12	11.75	8.51	7.23	6.52	6.07	5.76	5.52	5.35	5.20
13	11.37	8.19	6.93	6.23	5.79	5.48	5.25	5.08	4.94
14	11.06	7.92	6.68	6.00	5.56	5.26	5.03	4.86	4.72
15	10.80	7.70	6.48	5.80	5.37	5.07	4.85	4.67	4.54
16	10.58	7.51	6.30	5.64	5.21	4.91	4.69	4.52	4.38
17	10.38	7.35	6.16	5.50	5.07	4.78	4.56	4.39	4.25
18	10.22	7.21	6.03	5.37	4.96	4.66	4.44	4.28	4.14
19	10.07	7.09	5.92	5.27	4.85	4.56	4.34	4.18	4.04
20	9.94	6.99	5.82	5.17	4.76	4.47	4.26	4.09	3.96
21	9.83	6.89	5.73	5.09	4.68	4.39	4.18	4.01	3.88
22	9.73	6.81	5.65	5.02	4.61	4.32	4.11	3.94	3.81
23	9.63	6.73	5.58	4.95	4.54	4.26	4.05	3.88	3.75
24	9.55	6.66	5.52	4.89	4.49	4.20	3.99	3.83	3.69
25	9.48	6.60	5.46	4.84	4.43	4.15	3.94	3.78	3.64
26	9.41	6.54	5.41	4.79	4.38	4.10	3.89	3.73	3.60
27	9.34	6.49	5.36	4.74	4.34	4.06	3.85	3.69	3.56
28	9.28	6.44	5.32	4.70	4.30	4.02	3.81	3.65	3.52
29	9.23	6.40	5.28	4.66	4.26	3.98	3.77	3.61	3.48
30	9.18	6.35	5.24	4.62	4.23	3.95	3.74	3.58	3.45
40	8.83	6.07	4.98	4.37	3.99	3.71	3.51	3.35	3.22
60	8.49	5.79	4.73	4.14	3.76	3.49	3.29	3.13	3.01
120	8.18	5.54	4.50	3.92	3.55	3.28	3.09	2.93	2.81
∞	7.88	5.30	4.28	3.72	3.35	3.09	2.90	2.74	2.62

(Numerator degrees of freedom across columns.)

$F_{0.995}$

Denominator degrees of freedom	10	12	15	20	24	30	40	60	120	∞
1	24224	24426	24630	24836	24940	25044	25148	25253	25359	25465
2	199.4	199.4	199.4	199.4	199.5	199.5	199.5	199.5	199.5	199.5
3	43.69	43.39	43.08	42.78	42.62	42.47	42.31	42.15	41.99	41.83
4	20.97	20.70	20.44	20.17	20.03	19.89	19.75	19.61	19.47	19.32

(Numerator degrees of freedom across columns.)

Source: E. S. Pearson and H. O. Hartley, editors, *Biometrika Tables for Statisticians,* third edition, Volume 1, London: The Syndics of The Cambridge University Press, 1970. Used by permission of E. S. Pearson for the Biometrika Trustees.

Table H Percentiles of the F distribution (continued)

$F_{0.995}$

Denominator degrees of freedom	Numerator degrees of freedom 10	12	15	20	24	30	40	60	120	∞
5	13.62	13.38	13.15	12.90	12.78	12.66	12.53	12.40	12.27	12.14
6	10.25	10.03	9.81	9.59	9.47	9.36	9.24	9.12	9.00	8.88
7	8.38	8.18	7.97	7.75	7.65	7.53	7.42	7.31	7.19	7.08
8	7.21	7.01	6.81	6.61	6.50	6.40	6.29	6.18	6.06	5.95
9	6.42	6.23	6.03	5.83	5.73	5.62	5.52	5.41	5.30	5.19
10	5.85	5.66	5.47	5.27	5.17	5.07	4.97	4.86	4.75	4.64
11	5.42	5.24	5.05	4.86	4.76	4.65	4.55	4.44	4.34	4.23
12	5.09	4.91	4.72	4.53	4.43	4.33	4.23	4.12	4.01	3.90
13	4.82	4.64	4.46	4.27	4.17	4.07	3.97	3.87	3.76	3.65
14	4.60	4.43	4.25	4.06	3.96	3.86	3.76	3.66	3.55	3.44
15	4.42	4.25	4.07	3.88	3.79	3.69	3.58	3.48	3.37	3.26
16	4.27	4.10	3.92	3.73	3.64	3.54	3.44	3.33	3.22	3.11
17	4.14	3.97	3.79	3.61	3.51	3.41	3.31	3.21	3.10	2.98
18	4.03	3.86	3.68	3.50	3.40	3.30	3.20	3.10	2.99	2.87
19	3.93	3.76	3.59	3.40	3.31	3.21	3.11	3.00	2.89	2.78
20	3.85	3.68	3.50	3.32	3.22	3.12	3.02	2.92	2.81	2.69
21	3.77	3.60	3.43	3.24	3.15	3.05	2.95	2.84	2.73	2.61
22	3.70	3.54	3.36	3.18	3.08	2.98	2.88	2.77	2.66	2.55
23	3.64	3.47	3.30	3.12	3.02	2.92	2.82	2.71	2.60	2.48
24	3.59	3.42	3.25	3.06	2.97	2.87	2.77	2.66	2.55	2.43
25	3.54	3.37	3.20	3.01	2.92	2.82	2.72	2.61	2.50	2.38
26	3.49	3.33	3.15	2.97	2.87	2.77	2.67	2.56	2.45	2.33
27	3.45	3.28	3.11	2.93	2.83	2.73	2.63	2.52	2.41	2.29
28	3.41	3.25	3.07	2.89	2.79	2.69	2.59	2.48	2.37	2.25
29	3.38	3.21	3.04	2.86	2.76	2.66	2.56	2.45	2.33	2.21
30	3.34	3.18	3.01	2.82	2.73	2.63	2.52	2.42	2.30	2.18
40	3.12	2.95	2.78	2.60	2.50	2.40	2.30	2.18	2.06	1.93
60	2.90	2.74	2.57	2.39	2.29	2.19	2.08	1.96	1.83	1.69
120	2.71	2.54	2.37	2.19	2.09	1.98	1.87	1.75	1.61	1.43
∞	2.52	2.36	2.19	2.00	1.90	1.79	1.67	1.53	1.36	1.00

$F_{0.99}$

Denominator degrees of freedom	Numerator degrees of freedom 1	2	3	4	5	6	7	8	9
1	4052	4999.5	5403	5625	5764	5859	5928	5981	6022
2	98.50	99.00	99.17	99.25	99.30	99.33	99.36	99.37	99.39
3	34.12	30.82	29.46	28.71	28.24	27.91	27.67	27.49	27.35
4	21.20	18.00	16.69	15.98	15.52	15.21	14.98	14.80	14.66
5	16.26	13.27	12.06	11.39	10.97	10.67	10.46	10.29	10.16
6	13.75	10.92	9.78	9.15	8.75	8.47	8.26	8.10	7.98
7	12.25	9.55	8.45	7.85	7.46	7.19	6.99	6.84	6.72
8	11.26	8.65	7.59	7.01	6.63	6.37	6.18	6.03	5.91
9	10.56	8.02	6.99	6.42	6.06	5.80	5.61	5.47	5.35

Table H Percentiles of the F distribution (continued)

$F_{0.99}$

Denominator degrees of freedom	Numerator degrees of freedom 1	2	3	4	5	6	7	8	9
10	10.04	7.56	6.55	5.99	5.64	5.39	5.20	5.06	4.94
11	9.65	7.21	6.22	5.67	5.32	5.07	4.89	4.74	4.63
12	9.33	6.93	5.95	5.41	5.06	4.82	4.64	4.50	4.39
13	9.07	6.70	5.74	5.21	4.86	4.62	4.44	4.30	4.19
14	8.86	6.51	5.56	5.04	4.69	4.46	4.28	4.14	4.03
15	8.68	6.36	5.42	4.89	4.56	4.32	4.14	4.00	3.89
16	8.53	6.23	5.29	4.77	4.44	4.20	4.03	3.89	3.78
17	8.40	6.11	5.18	4.67	4.34	4.10	3.93	3.79	3.68
18	8.29	6.01	5.09	4.58	4.25	4.01	3.84	3.71	3.60
19	8.18	5.93	5.01	4.50	4.17	3.94	3.77	3.63	3.52
20	8.10	5.85	4.94	4.43	4.10	3.87	3.70	3.56	3.46
21	8.02	5.78	4.87	4.37	4.04	3.81	3.64	3.51	3.40
22	7.95	5.72	4.82	4.31	3.99	3.76	3.59	3.45	3.35
23	7.88	5.66	4.76	4.26	3.94	3.71	3.54	3.41	3.30
24	7.82	5.61	4.72	4.22	3.90	3.67	3.50	3.36	3.26
25	7.77	5.57	4.68	4.18	3.85	3.63	3.46	3.32	3.22
26	7.72	5.53	4.64	4.14	3.82	3.59	3.42	3.29	3.18
27	7.68	5.49	4.60	4.11	3.78	3.56	3.39	3.26	3.15
28	7.64	5.45	4.57	4.07	3.75	3.53	3.36	3.23	3.12
29	7.60	5.42	4.54	4.04	3.73	3.50	3.33	3.20	3.09
30	7.56	5.39	4.51	4.02	3.70	3.47	3.30	3.17	3.07
40	7.31	5.18	4.31	3.83	3.51	3.29	3.12	2.99	2.89
60	7.08	4.98	4.13	3.65	3.34	3.12	2.95	2.82	2.72
120	6.85	4.79	3.95	3.48	3.17	2.96	2.79	2.66	2.56
∞	6.63	4.61	3.78	3.32	3.02	2.80	2.64	2.51	2.41

$F_{0.99}$

Denominator degrees of freedom	Numerator degrees of freedom 10	12	15	20	24	30	40	60	120	∞
1	6056	6106	6157	6209	6235	6261	6287	6313	6339	6366
2	99.40	99.42	99.43	99.45	99.46	99.47	99.47	99.48	99.49	99.50
3	27.23	27.05	26.87	26.69	26.60	26.50	26.41	26.32	26.22	26.13
4	14.55	14.37	14.20	14.02	13.93	13.84	13.75	13.65	13.56	13.46
5	10.05	9.89	9.72	9.55	9.47	9.38	9.29	9.20	9.11	9.02
6	7.87	7.72	7.56	7.40	7.31	7.23	7.14	7.06	6.97	6.88
7	6.62	6.47	6.31	6.16	6.07	5.99	5.91	5.82	5.74	5.65
8	5.81	5.67	5.52	5.36	5.28	5.20	5.12	5.03	4.95	4.86
9	5.26	5.11	4.96	4.81	4.73	4.65	4.57	4.48	4.40	4.31
10	4.85	4.71	4.56	4.41	4.33	4.25	4.17	4.08	4.00	3.91
11	4.54	4.40	4.25	4.10	4.02	3.94	3.86	3.78	3.69	3.60
12	4.30	4.16	4.01	3.86	3.78	3.70	3.62	3.54	3.45	3.36
13	4.10	3.96	3.82	3.66	3.59	3.51	3.43	3.34	3.25	3.17
14	3.94	3.80	3.66	3.51	3.43	3.35	3.27	3.18	3.09	3.00

Table H Percentiles of the F distribution (continued)

$F_{0.99}$

Denominator degrees of freedom	Numerator degrees of freedom 10	12	15	20	24	30	40	60	120	∞
15	3.80	3.67	3.52	3.37	3.29	3.21	3.13	3.05	2.96	2.87
16	3.69	3.55	3.41	3.26	3.18	3.10	3.02	2.93	2.84	2.75
17	3.59	3.46	3.31	3.16	3.08	3.00	2.92	2.83	2.75	2.65
18	3.51	3.37	3.23	3.08	3.00	2.92	2.84	2.75	2.66	2.57
19	3.43	3.30	3.15	3.00	2.92	2.84	2.76	2.67	2.58	2.49
20	3.37	3.23	3.09	2.94	2.86	2.78	2.69	2.61	2.52	2.42
21	3.31	3.17	3.03	2.88	2.80	2.72	2.64	2.55	2.46	2.36
22	3.26	3.12	2.98	2.83	2.75	2.67	2.58	2.50	2.40	2.31
23	3.21	3.07	2.93	2.78	2.70	2.62	2.54	2.45	2.35	2.26
24	3.17	3.03	2.89	2.74	2.66	2.58	2.49	2.40	2.31	2.21
25	3.13	2.99	2.85	2.70	2.62	2.54	2.45	2.36	2.27	2.17
26	3.09	2.96	2.81	2.66	2.58	2.50	2.42	2.33	2.23	2.13
27	3.06	2.93	2.78	2.63	2.55	2.47	2.38	2.29	2.20	2.10
28	3.03	2.90	2.75	2.60	2.52	2.44	2.35	2.26	2.17	2.06
29	3.00	2.87	2.73	2.57	2.49	2.41	2.33	2.23	2.14	2.03
30	2.98	2.84	2.70	2.55	2.47	2.39	2.30	2.21	2.11	2.01
40	2.80	2.66	2.52	2.37	2.29	2.20	2.11	2.02	1.92	1.80
60	2.63	2.50	2.35	2.20	2.12	2.03	1.94	1.84	1.73	1.60
120	2.47	2.34	2.19	2.03	1.95	1.86	1.76	1.66	1.53	1.38
∞	2.32	2.18	2.04	1.88	1.79	1.70	1.59	1.47	1.32	1.00

$F_{0.975}$

Denominator degrees of freedom	Numerator degrees of freedom 1	2	3	4	5	6	7	8	9
1	647.8	799.5	864.2	899.6	921.8	937.1	948.2	956.7	963.3
2	38.51	39.00	39.17	39.25	39.30	39.33	39.36	39.37	39.39
3	17.44	16.04	15.44	15.10	14.88	14.73	14.62	14.54	14.47
4	12.22	10.65	9.98	9.60	9.36	9.20	9.07	8.98	8.90
5	10.01	8.43	7.76	7.39	7.15	6.98	6.85	6.76	6.68
6	8.81	7.26	6.60	6.23	5.99	5.82	5.70	5.60	5.52
7	8.07	6.54	5.89	5.52	5.29	5.12	4.99	4.90	4.82
8	7.57	6.06	5.42	5.05	4.82	4.65	4.53	4.43	4.36
9	7.21	5.71	5.08	4.72	4.48	4.32	4.20	4.10	4.03
10	6.94	5.46	4.83	4.47	4.24	4.07	3.95	3.85	3.78
11	6.72	5.26	4.63	4.28	4.04	3.88	3.76	3.66	3.59
12	6.55	5.10	4.47	4.12	3.89	3.73	3.61	3.51	3.44
13	6.41	4.97	4.35	4.00	3.77	3.60	3.48	3.39	3.31
14	6.30	4.86	4.24	3.89	3.66	3.50	3.38	3.29	3.21
15	6.20	4.77	4.15	3.80	3.58	3.41	3.29	3.20	3.12
16	6.12	4.69	4.08	3.73	3.50	3.34	3.22	3.12	3.05
17	6.04	4.62	4.01	3.66	3.44	3.28	3.16	3.06	2.98
18	5.98	4.56	3.95	3.61	3.38	3.22	3.10	3.01	2.93
19	5.92	4.51	3.90	3.56	3.33	3.17	3.05	2.96	2.88

Table H Percentiles of the F distribution (continued)

$F_{0.975}$

Denominator degrees of freedom	Numerator degrees of freedom 1	2	3	4	5	6	7	8	9
20	5.87	4.46	3.86	3.51	3.29	3.13	3.01	2.91	2.84
21	5.83	4.42	3.82	3.48	3.25	3.09	2.97	2.87	2.80
22	5.79	4.38	3.78	3.44	3.22	3.05	2.93	2.84	2.76
23	5.75	4.35	3.75	3.41	3.18	3.02	2.90	2.81	2.73
24	5.72	4.32	3.72	3.38	3.15	2.99	2.87	2.78	2.70
25	5.69	4.29	3.69	3.35	3.13	2.97	2.85	2.75	2.68
26	5.66	4.27	3.67	3.33	3.10	2.94	2.82	2.73	2.65
27	5.63	4.24	3.65	3.31	3.08	2.92	2.80	2.71	2.63
28	5.61	4.22	3.63	3.29	3.06	2.90	2.78	2.69	2.61
29	5.59	4.20	3.61	3.27	3.04	2.88	2.76	2.67	2.59
30	5.57	4.18	3.59	3.25	3.03	2.87	2.75	2.65	2.57
40	5.42	4.05	3.46	3.13	2.90	2.74	2.62	2.53	2.45
60	5.29	3.93	3.34	3.01	2.79	2.63	2.51	2.41	2.33
120	5.15	3.80	3.23	2.89	2.67	2.52	2.39	2.30	2.22
∞	5.02	3.69	3.12	2.79	2.57	2.41	2.29	2.19	2.11

$F_{0.975}$

Denominator degrees of freedom	Numerator degrees of freedom 10	12	15	20	24	30	40	60	120	∞
1	968.6	976.7	984.9	993.1	997.2	1001	1006	1010	1014	1018
2	39.40	39.41	39.43	39.45	39.46	39.46	39.47	39.48	39.49	39.50
3	14.42	14.34	14.25	14.17	14.12	14.08	14.04	13.99	13.95	13.90
4	8.84	8.75	8.66	8.56	8.51	8.46	8.41	8.36	8.31	8.26
5	6.62	6.52	6.43	6.33	6.28	6.23	6.18	6.12	6.07	6.02
6	5.46	5.37	5.27	5.17	5.12	5.07	5,01	4.96	4.90	4.85
7	4.76	4.67	4.57	4.47	4.42	4.36	4.31	4.25	4.20	4.14
8	4.30	4.20	4.10	4.00	3.95	3.89	3.84	3.78	3.73	3.67
9	3.96	3.87	3.77	3.67	3.61	3.56	3.51	3.45	3.39	3.33
10	3.72	3.62	3.52	3.42	3.37	3.31	3.26	3.20	3.14	3.08
11	3.53	3.43	3.33	3.23	3.17	3.12	3.06	3.00	2.94	2.88
12	3.37	3.28	3.18	3.07	3.02	2.96	2.91	2.85	2.79	2.72
13	3.25	3.15	3.05	2.95	2.89	2.84	2.78	2.72	2.66	2.60
14	3.15	3.05	2.95	2.84	2.79	2.73	2.67	2.61	2.55	2.49
15	3.06	2.96	2.86	2.76	2.70	2.64	2.59	2.52	2.46	2.40
16	2.99	2.89	2.79	2.68	2.63	2.57	2.51	2.45	2.38	2.32
17	2.92	2.82	2.72	2.62	2.56	2.50	2.44	2.38	2.32	2.25
18	2.87	2.77	2.67	2.56	2.50	2.44	2.38	2.32	2.26	2.19
19	2.82	2.72	2.62	2.51	2.45	2.39	2.33	2.27	2.20	2.13
20	2.77	2.68	2.57	2.46	2.41	2.35	2.29	2.22	2.16	2.09
21	2.73	2.64	2.53	2.42	2.37	2.31	2.25	2.18	2.11	2.04
22	2.70	2.60	2.50	2.39	2.33	2.27	2.21	2.14	2.08	2.00
23	2.67	2.57	2.47	2.36	2.30	2.24	2.18	2.11	2.04	1.97
24	2.64	2.54	2.44	2.33	2.27	2.21	2.15	2.08	2.01	1.94

Table H Percentiles of the F distribution (continued)

$F_{0.975}$

Denominator degrees of freedom	Numerator degrees of freedom 10	12	15	20	24	30	40	60	120	∞
25	2.61	2.51	2.41	2.30	2.24	2.18	2.12	2.05	1.98	1.91
26	2.59	2.49	2.39	2.28	2.22	2.16	2.09	2.03	1.95	1.88
27	2.57	2.47	2.36	2.25	2.19	2.13	2.07	2.00	1.93	1.85
28	2.55	2.45	2.34	2.23	2.17	2.11	2.05	1.98	1.91	1.83
29	2.53	2.43	2.32	2.21	2.15	2.09	2.03	1.96	1.89	1.81
30	2.51	2.41	2.31	2.20	2.14	2.07	2.01	1.94	1.87	1.79
40	2.39	2.29	2.18	2.07	2.01	1.94	1.88	1.80	1.72	1.64
60	2.27	2.17	2.06	1.94	1.88	1.82	1.74	1.67	1.58	1.48
120	2.16	2.05	1.94	1.82	1.76	1.69	1.61	1.53	1.43	1.31
∞	2.05	1.94	1.83	1.71	1.64	1.57	1.48	1.39	1.27	1.00

$F_{0.95}$

Denominator degrees of freedom	Numerator degrees of freedom 1	2	3	4	5	6	7	8	9
1	161.4	199.5	215.7	224.6	230.2	234.0	236.8	238.9	240.5
2	18.51	19.00	19.16	19.25	19.30	19.33	19.35	19.37	19.38
3	10.13	9.55	9.28	9.12	9.01	8.94	8.89	8.85	8.81
4	7.71	6.94	6.59	6.39	6.26	6.16	6.09	6.04	6.00
5	6.61	5.79	5.41	5.19	5.05	4.95	4.88	4.82	4.77
6	5.99	5.14	4.76	4.53	4.39	4.28	4.21	4.15	4.10
7	5.59	4.74	4.35	4.12	3.97	3.87	3.79	3.73	3.68
8	5.32	4.46	4.07	3.84	3.69	3.58	3.50	3.44	3.39
9	5.12	4.26	3.86	3.63	3.48	3.37	3.29	3.23	3.18
10	4.96	4.10	3.71	3.48	3.33	3.22	3.14	3.07	3.02
11	4.84	3.98	3.59	3.36	3.20	3.09	3.01	2.95	2.90
12	4.75	3.89	3.49	3.26	3.11	3.00	2.91	2.85	2.80
13	4.67	3.81	3.41	3.18	3.03	2.92	2.83	2.77	2.71
14	4.60	3.74	3.34	3.11	2.96	2.85	2.76	2.70	2.65
15	4.54	3.68	3.29	3.06	2.90	2.79	2.71	2.64	2.59
16	4.49	3.63	3.24	3.01	2.85	2.74	2.66	2.59	2.54
17	4.45	3.59	3.20	2.96	2.81	2.70	2.61	2.55	2.49
18	4.41	3.55	3.16	2.93	2.77	2.66	2.58	2.51	2.46
19	4.38	3.52	3.13	2.90	2.74	2.63	2.54	2.48	2.42
20	4.35	3.49	3.10	2.87	2.71	2.60	2.51	2.45	2.39
21	4.32	3.47	3.07	2.84	2.68	2.57	2.49	2.42	2.37
22	4.30	3.44	3.05	2.82	2.66	2.55	2.46	2.40	2.34
23	4.28	3.42	3.03	2.80	2.64	2.53	2.44	2.37	2.32
24	4.26	3.40	3.01	2.78	2.62	2.51	2.42	2.36	2.30
25	4.24	3.39	2.99	2.76	2.60	2.49	2.40	2.34	2.28
26	4.23	3.37	2.98	2.74	2.59	2.47	2.39	2.32	2.27
27	4.21	3.35	2.96	2.73	2.57	2.46	2.37	2.31	2.25
28	4.20	3.34	2.95	2.71	2.56	2.45	2.36	2.29	2.24
29	4.18	3.33	2.93	2.70	2.55	2.43	2.35	2.28	2.22

Table H Percentiles of the F distribution (continued)

$F_{0.95}$

Denominator degrees of freedom	Numerator degrees of freedom 1	2	3	4	5	6	7	8	9
30	4.17	3.32	2.92	2.69	2.53	2.42	2.33	2.27	2.21
40	4.08	3.23	2.84	2.61	2.45	2.34	2.25	2.18	2.12
60	4.00	3.15	2.76	2.53	2.37	2.25	2.17	2.10	2.04
120	3.92	3.07	2.68	2.45	2.29	2.17	2.09	2.02	1.96
∞	3.84	3.00	2.60	2.37	2.21	2.10	2.01	1.94	1.88

$F_{0.95}$

Denominator degrees of freedom	Numerator degrees of freedom 10	12	15	20	24	30	40	60	120	∞
1	241.9	243.9	245.9	248.0	249.1	250.1	251.1	252.2	253.3	254.3
2	19.40	19.41	19.43	19.45	19.45	19.46	19.47	19.48	19.49	19.50
3	8.79	8.74	8.70	8.66	8.64	8.62	8.59	8.57	8.55	8.53
4	5.96	5.91	5.86	5.80	5.77	5.75	5.72	5.69	5.66	5.63
5	4.74	4.68	4.62	4.56	4.53	4.50	4.46	4.43	4.40	4.36
6	4.06	4.00	3.94	3.87	3.84	3.81	3.77	3.74	3.70	3.67
7	3.64	3.57	3.51	3.44	3.41	3.38	3.34	3.30	3.27	3.23
8	3.35	3.28	3.22	3.15	3.12	3.08	3.04	3.01	2.97	2.93
9	3.14	3.07	3.01	2.94	2.90	2.86	2.83	2.79	2.75	2.71
10	2.98	2.91	2.85	2.77	2.74	2.70	2.66	2.62	2.58	2.54
11	2.85	2.79	2.72	2.65	2.61	2.57	2.53	2.49	2.45	2.40
12	2.75	2.69	2.62	2.54	2.51	2.47	2.43	2.38	2.34	2.30
13	2.67	2.60	2.53	2.46	2.42	2.38	2.34	2.30	2.25	2.21
14	2.60	2.53	2.46	2.39	2.35	2.31	2.27	2.22	2.18	2.13
15	2.54	2.48	2.40	2.33	2.29	2.25	2.20	2.16	2.11	2.07
16	2.49	2.42	2.35	2.28	2.24	2.19	2.15	2.11	2.06	2.01
17	2.45	2.38	2.31	2.23	2.19	2.15	2.10	2.06	2.01	1.96
18	2.41	2.34	2.27	2.19	2.15	2.11	2.06	2.02	1.97	1.92
19	2.38	2.31	2.23	2.16	2.11	2.07	2.03	1.98	1.93	1.88
20	2.35	2.28	2.20	2.12	2.08	2.04	1.99	1.95	1.90	1.84
21	2.32	2.25	2.18	2.10	2.05	2.01	1.96	1.92	1.87	1.81
22	2.30	2.23	2.15	2.07	2.03	1.98	1.94	1.89	1.84	1.78
23	2.27	2.20	2.13	2.05	2.01	1.96	1.91	1.86	1.81	1.76
24	2.25	2.18	2.11	2.03	1.98	1.94	1.89	1.84	1.79	1.73
25	2.24	2.16	2.09	2.01	1.96	1.92	1.87	1.82	1.77	1.71
26	2.22	2.15	2.07	1.99	1.95	1.90	1.85	1.80	1.75	1.69
27	2.20	2.13	2.06	1.97	1.93	1.88	1.84	1.79	1.73	1.67
28	2.19	2.12	2.04	1.96	1.91	1.87	1.82	1.77	1.71	1.65
29	2.18	2.10	2.03	1.94	1.90	1.85	1.81	1.75	1.70	1.64
30	2.16	2.09	2.01	1.93	1.89	1.84	1.79	1.74	1.68	1.62
40	2.08	2.00	1.92	1.84	1.79	1.74	1.69	1.64	1.58	1.51
60	1.99	1.92	1.84	1.75	1.70	1.65	1.59	1.53	1.47	1.39
120	1.91	1.83	1.75	1.66	1.61	1.55	1.50	1.43	1.35	1.25
∞	1.83	1.75	1.67	1.57	1.52	1.46	1.39	1.32	1.22	1.00

Table H Percentiles of the F distribution (continued)

$F_{0.90}$

Denominator degrees of Freedom	Numerator degrees of freedom 1	2	3	4	5	6	7	8	9
1	39.86	49.50	53.59	55.83	57.24	58.20	58.91	59.44	59.86
2	8.53	9.00	9.16	9.24	9.29	9.33	9.35	9.37	9.38
3	5.54	5.46	5.39	5.34	5.31	5.28	5.27	5.25	5.24
4	4.54	4.32	4.19	4.11	4.05	4.01	3.98	3.95	3.94
5	4.06	3.78	3.62	3.52	3.45	3.40	3.37	3.34	3.32
6	3.78	3.46	3.29	3.18	3.11	3.05	3.01	2.98	2.96
7	3.59	3.26	3.07	2.96	2.88	2.83	2.78	2.75	2.72
8	3.46	3.11	2.92	2.81	2.73	2.67	2.62	2.59	2.56
9	3.36	3.01	2.81	2.69	2.61	2.55	2.51	2.47	2.44
10	3.29	2.92	2.73	2.61	2.52	2.46	2.41	2.38	2.35
11	3.23	2.86	2.66	2.54	2.45	2.39	2.34	2.30	2.27
12	3.18	2.81	2.61	2.48	2.39	2.33	2.28	2.24	2.21
13	3.14	2.76	2.56	2.43	2.35	2.28	2.23	2.20	2.16
14	3.10	2.73	2.52	2.39	2.31	2.24	2.19	2.15	2.12
15	3.07	2.70	2.49	2.36	2.27	2.21	2.16	2.12	2.09
16	3.05	2.67	2.46	2.33	2.24	2.18	2.13	2.09	2.06
17	3.03	2.64	2.44	2.31	2.22	2.15	2.10	2.06	2.03
18	3.01	2.62	2.42	2.29	2.20	2.13	2.08	2.04	2.00
19	2.99	2.61	2.40	2.27	2.18	2.11	2.06	2.02	1.98
20	2.97	2.59	2.38	2.25	2.16	2.09	2.04	2.00	1.96
21	2.96	2.57	2.36	2.23	2.14	2.08	2.02	1.98	1.95
22	2.95	2.56	2.35	2.22	2.13	2.06	2.01	1.97	1.93
23	2.94	2.55	2.34	2.21	2.11	2.05	1.99	1.95	1.92
24	2.93	2.54	2.33	2.19	2.10	2.04	1.98	1.94	1.91
25	2.92	2.53	2.32	2.18	2.09	2.02	1.97	1.93	1.89
26	2.91	2.52	2.31	2.17	2.08	2.01	1.96	1.92	1.88
27	2.90	2.51	2.30	2.17	2.07	2.00	1.95	1.91	1.87
28	2.89	2.50	2.29	2.16	2.06	2.00	1.94	1.90	1.87
29	2.89	2.50	2.28	2.15	2.06	1.99	1.93	1.89	1.86
30	2.88	2.49	2.28	2.14	2.05	1.98	1.93	1.88	1.85
40	2.84	2.44	2.23	2.09	2.00	1.93	1.87	1.83	1.79
60	2.79	2.39	2.18	2.04	1.95	1.87	1.82	1.77	1.74
120	2.75	2.35	2.13	1.99	1.90	1.82	1.77	1.72	1.68
∞	2.71	2.30	2.08	1.94	1.85	1.77	1.72	1.67	1.63

Table H Percentiles of the F distribution (continued)

$F_{0.90}$

Denominator degrees of freedom	10	12	15	20	24	30	40	60	120	∞
				Numerator degrees of freedom						
1	60.19	60.71	61.22	61.74	62.00	62.26	62.53	62.79	63.06	63.33
2	9.39	9.41	9.42	9.44	9.45	9.46	9.47	9.47	9.48	9.49
3	5.23	5.22	5.20	5.18	5.18	5.17	5.16	5.15	5.14	5.13
4	3.92	3.90	3.87	3.84	3.83	3.82	3.80	3.79	3.78	3.76
5	3.30	3.27	3.24	3.21	3.19	3.17	3.16	3.14	3.12	3.10
6	2.94	2.90	2.87	2.84	2.82	2.80	2.78	2.76	2.74	2.72
7	2.70	2.67	2.63	2.59	2.58	2.56	2.54	2.51	2.49	2.47
8	2.54	2.50	2.46	2.42	2.40	2.38	2.36	2.34	2.32	2.29
9	2.42	2.38	2.34	2.30	2.28	2.25	2.23	2.21	2.18	2.16
10	2.32	2.28	2.24	2.20	2.18	2.16	2.13	2.11	2.08	2.06
11	2.25	2.21	2.17	2.12	2.10	2.08	2.05	2.03	2.00	1.97
12	2.19	2.15	2.10	2.06	2.04	2.01	1.99	1.96	1.93	1.90
13	2.14	2.10	2.05	2.01	1.98	1.96	1.93	1.90	1.88	1.85
14	2.10	2.05	2.01	1.96	1.94	1.91	1.89	1.86	1.83	1.80
15	2.06	2.02	1.97	1.92	1.90	1.87	1.85	1.82	1.79	1.76
16	2.03	1.99	1.94	1.89	1.87	1.84	1.81	1.78	1.75	1.72
17	2.00	1.96	1.91	1.86	1.84	1.81	1.78	1.75	1.72	1.69
18	1.98	1.93	1.89	1.84	1.81	1.78	1.75	1.72	1.69	1.66
19	1.96	1.91	1.86	1.81	1.79	1.76	1.73	1.70	1.67	1.63
20	1.94	1.89	1.84	1.79	1.77	1.74	1.71	1.68	1.64	1.61
21	1.92	1.87	1.83	1.78	1.75	1.72	1.69	1.66	1.62	1.59
22	1.90	1.86	1.81	1.76	1.73	1.70	1.67	1.64	1.60	1.57
23	1.89	1.84	1.80	1.74	1.72	1.69	1.66	1.62	1.59	1.55
24	1.88	1.83	1.78	1.73	1.70	1.67	1.64	1.61	1.57	1.53
25	1.87	1.82	1.77	1.72	1.69	1.66	1.63	1.59	1.56	1.52
26	1.86	1.81	1.76	1.71	1.68	1.65	1.61	1.58	1.54	1.50
27	1.85	1.80	1.75	1.70	1.67	1.64	1.60	1.57	1.53	1.49
28	1.84	1.79	1.74	1.69	1.66	1.63	1.59	1.56	1.52	1.48
29	1.83	1.78	1.73	1.68	1.65	1.62	1.58	1.55	1.51	1.47
30	1.82	1.77	1.72	1.67	1.64	1.61	1.57	1.54	1.50	1.46
40	1.76	1.71	1.66	1.61	1.57	1.54	1.51	1.47	1.42	1.38
60	1.71	1.66	1.60	1.54	1.51	1.48	1.44	1.40	1.35	1.29
120	1.65	1.60	1.55	1.48	1.45	1.41	1.37	1.32	1.26	1.19
∞	1.60	1.55	1.49	1.42	1.38	1.34	1.30	1.24	1.17	1.00

Table I Confidence limits for σ^2 (entries under 5% and 1% give 95% and 99% confidence limits, respectively).

Degrees of freedom	5%		1%	
	χ_1^2	χ_2^2	χ_1^2	χ_2^2
1	0.0^231593	7.8168	0.0^313422	11.345
2	0.084727	9.5303	0.017469	13.285
3	0.29624	11.191	0.101048	15.127
4	0.60700	12.802	0.26396	16.901
5	0.98923	14.369	0.49623	18.621
6	1.4250	15.897	0.78565	20.296
7	1.9026	17.392	1.1221	21.931
8	2.4139	18.860	1.4978	23.533
9	2.9532	20.305	1.9069	25.106
10	3.5162	21.729	2.3444	26.653
11	4.0995	23.135	2.8069	28.178
12	4.7005	24.525	3.2912	29.683
13	5.3171	25.900	3.7949	31.170
14	5.9477	27.263	4.3161	32.641
15	6.5908	28.614	4.8530	34.097
16	7.2453	29.955	5.4041	35.540
17	7.9100	31.285	5.9683	36.971
18	8.5842	32.607	6.5444	38.390
19	9.2670	33.921	7.1316	39.798
20	9.9579	35.227	7.7289	41.197
21	10.656	36.525	8.3358	42.586
22	11.361	37.818	8.9515	43.967
23	12.073	39.103	9.5755	45.340
24	12.791	40.383	10.2073	46.706
25	13.514	41.658	10.846	48.064
26	14.243	42.927	11.492	49.416
27	14.977	44.192	12.145	50.761
28	15.716	45.451	12.803	52.100
29	16.459	46.707	13.468	53.434
30	17.206	47.958	14.138	54.762
31	17.958	49.205	14.813	56.085
32	18.713	50.448	15.494	57.403
33	19.472	51.688	16.179	58.716
34	20.235	52.924	16.869	60.025
35	21.001	54.157	17.563	61.330
36	21.771	55.386	18.261	62.630
37	22.543	56.613	18.964	63.927
38	23.319	57.836	19.670	65.219
39	24.097	59.057	20.380	66.508
40	24.879	60.275	21.094	67.793
41	25.663	61.490	21.811	69.075
42	26.449	62.703	22.531	70.354
43	27.238	63.913	23.255	71.629
44	28.029	65.121	23.982	72.901
45	28.823	66.327	24.712	74.170
46	29.619	67.530	25.445	75.437
47	30.417	68.731	26.181	76.700
48	31.218	69.931	26.919	77.961
49	32.020	71.128	27.660	79.220
50	32.824	72.323	28.404	80.475

Source: D. V. Lindley, D. A. East, and P. A. Hamilton, "Tables for Making Inferences About the Variance of a Normal Distribution," *Biometrika*, 47 (1960), 433–437. Used by permission of E. S. Pearson for the Biometrika Trustees.

Table I (continued)

Degrees of freedom	5% χ_1^2	5% χ_2^2	1% χ_1^2	1% χ_2^2
51	33.630	73.516	29.150	81.729
52	34.439	74.708	29.898	82.979
53	35.248	75.897	30.649	84.228
54	36.060	77.085	31.403	85.474
55	36.873	78.271	32.158	86.718
56	37.689	79.456	32.916	87.960
57	38.505	80.639	33.675	89.200
58	39.323	81.820	34.437	90.437
59	40.143	83.000	35.201	91.673
60	40.965	84.178	35.967	92.907
61	41.787	85.355	36.735	94.139
62	42.612	86.531	37.504	95.369
63	43.437	87.705	38.276	96.597
64	44.264	88.878	39.049	97.823
65	45.092	90.049	39.824	99.048
66	45.922	91.219	40.600	100.271
67	46.753	92.388	41.379	101.492
68	47.585	93.555	42.159	102.71
69	48.418	94.722	42.940	103.93
70	49.253	95.887	43.723	105.15
71	50.089	97.051	44.508	106.36
72	50.926	98.214	45.294	107.58
73	51.764	99.376	46.081	108.79
74	52.603	100.536	46.870	110.00
75	53.443	101.696	47.661	111.21
76	54.284	102.85	48.452	112.42
77	55.126	104.01	49.245	113.62
78	55.969	105.17	50.040	114.83
79	56.814	106.32	50.836	116.03
80	57.659	107.48	51.633	117.23
81	58.505	108.63	52.431	118.44
82	59.352	109.79	53.230	119.64
83	60.200	110.94	54.031	120.84
84	61.049	112.09	54.833	122.03
85	61.899	113.24	55.636	123.23
86	62.750	114.39	56.440	124.43
87	63.601	115.54	57.245	125.62
88	64.454	116.68	58.052	126.81
89	65.307	117.83	58.859	128.01
90	66.161	118.98	59.668	129.20
91	67.016	120.12	60.477	130.39
92	67.871	121.26	61.288	131.58
93	68.728	122.41	62.100	132.76
94	69.585	123.55	62.912	133.95
95	70.443	124.69	63.726	135.14
96	71.302	125.83	64.540	136.32
97	72.161	126.97	65.356	137.51
98	73.021	128.11	66.172	138.69
99	73.882	129.25	66.990	139.87
100	74.744	130.39	67.808	141.05

Table J Percentage points of the Studentized range

Upper 5% points

Error df	2	3	4	5	6	7	8	9	10
1	17.97	26.98	32.82	37.08	40.41	43.12	45.40	47.36	49.07
2	6.08	8.33	9.80	10.88	11.74	12.44	13.03	13.54	13.99
3	4.50	5.91	6.82	7.50	8.04	8.48	8.85	9.18	9.46
4	3.93	5.04	5.76	6.29	6.71	7.05	7.35	7.60	7.83
5	3.64	4.60	5.22	5.67	6.03	6.33	6.58	6.80	6.99
6	3.46	4.34	4.90	5.30	5.63	5.90	6.12	6.32	6.49
7	3.34	4.16	4.68	5.06	5.36	5.61	5.82	6.00	6.16
8	3.26	4.04	4.53	4.89	5.17	5.40	5.60	5.77	5.92
9	3.20	3.95	4.41	4.76	5.02	5.24	5.43	5.59	5.74
10	3.15	3.88	4.33	4.65	4.91	5.12	5.30	5.46	5.60
11	3.11	3.82	4.26	4.57	4.82	5.03	5.20	5.35	5.49
12	3.08	3.77	4.20	4.51	4.75	4.95	5.12	5.27	5.39
13	3.06	3.73	4.15	4.45	4.69	4.88	5.05	5.19	5.32
14	3.03	3.70	4.11	4.41	4.64	4.83	4.99	5.13	5.25
15	3.01	3.67	4.08	4.37	4.59	4.78	4.94	5.08	5.20
16	3.00	3.65	4.05	4.33	4.56	4.74	4.90	5.03	5.15
17	2.98	3.63	4.02	4.30	4.52	4.70	4.86	4.99	5.11
18	2.97	3.61	4.00	4.28	4.49	4.67	4.82	4.96	5.07
19	2.96	3.59	3.98	4.25	4.47	4.65	4.79	4.92	5.04
20	2.95	3.58	3.96	4.23	4.45	4.62	4.77	4.90	5.01
24	2.92	3.53	3.90	4.17	4.37	4.54	4.68	4.81	4.92
30	2.89	3.49	3.85	4.10	4.30	4.46	4.60	4.72	4.82
40	2.86	3.44	3.79	4.04	4.23	4.39	4.52	4.63	4.73
60	2.83	3.40	3.74	3.98	4.16	4.31	4.44	4.55	4.65
120	2.80	3.36	3.68	3.92	4.10	4.24	4.36	4.47	4.56
∞	2.77	3.31	3.63	3.86	4.03	4.17	4.29	4.39	4.47

Error df	11	12	13	14	15	16	17	18	19	20
1	50.59	51.96	53.20	54.33	55.36	56.32	57.22	58.04	58.83	59.56
2	14.39	14.75	15.08	15.38	15.65	15.91	16.14	16.37	16.57	16.77
3	9.72	9.95	10.15	10.35	10.52	10.69	10.84	10.98	11.11	11.24
4	8.03	8.21	8.37	8.52	8.66	8.79	8.91	9.03	9.13	9.23
5	7.17	7.32	7.47	7.60	7.72	7.83	7.93	8.03	8.12	8.21
6	6.65	6.79	6.92	7.03	7.14	7.24	7.34	7.43	7.51	7.59
7	6.30	6.43	6.55	6.66	6.76	6.85	6.94	7.02	7.10	7.17
8	6.05	6.18	6.29	6.39	6.48	6.57	6.65	6.73	6.80	6.87
9	5.87	5.98	6.09	6.19	6.28	6.36	6.44	6.51	6.58	6.64
10	5.72	5.83	5.93	6.03	6.11	6.19	6.27	6.34	6.40	6.47
11	5.61	5.71	5.81	5.90	5.98	6.06	6.13	6.20	6.27	6.33
12	5.51	5.61	5.71	5.80	5.88	5.95	6.02	6.09	6.15	6.21
13	5.43	5.53	5.63	5.71	5.79	5.86	5.93	5.99	6.05	6.11
14	5.36	5.46	5.55	5.64	5.71	5.79	5.85	5.91	5.97	6.03
15	5.31	5.40	5.49	5.57	5.65	5.72	5.78	5.85	5.90	5.96
16	5.26	5.35	5.44	5.52	5.59	5.66	5.73	5.79	5.84	5.90
17	5.21	5.31	5.39	5.47	5.54	5.61	5.67	5.73	5.79	5.84
18	5.17	5.27	5.35	5.43	5.50	5.57	5.63	5.69	5.74	5.79
19	5.14	5.23	5.31	5.39	5.46	5.53	5.59	5.65	5.70	5.75
20	5.11	5.20	5.28	5.36	5.43	5.49	5.55	5.61	5.66	5.71
24	5.01	5.10	5.18	5.25	5.32	5.38	5.44	5.49	5.55	5.59
30	4.92	5.00	5.08	5.15	5.21	5.27	5.33	5.38	5.43	5.47
40	4.82	4.90	4.98	5.04	5.11	5.16	5.22	5.27	5.31	5.36
60	4.73	4.81	4.88	4.94	5.00	5.06	5.11	5.15	5.20	5.24
120	4.64	4.71	4.78	4.84	4.90	4.95	5.00	5.04	5.09	5.13
∞	4.55	4.62	4.68	4.74	4.80	4.85	4.89	4.93	4.97	5.01

Source: E. S. Pearson and H. O. Hartley, editors, *Biometrika Tables for Statisticians,* third edition, volume 1, London: The Syndics of the Cambridge University Press, 1970. Used by permission of E. S. Pearson for the Biometrika Trustees.

Table J (continued)

Upper 1% points

Error df	2	3	4	5	6	7	8	9	10
1	90.03	135.0	164.3	185.6	202.2	215.8	227.2	237.0	245.6
2	14.04	19.02	22.29	24.72	26.63	28.20	29.53	30.68	31.69
3	8.26	10.62	12.17	13.33	14.24	15.00	15.64	16.20	16.69
4	6.51	8.12	9.17	9.96	10.58	11.10	11.55	11.93	12.27
5	5.70	6.98	7.80	8.42	8.91	9.32	9.67	9.97	10.24
6	5.24	6.33	7.03	7.56	7.97	8.32	8.61	8.87	9.10
7	4.95	5.92	6.54	7.01	7.37	7.68	7.94	8.17	8.37
8	4.75	5.64	6.20	6.62	6.96	7.24	7.47	7.68	7.86
9	4.60	5.43	5.96	6.35	6.66	6.91	7.13	7.33	7.49
10	4.48	5.27	5.77	6.14	6.43	6.67	6.87	7.05	7.21
11	4.39	5.15	5.62	5.97	6.25	6.48	6.67	6.84	6.99
12	4.32	5.05	5.50	5.84	6.10	6.32	6.51	6.67	6.81
13	4.26	4.96	5.40	5.73	5.98	6.19	6.37	6.53	6.67
14	4.21	4.89	5.32	5.63	5.88	6.08	6.26	6.41	6.54
15	4.17	4.84	5.25	5.56	5.80	5.99	6.16	6.31	6.44
16	4.13	4.79	5.19	5.49	5.72	5.92	6.08	6.22	6.35
17	4.10	4.74	5.14	5.43	5.66	5.85	6.01	6.15	6.27
18	4.07	4.70	5.09	5.38	5.60	5.79	5.94	6.08	6.20
19	4.05	4.67	5.05	5.33	5.55	5.73	5.89	6.02	6.14
20	4.02	4.64	5.02	5.29	5.51	5.69	5.84	5.97	6.09
24	3.96	4.55	4.91	5.17	5.37	5.54	5.69	5.81	5.92
30	3.89	4.45	4.80	5.05	5.24	5.40	5.54	5.65	5.76
40	3.82	4.37	4.70	4.93	5.11	5.26	5.39	5.50	5.60
60	3.76	4.28	4.59	4.82	4.99	5.13	5.25	5.36	5.45
120	3.70	4.20	4.50	4.71	4.87	5.01	5.12	5.21	5.30
∞	3.64	4.12	4.40	4.60	4.76	4.88	4.99	5.08	5.16

Error df	11	12	13	14	15	16	17	18	19	20
1	253.2	260.0	266.2	271.8	277.0	281.8	286.3	290.4	294.3	298.0
2	32.59	33.40	34.13	34.81	35.43	36.00	36.53	37.03	37.50	37.95
3	17.13	17.53	17.89	18.22	18.52	18.81	19.07	19.32	19.55	19.77
4	12.57	12.84	13.09	13.32	13.53	13.73	13.91	14.08	14.24	14.40
5	10.48	10.70	10.89	11.08	11.24	11.40	11.55	11.68	11.81	11.93
6	9.30	9.48	9.65	9.81	9.95	10.08	10.21	10.32	10.43	10.54
7	8.55	8.71	8.86	9.00	9.12	9.24	9.35	9.46	9.55	9.65
8	8.03	8.18	8.31	8.44	8.55	8.66	8.76	8.85	8.94	9.03
9	7.65	7.78	7.91	8.03	8.13	8.23	8.33	8.41	8.49	8.57
10	7.36	7.49	7.60	7.71	7.81	7.91	7.99	8.08	8.15	8.23
11	7.13	7.25	7.36	7.46	7.56	7.65	7.73	7.81	7.88	7.95
12	6.94	7.06	7.17	7.26	7.36	7.44	7.52	7.59	7.66	7.73
13	6.79	6.90	7.01	7.10	7.19	7.27	7.35	7.42	7.48	7.55
14	6.66	6.77	6.87	6.96	7.05	7.13	7.20	7.27	7.33	7.39
15	6.55	6.66	6.76	6.84	6.93	7.00	7.07	7.14	7.20	7.26
16	6.46	6.56	6.66	6.74	6.82	6.90	6.97	7.03	7.09	7.15
17	6.38	6.48	6.57	6.66	6.73	6.81	6.87	6.94	7.00	7.05
18	6.31	6.41	6.50	6.58	6.65	6.73	6.79	6.85	6.91	6.97
19	6.25	6.34	6.43	6.51	6.58	6.65	6.72	6.78	6.84	6.89
20	6.19	6.28	6.37	6.45	6.52	6.59	6.65	6.71	6.77	6.82
24	6.02	6.11	6.19	6.26	6.33	6.39	6.45	6.51	6.56	6.61
30	5.85	5.93	6.01	6.08	6.14	6.20	6.26	6.31	6.36	6.41
40	5.69	5.76	5.83	5.90	5.96	6.02	6.07	6.12	6.16	6.21
60	5.53	5.60	5.67	5.73	5.78	5.84	5.89	5.93	5.97	6.01
120	5.37	5.44	5.50	5.56	5.61	5.66	5.71	5.75	5.79	5.83
∞	5.23	5.29	5.35	5.40	5.45	5.49	5.54	5.57	5.61	5.65

Table K Power function for analysis of variance

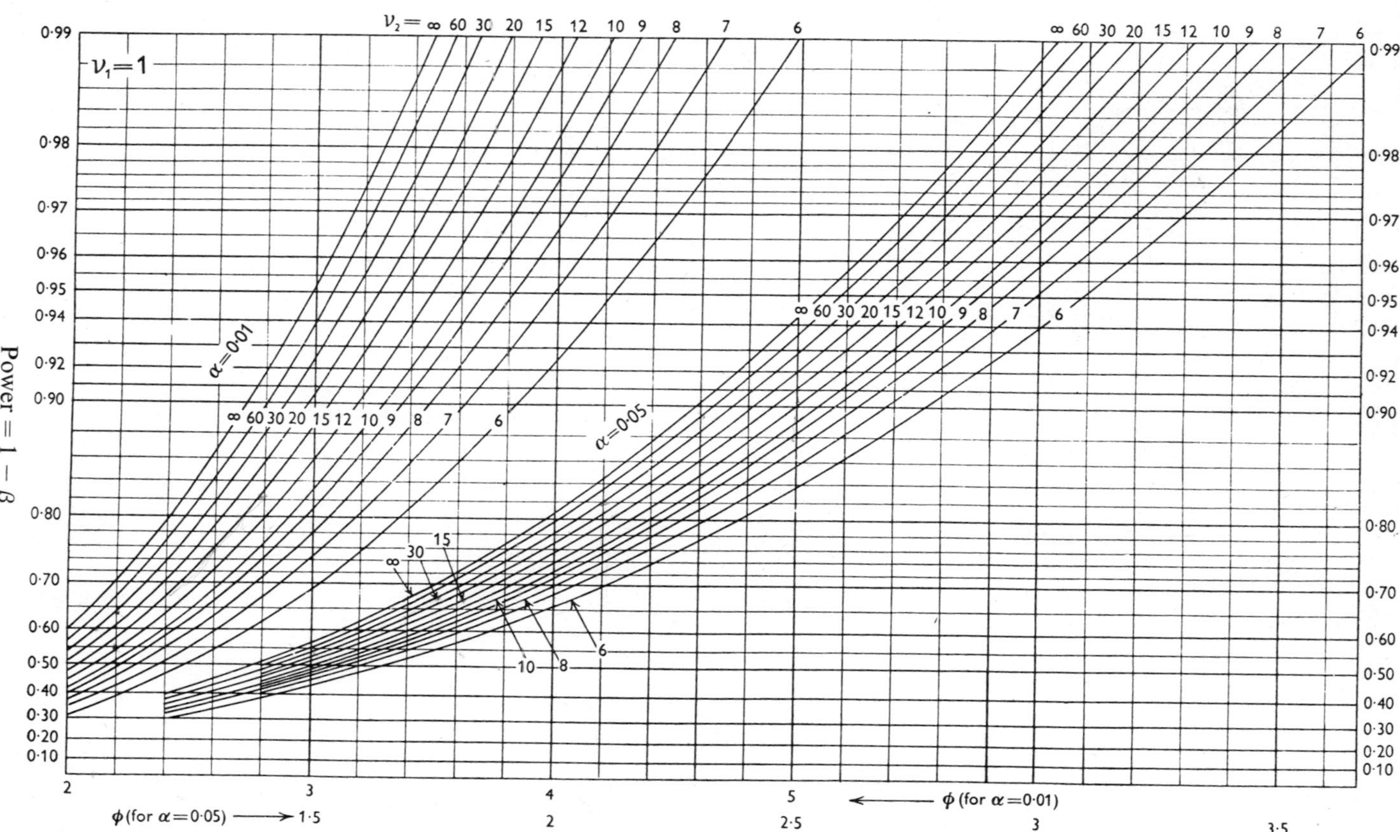

Source: E. S. Pearson and H. O. Hartley, "Charts of the Power Function for Analysis of Variance Tests, Derived from the Non-Central *F*-Distribution," *Biometrika,* 38 (1951), 112–130. Used by permission of E. S. Pearson for the Biometrika Trustees. [*Note:* According to the original Neyman and Pearson practice, the power was called β.]

Table K (continued)

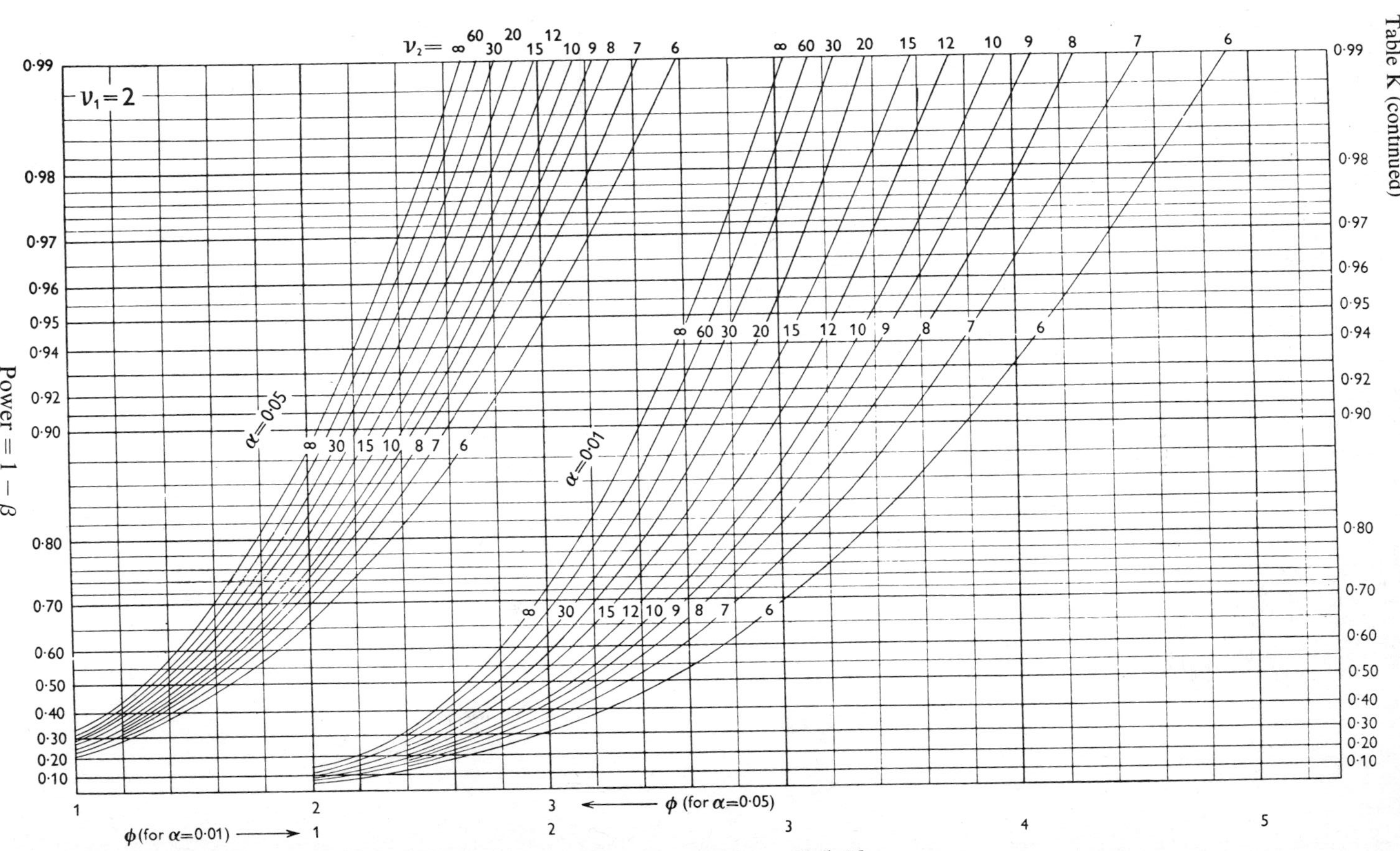

[*Note:* According to the original Neyman and Pearson practice, the power was called β.]

Table K (continued)

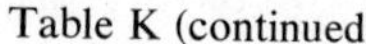

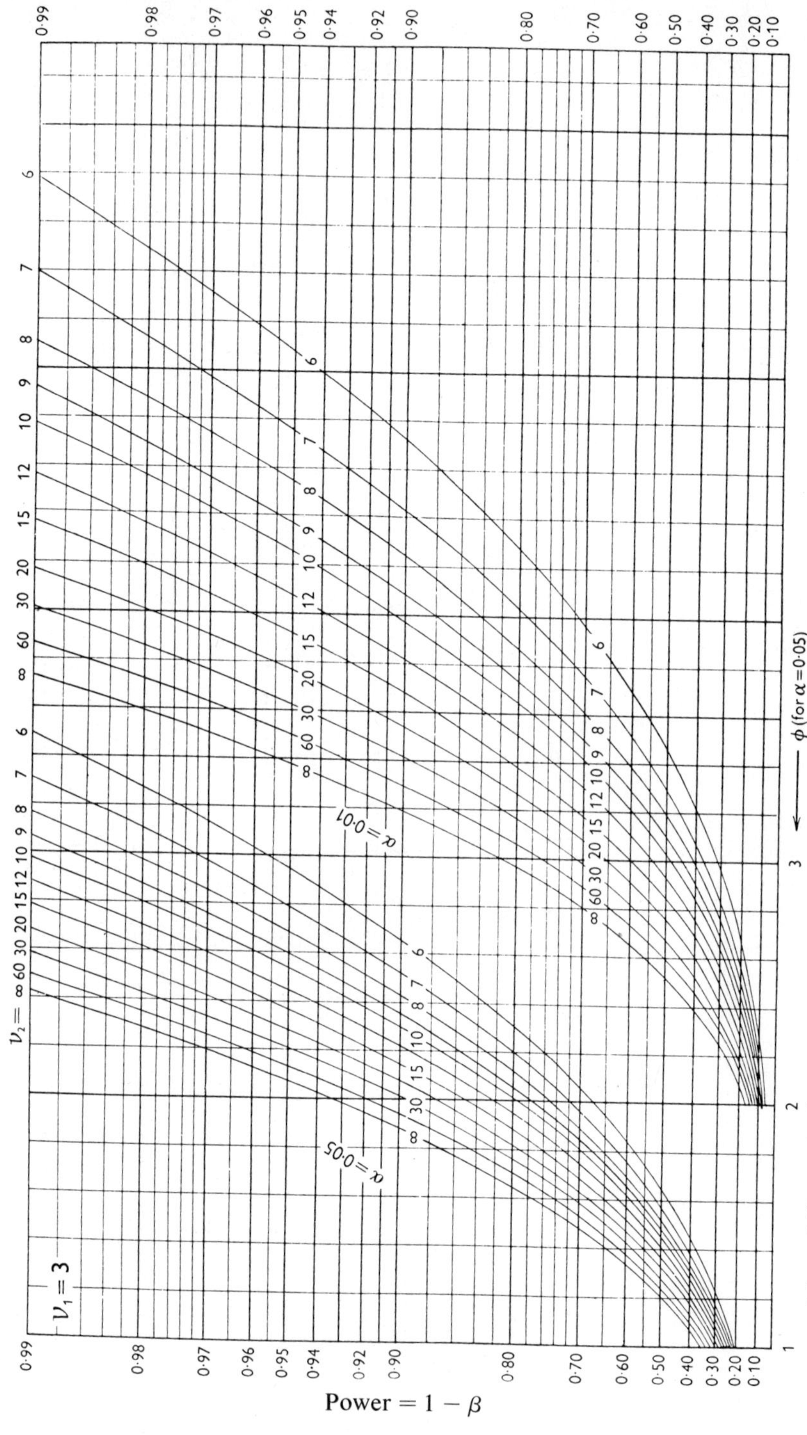

$\nu_1 = 3$

$\nu_2 = \infty$ 60 30 20 15 12 10 9 8 7 6 ∞ 60 30 20 15 12 10 9 8 7 6

$\alpha = 0{\cdot}05$: ∞ 30 15 10 8 7 6

$\alpha = 0{\cdot}01$: ∞ 60 30 20 15 12 10 9 8 7 6

Power $= 1 - \beta$

0·99, 0·98, 0·97, 0·96, 0·95, 0·94, 0·92, 0·90, 0·80, 0·70, 0·60, 0·50, 0·40, 0·30, 0·20, 0·10

ϕ (for $\alpha = 0{\cdot}01$) ⟶ 1 2 3

ϕ (for $\alpha = 0{\cdot}05$) ⟵ 1 2 3 4 5

[*Note:* According to the original Neyman and Pearson practice, the power was called β.]

Table K (continued)

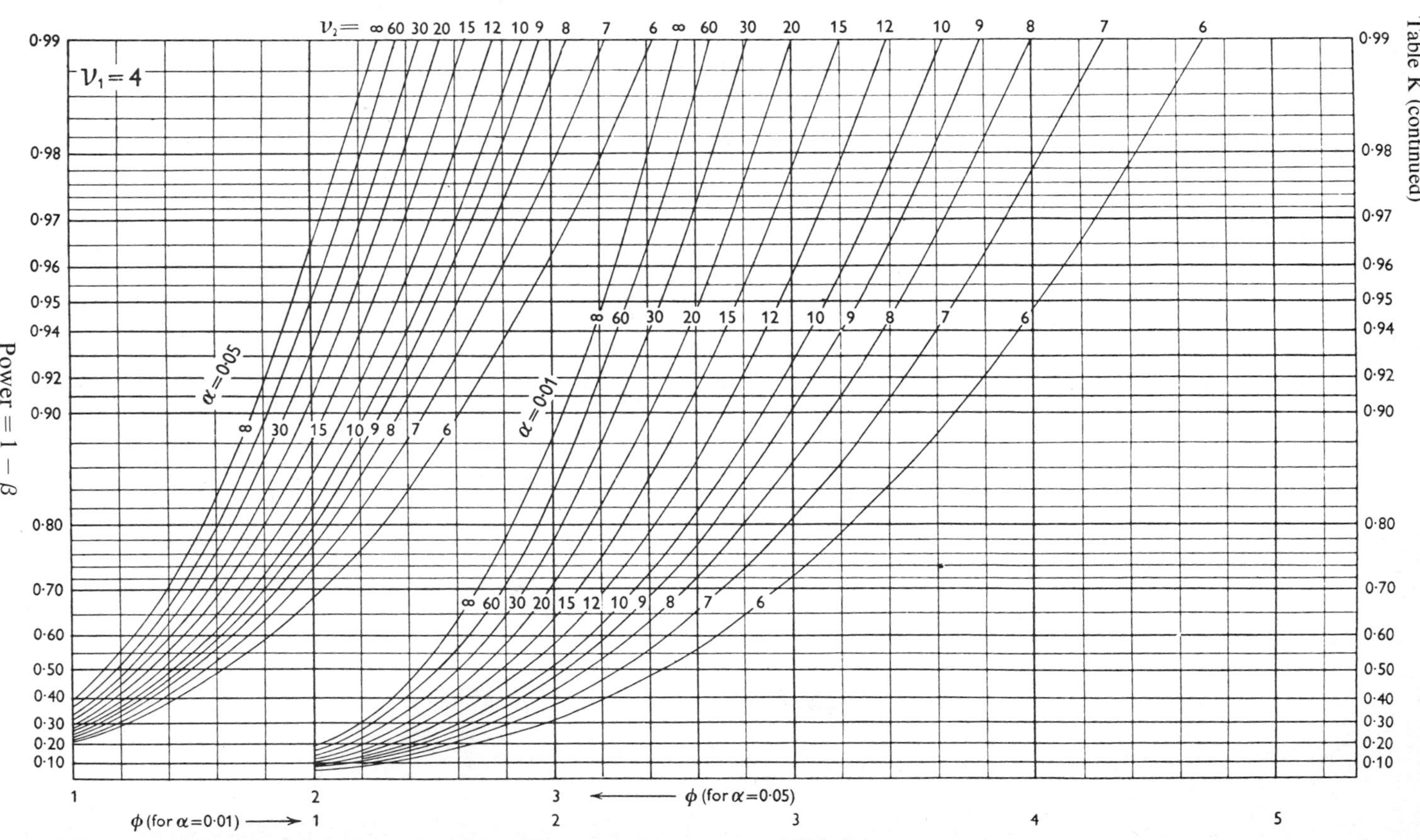

[*Note:* According to the original Neyman and Pearson practice, the power was called β.]

Table K (continued)

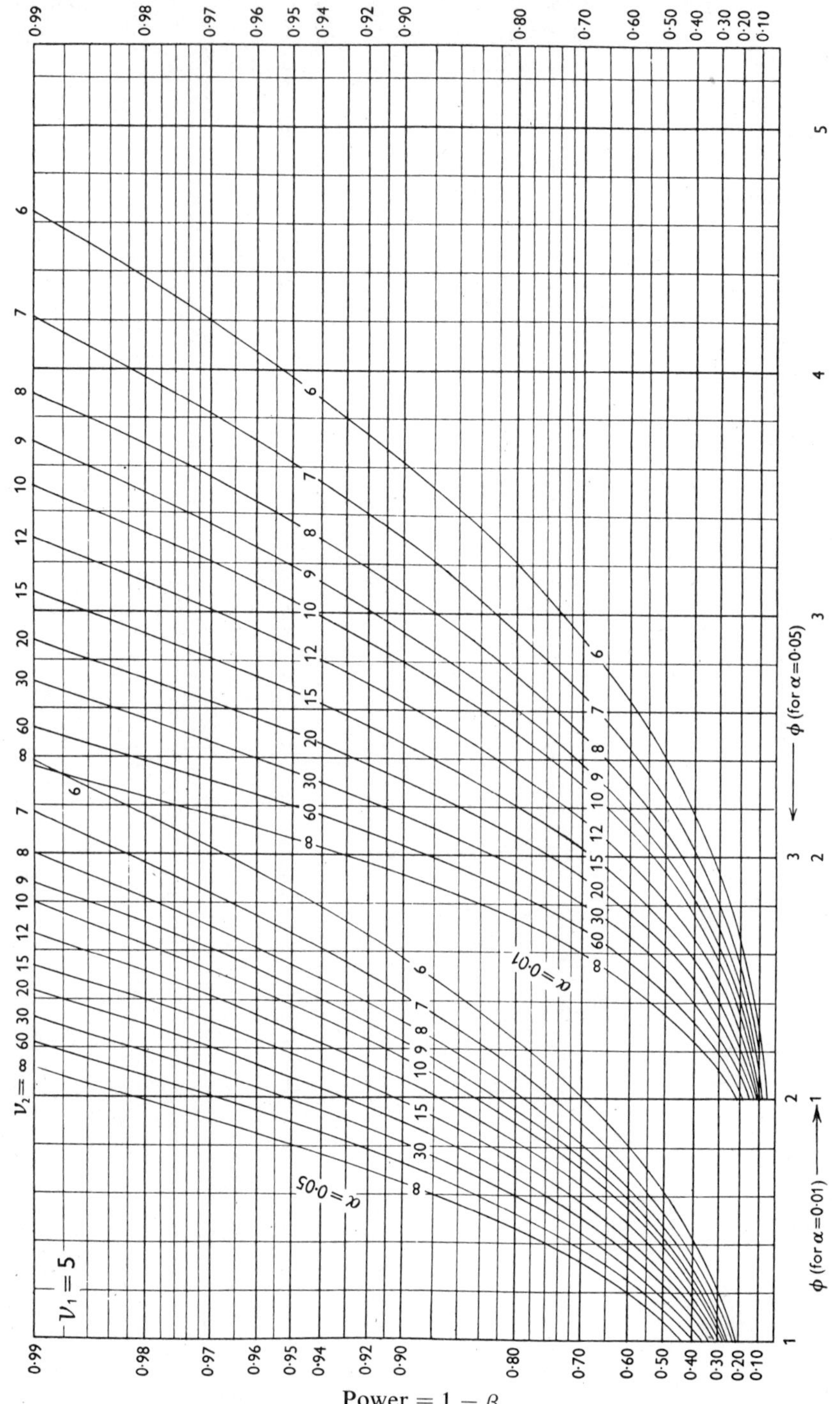

[*Note:* According to the original Neyman and Pearson practice, the power was called β.]

Table K (continued)

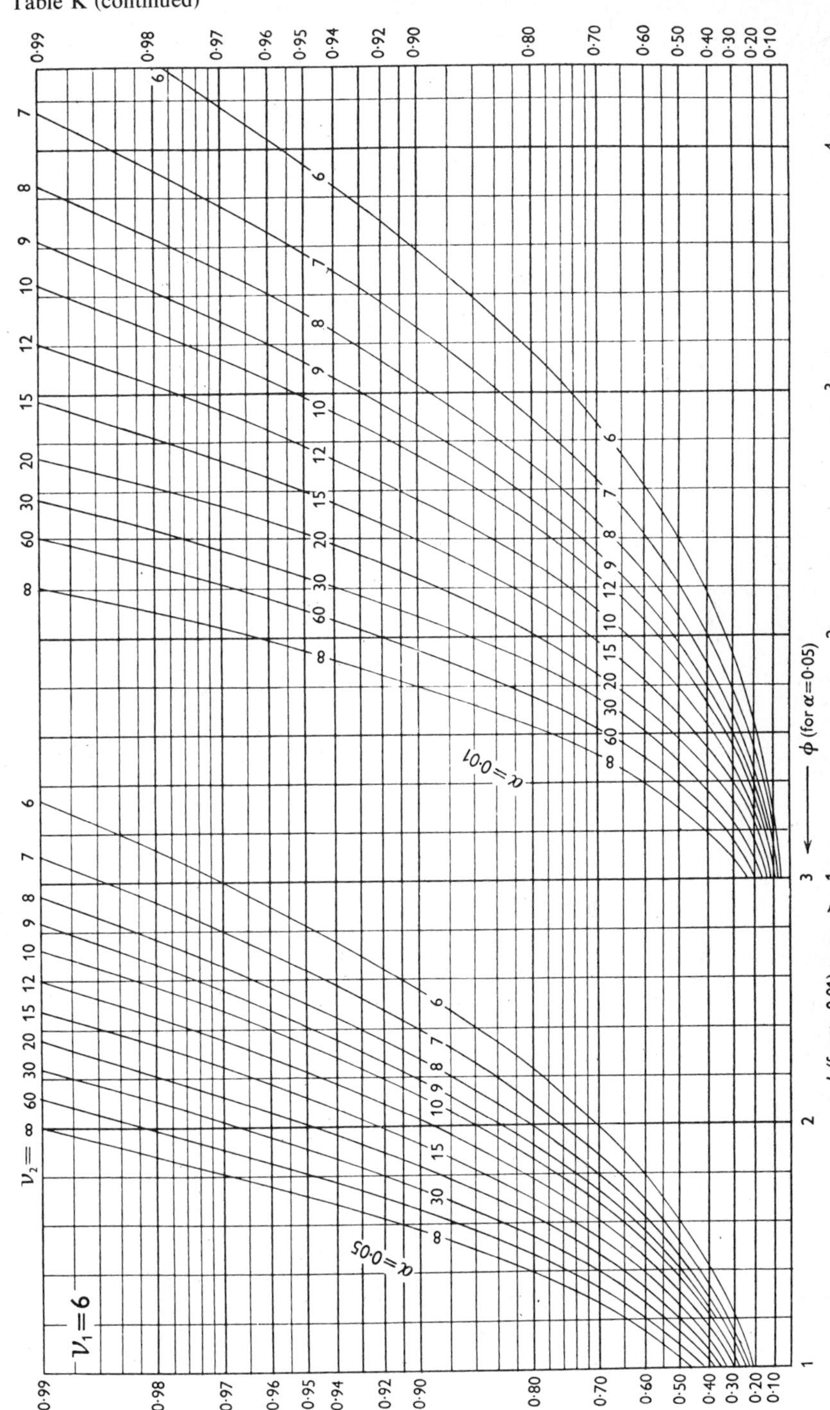

[*Note*: According to the original Neyman and Pearson practice, the power was called β.]

Table K (continued)

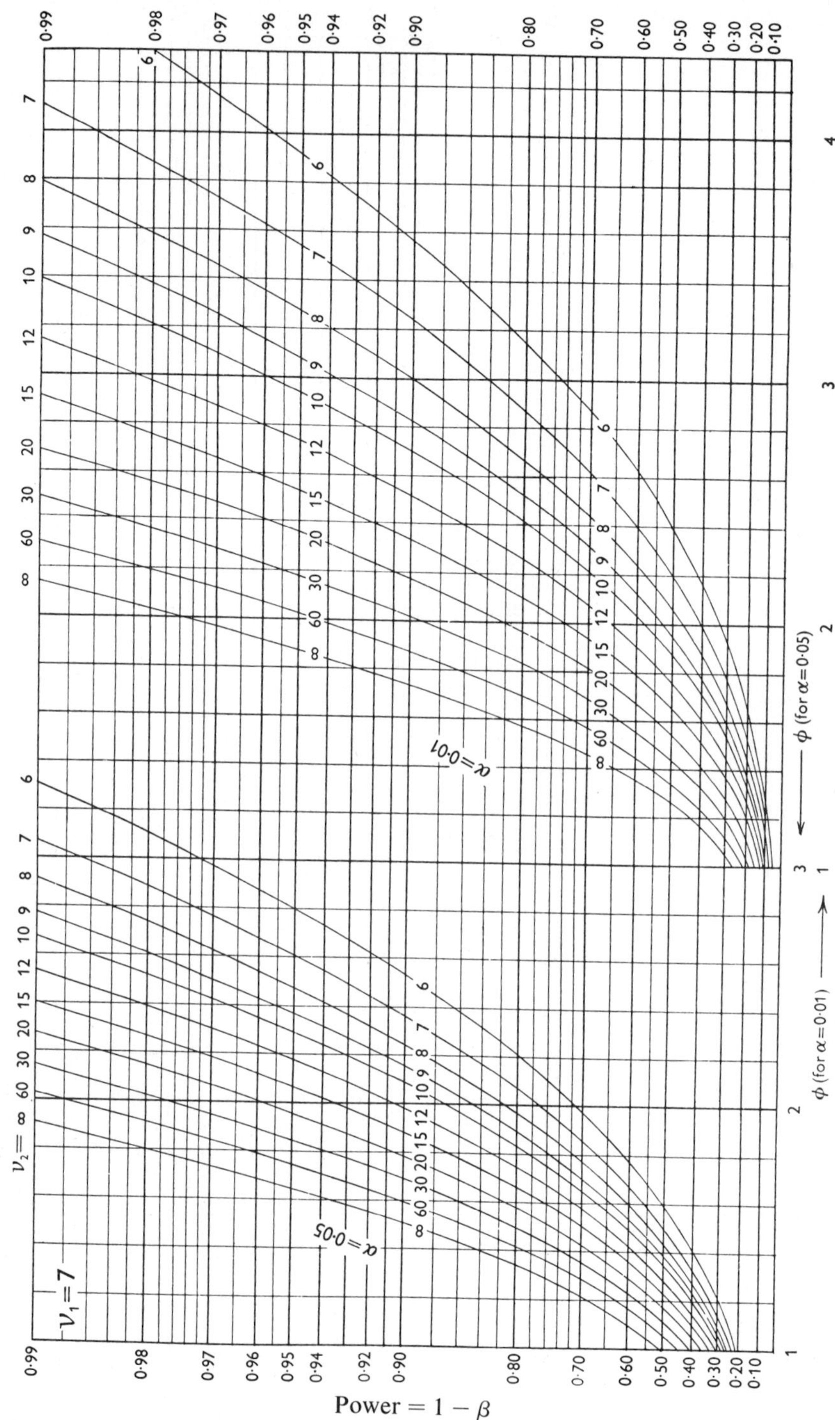

[*Note*: According to the original Neyman and Pearson practice, the power was called β.]

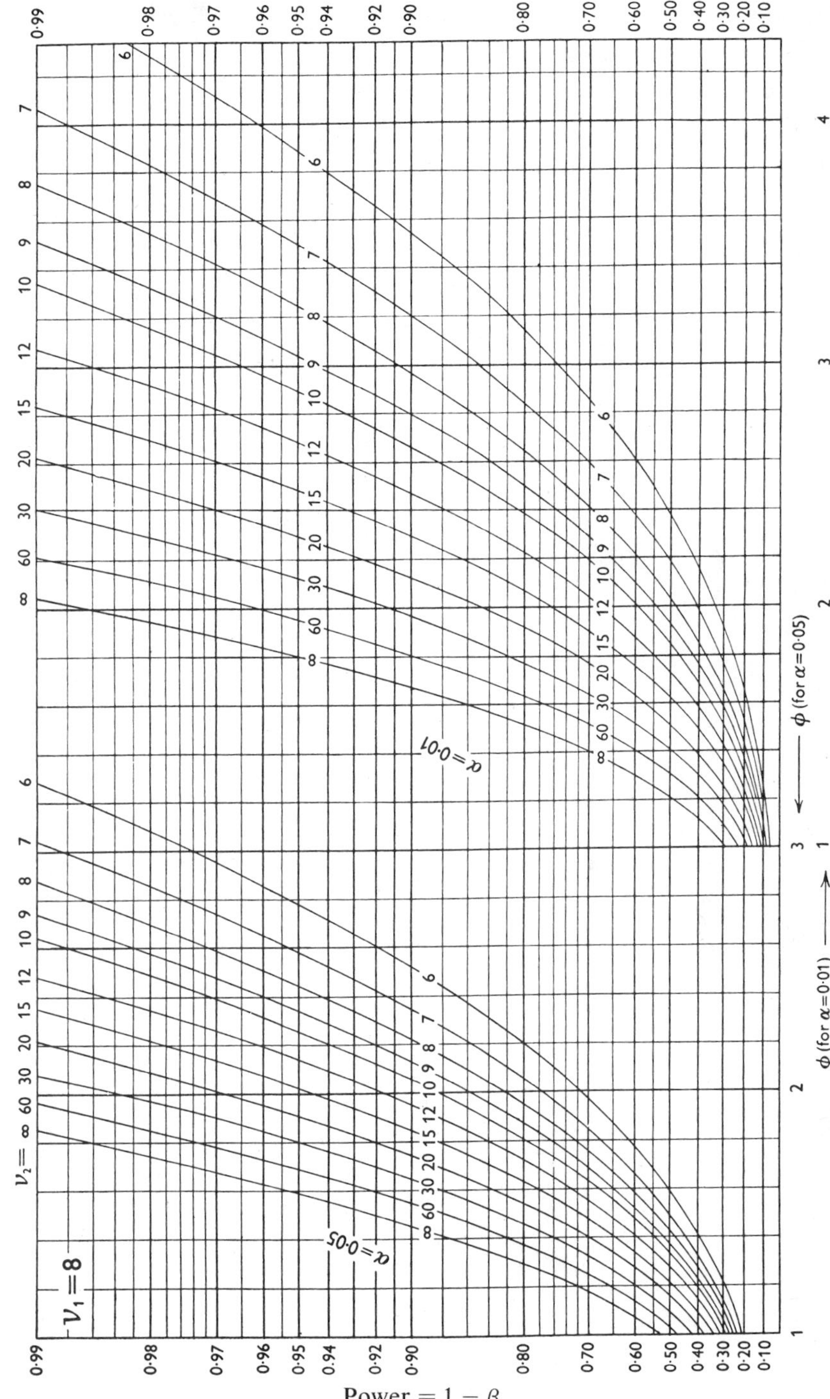

[*Note:* According to the original Neyman and Pearson practice, the power was called β.]

Table L Transformation of r to z

The body of the table contains values of $z = 0.5 \ln [(1 + r)/(1 - r)] = \tanh^{-1}$ for corresponding values of r, the correlation coefficient. For negative values of r, put a minus sign in front of the tabled numbers.

r	0.00	0.01	0.02	0.03	0.04	0.05	0.06	0.07	0.08	0.09
0.0	0.00000	0.01000	0.02000	0.03001	0.04002	0.05004	0.06007	0.07011	0.08017	0.09024
0.1	0.10034	0.11045	0.12058	0.13074	0.14093	0.15114	0.16139	0.17167	0.18198	0.19234
0.2	0.20273	0.21317	0.22366	0.23419	0.24477	0.25541	0.26611	0.27686	0.28768	0.29857
0.3	0.30952	0.32055	0.33165	0.34283	0.35409	0.36544	0.37689	0.38842	0.40006	0.41180
0.4	0.42365	0.43561	0.44769	0.45990	0.47223	0.48470	0.49731	0.51007	0.52298	0.53606
0.5	0.54931	0.56273	0.57634	0.59015	0.60416	0.61838	0.63283	0.64752	0.66246	0.67767
0.6	0.69315	0.70892	0.72501	0.74142	0.75817	0.77530	0.79281	0.81074	0.82911	0.84796
0.7	0.86730	0.88718	0.90764	0.92873	0.95048	0.97296	0.99622	1.02033	1.04537	1.07143
0.8	1.09861	1.12703	1.15682	1.18814	1.22117	1.25615	1.29334	1.33308	1.37577	1.42193
0.9	1.47222	1.52752	1.58903	1.65839	1.73805	1.83178	1.94591	2.09230	2.29756	2.64665

Table M(a) Table of critical values of r in the runs test

Tables M(a) and M(b) contain various critical values of r for various values of n_1 and n_2. For the one-sample runs test, any value of r that is equal to or less than that shown in Table M(a) or equal to or greater than that shown in Table M(b) is significant at the 0.05 level.

n_1 \ n_2	2	3	4	5	6	7	8	9	10	11	12	13	14	15	16	17	18	19	20
2											2	2	2	2	2	2	2	2	2
3					2	2	2	2	2	2	2	2	2	3	3	3	3	3	3
4				2	2	2	3	3	3	3	3	3	3	3	4	4	4	4	4
5			2	2	3	3	3	3	3	4	4	4	4	4	4	4	5	5	5
6		2	2	3	3	3	3	4	4	4	4	5	5	5	5	5	5	6	6
7		2	2	3	3	3	4	4	5	5	5	5	5	6	6	6	6	6	6
8		2	3	3	3	4	4	5	5	5	6	6	6	6	6	7	7	7	7
9		2	3	3	4	4	5	5	5	6	6	6	7	7	7	7	8	8	8
10		2	3	3	4	5	5	5	6	6	7	7	7	7	8	8	8	8	9
11		2	3	4	4	5	5	6	6	7	7	7	8	8	8	9	9	9	9
12	2	2	3	4	4	5	6	6	7	7	7	8	8	8	9	9	9	10	10
13	2	2	3	4	5	5	6	6	7	7	8	8	9	9	9	10	10	10	10
14	2	2	3	4	5	5	6	7	7	8	8	9	9	9	10	10	10	11	11
15	2	3	3	4	5	6	6	7	7	8	8	9	9	10	10	11	11	11	12
16	2	3	4	4	5	6	6	7	8	8	9	9	10	10	11	11	11	12	12
17	2	3	4	4	5	6	7	7	8	9	9	10	10	11	11	11	12	12	13
18	2	3	4	5	5	6	7	8	8	9	9	10	10	11	11	12	12	13	13
19	2	3	4	5	6	6	7	8	8	9	10	10	11	11	12	12	13	13	13
20	2	3	4	5	6	6	7	8	9	9	10	10	11	12	12	13	13	13	14

Table M(b) Table of critical values of r in the runs test

n_1 \ n_2	2	3	4	5	6	7	8	9	10	11	12	13	14	15	16	17	18	19	20
2																			
3																			
4				9	9														
5			9	10	10	11	11												
6			9	10	11	12	12	13	13	13	13								
7				11	12	13	13	14	14	14	14	15	15	15					
8				11	12	13	14	14	15	15	16	16	16	16	17	17	17	17	17
9					13	14	14	15	16	16	16	17	17	18	18	18	18	18	18
10					13	14	15	16	16	17	17	18	18	18	19	19	19	20	20
11					13	14	15	16	17	17	18	19	19	19	20	20	20	21	21
12					13	14	16	16	17	18	19	19	20	20	21	21	21	22	22
13						15	16	17	18	19	19	20	20	21	21	22	22	23	23
14						15	16	17	18	19	20	20	21	22	22	23	23	23	24
15						15	16	18	18	19	20	21	22	22	23	23	24	24	25
16							17	18	19	20	21	21	22	23	23	24	25	25	25
17							17	18	19	20	21	22	23	23	24	25	25	26	26
18							17	18	19	20	21	22	23	24	25	25	26	26	27
19							17	18	20	21	22	23	23	24	25	26	26	27	27
20							17	18	20	21	22	23	24	25	25	26	27	27	28

These tables are adapted from Frieda S. Swed and C. Eisenhart, "Tables for Testing Randomness of Grouping in a Sequence of Alternatives," in *Annals of Mathematical Statistics*, vol. 14 (1943), pages 66–87.

Table N Critical values of the Spearman test statistic
Approximate upper-tail critical values, r_s^*, where $P(r_s > r_s^*) \leq \alpha$, $n = 4(1)30$

Significance Level, α

n	0.001	0.005	0.010	0.025	0.050	0.100
4	—	—	—	—	.8000	.8000
5	—	—	.9000	.9000	.8000	.7000
6	—	.9429	.8857	.8286	.7714	.6000
7	.9643	.8929	.8571	.7450	.6786	.5357
8	.9286	.8571	.8095	.7143	.6190	.5000
9	.9000	.8167	.7667	.6833	.5833	.4667
10	.8667	.7818	.7333	.6364	.5515	.4424
11	.8364	.7545	.7000	.6091	.5273	.4182
12	.8182	.7273	.6713	.5804	.4965	.3986
13	.7912	.6978	.6429	.5549	.4780	.3791
14	.7670	.6747	.6220	.5341	.4593	.3626
15	.7464	.6536	.6000	.5179	.4429	.3500
16	.7265	.6324	.5824	.5000	.4265	.3382
17	.7083	.6152	.5637	.4853	.4118	.3260
18	.6904	.5975	.5480	.4716	.3994	.3148
19	.6737	.5825	.5333	.4579	.3895	.3070
20	.6586	.5684	.5203	.4451	.3789	.2977
21	.6455	.5545	.5078	.4351	.3688	.2909
22	.6318	.5426	.4963	.4241	.3597	.2829
23	.6186	.5306	.4852	.4150	.3518	.2767
24	.6070	.5200	.4748	.4061	.3435	.2704
25	.5962	.5100	.4654	.3977	.3362	.2646
26	.5856	.5002	.4564	.3894	.3299	.2588
27	.5757	.4915	.4481	.3822	.3236	.2540
28	.5660	.4828	.4401	.3749	.3175	.2490
29	.5567	.4744	.4320	.3685	.3113	.2443
30	.5479	.4665	.4251	.3620	.3059	.2400

Note: The corresponding lower-tail critical value for r_s is $-r_s^*$.

Source: Gerald J. Glasser and Robert Winter, "Critical Values of the Coefficient of Rank Correlation for Testing the Hypothesis of Independence," *Biometrkia*, 48 (1961) 444–448. Used by permission of the authors and Professor E. S. Pearson on behalf of the Biometrika Trustees. The table as reprinted here contains corrections given in W. J. Conover, *Practical Nonparametric Statistics*, New York: John Wiley, 1971.

Table O Quantiles of the Wilcoxon matched-pairs signed-ranks test statistic

	$w_{0.005}$	$w_{0.01}$	$w_{0.025}$	$w_{0.05}$	$w_{0.10}$	$w_{0.20}$	$w_{0.30}$	$w_{0.40}$	$\frac{n(n+1)}{2}$
$n=4$	0	0	0	0	1	3	3	4	10
5	0	0	0	1	3	4	5	6	15
6	0	0	1	3	4	6	8	9	21
7	0	1	3	4	6	9	11	12	28
8	1	2	4	6	9	12	14	16	36
9	2	4	6	9	11	15	18	20	45
10	4	6	9	11	15	19	22	25	55
11	6	8	11	14	18	23	27	30	66
12	8	10	14	18	22	28	32	36	78
13	10	13	18	22	27	33	38	42	91
14	13	16	22	26	32	39	44	48	105
15	16	20	26	31	37	45	51	55	120
16	20	24	30	36	43	51	58	63	136
17	24	28	35	42	49	58	65	71	153
18	28	33	41	48	56	66	73	80	171
19	33	38	47	54	63	74	82	89	190
20	38	44	53	61	70	82	91	98	210

Source: Robert L. McCornack, "Extended Tables of the Wilcoxon Matched-Pair Signed-Ranks Statistic," *Journal of the American Statistical Association,* 60 (1965), 864–871. Used by permission.

Table P Quantiles of the Mann-Whitney test statistic

n	p	$m=2$	3	4	5	6	7	8	9	10	11	12	13	14	15	16	17	18	19	20
	.001	0	0	0	0	0	0	0	0	0	0	0	0	0	0	0	0	0	0	0
	.005	0	0	0	0	0	0	0	0	0	0	0	0	0	0	0	0	0	1	1
2	.01	0	0	0	0	0	0	0	0	0	0	0	1	1	1	1	1	1	2	2
	.025	0	0	0	0	0	0	1	1	1	1	2	2	2	2	2	3	3	3	3
	.05	0	0	0	1	1	1	2	2	2	2	3	3	4	4	4	4	5	5	5
	.10	0	1	1	2	2	2	3	3	4	4	5	5	5	6	6	7	7	8	8
	.001	0	0	0	0	0	0	0	0	0	0	0	0	0	0	0	1	1	1	1
	.005	0	0	0	0	0	0	0	1	1	1	2	2	2	3	3	3	3	4	4
3	.01	0	0	0	0	0	1	1	2	2	2	3	3	3	4	4	5	5	5	6
	.025	0	0	0	1	2	2	3	3	4	4	5	5	6	6	7	7	8	8	9
	.05	0	1	1	2	3	3	4	5	5	6	6	7	8	8	9	10	10	11	12
	.10	1	2	2	3	4	5	6	6	7	8	9	10	11	11	12	13	14	15	16
	.001	0	0	0	0	0	0	0	0	1	1	1	2	2	2	3	3	4	4	4
	.005	0	0	0	0	1	1	2	2	3	3	4	4	5	6	6	7	7	8	9
4	.01	0	0	0	1	2	2	3	4	4	5	6	6	7	9	8	9	10	10	11
	.025	0	0	1	2	3	4	5	5	6	7	8	9	10	11	12	12	13	14	15
	.05	0	1	2	3	4	5	6	7	8	9	10	11	12	13	15	16	17	18	19
	.10	1	2	4	5	6	7	8	10	11	12	13	14	16	17	18	19	21	22	23
	.001	0	0	0	0	0	0	1	2	2	3	3	4	4	5	6	6	7	8	8
	.005	0	0	0	1	2	2	3	4	5	6	7	8	8	9	10	11	12	13	14
5	.01	0	0	1	2	3	4	5	6	7	8	9	10	11	12	13	14	15	16	17
	.025	0	1	2	3	4	6	7	8	9	10	12	13	14	15	16	18	19	20	21
	.05	1	2	3	5	6	7	9	10	12	13	14	16	17	19	20	21	23	24	26
	.10	2	3	5	6	8	9	11	13	14	16	18	19	21	23	24	26	28	29	31
	.001	0	0	0	0	0	0	2	3	4	5	5	6	7	8	9	10	11	12	13
	.005	0	0	1	2	3	4	5	6	7	8	10	11	12	13	14	16	17	18	19
6	.01	0	0	2	3	4	5	7	8	9	10	12	13	14	16	17	19	20	21	23
	.025	0	2	3	4	6	7	9	11	12	14	15	17	18	20	22	23	25	26	28
	.05	1	3	4	6	8	9	11	13	15	17	18	20	22	24	26	27	29	31	33
	.10	2	4	6	8	10	12	14	16	18	20	22	24	26	28	30	32	35	37	39
	.001	0	0	0	0	1	2	3	4	6	7	8	9	10	11	12	14	15	16	17
	.005	0	0	1	2	4	5	7	8	10	11	13	14	16	17	19	20	22	23	25
7	.01	0	1	2	4	5	7	8	10	12	13	15	17	18	20	22	24	25	27	29
	.025	0	2	4	6	7	9	11	13	15	17	19	21	23	25	27	29	31	33	35
	.05	1	3	5	7	9	12	14	16	18	20	22	25	27	29	31	34	36	38	40
	.10	2	5	7	9	12	14	17	19	22	24	27	29	32	34	37	39	42	44	47
	.001	0	0	0	1	2	3	5	6	7	9	10	12	13	15	16	18	19	21	22
	.005	0	0	2	3	5	7	8	10	12	14	16	18	19	21	23	25	27	29	31
8	.01	0	1	3	5	7	8	10	12	14	16	18	21	23	25	27	29	31	33	35
	.025	1	3	5	7	9	11	14	16	18	20	23	25	27	30	32	35	37	39	42
	.05	2	4	6	9	11	14	16	19	21	24	27	29	32	34	37	40	42	45	48
	.10	3	6	8	11	14	17	20	23	25	28	31	34	37	40	43	46	49	52	55

Source: Adapted from L. R. Verdooren, "Extended Tables of Critical Values for Wilcoxon's Test Statistic," *Biometrika,* 50 (1963), 177–186. Used by permission of the author and E. S. Pearson on behalf of the Biometrika Trustees. The adaptation is due to W. J. Conover, *Practical Nonparametric Statistics,* New York: John Wiley, 1971, 384–388.

Table P Quantiles of the Mann-Whitney test statistic (continued)

n	p	$m=2$	3	4	5	6	7	8	9	10	11	12	13	14	15	16	17	18	19	20
	.001	0	0	0	2	3	4	6	8	9	11	13	15	16	18	20	22	24	26	27
	.005	0	1	2	4	6	8	10	12	14	17	19	21	23	25	28	30	32	34	37
9	.01	0	2	4	6	8	10	12	15	17	19	22	24	27	29	32	34	37	39	41
	.025	1	3	5	8	11	13	16	18	21	24	27	29	32	35	38	40	43	46	49
	.05	2	5	7	10	13	16	19	22	25	28	31	34	37	40	43	46	49	52	55
	.10	3	6	10	13	16	19	23	26	29	32	36	39	42	46	49	53	56	59	63
	.001	0	0	1	2	4	6	7	9	11	13	15	18	20	22	24	26	28	30	33
	.005	0	1	3	5	7	10	12	14	17	19	22	25	27	30	32	35	38	40	43
10	.01	0	2	4	7	9	12	14	17	20	23	25	28	31	34	37	39	42	45	48
	.025	1	4	6	9	12	15	18	21	24	27	30	34	37	40	43	46	49	53	56
	.05	2	5	8	12	15	18	21	25	28	32	35	38	42	45	49	52	56	59	63
	.10	4	7	11	14	18	22	25	29	33	37	40	44	48	52	55	59	63	67	71
	.001	0	0	1	3	5	7	9	11	13	16	18	21	23	25	28	30	33	35	38
	.005	0	1	3	6	8	11	14	17	19	22	25	28	31	34	37	40	43	46	49
11	.01	0	2	5	8	10	13	16	19	23	26	29	32	35	38	42	45	48	51	54
	.025	1	4	7	10	14	17	20	24	27	31	34	38	41	45	48	52	56	59	63
	.05	2	6	9	13	17	20	24	28	32	35	39	43	47	51	55	58	62	66	70
	.10	4	8	12	16	20	24	28	32	37	41	45	49	53	58	62	66	70	74	79
	.001	0	0	1	3	5	8	10	13	15	18	21	24	26	29	32	35	38	41	43
	.005	0	2	4	7	10	13	16	19	22	25	28	32	35	38	42	45	48	52	55
12	.01	0	3	6	9	12	15	18	22	25	29	32	36	39	43	47	50	54	57	61
	.025	2	5	8	12	15	19	23	27	30	34	38	42	46	50	54	58	62	66	70
	.05	3	6	10	14	18	22	27	31	35	39	43	48	52	56	61	65	69	73	78
	.10	5	9	13	18	22	27	31	36	40	45	50	54	59	64	68	73	78	82	87
	.001	0	0	2	4	6	9	12	15	18	21	24	27	30	33	36	39	43	46	49
	.005	0	2	4	8	11	14	18	21	25	28	32	35	39	43	46	50	54	58	61
13	.01	1	3	6	10	13	17	21	24	28	32	36	40	44	48	52	56	60	64	68
	.025	2	5	9	13	17	21	25	29	34	38	42	46	51	55	60	64	68	73	77
	.05	3	7	11	16	20	25	29	34	38	43	48	52	57	62	66	71	76	81	85
	.10	5	10	14	19	24	29	34	39	44	49	54	59	64	69	75	80	85	90	95
	.001	0	0	2	4	7	10	13	16	20	23	26	30	33	37	40	44	47	51	55
	.005	0	2	5	8	12	16	19	23	27	31	35	39	43	47	51	55	59	64	68
14	.01	1	3	7	11	14	18	23	27	31	35	39	44	48	52	57	61	66	70	74
	.025	2	6	10	14	18	23	27	32	37	41	46	51	56	60	65	70	75	79	84
	.05	4	8	12	17	22	27	32	37	42	47	52	57	62	67	72	78	83	88	93
	.10	5	11	16	21	26	32	37	42	48	53	59	64	70	75	81	86	92	98	103
	.001	0	0	2	5	8	11	15	18	22	25	29	33	37	41	44	48	52	56	60
	.005	0	3	6	9	13	17	21	25	30	34	38	43	47	52	56	61	65	70	74
15	.01	1	4	8	12	16	20	25	29	34	38	43	48	52	57	62	67	71	76	81
	.025	2	6	11	15	20	25	30	35	40	45	50	55	60	65	71	76	81	86	91
	.05	4	8	13	19	24	29	34	40	45	51	56	62	67	73	78	84	89	95	101
	.10	6	11	17	23	28	34	40	46	52	58	64	69	75	81	87	93	99	105	111
	.001	0	0	3	6	9	12	16	20	24	28	32	36	40	44	49	53	57	61	66
	.005	0	3	6	10	14	19	23	28	32	37	42	46	51	56	61	66	71	75	80
16	.01	1	4	8	13	17	22	27	32	37	42	47	52	57	62	67	72	77	83	88
	.025	2	7	12	16	22	27	32	38	43	48	54	60	65	71	76	82	87	93	99
	.05	4	9	15	20	26	31	37	43	49	55	61	66	72	78	84	90	96	102	108
	.10	6	12	18	24	30	37	43	49	55	62	68	75	81	87	94	100	107	113	120

Table P Quantiles of the Mann-Whitney test statistic (continued)

n	p	$m=2$	3	4	5	6	7	8	9	10	11	12	13	14	15	16	17	18	19	20
	.001	0	1	3	6	10	14	18	22	26	30	35	39	44	48	53	58	62	67	71
	.005	0	3	7	11	16	20	25	30	35	40	45	50	55	61	66	71	76	82	87
17	.01	1	5	9	14	19	24	29	34	39	45	50	56	61	67	72	78	83	89	94
	.025	3	7	12	18	23	29	35	40	46	52	58	64	70	76	82	88	94	100	106
	.05	4	10	16	21	27	34	40	46	52	58	65	71	78	84	90	97	103	110	116
	.10	7	13	19	26	32	39	46	53	59	66	73	80	86	93	100	107	114	121	128
	.001	0	1	4	7	11	15	19	24	28	33	38	43	47	52	57	62	67	72	77
	.005	0	3	7	12	17	22	27	32	38	43	48	54	59	65	71	76	82	88	93
18	.01	1	5	10	15	20	25	31	37	42	48	54	60	66	71	77	83	89	95	101
	.025	3	8	13	19	25	31	37	43	49	56	62	68	75	81	87	94	100	107	113
	.05	5	10	17	23	29	36	42	49	56	62	69	76	83	89	96	103	110	117	124
	.10	7	14	21	28	35	42	49	56	63	70	78	85	92	99	107	114	121	129	136
	.001	0	1	4	8	12	16	21	26	30	35	41	46	51	56	61	67	72	78	83
	.005	1	4	8	13	18	23	29	34	40	46	52	58	64	70	75	82	88	94	100
19	.01	2	5	10	16	21	27	33	39	45	51	57	64	70	76	83	89	95	102	108
	.025	3	8	14	20	26	33	39	46	53	59	66	73	79	86	93	100	107	114	120
	.05	5	11	18	24	31	38	45	52	59	66	73	81	88	95	102	110	117	124	131
	.10	8	15	22	29	37	44	52	59	67	74	82	90	98	105	113	121	129	136	144
	.001	0	1	4	8	13	17	22	27	33	38	43	49	55	60	66	71	77	83	89
	.005	1	4	9	14	19	25	31	37	43	49	55	61	68	74	80	87	93	100	106
20	.01	2	6	11	17	23	29	35	41	48	54	61	68	74	81	88	94	101	108	115
	.025	3	9	15	21	28	35	42	49	56	63	70	77	84	91	99	106	113	120	128
	.05	5	12	19	26	33	40	48	55	63	70	78	85	93	101	108	116	124	131	139
	.10	8	16	23	31	39	47	55	63	71	79	87	95	103	111	120	128	136	144	152

Index

ABCDEFGHIJ–VB–7987

SOME FREQUENTLY USED FORMULAS

Sample mean

$$\bar{x} = \sum_{i=1}^{n} \frac{x_i}{n}$$

Sample variance

$$s^2 = \sum_{i=1}^{n} \frac{(x_i - \bar{x})^2}{n-1}$$

Number of permutations of n objects taken r at a time

$${}_nP_r = \frac{n!}{(n-r)!}$$

Number of combinations of n objects taken r at a time

$$\binom{n}{r} = \frac{n!}{r!\,(n-r)!}$$

Confidence interval for μ, unknown population variance

$$C\left[\bar{x} - t\left(\frac{s}{\sqrt{n}}\right) \le \mu \le \bar{x} + t\left(\frac{s}{\sqrt{n}}\right)\right] = 1 - \alpha$$

Sample size for estimating μ

$$n = \frac{z^2\sigma^2}{d^2}$$

Sample size for estimating a population proportion

$$n = \frac{z^2 pq}{d^2}$$

Pooled estimate of common population variance

$$s_p^2 = \frac{(n_1 - 1)s_1^2 + (n_2 - 1)s_2^2}{n_1 + n_2 - 2}$$

Test statistic for testing H_0: $\mu = \mu_0$, population variance unknown

$$t = \frac{\bar{x} - \mu_0}{s/\sqrt{n}}$$

Test statistic for testing H_0: $\mu_1 = \mu_2$, unknown, but equal, population variances

$$t = \frac{\bar{x}_1 - \bar{x}_2}{\sqrt{\dfrac{s_p^2}{n_1} + \dfrac{s_p^2}{n_2}}}$$

Confidence interval for a population proportion

$$C\left[\hat{p} - z\sqrt{\frac{\hat{p}(1-\hat{p})}{n}} \le p \le \hat{p} + z\sqrt{\frac{\hat{p}(1-\hat{p})}{n}}\right] = 1 - \alpha$$